S0-FJU-798

WITHDRAWN
UNIVERSITY OF ST. THOMAS LIBRARIES
UST
Libraries

The Conservation Movement:
A History of Architectural Preservation

In many places across the world, particularly in Europe, old buildings form a prominent part of the built environment, and we often take it for granted that their contribution is intrinsically positive. How has that widely-shared belief come about, and is its continued general acceptance inevitable?

Certainly, ancient structures have long been treated with care and reverence in many societies, including classical Rome and Greece. But only in modern Europe and America, in the last two centuries, has this care been elaborated and energised into a forceful, dynamic ideology: a 'Conservation Movement', infused with a sense of historical destiny and loss, that paradoxically shared many of the characteristics of Enlightenment modernity. The close inter-relationship between conservation and modern civilisation was most dramatically heightened in periods of war or social upheaval, beginning with the French Revolution, and rising to a tragic climax in the twentieth-century age of totalitarian extremism; more recently the troubled relationship of 'heritage' and global commercialism has become dominant.

Miles Glendinning's new book authoritatively presents, for the first time, the entire history of architectural conservation, and traces its dramatic fluctuations in ideas and popularity, ending by questioning whether its recent international ascendancy can last indefinitely.

Miles Glendinning is Professor of Architectural Conservation at the University of Edinburgh and Director of the Scottish Centre for Conservation Studies. He has published extensively on modernist and contemporary architecture and housing: his books include the award-winning *Tower Block* (with Stefan Muthesius), *Modern Architect*, on the life and times of Sir Robert Matthew, and *Architecture's Evil Empire*, a polemical evaluation of contemporary 'iconic modernism'. His current research projects include an international history of mass social housing, focusing in particular on the experience of Hong Kong and Singapore.

The Conservation Movement: A History of Architectural Preservation

Antiquity to modernity

Miles Glendinning

Routledge
Taylor & Francis Group

LONDON AND NEW YORK

First published 2013
by Routledge
2 Park Square, Milton Park, Abingdon, Oxon OX14 4RN

Simultaneously published in the USA and Canada
by Routledge
711 Third Avenue, New York, NY 10017

Routledge is an imprint of the Taylor & Francis Group, an informa business

© 2013 Miles Glendinning

The right of Miles Glendinning to be identified as author of this work has been asserted by him/her in accordance with sections 77 and 78 of the Copyright, Designs and Patents Act 1988.

All rights reserved. No part of this book may be reprinted or reproduced or utilised in any form or by any electronic, mechanical, or other means, now known or hereafter invented, including photocopying and recording, or in any information storage or retrieval system, without permission in writing from the publishers.

Trademark notice: Product or corporate names may be trademarks or registered trademarks, and are used only for identification and explanation without intent to infringe.

Every effort has been made to contact and acknowledge copyright owners. If any material has been included without permission, the publishers offer their apologies. The publishers would be pleased to have any errors or omissions brought to their attention so that corrections may be published at later printing.

British Library Cataloguing in Publication Data
A catalogue record for this book is available from the British Library

Library of Congress Cataloging in Publication Data
Glendinning, Miles, 1956–
    The conservation movement: a history of architectural preservation: antiquity to modernity/Miles Glendinning.
        pages cm
    (print) 1. Architecture–Conservation and restoration–History. I. Title.
    NA105.G49 2013
    720.28'8–dc23   2012022396

ISBN: 978–0–415–49999–6 (hbk)
ISBN: 978–0–415–54322–4 (pbk)
ISBN: 978–0–203–08039–9 (ebk)

Typeset in Minion
by Keystroke, Station Road, Codsall, Wolverhampton

– Contents –

— Acknowledgements —

I would like to thank the following individuals and organisations for helping me in this project. In some cases, the help was immediately and specifically related to the text or illustrations, including commenting on drafts or bringing new points to my attention. In other cases, the input was more indirect, including organisational assistance, general advice on conservation-history matters, or more general support, sometimes over a period of years. Those whose names are asterisked contributed through participation in historical interviews, in some cases up to two decades ago.

Jane Anderson; Kerstin Barup; Ivan Blasi; Jonathan Blower; Allison Borden; Brownstone Coalition; Stephen Cairns; Ian Campbell; Louise Campbell; Maristella Casciato; Mark Cousins; Ellen Creighton; Catherine Croft; *Theo Crombie; Kitty Cruft; Audrey Dakin; Wessel de Jonge; Lynne DiStefano; DOCOMOMO-International; DOCOMOMO-Scotland; Emilie D'Orgeix; Marija Drėmaitė; Stuart Eydmann; Clive Fenton; Leslie Forsyth; Colum Giles; Margaret Gilfillan; Isla Glendinning; Ian Gow; Giovanna Guidicini; Peter Guillery; Ewan Harrison; Elain Harwood; Hubert-Jan Henket; Carsten Hermann; Amy Hickman; Yvonne Hillyard; Julian Holder; Alison, Amy-Felicity, Kitty, Margaret and Sali Horsey; John Hume; Jane Jacobs; Hazel Johnson; ICOMOS Australia; Jukka Jokilehto; Mart Kalm; *Lord Kennet; Janis Krastins; Gus Lamb; Leticia Leitão; Hannah Lewi; Sian Loftus; Elizabeth and Nancy Love; Aonghus MacKechnie; Kirsten McKee; Colin McWilliam; Debbie Mays; Suzannah Meade; Kasia Murawska; Michał Murawski; Stefan Muthesius; Miles Oglethorpe; Ali and Matthew Owens; David Page; Brian Park; Anne Raines; Lukas Rekevičius; Una Richards; Peter Robinson; Johnny Rodger; Richard Rodger; Dennis Rodwell; Joanna Roscoe; Indrė Ruseckaitė; Antonello Sanna; Achim Schröer; Moira Seftor; *Konrad Smigielski; Wolfgang Sonne; Chris Speed; Gavin Stamp; Geoffrey Stell; Fiona Stenke; Graeme Stewart; Ruxandra Stoica; *Herb Stovel; Ian Tan Yuk-Hong; Jessica Taylor; Dimitris Theodossopoulos; *Mary and Frank Tindall; Emily Tracey; Mark Tripney; Ola Uduku; Florian Urban; Rosman Wai; *David Walker; Diane Watters; Vicky Webster; Ola Wedebrunn; *David and Enid Whitham; Alfred Wood; *Roy Worskett; *Raymond Young.

– Abbreviations –

AHSS	Architectural Heritage Society of Scotland
AR	*Architectural Review*
ARQ	*Architectural Research Quarterly*
ASHPS	American Scenic and Historic Preservation Society
CA	conservation area
CHAC	Central Housing Advisory Committee
CIAM	International Congresses of Modern Architecture
CMH	Historic Monuments Commission (France)
COE	Council of Europe
CPS	Commons Preservation Society
CT	Civic Trust
DD	*Die Denkmalpflege*
DKD	*Deutsche Kunst und Denkmalpflege*
DHS	Department of Health for Scotland
DOCOMOMO	Documentation and Conservation of the Modern Movement (Modern architecture heritage group)
EAHY	European Architectural Heritage Year (1975)
ENTCC	Edinburgh New Town Conservation Committee
FDH	Freies Deutsches Hochstift
GDR	German Democratic Republic
GHND	Gesellschaft Historischer Neumarkt Dresden
GIA	General Improvement Area
GLC	Greater London Council
GRKG	*Geschichte der Rekonstruktion: Konstruktion der Geschichte*
HAA	Housing Action Area
HABS	Historic American Buildings Survey
HBCs	Historic Buildings Councils (for England and Scotland)
HTA	Housing Treatment Area
HUD	US Department of Housing and Urban Development
IBA	Internationale Bauaustellung Berlin
ICCROM	International Centre for the Study of the Preservation and Restoration of Cultural Property
ICIC	International Committee on Intellectual Cooperation
ICOMOS	International Council on Monuments and Sites
IFHTP	International Federation for Housing and Town Planning

IMO	International Museums Office
INTACH	Indian National Trust for Art and Cultural Heritage
IPHAN	National Institute for the Preservation of Historical and Artistic Heritage(Brazil)
ISMEO	Italian Middle and Far Eastern Institute
ISOCARP	International Society of City and Regional Planners
IUCN	International Union for the Conservation of Nature and Natural Resources
JSAH	*Journal of the Society of Architectural Historians*
LCC	London County Council
MH	*Les Monuments Historiques*
MHLG	Ministry of Housing and Local Government
MOMA	Museum of Modern Art
MoMo	sometimes used for Modern Movement
NHS	Neighborhood Housing Services (USA)
NPS	National Park Service
NSDAP	National Socialist German Workers' Party (Nazi Party)
NT	National Trust
NTS	National Trust for Scotland
OIRU	Society for Study of Russian Mansions
OPAH	Opérations programmes d'amélioration de l'habitat
OUV	[of] outstanding universal value
ÖZKD	*Österreichische Zeitschrift für Kunst und Denkmalpflege*
PB	*Planning Bulletin*
PEEP	Piano per l'Edilizia Economica Popolare
PKZ	Monument Restoration Atelier (Poland)
PNF	National Fascist Party (Italy)
RCAHMS	Royal Commission on the Ancient and Historical Monuments of Scotland
RFAC	Royal Fine Art Commission (England)
RFACS	Royal Fine Art Commission (Scotland)
RIAS	Royal Incorporation of Architects in Scotland
RIBA	Royal Institute of British Architects
SAHGB	Society of Architectural Historians of Great Britain
SCAPA	Society for the Checking of the Abuses of Public Advertising
SCT	Scottish Civic Trust
SDD	Scottish Development Department
SEM	Sociétés d'économie mixte
SHAC	Scottish Housing Advisory Committee
SLASH	Scottish Local Authorities Special Housing Group
SPAB	Society for the Protection of Ancient Buildings
SPRND	School of Planning and Research for National Development
SÚRPMO	State Institute for the Reconstruction of Historic Towns and Monuments (Czechoslovakia)
TC	*The Theodosian Code*
TPRM	Trustees of Public Reservations in Massachusetts

TSNRM	Central Scientific Restorative Workshop (USSR)
UNESCO	United Nations Educational, Scientific and Cultural Organisation
USSR	Union of Soviet Socialist Republics
Vixoc	Victorian Society
VOOPIK	All-Russian Society for the Safeguarding of Historical and Cultural Monuments

Introduction

The Conservation Movement: Stepchild of progress

*'I never saw anything like you clergymen,' said Eleanor; 'you are always thinking of
fighting each other.' 'Either that,' said he, 'or else supporting each other. The pity is that
we cannot do the one without the other. But are we not here to fight? Is not ours a church
militant? What is all our work but fighting, and hard fighting, if it be well done?'*
Anthony Trollope, *Barchester Towers*, 1857[1]

Today, in many developed societies, especially in Europe, the built environment is dominated
not by outright modernity but by 'old' buildings and ensembles. These are known under a vast
range of names: conservation, historic preservation, listed buildings, heritage, historic monuments,
Denkmalpflege, patrimoine, Altstadt, centro storico, monuments classés, World Heritage Site, and so
forth. Most people simply take this situation for granted and, in the most general terms, assume it
to be not only a natural but a good thing. Any debates and controversies tend to be confined to
relatively narrow circles of interest groups. These tend to focus either on the problems of individual
monuments, or on hackneyed confrontations between intransigent conservationists railing against
'vandalism' and 'threat', and developer-led groups, indignant at hole-in-corner obstruction of
'progress', or at towns turned to 'museums' or 'Disney pastiche'. Yet change is unavoidable, even in
the most cherished places. And conservation clearly plays an intimate part in that wider process of
development and change in the existing built environment – a process that affects everyone in
society.

Architectural conservation, in fact, is something that embraces not just architecture in all its
various forms, but a vast range of other subjects – environmental politics, urban planning, housing,
urban economics and tourism, and even wartime destruction and renewal. And it has attracted, and
continues to attract, intense if intermittent attention not just from narrow interest groups, but from
national politicians and media, who have at times seen it as a part of their own ideological self
projection, as well as from increasingly influential international cultural organisations. Conservation
is, and has always been, an integral part of modern society, and its environments, like all modern
environments, did not just happen. They were 'made', chiefly during the 19th and 20th centuries, and
sometimes also remade, or destroyed, several times over.

Heritage is not something that is just 'there' and has always been 'there'. It has a story, a
dramatic history, firmly knitted into the wider trajectory of European, or Western, modernity. It is
that story, of conservation as a constantly changing modern phenomenon, a future-oriented

'Movement' drawing on the past, which this book sets out to tell. Of course, societies throughout the world and across the centuries and millennia, from China to Central America and Africa, have treated old buildings and structures with care and reverence; but only in the societies dealt with in this account, the societies shaped by Western modernity, has that sense of respect been transformed into a dynamic historical narrative, a conscious ideology, a Movement.

The history of this Conservation Movement is a tale dominated, on the whole, not by high-flown, intellectually abstruse theories, but by more middle-brow, collective, sometimes even bureaucratic ways of thinking – ideas that are quite circumscribed in some ways, and yet are in other respects unusually eclectic, wide-ranging, even political. This is a narrative that combines broad, overarching themes with a great plethora of facts from many disciplines, including laws, policies and technical data. And it is a story that is not yet finished – which makes the historian's task especially relevant. The ancient Athenian historian, Thucydides, hoped that his *History* of the rise, excesses and final calamitous collapse of the 5th-century-BC Athenian Empire 'will be judged useful by those inquirers who desire an exact knowledge of the past as an aid to the interpretation of the future, which, human nature being what it is, must resemble it, if it does not actually reflect it'.[2] We will return later to the issue of whether the story contained in this book is a straightforward narrative of origins, growth and triumphant, open-ended ascendancy, or whether it also contains some early elements of 'decline and fall'.

Naturally, the story of the Conservation Movement is one that has long roots, stretching back to Western antiquity, in other words to Greece and Rome. Care for old structures in those ages stemmed from both practical and symbolic motives, including the diverse demands of polytheistic religion and the salvation narratives of Christianity; and the 15th–17th centuries saw early harbingers of more modern heritage values, for example in the Renaissance revival of the classical world, the work of the first antiquarians and the first concerted schemes of postwar reconstruction. As the modern world, with its unique sense of historical destiny, emerged out of the old timeless traditional religious society – with monotheistic Christianity as a kind of halfway house, partly embracing the idea of progress and human control and partly resisting it – old buildings gradually came to occupy a more important role, both rationally and emotionally.

But conservation as a concerted, modern phenomenon only really emerged in the late 18th century, in the violent political, social and economic modernisations in Europe, especially the French Revolution of 1789. Conservation, as an ideology, only became seen as a necessity once the juggernaut of Enlightenment Progress got underway, unleashing an all-embracing upheaval that, many felt, urgently demanded stabilisation. Like many other children and stepchildren of the Enlightenment, the Conservation Movement was imprinted with its ideas, stamped from one end to the other, like a stick of seaside rock, with the values of historical Progress. Stimulated by these, it developed its own specific values and ideologies – values that were furiously contested by rival groups, and changed radically over time. From this point, conservation became a Movement in the broad modern sense, just like socialism, nationalism, environmentalism, or other more issue-specific cultural or political groupings, such as the anti-slavery or Prohibition movements – or, of course, the Modern Movement in architecture and planning, which became almost an *alter ego* to the Conservation Movement during the course of the 20th century.

The story of the Conservation Movement, it should be stressed, is a specifically 'Western' story, not least because it has been so strongly bound up with the Western drive for a codified, rational, secular exercise of power and knowledge. Although conservation often presents itself as a non-modern, even anti-modern phenomenon, and certainly as something 'traditional', it also in many ways *exemplifies* Western modernity. The history of conservation, above all, has been about exploiting the past for useful modern purposes, especially as a way of underpinning or counterbalancing the drive for modern progress and for rational control of the world. It was no coincidence that modern conservation really came into its own after the French Revolution overthrew the last significant bastions of traditional religious society. It helped provide an anchor against revolutionary chaos, or, occasionally, even itself participated in modern upheavals, as in the case of its deep complicity in the late-19th- and early-20th-century craze of militaristic, aggressive nationalism.

This, therefore, is a story that in some ways becomes more coherent and unified in more recent centuries, as we move from the unstructured, confused, *ad hoc* practices of the Middle Ages or even the early 18th century, to the ideologies and organised structures of the modern age. But that increasing cohesion was disguised by the apparent chaos and conflict of the many competing ideological factions in the modern history of the Movement. This paradoxical character emerged most forcibly in the closest external relationship of the Conservation Movement, at least in the late 19th and 20th centuries: its intimate but troubled relationship with modern nationalism. Theirs was a relationship that brought with it a combination of bitter yet superficial antagonism between separate national traditions, with unifying ideological currents below the surface. Later in the 20th century, however, the unifying pressures began gradually to emerge into the open and gain the upper hand, as the rival conservation nationalisms were gradually outweighed and replaced by an overt internationalism. Alongside this geopolitical development of conservation, our account will also trace a variety of long-running, relatively autonomous intellectual debates within the Movement, often championed by bitterly opposed factions – including the centuries-old controversy over the rights and wrongs of restoration, the post-1918 demand for building of facsimiles of war-destroyed buildings and the rumbling mid-20th-century architectural tensions between the Conservation Movement and the Modern Movement.

These debates were often conducted with extreme passion, focusing intensely on specific issues such as authenticity, or on condemnation of 'fakes' and 'pastiche', the proper relationship between conservation and community, or the constantly fluctuating relationship between conservation and new architecture. At some times in past centuries, certain strands of conservation became almost identical to trends within contemporary architecture: for instance, the early history of the Gothic Revival in England and the role within it of church restoration, from the late 18th to the mid-19th centuries. In general, however, by comparison with the often highly legible narratives of new architecture, the progression of conservation theories and values, and their relation to the built fabric, were more oblique, roundabout and even circular.

The lack of 'legibility' of the history of conservation, in today's built fabric, can be illustrated by just one example, from the North German town of Hildesheim, which has at its core a tightly planned market square, dominated by a spectacularly gabled, timber-framed building, the Knochenhauer-Amtshaus. Historic photographs from the 19th century appear exactly the same as today, suggesting an exceptional case of heritage continuity; only minor details of today's ensemble hint at any date more modern than the 16th or 17th centuries. There is nothing at all to suggest that almost the whole of today's square, including the Knochenhauer-Amtshaus, in fact only dates from the 1980s, when a facsimile of an 'original' building was constructed, and that during the 40 preceding years since 1945, the site was initially an empty bomb-site, and then was occupied by a completely different street layout, including a much larger square and a modern multi-storey hotel

Figure 0.1 The inscrutability of conservation: from 'new' to 'old' in the Hildesheim Marktplatz

(a) 1972 view of the west side of the Marktplatz, showing the Hotel Rose (opened 1963 and demolished 1985); (b) 2005 view of the same site, now occupied by the facsimile Knochenhauer-Amtshaus (completed 1989)

block. We will return in a later chapter to that specific, extraordinary case, together with its implications for the 20th and early 21st centuries' increasingly passionate debates about facsimile-building and authenticity. Its importance here is to emphasise the frequent impossibility of 'reading' the history of conservation itself in the fabric and appearance of the built environment.

Conservation was a part of modernity that travelled not at the front but at the side or the back, shifting and moulding itself, chameleon-like, in reaction to developments elsewhere. Often, it served as a mirror of modernity, developing its values in reaction to the mainstream – old as opposed to new, static as opposed to dynamic, mixed as opposed to segregated, and so forth. Yet, often, that opposition-principle became muddied: in the mid-20th century, for example, modern architects and conservation leaders were in agreement on the need for absolute segregation and legibility of new and old. But at other times the Conservation Movement directly reproduced the most driving modern values, as in the mania for clearing wide open spaces around monuments, or the sharp 'newness-value' of the radical restorations by architect Viollet-le-Duc.

It goes without saying, of course, that this book takes no sides in any of these debates – in contrast to many conservation texts that declare certain key values (such as the opposition to restoration or facsimiles) to be sacrosanct, or one building to be superior to another in conservation terms. For example, Coventry Cathedral (1951–62, by Basil Spence) and the Dresden Frauenkirche (1993–2005) present two very different responses, by two different generations, to the task of building a new replacement for a cathedral bombed in World War II, incorporating fragments of the old in the new fabric. Coventry embraces the stabilised ruin of the old cathedral as a memorial annexe to a large new building in a 'modern-traditional' style. The Frauenkirche takes the form of a facsimile of the old building on exactly the same site, in which are physically embedded its surviving fragments, discernible at the moment by their darker colour. Which is the better work of conservation? Which is the more 'authentic'? This book makes no attempt to answer questions such as these. Its task is instead to record the often extreme opinions held about them by others. One of the Conservation Movement's foremost and most pugnacious theorists, John Ruskin, spoke of old buildings as having a 'life' of their own. In this book, the Conservation Movement emerges as a phenomenon with its own life, interacting constantly and at times passionately with architecture, politics, culture and society, but never quite tied to any of them.

How, then, is the book laid out? In order to keep constantly before the reader the fact that the Conservation Movement was a modern, dynamic, historical movement, it takes the form of an overarching narrative, within which a variety of complex strands and ideas are traced over time. Overall, the book is arranged chronologically, with a special emphasis on the 20th century, the prime century of mass movements and thus, arguably, the spiritual heartland and climax of the Conservation Movement itself. The story of conservation in the 18th and 19th centuries, and earlier, has already been covered extensively in previous accounts, including those of Erder, Choay, and Murtagh – and above all in Jukka Jokilehto's monumental *History of Architectural Conservation* (1999/2004).[3] As the numerous note references in the earlier chapters (especially 1–4) make clear,

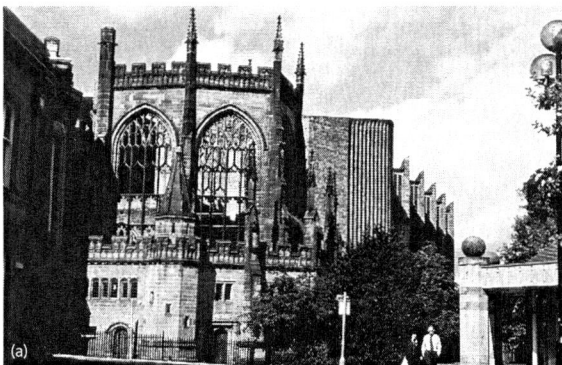

Figure 0.2 How to rebuild a war-damaged cathedral

(a), (b) Sir Basil Spence's solution at Coventry in 1951–62: new cathedral adjoining preserved ruins; (c) The Dresden Frauenkirche, 1993–2005: facsimile building incorporating ruin fragments

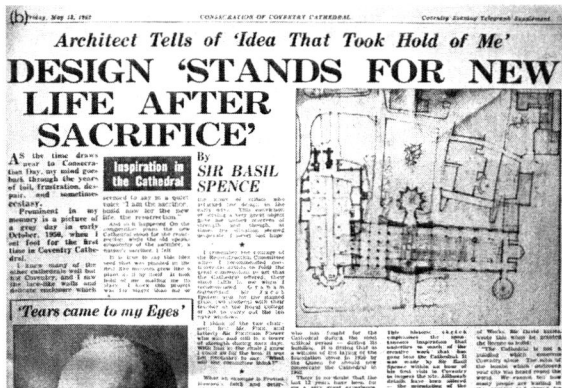

my account of those earlier centuries draws extensively on these works. The chapters dealing with the 20th century, however, are completely new, and deal with the story at comparatively greater length. Chapters 1–7 follow a single, linear narrative but, in the most complex phase of our story, the years 1945–89, when conservation rose to ascendancy in the built environment of many countries, the text is divided into parallel narratives, reflecting the sharp geopolitical division into competing blocs, and allowing both the competitive differences and the underlying commonalities to be expressed.

Geographically, the arrangement of the book reflects the especially strong interactions between the Conservation Movement and the modern ideology of nationalism. As we noted above, there was a constant tension between conservation theories common to most countries – such as the emphasis on 'rediscovery' of neglected masterpieces, or the restoration-versus-repair debate – and the forcibly competing national variations and trends. The book's layout is also shaped by the need to present a balanced variety of those many national trends and traditions in the 19th and 20th centuries, and to supply readers with the basic facts about key nations, bearing in mind also that the primary readership will be in the UK and the anglophone countries. For that latter reason, the story in Britain is given consistent prominence: each chapter from 2 onwards (except 10 and 11) presents not only an international overview of developments throughout the European and American territory of the Conservation Movement, but also a specific account of developments in Britain – including a proper acknowledgement of the strong and persisting differences in ideas and policies between England and Scotland. Among the other principal centres of conservation, the developments in Italy, France and the USA have been very fully covered in previous accounts. Although they, too, are dealt with at some length in this account, there is also a compensatory additional emphasis on the often seminal developments and debates in the German-speaking world – an area of conservation history distinctly under-represented in English-language publications.

The book finishes with a more speculative chapter, tracing developments in the two decades since the end of the Cold War, drawing strands together and looking to the future – as conservation has always done throughout its history. But here we face something of a paradox. If the Conservation Movement has always, despite its own propaganda, been a child of Progress and Western modernity, a product of the ideal of using the past for useful modern progress, then where does it stand today, in an era of postmodern relativism that has deconstructed the old modern certainties of grand narratives and normative values? We will return again to that unsettling question in Chapter 12 and the 'Epilogue'.

Foundations of the Movement

Care for old buildings in the pre-modern age

Harbingers of heritage

Antiquity, Christendom, Renaissance

Some people were talking about the temple and the fine stones and votive offerings with which it was adorned. He said, 'Those things you are gazing at – the time will come when not one stone of them will be left upon another; all will be thrown down.'

Luke 21, 5–7[1]

THE story of the modern Conservation Movement that dominates this book belongs over-whelmingly to Europe and America, and to the two centuries following the French and Industrial Revolutions. But the Movement's roots are long, stretching right back to classical Greece and Rome. Indeed, the very fact that the Conservation Movement can be interpreted as a 'movement' at all, is a consequence of the momentous changes that began in classical antiquity, in how people saw the relationship of past, present and future – a relationship within which the built environment played a central role.

In pre-classical civilisations, in the Middle East, India or China, there was little concept of historical progression, and the combination of ingrained social hierarchies and polytheistic religion encouraged a tremendous stability, often symbolised through monumental architecture: many sacred sites experienced multiple layering and reappropriation over millennia. In Ancient Egypt, everything was assumed to stay exactly the same, including the monarchy, with its dynasties spanning over two millennia and its pharaonic religion, interrupted only by the brief, monotheistic aberration of the reign of Akhenaten (1364–47 BCE). This unchanging cultural character was expressed in the stone-built monumentality of its architectural set pieces, such as the vast temple complex of Western Thebes, revered from 2000 BC to about AD 500 as the seat of Amun-Re, king of the gods. Reflecting Egyptian belief in the circular character of time, Amun was thought to return annually to be reborn in the Luxor temples (built cumulatively over several hundred years in c.1500–1230 BC). To the north was the massive Karnak Temple where Amun 'lived' for the rest of the year. Across the river in the 'land of the dead', the Valley of the Kings and of the Queens housed the royal dead of the New Kingdom in a succession of grand mortuary temples, such as that of Amenhotep III (builder of most of Luxor Temple) – a landscape that fused natural and built environments. Although these mortuary precincts often fell into ruin or were looted, their rise and fall followed no conscious framework of 'progress': the monumental commemorative statues that dotted them were not monuments in the modern sense, but religious objects, invested with unchanging divine force, like the buildings themselves. In the Mesopotamian civilisations, a similar theocratic traditionalism prevailed and sites

Figure 1.1 David Roberts, grand portico of the Temple of Philae, Nubia (from *Views of Egypt and Nubia*, c.1847–9)

such as the multi-layered ziggurats of Babylon were constantly reused. Later and more advanced civilisations elsewhere, such as China in the age of Confucius (551–479 BC), also emphasised stability and deference to ancestral practices and to the family unit.[2]

In all these eras and places, there was often an intense care for old structures, varying in character depending on whether local building tradition stressed masonry (as in the Middle East) or timber (requiring periodical renewal). Nowhere, however, was there conscious conservation in the modern sense, still less historically informed concepts such as 'authenticity'. Polytheistic religion encouraged a static view of a world in which the sacred was completely intermingled with perceptible reality, and in which everyday objects or places could be invested by gods or ancestral spirits. These had a real call on the respect of the living – and, in turn, their sacred aura could help bolster the ruling order among the living.

This situation only began to alter a little in Mycenean Greece – one of the first societies where *change*, rather than stability, became a norm. Traces of this process, the first 'fashions' in artefacts, are discernible through archaeological investigation, with pot-sherds traceable to a specific century. The restless, seafaring Greeks no longer assumed that the best thing was that which had existed since time immemorial: civilisation began to 'go faster'. And inevitably, in due course, that new restlessness was consciously articulated, in 5th-century-BC Athens's outpouring of written and built celebrations of the rise of its civilisation. In classical Athens, we have almost arrived at a concept of historical progress, with a semi-secular nation-state commissioning built monuments to its own advances. In 5th-century Athens, concepts, fashions and leaders followed each other with bewildering rapidity:

condemnation hot on the heels of praise. More happened here in a hundred years than in a millennium and a half in Egypt, and the Athenian Oath looked on to the future, pledging that 'we will transmit the city, not only not less, but greater and more beautiful than when transmitted to us'; the reconstruction of the Acropolis Wall after the Persian War emphasised the visible heritage of surviving masonry.[3]

But classical Greece, including the diverse Hellenistic world established by Alexander's descendants, was still a polytheistic society. Its most costly and monumental constructions were almost invariably associated with religion, and thus were mixed together with the need for respect to the gods and ancestors. Within the house, that meant maintaining household shrines – whose aura underlined the authority of the master over family and slaves. The same principle was displayed publicly, on a greater scale, in temples of priestly cults, supplemented by the monuments erected by great men to their achievements and conquests, including statues, triumphal arches, inscriptions, great public buildings and their own tombs. It was the duty of subsequent generations to honour and maintain these, as it was their duty to honour the gods.

In Greek art, the principle of imitation generally prevailed, and there was a strong concern for repair and restoration of old statues and temples. The intermixing of religion and secular life bolstered the ruling order, which invariably claimed the sanction of the gods, through founding myths or the claim of monarchies that their leaders were semi-divine themselves. In Greece, the word for the public built evocation of religious respect was *mnema* or *mnemeton* – literally, a spur to memory. In the built environment, this represented an ethical codification of the universal principle of respect and reappropriation of sacred sites. Destruction of temples was universally seen as a violation of morality. But the same could apply to secular monuments too. For example, in the 4th century BC, when Alexander the Great overthrew the Achaemenid Persian empire, he showed careful respect to the tomb of Cyrus the Great, 6th-century BC founder of the Persian empire. Shocked that it had been plundered by robbers, he had it repaired and had Cyrus's memorial inscription reproduced in Greek.[4] Destruction of temples, a much greater sacrilege, could be commemorated through leaving 'intentional ruins'. Following the second Persian invasion of Greece in 480–79 BC, an inscription at Plataea recorded that 'I will not rebuild any of the temples that have been burnt and destroyed by the barbarians, but I will let them be left as a memorial to those who come after, of the sacrilege of the barbarians'. And indeed, very few temples were built in Greece in the 30 years after the Persian War.[5]

Classical antiquity also saw the first restorations, not only of buildings but also of paintings and sculpture – for example, of the Parthenon in 398 BC after the Peloponnesian War. But by the Hellenistic period and later antiquity, sites such as the Athenian Agora, and the Delos and Olympia complexes had become vast open-air museums. In the description of Greece by Pausanias, a traveller of the 2nd century AD, renowned ancient monuments were recorded as still in use, often through like-for-like replacement of decayed elements – as with the wooden columns of the 500-year-old Temple of Hera at Olympia. The Temple of Zeus in that complex, originally built in c.470–50 BC, was repeatedly rebuilt in the following centuries, until finally the statue of Zeus was removed by a private collector to Constantinople in the 5th century AD. Plutarch, writing around AD 75, related the story of the boat of Theseus, allegedly preserved by the Athenians until 310 BC by successive replacement of rotted planks: contemporaries had speculated whether the resulting object was the 'real' boat or not. The first secular relics were bound up with religious sites: Vitruvius recorded that 'in Athens, there is on the Areopagus an example of ancient building roofed with clay to this day'.[6]

Figure 1.2 The Temple of Hera at Olympia, following the German excavations and reconstructions of the late 19th and early 20th centuries (1975 view)

Pietas and polytheistic heritage in classical Rome

As in other fields, the Romans vastly elaborated these Greek themes. Rome's relationship with Greece, however, was quite different from that of modern civilisation to classical antiquity. Certainly, Rome embraced not only Greek classical architecture but also Greek art-collecting. This had begun in the Hellenistic period with the Attalids and Pergamon, and influential Romans followed suit: beginning with the destruction of Corinth in 146 BC and culminating in the systematic plunder by the dictator Sulla (from 82 BC) and the corrupt governor Verres, a tradition became established of the accumulation of private collections of looted Greek art. Equally, Greek civilisation was seen as a model, to be admired and protected: under the rule of Augustus, the Athenian Agora was augmented by a 5th-century-BC Temple of Ares, relocated from Acharnai, and the Hellenophile, 2nd-century-AD emperor Hadrian collected Greek art on a vast scale, especially in his Tivoli villa complex, and helped to preserve Greek cities themselves.[7]

But there was no sense of separation from Greece: its monuments were not seen as 'historic'. Rome's attitude to Greek religion was one of unstructured continuity; building on the same pantheon, they recognised a similar duty of self-interested respect to the world of the sacred, including gods and ancestors, a duty for which the Latin word was *pietas*. This meant something very different to its modern derivative, piety. The Latin equivalent of *mnemeton*, *monumentum*, carried the added overtone of warning, from *monere*, to warn. In the Roman world, a monument was a physical focus of *pietas*. The scope of the term was surprisingly wide, including not just statues or individual buildings but also 'cultural property' in almost a modern sense, including even entire towns. Monumentality in imperial Rome implied not only durable and imposing structures capable of evoking the past, but also a degree of sumptuousness that would perpetuate the fame of the patron and help uphold wider social norms and power structures.[8]

In Rome, with its driving sense of imperial mission, the public demands of *pietas* also encompassed the glorification of the city's power and history, and of its '*dignitas*' in general. The power of the empire and the emperors was bound up with the gods and their temples. The goddess Vesta in Rome was the goddess of hearth and home *and* the protector of the Roman nation. From the founding of the principate (imperial system) by Augustus at the end of the first century BC, there was an increasing shift towards an explicitly historical and secular approach to *pietas*, including ideas of continuity between past and future.

This was most explicitly articulated in literature, in the *Aeneid*, the epic written by the poet Virgil to exalt Rome's destiny. Here, symbolic objects, and the *pietas* of the poem's hero, are shown as bound up with the historicised future of the city, and the epic literary heritage of Greece and Homer is appropriated for the Roman cause. In Book 8, echoing the *Iliad*, Aeneas' mother, the goddess Venus, gives him a special shield on which Vulcan had 'wrought the story of Italy and the triumphs of Rome', including, of course, those of Augustus; Aeneas looked in wonder at these scenes and then 'lifted on to his shoulder the glory and destiny of his heirs' (*atollens umero famamque et fata nepotum*). Earlier, in Book 6, he had witnessed in the underworld Rome's unborn heroes, 'souls of renown now awaiting life, who shall succeed to our name' and build great cities 'which are now nameless places, but whose names shall be famous one day'.[9] Although expressed in the semi-religious language of classical mythology and epic poetry, the *Aeneid* was actually a secular manifesto in all but name. We should remember its language of destiny, and of the conscious interrelationship

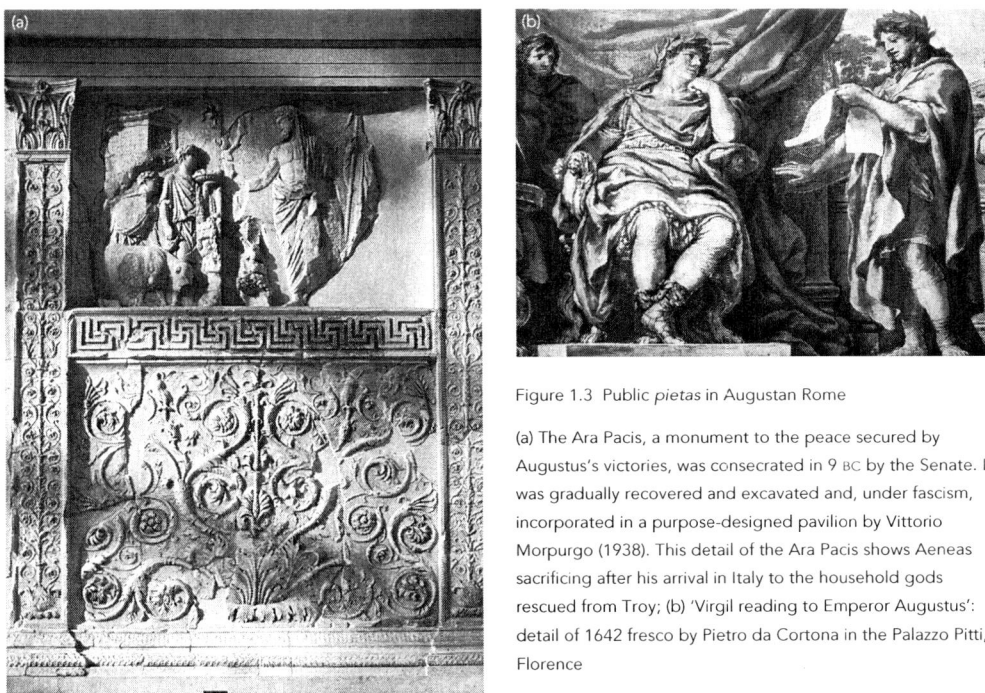

Figure 1.3 Public *pietas* in Augustan Rome

(a) The Ara Pacis, a monument to the peace secured by Augustus's victories, was consecrated in 9 BC by the Senate. It was gradually recovered and excavated and, under fascism, incorporated in a purpose-designed pavilion by Vittorio Morpurgo (1938). This detail of the Ara Pacis shows Aeneas sacrificing after his arrival in Italy to the household gods rescued from Troy; (b) 'Virgil reading to Emperor Augustus': detail of 1642 fresco by Pietro da Cortona in the Palazzo Pitti, Florence

of past and future, when tracing the dynamic ideologies of more recent centuries – including the Conservation Movement.

Rome's own fabric, although looked on with reverence, was highly unstructured in character. There was no concept of public land to allow concerted urban planning. Unlike the rigid Roman grid plan of smaller provincial cities and military settlements, or the memorable 'iconic' structures that dominated some smaller centres, such as Alexandria or Athens, within the *caput mundi* itself monumental, processional splendour was informally jumbled together with private squalor. The city's rich complexity was encapsulated in every fragment, but, by the 4th century AD, it was considered sacred as a whole, and visitors tended to follow standard itineraries, beginning at the Palatine and proceeding to the Forum complexes. That veneration of cities as organic amalgams of public and private applied throughout the empire: Plutarch argued that 'A city, like a living thing, is a united and continuous whole. It does not cease to be itself as it changes in growing older, nor does it become one thing after another . . . but is always at one with its former self in feeling and identity'.[10]

Within this framework (misleadingly resembling 20th-century conservation's rhetoric of the informal, 'living' town), the individual elements of the urban fabric were of lesser importance, especially given the frequent fires and destructions that afflicted public and domestic buildings during peacetime: Pompeii was only the most dramatic example of this. The built substance of the shrines around Rome's Forum was frequently renewed, as Augustus boasted in his *Res Gestae*; in turn, Septimius Severus restored the Portico of Octavian in 203, and in 203 and 250 the upper storeys of the Colosseum. During the empire, formal provisions for protection of the built fabric of Rome and other urban centres coalesced. Hadrian's principate saw a 'code of private building', prescribing stringent penalties for sale of private houses for demolition; similar penalties for demolition were recorded earlier in Herculaneum.[11] The ahistoric character of *pietas* meant that repairs and alterations to older buildings, however respectful, were not invested with any kind of sense of historical rigour or legibility, as modern conservation ethics demands: for example, when Hadrian built the Pantheon in the 2nd century AD, he crowned it with an inscription referring to the predecessor building on the site, built by Augustus's general, Agrippa.

The extreme mutability of symbolically venerated built fabric in Rome was highlighted in the fate of one idiosyncratic group of structures on the Palatine Hill, residence of Augustus and his successors. This group embodied both humble-family and grand-national *pietas*, and they were preserved not for their grandeur but as exemplars of spartan modesty – a cherished element of Roman traditional morality. By the late Republic, the focus of the group was a circular, thatch-roofed structure, which supposedly represented the hut of Faustulus, foster-father of Romulus and Remus. In its form, the hut resembled the small circular buildings used to house funerary urns during the first millennium BC in central Italy; it was constantly hailed by moralists as an example of the humble origins of Rome's glory. However, its substance was several times renewed following fire – first in 38 BC during a ritual, when burning altar sacrifices landed on its roof, and then again in 12 BC. In each case, the hut was 'restored' again at once and, throughout the early Christian era, it survived in good repair. Also on the Palatine, Augustus's own ostentatiously modest home was preserved alongside the palaces of later emperors. On the Capitoline Hill, a further thatched 'hut of Romulus' was preserved, to remind citizens (in Vitruvius's words) of 'the ancient ways and their memory'.[12] Although these structures seemed to prefigure the modern age's preserved 'houses of great men' in their mixture of humble-family and grand-national 'memory', they were still rooted in the religio-secular world of *pietas*.

Decay, recycling and *spolia* in early Christian Rome

BEGINNING with the emperor Constantine in the 4th century AD, a shift began in the interwoven relationship of old and new in Rome. Grand buildings were no longer constructed out of fresh materials, but made out of stone procured by demolishing older structures – a predatory relationship that would continue, in various forms, down the centuries to the present. People were in two minds about this process, condemning it in others if it went too far or too fast, while often practising it themselves. As early as the 1st century BC, Augustus initiated one of the first systematic campaigns of 'international monument plunder', following his annexation of Egypt, when he ordered the transporting of several obelisks to Italy (in 13–10 BC), at massive cost and effort.[13] From the reign of Constantine onwards, it became perfectly respectable to display parts of venerated older buildings as '*spolia*' on new buildings: *pietas* was compatible with cannibalising other buildings for approved ideological purposes. The Arch of Constantine (AD 315) has columns and part of an entablature from an Antonine building, and reliefs and statues from Trajanic, Hadrianic and Antonine monuments, all built by 'good' emperors. These *spolia* were freely doctored where required: for example, by removing a head of Hadrian from a relief, and substituting one of Constantine. Following damage or fire, surviving elements were reassembled at will, or incorporated into successor buildings – as with the Temple of Saturn in Rome, after a fire in the 5th century AD.[14]

Away from the capital, the fate of public buildings was a more haphazard affair. Within Italy, even small towns were normally equipped with a standard collection of public buildings, including town walls, paved streets, baths, temples, forum buildings and statues. For these, though, there was no systematic provision of maintenance, leading to cycles of neglect followed by major civic repairs to structures that were recorded as 'collapsed from age' or 'derelict from long neglect': especially vulnerable were fragile structures such as aqueducts.[15] By the 2nd century AD, a law of Antoninus Pius exhorted repair rather than unnecessary new building – a preference which would be repeatedly echoed in later centuries.

From the 4th century AD, the monumental legacy of past Roman glory was ever more disrupted by the irresistible forces of imperial disintegration and monotheistic religion, rising to a climax in the rule of Constantine. He decided in 330 to move the capital to the renamed Constantinople – a move followed by the usual mass migration of works of art from across the empire to the new capital. The secular pride in Rome, the Universal City, continued, and Christianity strove to appropriate it. But with most temples rendered redundant, the old supporting networks of public shrines and household altars began decaying away.

Christian theologians, whose own discourse and sacred book were also arranged in strongly narrative form, wrestled with the problem of human history. In the early 5th century, St Augustine argued that time was continuous and irreversible, stretching from the creation to the present, and in the 19th century, John Henry Newman would argue that 'in a higher world it is otherwise, but here below to live is to change, and to be perfect is to have changed often'.[16] But there was a profound ambivalence: history, or *tempus*, time, was a reality, but negative and transitory, a litany of trials: Christians, looking towards God's glory and the end of the world at any time, had no real need of history. In terms of secular conceptions of the monument, Christianity was a step back by comparison with the *Aeneid* and the Age of Augustus. The practical effects of this on the integrated public-private infrastructure ethos of *pietas* were very destructive. This seemed puzzling to Christian Rome, which had appropriated the imperial chosen-city rhetoric, but now faced the fact that it was

actually declining under Christianity where it had flourished under the 'pagans'. In the long run, however, the crusading passion of monotheistic religion would enhance the force of all 'missionary' secular movements – including modern conservation.[17]

Within a monotheistic system, inanimate objects lost much of their potential aura of sacredness – although ordinary objects could still be invested with a symbolic charge: in the Book of Exodus, an ordinary object – manna – symbolised the continuity of 'chosen' status. Within Judaism, the sharp polarisation between oppressive everyday reality and heavenly promise had assigned a metaphoric role to architecture: 'Let us build a city and tower that reaches to the heavens, to make a name for ourselves.' This language was reinforced by Christianity. Rather than the literal Rome, the *caput mundi*, now the aspiration was the heavenly city: 'the holy city, new Jerusalem, coming down out of heaven from God'. And buildings' ruination and decay could be seen as a metaphor for the transience of human existence and mortal pride.[18]

Two tendencies now began running in parallel, and at an ever increasing pace and scale: attempts to stave off the decline and disintegration of old structures, and construction of new buildings using materials from old. This interplay between protection and active destruction would continue into the modern age, mutating eventually into the 19th and 20th centuries' combination of mass preservation and destruction.

In the matter of simple decay and destruction, one particular building type, the temple, experienced particularly precarious fortunes in later imperial Rome. The classical gods were officially banned, and labelled 'pagan', in 341.[19] At that point, in a harbinger of later 'reformations', the temples were simply closed, but remained state property. As with the later mass redundancy of Catholic religious buildings following the Reformation, the question of what to do with them was far from straightforward. They could not be converted wholesale to churches: their grand scale and orientation towards external processions and display of statuary made them unsuited to the more private, introverted character of Christian worship. Likewise, with the decline of public *pietas* and the emergence of church art, the collecting of works of art went into decline: some statues were removed from temples and put on public display.

In the late 4th century, the emperors Julian the Apostate and Symmachus briefly revived tolerance towards the older religion; in most of Italy, from the 380s, temples were allowed to rot slowly. But this was followed by a fresh burst of Christian zeal; in 399 Arcadius and Honorius demanded fresh efforts to root out 'idols' and sacrificial altars, while stressing that the buildings themselves must 'remain unimpaired', becoming public property and transferred to other uses where possible. Elsewhere in the empire, 'pagan' remains from other civilisations suffered similarly: in Egypt, the replacement of the pharaonic religion by Christianity, and later Islam, led to systematic recycling of tomb and monument structures, to remove all traces of the former cults. Some were converted to hermitages, houses or churches; others were dismantled for building materials. As a result, the religious complexes of northern Egypt largely disappeared. In other provinces, overrun by 'barbarians' with no concept of *pietas*, the built fabric was simply abandoned, as with the towns of Roman Britain.[20]

facing page

Figure 1.4 Reuse of classical monuments in Rome

(a) The Arch of Constantine (AD 315), faced with sculptural *spolia* from Trajanic, Hadrianic and Antonine monuments; (b) The church of SS. Cosmas and Damian (converted AD 526–30); (c) The Pantheon (2nd century AD; converted into a church after 608); (d) The church of S. Lorenzo in Miranda (built in the 8th century AD inside the cella of the temple of Antoninus and Faustina)

During the early empire, the gradual appropriation process continued the long-standing practice of recycling building materials and *spolia*. Within 40 years of the removal of the capital to Constantinople, decrees were issued forbidding the construction of new buildings in the 'Eternal City of Rome' out of old materials. Much of this destruction was clearly being carried out by public officials. In 376, the emperors Valentinian, Gratian and Valens forbade public magistrates to

> undertake any new structure in the renowned City of Rome [rather than] improving the old. If any person should wish to undertake any new building in the city, he must complete it with his own money and labour, without quarrying out of old buildings, digging up the foundations of noble buildings, obtaining renovated stones from the public, or tearing away pieces of marble by the mutilation of despoiled buildings.[21]

A series of similar laws followed, usually forbidding despoliation of old buildings for new or appropriation of monumental spaces by private residences, and stipulating removal of shoddy private infill structures from grand public buildings – a plague that equally affected the new capital of Constantinople. To combat decay, building maintenance in Rome was stepped up, under the control of the urban prefect. There were also sporadic efforts at repair outside Rome, for example of the baths at Antium (Arezzo) in 379–82. But, by the 5th century, several massive church-building projects were underway – the Basilicas of St John Lateran and St Peter – requiring large quantities of antique materials: both incorporated over 100 reused columns, not counting courtyards (for comparison, one of the largest of the old temples, that of Venus and Rome, only had 58 peristyle columns). From the 5th century, Rome was above all famed as the seat of the see of St Peter, with pilgrims flocking to the sites of 'martyrdom'.[22]

The following century, following the Visigothic sack of 410, the pace of destruction in Rome accelerated considerably, and a law of the emperors Leo and Maiorian of 458, issued in Ravenna, forbade plundering of stone and bricks from old buildings, arguing that, 'in order that something small may be repaired, great things are being destroyed'.[23] After the fall of the western empire, a determined effort to stem the tide of decay was made by the Ostrogoth king, Theodoric (493–526), whose outlook was shaped less by Christianity than by his classical education in Constantinople. However, he venerated Rome chiefly for its past glory, rather than as a living organism, arguing that

> these excellent buildings are my delight, the noble image of the empire's power and the witnesses of its grandeur and its glory. It is my wish that you shall preserve in its original splendour all that is excellent, and that whatever you may add will conform to it in style.[24]

In 500, Theodoric appointed an *architectus publicorum* and *curator statuarum* and began patching up major monuments such as the Colosseum and Theatre of Pompeius; statues were moved from outlying temples to the Forum to protect them from theft. Theodoric's chief minister, Cassiodorus, records a special concern for commemorative monuments and inscriptions: 'the forest of walls and the population of statues which make up Rome' were under constant threat from evil-doers, despite penalties rather more severe than those of modern conservation laws: 'Rightly does the public grief punish those who mar the beauty of the ancients with amputation of limbs, inflicting on them that which they have made our monuments to suffer.' At night, when the possibilities of theft were most tempting, the statues gave forth a tell-tale 'ringing sound . . . under the blows of the thief'. Increasingly, permission was granted for disused porticoes in the Forum to be converted into private houses, or churches: in 526–30, Pope Felix IV converted a public building on the north of the Forum into the church of SS. Cosmas and Damian, adding an elaborate new Christian mosaic. But any care and protection was always precarious: by 536, during a fresh sack of Rome by Goths, the city's defenders were reduced to pelting the attackers with Greek statues from the top of the Mausoleum of Hadrian.[25]

The dispersal of 'monumental Rome'

BY THE 7th and 8th centuries, the general replacement of the Roman empire by the spiritual empire of 'Christendom' (or Islam) had stimulated a growing stress on local diversity, in the burgeoning national monarchies. Alongside the political successor empires (Byzantine, Holy Roman), Christendom appropriated the Roman empire's sense of destiny and heritage, beginning to cast Rome itself as a seductive lost vision that could inspire attempts at revival, or 'renaissance'. During the period of Byzantine control after the fall of the Ostrogothic kingdom, a degree of public control survived, with attitudes to historic fabric wildly fluctuating between care and plunder. In Rome, when Pope Boniface IV (608–15) proposed to convert the Pantheon into a church, he had to petition the emperor Phocas for permission; but in 663/4, a personal visit by the emperor Constans II resulted in the pillaging of all bronze roofing from city monuments, including the Pantheon. After the 7th century, control of Rome's ruinous public buildings passed decisively to the popes. Building resources were diverted from traditional secular patronage to the construction of churches, especially in Rome, Milan and Ravenna, while security worries encouraged systematic adaptation of Roman town walls.[26]

Between the 8th and 12th centuries, the antique seemed both impenetrable and near at hand. With the lack of any systematic concept of historical age, sections of ancient Roman buildings were gradually incorporated into churches in Rome or exported throughout Italy and beyond; in the beginnings of 'archaeology', Christian *loca sancta* ('holy places' connected with saints) were often excavated. Antique building fragments could be very valuable: in 821, for instance, Aripert, rector of San Donato in Lucca, exchanged four stone columns for a piece of land. For much of the Middle Ages, Rome was basically a huge quarry of raw material, its population ebbing away. By the mid-8th-century visit of the unknown Swiss monk who compiled the *Einsiedeln Itinerarium* – the first 'guide' to the city – the Forum was largely a field of ruins. The few major surviving buildings were concentrated around the steps of the Capitol: the Arch of Septimius Severus, the temples of Concord and Vespasian. Others had disappeared or become churches; in 625–38 the Curia Iulia (Julian Senate House), repaired by Diocletian following a fire in 283, became the church of S. Adriano: this was probably a *pietas*-driven act of conservation, safeguarding the core building of Roman government. In the 8th century, the church of S. Lorenzo in Miranda was built inside the cella of the temple of Antoninus and Faustina in the Roman Forum. Other monuments had been encased in fortified complexes, such as the Arch of Titus, incorporated in the castle of the Frangipane barons, or Hadrian's Mausoleum, which was fortified from the mid-6th century by the Byzantines and the Goths and then, from the 9th century, used as a redoubt by the popes in opposition to the city. By 1000, only 25 per cent of the area of the ancient city was still inhabited.[27]

Frequently, classical architectural *spolia* were reused in new buildings, although this was seen as a slightly inferior solution to newly cut masonry. Theodoric's new palace in Ravenna comprised a mixture of newly carved marble and *spolia* from Rome; Charlemagne's new palace chapel at Aachen (*c.*AD 800), with its resemblance to Theodoric's church of San Vitale in Ravenna, was internally lined with columns, mosaics and marble slabs from Rome and Ravenna, as local stone was not available. In a series of 8th- and 9th-century churches in Rome and elsewhere in Italy, *spolia* predominated in the building materials. One of Charlemagne's scholars, Alcuin, bemoaned that Rome was 'now a pitiful ruin, the wreck of its glory of old'. And the Byzantine emperor Justinian's grand new church of Hagia Sophia in Constantinople (532–7) was adorned with lavish *spolia* from across the eastern empire – although, subsequently, this showpiece was repeatedly wrecked by earthquakes and fires.[28]

Outside Rome, the substance of towns gradually mutated through decay or adaptation. In Verona, a description of *c.*800 and a 10th-century drawing indicated the great amphitheatre still intact. In Spoleto, the amphitheatre was converted into a fortress by the Goths during their mid-6th-century war with the Byzantines, while, in Arles, the amphitheatre was infilled with a huddle of later dwellings. In 7th-century Syracuse, the Doric columns of the 5th-century-BC Temple of Athena were encased in a new cathedral (whose present Baroque facade dates from 1725–53). Especially persistent were the orthogonal town plans of the Roman provincial settlements, in places such as Pavia, Florence and Verona, through public control of the alignments of the *via publica*, even as the magnificent Roman paving was allowed to decay. Two-thirds of the Augustan towns of Italy survived late antiquity and the early Middle Ages in some form. However, there were still acts of systematic destruction, as when Pope Boniface VIII in 1298 ordered the destruction of the classical complex of Palestrina, near Rome.[29]

In Greece, as with Rome under Theodoric, the Byzantine period saw determined attempts to revive the past glories of classical Athens in Christianised form, including conversion of the Parthenon into a church (later, cathedral) and centre of pilgrimage. Restored following severe damage in AD 267 and 396 (when it was sacked by the Goths), the Parthenon reached the apogee of its Christian fortunes in the 12th century, with an interior richly glowing with mosaics. After the conquest of Athens (and Constantinople) by Crusaders in 1204, the exiled bishop Michael Choniates bemoaned the despoliation of 'the holy Acropolis of Athens . . . and the most holy Parthenon of the Mother of God upon it, which has now become a den of thieves'. Subsequently, under Ottoman rule, the building underwent an extended period as a mosque prior to its destruction by the Venetians in the late 17th century and its post-1800 elevation into an object of international secular veneration.[30]

Outside the Mediterranean core of the empire, the old universal narrative of classical Rome faded from view, and the concept of *monumentum* narrowed again to a restricted definition of inscriptions and statues. The principle of *pars pro toto* became very important, with relics and fragments symbolising continuity in tradition. In Charlemagne's Aachen chapel of *c.*AD 800, the royal precedent of Theodoric was evoked by incorporating antique columns from Ravenna. Conversely, the early-11th-century St Michael, Hildesheim, incorporated small relics set into the capitals of new columns – the objects being far 'holier' than any entire old building could be.[31]

facing page

Figure 1.5 Decline and fall

(a), (b) Reconstruction views of the Forum of Nerva, Rome, when
newly completed, *c.*AD 100, and during the early Middle Ages

Figure 1.6 Hagia Sophia, Constantinople: 1984 interior view

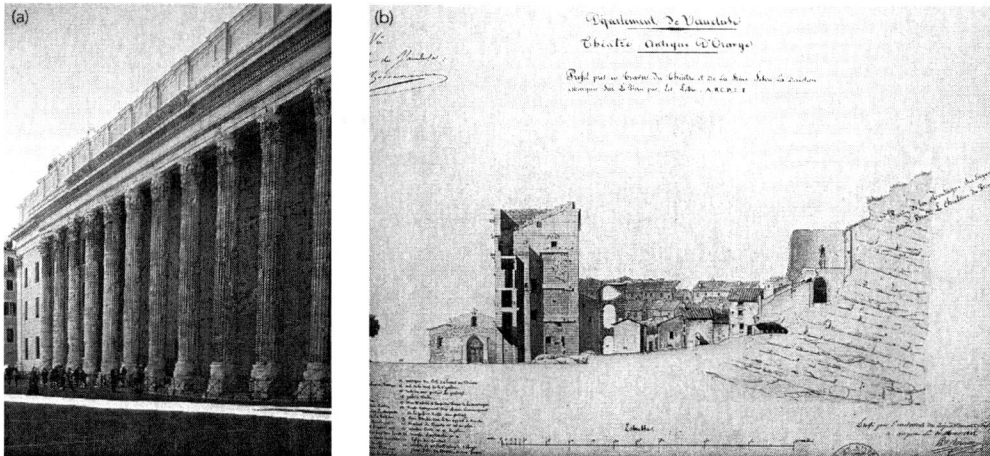

Figure 1.7 The embedding of classical remains

(a) Colonnade of the Temple of Hadrian, built in AD 145 by Antoninus Pius, incorporated in Carlo Fontana's late-17th-century Borsa Valori di Roma (Stock Exchange); (b) 19th-century cross-section of the Roman Theatre at Orange, France (built in the 1st century AD and closed in 391), showing medieval infilling with houses. The theatre was steadily restored during the 19th century, beginning in 1825 at the instigation of Prosper Mérimée; a 'Roman Festival' was inaugurated in 1869, and the tiered seating was installed by the end of century

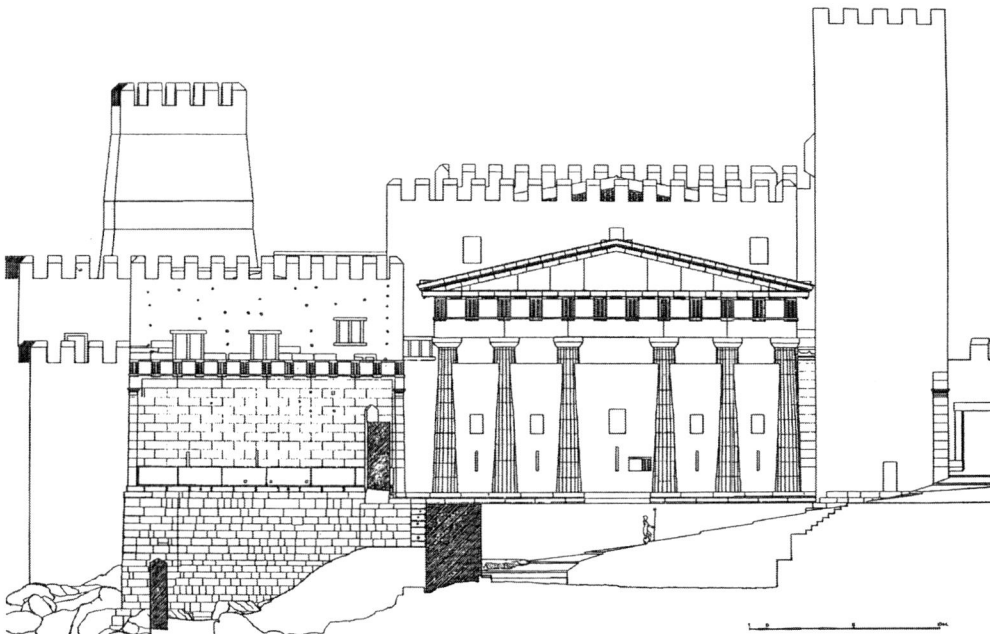

Figure 1.8 Reconstruction elevation of the Propylaea of the Athenian Acropolis as at c.1450

Figure 1.9 Aachen Cathedral

(a) External view, showing the Chapel of c.AD 800 at the centre; (b) Interior of the Chapel (taken c. 1900), showing reused antique columns from Ravenna

By the 10th century, Rome's remaining cultural prestige was sustained through the power of the papacy. Every emperor of the German Holy Roman Empire from Otto I in AD 962 to Karl IV in 1355 had to make the long journey south to be crowned by the Pope, and even after the worst-ever sack of Rome in 1084 by the Normans, it remained the destination for large numbers of pilgrims and other visitors. For their benefit, the surviving columns of Marcus Aurelius and Trajan were repaired in the early and mid-12th century; and the *Mirabilia Urbis Romae* – a 'guidebook' based on the Einsiedeln Itinerary – was produced.[32]

The Middle Ages: local narratives of care and conservation

DURING the early Christian centuries, Christianity's spread to outlying parts of Europe was generally respectful of local cultural traditions. In 601, Gregory the Great instructed St Augustine's mission to convert the Anglo-Saxons that they should avoid destroying shrines except where they were actively anti-Christian. Just as in the prehistoric era, with its repeated appropriations of older sites, there were many instances of churches being built at holy wells, stone circles or sacred hills, or of standing stones being converted to crosses: the Pictish standing stone tradition, for instance, was respectfully Christianised with cross motifs. But there are also reports in Bede of shrines being destroyed. There was also, throughout the Middle Ages, a trend to reuse earlier Christian holy sites, as well as the cult of saints' relics – as with the preservation for a time of the stake on which St Cuthbert used to tie himself up and pray in the sea at Inner Farne, north-east England. The concept of national heritage was only at an embryonic stage: Geoffrey of Monmouth's 12th-century *History of the Kings of Britain* claimed Stonehenge had been magically brought from Ireland in 483 by the wizard Merlin. But the pattern of local initiatives of care and reuse of ancient structures had been decisively set.[33]

In medieval Italy, the processes of appropriation and adaptation were more complex – as epitomised in the case of Syracuse Cathedral. Many churches comprised successive layers of *spolia*, used to add antique authority. For example, in the 12th-century Santa Maria in Trastevere, the nave arcade incorporates materials from the Baths of Caracalla. Individual works of public art often underwent complicated sequences of ownership: the 'Horses of San Marco', originally Roman 2nd- and 3rd-century (AD) statues, passed to Constantinople in the 4th century, Venice in the 13th and, briefly, later in the 19th century to the Arc du Carousel in Paris, before their eventual 'return' to Venice.[34]

Medieval Italian towns presented a radically different image from those of classical antiquity: Rome, Florence and others were dominated by forests of fortified noblemen's towers. With its perpetuation of a strong tradition of civic life, it is unsurprising that 13th-century Italy saw the first attempts at civic control of the built heritage. In San Gimignano a 1282 law made it illegal to demolish anything, except as part of an enhancement scheme. In Siena, 1295 saw a pioneering urban planning measure, regulating shape of windows and distance between buildings. The concept that a city's political strength was tied to its physical condition, originating in the early 14th century if not before, was expressed in Ambrogio Lorenzetti's 1338–40 'Allegory of Good and Bad Government' murals in the Palazzo Pubblico in Siena, with declining morals symbolised by dilapidated buildings.[35]

Figure 1.10 The nave arcade of the 12th-century St Maria Trastevere, Rome, showing *spolia* from the Baths of Caracalla

With the fading in power of the Roman centre and growth in national and local diversity in the Middle Ages, local narratives of care and conservation emerged and became more prominent. Owing to the difficulty of travel, contemporary medieval architecture, with its local variations and rapid fluctuations in styles, was often mistakenly seen by contemporaries as offering an accurate reflection of the Christian heritage of Rome or Jerusalem. From Abelard onwards, the Temple of Solomon was an ever-recurring ideal image, even in its proportions. But these constant changes and variations in the styles of medieval architecture – so different from the essentially unitary tradition of classical antiquity – posed designers new problems in reconciling the styles of new and old work, completing unfinished structures, or putting right damage or destruction.[36] These problems of reconciling new and old, and the diversity of architectural responses, have continued ever since. We should not, however, anachronistically project modern culture's sharply defined historical awareness onto the medieval world, where the old-new relationship was very imprecise – not least because the most monumental structures of the Middle Ages were only achieved through building programmes protracted over centuries, with little or no concept or expectation of a unified design. Eventually, Renaissance figures such as Alberti would turn that system upside down.

Within 'completion' schemes, some designers did straightforwardly use the style of their day, disregarding the existing buildings. For example in England, at Bolton Abbey, in 1520, a tower in the latest Perpendicular style was planted by Prior Richard Moon in front of the 13th-century Early Pointed nave: his scheme was abandoned uncompleted in 1539 at the Dissolution of the Monasteries. A frequent medieval pattern was the fusion of a later church in a new style, with an older tower. In the Champagne church of Notre-Dame de Donnemarie-en-Montois, a tall 12th-century tower was embedded in an early-13th-century church in the modish style of Reims Cathedral. Likewise, in the 14th century, Abbe Suger demolished part of the abbey of St Denis as part of his building scheme.[37]

In other cases, old styles were perpetuated for ideological reasons that can only be conjectured. In 1377, Henry Yevele rebuilt the nave of Canterbury Cathedral in the latest Perpendicular style, whereas in 1387–1400, continuing works by John Palterton commissioned in 1375–6 by Abbot Litlyngton to complete the 13th-century nave of Westminster Abbey, he copied the by-then-unfashionable style of Henry III's church. Some extension works evoked a more archaic period of architecture, such as the Romanesque: for example, in the 14th-century crossing arcade of Bamberg's Carmelite Abbey, or the south tower of Xanten Cathedral, or the much later (16th-century) additions to the Vienna Stephansdom, and a number of buildings in the Augsburg/Regensburg area. In 15th-century Scotland, at Dunfermline Abbey, upgrading of the nave combined contemporary late Gothic with a 'neo-Romanesque' matching the older parts of the building. Perhaps the neo-Romanesque of later medieval Scotland was a reaction to the 'Perpendicular' fashion of contemporary England – but there was not yet, in the late Middle Ages, anything resembling modern nationalism. Nor was there any question of specific ideas of historical authenticity, although, in the frequent efforts at recon-struction or restitution of damage, there was certainly some inchoate awareness of historical context.[38]

What is difficult for us, today, is to differentiate between practical conservatism and ideo-logically charged love of the past, given the diversity of individual approaches and lack of documentary evidence. During the Middle Ages, there are plenty of examples of what at first glance seem to be conservative repairs. For example, in a Byzantine church just outside Constantinople, damaged in an earthquake in 869, which the Emperor Basil wanted to completely replace, he was dissuaded and restored only the parts which had fallen down – such as the dome. Later, when the choir at Canterbury Cathedral was rebuilt after a fire in 1174, there were attempts to salvage the

Figure 1.11 Westminster Abbey's 13th-century nave, extended 'in keeping' from 1375 (by masons John Palterton and Henry Yevele) and western towers (completed in 1735–45 to designs by Thomas Hawksmoor)

remnants of the Romanesque 'Conrad's Choir': William of Sens was engaged to do the job, as he claimed he would retain much of the fabric. In the event, he failed to do so, causing consternation when the scaffolding came down.[39]

In Germany, two post-fire rebuilding schemes of *c.*1200 were both seemingly shaped by symbolic considerations, with strongly contrasting built outcomes. In Bamberg, a fire of 1185 gutted the cathedral, built in 1012 by Heinrich II (who was buried there). Following protracted debates about replacement, eventually it was decided to replace the ruins with an entirely new structure, evoking the old plan-form, with both an east and a west choir, and incorporating saints' relics: heated debate within the chapter about whether the new roof should be flat (like the old cathedral) or stone vaulted (in the contemporary fashion) was resolved in favour of the latter, 'modernising' option. At Magdeburg, by contrast, a 1207 fire, which destroyed Otto the Great's 10th-century cathedral, with its *spolia* columns from Ravenna, was followed by a rebuilding scheme that physically incorporated surviving antique columns in the apse, but within a completely new, Gothic design and layout. Both extremes showed respect for the past, but within a *pars pro toto* approach not requiring preservation or restoration of old buildings. As earlier at Aachen and Hildesheim, either the incorporation of small fragments in new fabric, or generalised evocation, was enough.[40]

The tension between conservatism and 'modernity' in rebuilding projects continued into the later Middle Ages. At Reims, the 15th century saw completion of the west towers in a style matching

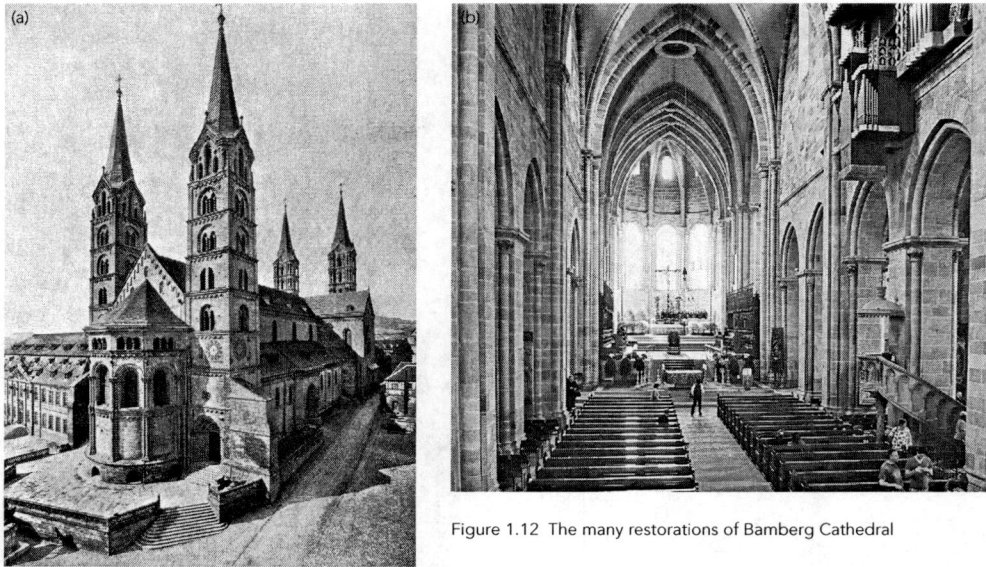

Figure 1.12 The many restorations of Bamberg Cathedral

(a) Exterior view of the cathedral as rebuilt following the 1185 fire; (b) Present-day interior view, 2010, showing the outcome of a succession of 19th-century 'purifying' restorations (by F K Rupprecht, 1828–31, K A von Heideloff, 1831–4 and Friedrich von Gärtner, 1834–7) which first banished 17th-century Baroque decorations and then vied with each others' interpretations of authenticity

the *c.*1250 body of the cathedral, and a major repair scheme to the arcading around the base of the roof, conforming in general massing to the original, but with late medieval detailing. Several other French cathedrals, such as St Quentin, Evreux and Auxerre, also underwent late medieval restorations that balanced conformity and innovation. At the abbey church of St Pierre de Corbie in Picardy, demolition of an old Romanesque church in 1502 was followed by an extraordinarily protracted replacement project, beginning with a 'contemporary' late Gothic choir and crossing and only completed with a neo-Gothic nave and towered west facade (in the general style of Notre-Dame de Paris) in the 18th and 19th centuries. In detail, too, care was often taken to respect existing fabric: an early-15th-century repair of the 12th/13th-century stained glass of Chartres Cathedral attempted to repair damaged sections in a '13th-century' style.[41] We will see in Chapter 2 how the challenges of reconstruction became more urgent from the 16th and 17th centuries, with widespread war damage to architecturally significant monuments – beginning with the Huguenot wars of 1562–89, when dozens of cathedrals and hundreds of abbeys and parish churches were destroyed. Indeed, the interrelationship of conservation and mass warfare is one of the most enduring themes of our story, forming one of the chief stimuli for the emergence of the modern Conservation Movement.

The Renaissance: antiquity as a 'mirror of modernity'

THE late Middle Ages saw a burgeoning diversity of built solutions to the challenge of care and reuse of old buildings, following the break-up of antiquity. But we need now to return again to the 'centre', where a momentous cultural movement was brewing, that would take the narrative of conservation a stage further forward, by setting up a more deliberate, intellectualised tension between 'local' and 'universal', that would continue through to the present and shape the development of the modern Conservation Movement from the late 18th century.

Within Italy, the severance of the awareness of a unitary classical civilisation was almost complete by the 12th and 13th centuries. The 12th-century *Mirabilia Urbis Romae* described imperial monuments as if they were the magical remains of a distant age, with classical and Christian martyrdom associations freely mixed together. Now, the way was clear for a more calculated, distanced 'revival' of classical antiquity. Partly, the stage was set in 1420 by Martin V's return to Rome after the papacy's prolonged exile in Avignon: the city's population duly recovered from 17,000 in 1400 to 110,000 in 1600. Other Italian cities, such as Florence, would also play a key role in the Renaissance: an essential element was, indeed, a competitive sense of civic pride, as organisations and individuals throughout Italy built proud monuments and palaces to express and extend their influence.[42]

Within this new world-outlook, monuments of antiquity could begin to play the role of a consciously defined, reflexive mirror of modernity, a role that has continued in differing forms to today. Thinkers like Francesco Petrarch (1304–74) were beginning to argue that secular history had its own autonomous validity, as distinct from universal, salvation-oriented religious history, and that it had its own internal narrative of evolution and change, focused on the three eras of the ancient, middle and modern ages. Petrarch played a key role in reviving respect for Rome as a centre of cultural and political life: the 1340s saw a short-lived movement for restoration of a Roman 'republic' in opposition to the nobility, led by Petrarch's friend and fellow-admirer of antiquity, Cola di Rienzo.[43]

On his first visit to decayed Rome in 1337, Petrarch was moved to tears. A new, intense regret at the '*deploratio urbis*' (ruination of the city) began to establish itself; something detached from the ahistorical antique idea of *pietas*, and which pointed forward to the Romantic Movement of the 18th century. In an almost proto-Romantic gesture, in 1341 Petrarch was presented with a laurel wreath on the Capitoline Hill. During the 1340s, Cola di Rienzo began systematically recording the city's classical monuments on a plan, collecting and deciphering inscriptions; and in 1375 Giovanni Dondi, humanist and friend of Petrarch, was also busy measuring antique remains. After 1500, there was a fashion for a neo-Latin literature of ruins, and sculptor-architect Jacopo Sansovino – driven by the 1527 sack of Rome to flee to Venice, which he helped build up as a 'new Rome' – began to interpret classical ruins as a metaphor of the frailty of human existence. The new historical sensibility highlighted the polarity between universal and local/national: one could research the culture and remains of Roman antiquity, as a universal legacy of lost grandeur and a testimony of potential regeneration but, at the same time, one could develop distinct national narratives.[44]

A further strand of the new 'universal narrative' of the Renaissance was a new, intense concept of Art, based partly on neo-Platonic ideas of innate ideal beauty, and partly on a new socio-economic demand for art-collecting. This, in turn, created a new standing for the individual artist – with Raphael in the early 16th century the first to be accepted on the same level as the aristocracy. The new cult of Art and the artist-creator – unlike the cult of *deploratio urbis* – was something completely modern, as nothing was known about artists or art-collecting in antiquity. Following the literary efforts of the 14th century, the early 15th saw a new emphasis on aesthetic connoisseurship, and galleries and collections began to appear on an informal basis. In Florence, the 'Uffizi' were adapted in 1581 to a gallery, and the first gallery in England was created in 1615.[45]

29

From the early Renaissance onwards, influential people began collecting antique figures for study and, by the end of the 15th century, there were some 40 collections in Rome alone. They included not just statues and freestanding sculpture, but also, as in the age of Constantine, cannibalised reliefs – for example fragments of Augustus' *Ara Pacis*, incorporated into the Villa Medici. The 16th century saw a huge increase in collecting, fuelled by demand from abroad, as other up-and-coming countries sought to appropriate a chunk of universal antique prestige for themselves. This internationalisation of the cult of the antique, and the new, systematic art-collecting, began to take on distinct overtones of 'tourism'. Already, by the 16th century, it was common for educated northern Europeans to visit Rome to make engravings – as in the case of Hieronymus Cock's 1540s' drawings or those of Heemskerck the previous decade. In response, the first 'tourist guidebooks' began to appear. Following Cola di Rienzo's pioneering surveys, 1444–6 saw the first, inaccurate attempts to inventorise ancient monuments in print, in Flavio Biondo's *Roma Instaurata* (Rome revised). By the 16th century, a range of publications was available: for instance, the architect Palladio's enormously popular 1554 guide to Rome, or Pirro Ligorio's 1553 *Libro delle Antichita di Roma*, with its categorised accounts of amphitheatres, circuses and theatres. By this time, too, earlier misconceptions about classical architecture were beginning to clear: for example, the supposition,

Figure 1.13 Heritage depictions in the Renaissance

(a) Part of Pirro Ligorio's 1562 reconstruction plan of classical Rome; (b) Marten van Heemskerck's 1536 drawing of the progress of the St Peter's demolition and reconstruction project

(a)

(b)

based on the Pantheon, that antique temples had generally been circular in plan.[46] The beginnings of cultural tourism to Rome were unambiguously elitist, culminating in the 18th-century Grand Tours by rich northern Europeans, but eventually the interrelationship of heritage and tourism would become bound up with the upsurge of 'mass society'.

It was from the Renaissance cult of Art, and from the collecting of classical sculpture – itself only indirectly 'architectural' – that there now arose another of the enduring narratives of architectural conservation: the conflict between restoration and conservative repair. From around 1500, there were the first systematic efforts at completion of antique statues lacking arms and legs. The discovery in 1506 of a statue of Laocoon and his sons being attacked by snakes led to a vigorous debate about completion of a truncated arm – should it be straight or bent? – with Michelangelo and Giulio de Sangallo both proposing 'completions'. In 1550, Vasari argued that restored antiquities had more 'grace' than truncated ones.[47]

How did all this connect with the conservation of old buildings? As a matter of principle, the building of new architectural projects in overtly classical forms – beginning with Brunelleschi's Ospedale dei Innocenti, Florence (from 1421) – went hand in hand with a growing respect for ancient buildings: the rediscovery of Vitruvius in 1414 spawned a host of imitation treatises, including Alberti's ten-book *De re aedificatoria*, published in 1485. Echoing Vitruvius's distinction between the three architectural values of beauty (*uenustas*), practical utility (*utilitas)* and good construction (*firmitas*), Alberti hailed antique buildings as worthy of protection both for their resilience and their beauty. The planning of the many new noblemen's *palazzi* was also profoundly influenced by the vague ideas of what ancient houses actually looked like. Ruins stimulated architects, not just as a spur to laments of lost glory, but through their fragmented or incomplete character, which allowed scope for exercise of creative imagination.[48]

The revival and destruction of classical Rome

THE experience 'on the ground' of the relationship between antique survivals and new classical architecture was to be much more fraught and confrontational. The Renaissance potentially revitalised conservation values, opening up ways of interpreting old buildings based on the new artistic and historic self-consciousness rather than the old symbolic values of Christendom. But this growth of interest in the past also coincided with a vast programme of building to celebrate the glory of the resurgent papacy. It was here, in 16th-century Rome, that we see the roots of the intense interrelationship of destruction and preservation that would continue into modern centuries. The overtones of hypocrisy that attended this relationship, with the same agency protecting and destroying, were already displayed in 16th-century Rome, where a succession of popes and their officials trumpeted classical humanism while pillaging the surviving built legacy of the emperors for their own construction programmes.

At first, in the 'Renewal of Rome' following the return of the papacy in 1420, the emphasis was largely on restoration of the old churches, including St John Lateran and St Peter, which was in a decayed state, with nave walls leaning dangerously outwards. For this, the most convenient source of building materials, as always, was the ancient monuments. Pope Martin V, inaugurated in 1425, was the first to try to bring order back to the city. He releaded the roof of the Pantheon – which was, of course, now a church. His successor, Eugenius IV (1431–47), issued an edict to protect the Colosseum from stone thieves, while himself using stones from it to repair the Lateran Basilica, and took marbles from the Senate House and other monuments for the Lateran Palace. Marble was excavated for use both as a building material and also for burning for lime – as in 1426, when the paving of the Basilica Iulia in the Forum was removed for burning. When early-15ᵗʰ-century Romans discovered even marble sculptures and inscriptions, they usually melted them down into mortar. A century later, all that had changed: these antique marbles had assumed their revered modern role as works of art in dozens of private collections, above all the Vatican Belvedere. This first generation of antique art collections was, in turn, dispersed in the devastating 1527 sack of Rome by the troops of Charles V – the worst since the Normans in 1084 – but the precedent for classical art-collecting was now set.[49]

As the pace of the classical revival increased, so did the threat of destruction of antiquities. Poggio Bracciolini recorded that, when he first went to Rome in 1431, the temple of Saturn (as rearranged in the 5ᵗʰ century) was virtually intact, whereas by 1447 only the portico was left – as today. Under the 'restoring' regime of Pope Nicholas V (1447–55), attempts at protection were matched by destruction, for example, of the Arch of Valentinian and Gratian; a contractor was allowed to remove 2,522 cartloads of travertine from the Colosseum in nine months. Some destruction was caused directly by war or conflict – notably in 1527. But more usually there was a circular process, in which destruction was stimulated by the classical revival then, in turn, provoked measures of protection, often by the agents most energetically pursuing the destruction! The most basic level of protection was to direct the destructions along the least damaging channels. For example, it was the south side of the Colosseum, and the interior, that suffered the most extensive quarrying and damage, whereas the north side, facing the main ceremonial route from St Peter's to St John Lateran, was left relatively intact – an asymmetrical profile that had established itself as early as the 8ᵗʰ century.[50]

Active protection regulations were also gradually introduced, especially once the old aristocratic barons like the Frangipane were squeezed out by resurgent civic spirit. The year 1363 saw the first recorded civic statutes in medieval Rome forbidding destruction of ancient remains. In the Colosseum, as early as the 12ᵗʰ century, civic magistrates had established control over the intact northern half; by the 15ᵗʰ century, officers of the *Comune* such as Lorenzo Caffarelli were rigorously controlling any excavations and protecting the building's external arcades, while a religious community (the Salvatore confraternity) occupied the ruinous southern third. Alberti carried out numerous remodellings of early Christian and medieval buildings, preserving older cores behind classical facades, as in his Tempio Malatestiano, Rimini. And, from the 15ᵗʰ century, the returned papacy, which enjoyed much stronger powers than the civic *Comune* (via the *curia apostolica*), attempted to introduce protective measures to moderate the destructive effects of the rebuilding campaigns. These, however, were generally as ineffective as the edicts of the later Roman empire. For example, Pope Pius II issued an edict in 1462 protecting ancient monuments, but broke it himself in the collection of building material for the Vatican Benediction Loggia – whose design, ironically, was inspired by the Colosseum! In 1466 there were repairs to the Arch of Titus by Florentine masons, and Pope Sixtus IV (1471–84), self-styled *Restaurator Urbis*, repaired many palaces and churches (ancient and modern) and issued a bull of 1474 against damage and despoliation of religious buildings – while in 1471 allowing the Vatican architects to dig where they wanted for building stones.[51]

The dichotomy between destruction and protection culminated in the event which unleashed more destruction than any other – the decision taken by Julius II in 1506 to rebuild Old St Peter. This decision, by a pope who considered himself a new 'Caesar', to destroy the most venerable

Christian monument outside the Holy Land, albeit through demolition by gradual stealth, led to an enormous demand for building materials during his lifetime – quite apart from the destructive effects on church unity of the insatiable financial demands made by the papacy. The project was so large that it could only progress very gradually, and the next Pope, Leo X (1513–22), a significant patron of antiquarian research, slowed down its progress. In 1515, Raphael, who was also chief architect of St Peter's, wrote to Leo X, complaining generally of the pace of destruction, but especially of the use of marble for lime-burnings. In part-compensation for the destruction, he was appointed 'curator of monuments', but as his title (Prefect of Marbles and Statues) implied, protection only extended to stones bearing inscriptions or sculpture – an interesting order of priorities that implied the persistence of the old ideas of *pietas*. Raphael prepared a map of classical monuments in the city, including excavation sites; in 1521, he issued the first list of protected inscriptions, and in 1527 his colleagues published the first study of Rome's antiquities.[52]

Protests at the destruction came not just from Raphael but also from the municipal *Comune*, frustrated at its weakness in relation to the Pope. When a senator wanted to complete the loggia on the Campidoglio, he was granted permission to take stone from the Arch of Septimius Severus, but the *Comune* stipulated that work should be inspected by ten citizens to ensure the structure of the arch was safeguarded. The attitude of the *Comune* was contradicted by that of the *Fabbrica di S. Pietro* – the papal-controlled building organisation. Around 1540, Paul III decided to get the St Peter's project moving again, with Antonio di Sangallo as architect, and gave the *Fabbrica* permission to dismantle more classical remains, despite protests from the *Comune* at removal of stones from the Forum and Via Sacra.[53]

The Forum regained its old splendour momentarily in 1536, when Paul III celebrated the entry of Charles V into Rome after his victory over the Turks: he created a temporary triumphal way from the Arch of Septimius Severus to the Arch of Titus – prototype of countless processional tableaux until the 18th century. But the destruction only accelerated after this, expanding from opportunistic plunder to methodical excavation, almost archaeological in its systematic character, but destructive in its intentions. Large areas around the Forum were excavated during the 1540s and '50s, in preparation for the building of St Peter's, and countless monuments were destroyed: white marble was usually burnt down for lime, only inscriptions being spared. The areas round the temples of Saturn and Vespasian, the Curia Iulia, the Basilica Aemilia, the Temple of the Castores and the road from the Temple of Antoninus and Faustina to the Arch of Titus were all turned into quarries. Structures such as the Temple of Divus Iulius were simply obliterated for building material, sometimes in the course of a single month.[54]

In some of the original projects, such as the enlargement of St John Lateran, the age-old practice of incorporating classical columns as *spolia* continued, as part of the process of appropriation of antique forms for modern progress. In 1537, Paul III initiated the systematisation of the hitherto informally planned Capitoline, with its towered, medieval Palazzo Senatorio built on the foundations of the Roman Tablinum (public records office). He commissioned Michelangelo to transfer the classical equestrian statue of Marcus Aurelius to the Capitoline, and began the regularisation of the piazza around it into a classical, symmetrical, three-sided composition of civic buildings and an art gallery, the Palazzi Senatorio, dei Conservatori, and Nuovo – a project doggedly pursued by successive popes, and not definitively completed until the 18th century.[55]

The tension between Roman revival and destruction continued into the late 16th century. The Counter-Reformation fuelled further aggressive surgery of 'pagan' monuments, and under Sixtus V

(1585–99), the papal architect/planner Domenico Fontana implemented a grand plan for the stately systematisation of the city fabric, in the face of fierce citizen protests. His plans involved the construction of wide new axes linking key religious sites. Principal intersections were marked by open spaces in which were set existing monuments (such as the columns of Marcus Aurelius and Trajan) or relocated structures, including four obelisks. Here we see the earliest roots of the concept of enhancing the impact and status of historic monuments by clearing vast spaces around them – a philosophy that would carry on until the 20th century, under various names in various countries. More generally, Sixtus's strategy required sweeping demolition: Fontana was to 'tear down the ugly old and repair the worthwhile'.[56] Yet, despite threatening demolition of 'pagan' structures, including the tomb of Caecilia Metella, Sixtus also commissioned Fontana to restore principal monuments.

Increasingly, key structures were restored rather than totally rebuilt. Michelangelo had designed a conversion of the *frigidarium* of the Baths of Diocletian into a church for Pius IV in 1563–4 with minimum alteration, and Borromini drastically remodelled the Lateran Basilica while incorporating the old structure underneath, reusing antique columns as *spolia*. In the 1650s, Alexander VII changed the Pantheon into a mausoleum for his family, without major reconstruction.[57] But, by then, as we will see in the next chapter, a much more explicit discourse of care for antiquities was beginning to evolve, not just in 'universal' Rome but throughout the emergent countries of western and northern Europe. In the process, the foundations were being laid from which the modern Conservation Movement could develop, in those countries, in the late 18th and 19th centuries.

(a) (b)

Figure 1.14 *Spolia* and shells

(a) The Villa Medici in Rome, showing the classical *spolia* incorporated following its 1576 acquisition by Cardinal Ferdinando de' Medici; (b) Michelangelo's 1563–4 reconstruction of part of the Baths of Diocletian into the church of Santa Maria degli Angeli

Antiquarian antecedents

17th and 18th centuries

Quod non fecerunt barbari, fecerunt Barberini.[1]
(The Barberini did what the barbarians failed to do)

Warfare and restoration in the 'Confessional Age'

DURING the 17th and 18th centuries, the old universal narrative of heritage gradually fragmented into a constellation of competing narratives. There was a new diversity in interpretations of classical antiquity; but there were also emergent national heritages, each a microcosm of the universal tradition. This was not the same as the aggressive, disciplined nationalisms whose competition in the 19th and 20th centuries would profoundly affect the world of monuments. The conflicts of this earlier age were focused not on nationality but on religion: the Reformation broke up the old unity of western Christendom, provoking a crescendo of conflicts culminating in the Thirty Years War of 1618–48. Throughout northern and western Europe, these inflicted a repeated and massive devastation on the urban and rural landscape. In reaction, the built fabric underwent a sharp collective 'valorisation', as its set pieces became charged with cultural significance and valued as subjects both of loss and of potential restoration – sentiments previously commanded only by the ruins of antiquity. This valorisation was focused on the great medieval religious buildings, numbers of which suffered severe damage, sometimes amounting to virtual destruction, as the locus of conflict moved around Europe. Nor was it just wars that caused damage: peacetime convulsions such as the 1536–41 Dissolution of the Monasteries in England could render an entire building-class redundant at a stroke.

Although previous conflicts had endangered particular building-types, such as the over 1,000 castles and abbeys attacked by peasants in the 1524–6 *Bauernkrieg* in present-day southern Germany and Austria, the first war in which monumental architecture was systematically targeted for ideological reasons was the Huguenot conflict of 1562–89.[2] Dozens of cathedrals and hundreds of monasteries and parish churches across France were ruined and stripped of ornamentation, in a deliberate campaign of cultural decapitation motivated by the belief that (in the words of a Swiss Calvinist) 'once their nest is destroyed, the storks will not come back'. At the same time, the northern Netherlands witnessed in 1566 a so-called *Beeldenstorm* of Calvinist iconoclasm, targeting mainly sculpture and relics rather than buildings, and leaving churches stripped of fittings and stained glass, and whitewashed internally.[3]

From 1600 or so, under Henri IV, a post-Huguenot wave of restoration began across France, in some cases immediately, in others after a lapse of time beyond the lifespan of the generation that had experienced the destruction. In cases of only partial destruction the task was often confined to details, for example in Auxerre Cathedral, where much 13th-century stained glass had been smashed, and surviving fragments were concentrated in the choir, mixed together with little regard for iconographic coherence. Where the entire church was destroyed, the demands of reconstruction were naturally more far-reaching and the potential choices more varied, although in most cases rebuilding in practice was in a late Flamboyant Gothic even if the original churches had been largely Romanesque. In the case of Mende, for example, where the Huguenots had blown up the entire cathedral in 1580 except for the towers, the building was reconstructed in 1600–20 exactly as it had been, in a strict late-Gothic style, to demonstrate Catholic continuity. In the abbey of St Maixent l'Ecole, in Poitou, the pre-destruction church was a mixture of Romanesque and 13th-century Gothic. Almost like 19th-century architects, the monks debated whether the rebuilding should be in a Gothic, Romanesque or contemporary classical Baroque style. Eventually the first was selected, and so in 1670–82 a church of relatively strict 15th-century-Gothic style was built, showing its modernity only in details. At the cathedral of St Apollinaire in Valence, conversely, the pre-destruction church was largely 12th century in date. So the 1604–19 reconstruction, by architect Jean Thuillier and mason Jacques Blanc, partly financed by Henri IV, followed a tolerably exact Romanesque style, incorporating surviving capitals as *spolia*.[4]

The most remarkable, and protracted, of all the post-Huguenot restitution schemes was the rebuilding of Sainte-Croix Cathedral in Orleans, targeted by retreating Protestants in 1568 and so comprehensively blasted by mines placed at the crossing that there survived only two bays of the nave and fragments of choir and transepts, together with the Porte de l'Évêque. The pre-1568 cathedral was an extensive, multi-period structure, intermittently under construction since 1287. With uncanny symmetry, the post-1568 reconstruction took almost as long, despite the enthusiastic patronage of successive kings, including Louis XIV, anxious to showcase their Catholic fidelity. The first stage of rebuilding was planned in 1601 by Henri IV, and the structure was only finished in 1829. Despite this protracted process, the rebuilding followed a remarkably unified late-Gothic style, derived from the details of the surviving nave bays. The surviving choir fragments were earlier in date (13th and 14th centuries), but were integrated carefully into the 16th-century image, to reinforce the impression of a perfect Flamboyant cathedral. The late-17th-century work focused on the north and south transepts, the former started in 1636 under Louis XIII, the latter in 1675 under Louis XIV. Both combined an overall Gothic form with intriguing classical details (including fully fledged pedimented doorways of 1693–6 by Claude Godard): the tracery of the rose windows featured Louis XIV's sun-motif, repeated inside the building.[5] The 18th-century work, pushed forward by Louis XV, culminated in the completion of the western towers, designed in an elaborately tiered, intricately arcaded neo-Gothic style by a succession of architects including Jacques Gabriel and built in 1773–90. This reconstruction was not motivated by romantic eclecticism in the 19th-century manner:

facing page

Figure 2.1 Pre-modern cultural destruction

(a) Plundering of castles in Upper Franconia during the 1525–6 'Peasants' War' in southern Germany; (b) Orleans Cathedral, present-day view of the exterior as reconstructed after 1568, showing the south portal (started 1675); (c) Interior view of Orleans, rebuilt as a 'perfect Flamboyant cathedral': the nave design was extrapolated from the two surviving pre-1568 bays

indeed, Victor Hugo attacked it as an 'odious church that makes so many promises from afar and breaks them all, close-up'.[6] Rather, it was driven by a determination to turn the clock back to the *status quo ante* and emphasise the evils of the Reformation.

But the Huguenot conflict was only a curtain-raiser to successive decades of destruction during the 17ᵗʰ century, including the continent-wide Thirty Years War and the later campaigns of Louis XIV. Even within peacetime France, Louis's grandiose urban visions prompted the systematic destruction of much medieval fabric: in Paris, houses were cleared from several city bridges, as were 17 small churches from around Notre-Dame.[7] On campaigns abroad, the destruction was yet more radical, with widespread devastation in the Palatinate and the Low Countries, and systematic demolitions not unlike those perpetrated by the Huguenots. At Speyer, the expulsion of the townspeople and the burning of the town in 1689 was followed by a systematic attempt by the French army to demolish the Romanesque cathedral by explosive mines: by the time the cathedral dean persuaded Marshal Duras to stop the demolition, the western half of the cathedral was a heap of rubble – a loss compounded by further demolition in the 1750s. This French destruction of a national imperial German shrine (the largest Romanesque church to the east of France) had long-lasting consequences stretching over several centuries, not only in provoking a succession of attempted restorations, but also in exacerbating the feelings of resentment that would fuel the eventual military expansionism of reunified Germany.[8]

In the late 17ᵗʰ century, we encounter the first conscious replica-rebuilding of a destroyed urban ensemble (a task that became a dominant theme of conservation in the 20ᵗʰ century), when Louis XIV's Marshal de Villeroy, on a punitive campaign in the Netherlands in 1695, shelled and burned to the ground the old centre of Brussels, including the already famous Grand' Place. Although few civilians were killed, the area was reduced to rubble. However, the Grand' Place was immediately (1697–1700) rebuilt as a partial copy of the spiky Northern Renaissance gabled facades that had existed before, with no change to the street plan other than modest street-widening. Authorised by the City Council, the reconstruction comprised individual houses whose details were ingeniously derived from those of two surviving house-facades; it was proudly commemorated in date inscriptions on the ornate gables. When discussing the 20ᵗʰ century's many postwar reconstructions, we should not forget this pioneering example.[9] Only 30 years earlier, the greater destruction inflicted on the non-monumental fabric of London by the Great Fire had provoked a very different response, in Sir Christopher Wren's proposal to sweep away the medieval street plan and substitute an ordered classical layout more reminiscent of absolutist France. The compromise eventually implemented retained the medieval layout, but stocked this with completely new buildings, including a new, classical St Paul's Cathedral, and an array of new parish churches.

Iconoclasm and rescue in 17ᵗʰ-century Britain and Ireland

APPROPRIATELY enough, it was at this time of conflict that many modern heritage words began to assume their present meanings, including the word 'monument' in England and France, and '*Denkmal*' in Germany – a word said to have been originally coined by Martin Luther.[10] The concept of antiquarianism, too, was emerging in a number of countries, especially those involved in fighting and expansionism. In aggressively imperialist Sweden, 1630 saw the appointment of a royal Director General of Antiquities and 1666 the passing of an antiquities ordinance (the first outside Italy) to

Figure 2.2 The Grand' Place in Bruxelles

(a) View during the 1695 bombardment by Marshal Villeroy's forces;
(b) Present-day view of the same frontage as rebuilt, including 1690s'
commemorative plaques

protect old objects and remains, both moveable and fixed – although little preservation activity actually ensued. A year later an antiquarian and archaeological studies institute (*Collegium Antiquitatum*) was founded at Uppsala.[11]

It was, however, chiefly in England, a country that experienced relative peace during the 16ᵗʰ and 17ᵗʰ centuries, that the principles of modern antiquarianism, an expertise that would be a precondition for effective heritage protection, were developed. At this stage, as with art-collecting and proto-tourism in Italy, this was a strictly small-scale, elite practice, unlike the mass public enthusiasm that would swell up in the 19ᵗʰ century.

In England, the late 16ᵗʰ and 17ᵗʰ centuries saw the emergence of various competing strands of 'national meaning', reflecting the dynastic-religious tensions of the post-Reformation years, and often somewhat pessimistic in character.[12] The mid-17ᵗʰ-century civil wars in Britain, although less destructive than the contemporary continental conflicts, still claimed numerous prominent architectural casualties, such as Basing House, Hampshire, one of England's largest country houses. A vast, double-courtyard complex of 1535, Hampton Court-like in scale and owned by the royalist Marquess of Winchester, Basing was besieged and razed by Cromwell's parliamentary troops in 1645. Roman masonry in Colchester Castle was bombarded by Parliamentary forces, and London's Roman wall was plundered for stone. By the end of the 17ᵗʰ century, only 42 out of 395 manor houses in Hertfordshire, for example, were still as intact as in 1540.[13] To this was added a continuing rumble of Protestant iconoclasm, admittedly milder than the militancy of the Huguenots. The fundamentalist Protestant position on the First Commandment resembled the early Christian attacks on 'pagan idols'. Here, though, images of anything whatever were destroyed – statues, glass, or pictures of saints. In England, attitudes were more ambiguous and potentially conservative than in Calvinist Scotland, where many churches were simply abandoned, such as the vast St Andrews Cathedral, or drastically remodelled for preaching-centred worship, as in the case of St Giles, in Edinburgh (subdivided not only into several churches but also, ultimately, municipal chambers and a fire station). In England, the Dissolution of the Monasteries of 1536–41 had been followed by an initial wave of organised destruction, with detailed records, for example, chronicling the 1537 dismantling of the Cluniac Priory of St Pancras in Lewes (on which see also Chapter 3). This campaign of iconoclasm and abandonment provoked a September 1560 decree from Elizabeth I, forbidding the 'defacing of Monuments of antiquity, being set up in the churches or other public places for memory, and not for superstition', and instructed 'that no such barbarous disorder be hereafter used, and to repair as much of the said Monuments as conveniently may be'.[14]

Ironically, unlike the widespread removal of screens in Counter-Reformation Catholic churches, following the Council of Trent, Anglican churches, generally conservative regarding ecclesiastical institutions and liturgy, often deliberately preserved their late medieval interior

Figure 2.3 Iconoclasm in Britain

(a) The English Civil War: Victorian painting of the 1645 siege and destruction by Parliamentary forces of the early-16th-century Basing House, Hampshire ('Cromwell at the Storming of Basing House', 1900, by Ernest Crofts); (b) The Scottish Reformation: 19th-century painting of the 1579 defence of Glasgow Cathedral from Calvinist iconoclasts, by members of the town's Trades House

(a)

(b)

arrangements, including the '*pulpitum*' or choir screen blocking off an open view of the choir, as in York Minster. The 16th-century damage was mostly confined to abbey and priory churches, although there was insidious neglect of the two dozen or so cathedral-size churches. The first Stuart kings of England, James I and Charles I, addressed this neglect through a programme of restorations, culminating in that of Old St Paul's, London, on which around £100,000 was spent in 1634–44 on improvements authorised by Charles I. Inigo Jones encased the Romanesque nave and transepts in a classical shell, while the Gothic choir was left as it stood. The mid-century conflicts took a further toll on the cathedral: William Dugdale bemoaned its 'lamentable condition' in 1660,

> being made a horse quarter for soldiers during the whole time of the late Usurpation; the stately Portico, with beautiful Corinthian pillars, being converted into shops for the seamstresses, and other trades . . . for the fitting whereof to that purpose those stately pillars were shamefully hewed and defaced for the support of timber work.[15]

More generally, the 1640s' civil wars in Britain unleashed a fresh wave of iconoclasm and damage to churches. In 1641 the House of Commons ordered 'Commissions to be sent into all Counties, for the Defacing, Demolition, and quite taking away of all Images, Altars or Tables turned Altarwise, Crucifixes, superstitious Pictures, Monuments, and Relicts of idolatry, out of all Churches and Chapels.' A lengthy account records the activities of William Dowsing, an official commissioner appointed under this ordinance, who toured churches in Norfolk in 1643–4 smashing images. The iconoclasm of the French Revolution here seems a relatively short distance away. Churches suffered war damage, too: Lichfield Cathedral was besieged three times, losing its spire, roofs and much vaulting. Extreme Protestants were vehemently opposed to any maintenance of the great churches. With the Restoration of King Charles II, conversely, once more the stress was on identification of the monarch with the Church of England, and renewed attention to restoration projects such as Westminster Abbey.[16]

The sharper ethnic and religious confrontations in contemporary Ireland were reflected in a more diverse instrumentalisation of heritage. During the Stuart age, key Protestant and Catholic building patrons systematically reused Romanesque building fragments in new projects, like Roman *spolia*, as ideological devices to emphasise either their patriotic zeal or the antiquity of their ancestry and the legitimacy of their presence in Ireland. Examples include a gateway erected in 1611 at Lismore Castle, Co. Waterford, by the Protestant Richard Boyle, 1st Earl of Cork, incorporating Romanesque carved stones, or the 'Main Guard' at Clonmel, Co. Tipperary, *c*.1675, one of Ireland's first fully classical buildings, which also contains quantities of salvaged Romanesque piers (possibly from the 12th-century Cistercian abbey of Inislounaght). Early Christian monuments were also appropriated by devout Catholics in the late 17th century, sometimes reusing ancient sites.[17]

Early antiquarians and medievalists in northern Europe

ALONGSIDE all this conflict, there was a growing practical consensus in England over a canon of patriotic relics. By the Restoration in 1660, the armouries of the Tower of London – previously an arsenal for the 14th-century White Tower, opened occasionally to important visitors since 1489 – were accessible to paying visitors. These domestic tourists could view an array of new displays commemorating historic English victories, including a 'Spanish armoury' supposedly salvaged from wrecks of the Armada, and the 'Line of Kings' – a row of mounted, armoured effigies on horses, with associated portrait paintings of monarchs. From 1696 onwards, the Grand Storehouse at the Tower was used for exhibitions of historic weapons.[18]

In response to all this, the phenomenon of the 'antiquarian', studying the post-classical past, emerged in England, rather earlier than elsewhere. These antiquarians tended to be royalist in their political sympathies and compensated for lack of scholarly precision with a nostalgic zeal echoing Petrarch's '*deploratio urbis*' language. In their work, revulsion against religious iconoclasm shaded into criticism of aggressive modernity, anticipating the 19th century's fierce attacks against demolitions of old buildings and the bewildering fluctuations of opinion about whether restoration was a creative or destructive practice.[19] As early as the late 15th century, educated topographer-travellers, such as William Worcestre (in 1478–80), had begun touring historic relics. In 1535–43, the *Itinerary* of John Leland (King's Antiquary since 1533) signalled the first systematic search for antiquities – ironically, just as the Dissolution of the Monasteries was creating new 'monuments' for the future. And in 1586, William Camden's *Britannica* attempted to classify prehistoric relics. In 1620, reflecting the growing interest in ancient monuments, Inigo Jones surveyed Stonehenge, but declared it was of Roman origin.[20]

Renaissance humanism in England could not cope adequately with non-classical remains and sites, rejecting the proposition that Aeneas and Brutus were the first British rulers, but providing no alternative to fill the gap between Old Testament and Renaissance: even in the early 18th century, William Stukely's research still envisaged the great field monuments as the work of Druids. The Restoration was followed by the first upsurge of archaeological discovery in England, including surveying of medieval structures, by scholars such as John Aubrey, Roger Dodsworth and William Dugdale, Elias Ashmole and William Drydale. In 1657, Anthony Wood at Evesham was 'wonderfully struck with a veneration of the stately, yet much lamented ruins of the abbey', and Aubrey's 'Chronologia Architectonica' of 1656–86 pioneered the chronological analysis of medieval styles. Edward Lhuyd and Robert Sibbald charted Celtic remains in Wales and Scotland. Destruction continued alongside this: in the early 18th century Stukely recorded that hundreds of cartloads of Roman brick were removed from St Albans for road-making.[21]

Only in the late 17th century did similar antiquarian tendencies emerge across Europe, including Michel Germain's *Monasticon Gallicanum*, and the 1690s' studies by Giovanni Giustino Ciampini; in Germany, 1692–1715 saw the first 'castle guidebook' published, in four editions (to the Saxon stronghold of Königstein).[22] But by the early 18th century, echoing the late 17th-century Swedish innovations, England had moved on again, towards a more systematic 'national' apparatus of antiquarianism, focused on a new Society of Antiquaries of London, founded in 1707: this claimed to be a revival of an Elizabethan society of 1585 and focused on English archaeological remains. Although Protestant public opinion in the early 18th century was still suspicious of preservation of medieval monuments, seeing antiquarians as Catholic and Jacobite sympathisers, that opinion was shifting. While, in 1733, Bristol citizens advocated demolition of the town's High Cross as a 'ruinous and superstitious Relick' symbolising 'Popery', in 1771 the Society of Antiquaries paid ten shillings 'for setting down two oak posts to save Waltham Cross from injury by carriages'.[23]

During the 17th and 18th centuries, there was still a lack of consensus concerning the merits of English medieval architecture. There was, as in most European countries, still a distrust of its irregularity. In 1664, diarist John Evelyn, in his *Account of Architects and Architecture*, decried Gothic buildings as 'congestions of heavy, dark, melancholy and Monkish Piles', and as late as 1771, Tobias Smollett's old-fashioned Mrs Bramble in *The Expedition of Humphrey Clinker* argued that 'the external appearance of an old cathedral cannot but be displeasing to the eye of every man who has any idea of propriety and proportion'.[24]

Yet medieval architecture had never been rejected absolutely in any country. When the 1689 war-damaged Speyer Cathedral was eventually rebuilt in 1772–8, nearly a century after its part-destruction, the bishop decided it should be reconstructed in facsimile, and engaged Franz Ignaz Neumann, son of southern Germany's foremost Baroque architect, to draw up a scheme. Neumann's design copied the surviving 11th-century nave bays so exactly that one cannot tell the two apart. Shortage of money made it impracticable to rebuild the towered *Westbau* at full scale. That had to await a later, neo-medieval scheme of 1854–8 by Heinrich Hübsch. However, the enduring resentment over the French destruction of the cathedral left each generation of Speyer restorers unhappy with its predecessors' contribution, and ultimately provoked a 'purifying' 20th-century restoration that expunged Hübsch's rich 19th-century decoration. In Bologna, the 250-year part-completion saga of San Petronio Church saw Gothic rib vaults erected as late as 1658, while the early 1700s witnessed pioneering eclecticism and 'Gothic revival' in Ferdinando Galli Bibbiena's theatre designs.[25]

Figure 2.4 Speyer Cathedral, destroyed in 1689 by French forces and rebuilt in 1772–8 by F I Neumann

(a) Present-day external view showing the five surviving 11th-century nave bays on the right and Neumann's work on the left (in lighter masonry); (b) Inside, a richly polychromatic scheme of 1846–58 by Heinrich Hübsch was expunged in a 'purifying' restoration of 1957–63 by Rudolf Esterer

In England, too, medieval buildings were never completely reviled, and there was a long, complicated phase of Gothic survival, including, as in France, repair of damaged buildings. The war-damaged Lichfield Cathedral was carefully reinstated after 1660 in only seven years, its Perpendicular Gothic tracery and vaulting carefully replaced by local masons, and enhanced by a vast new west window of eccentric design. When part of the Romanesque north transept of Ely Cathedral collapsed in 1699, the authorities rebuilt it 'exactly in the same manner and on the same foundation it stood before'.[26] Gothic churches were still built in England throughout the 17th and early 18th centuries, the last significant example being the nave of St Margaret's, King's Lynn, from 1742. By then, as we will see shortly, the Gothic survival had become revival, aided by the first Gothic pattern books; from here, the story of conservation and restoration overlaps substantially with the very well-known mainstream architectural history of the early Gothic Revival in England.

The conflicting English values of the age were exemplified in the work of the leading (royalist) architect of the day, Sir Christopher Wren – whose advice had prompted the Ely rebuilding plan. Before the Great Fire of London in 1666, Wren proposed a conservative restoration of Old St Paul's,

Figure 2.5 Gothic survival or revival? Christopher Wren's Gothic entrance tower at Christ Church, Oxford, 1681–2. Wren had successfully argued that its design 'ought to be Gothick to agree with the Founders worke'

and even his all-new, post-fire design adopted a 'Gothic' plan complete with concealed flying buttresses, arguing that its medieval predecessor had been a 'monument of power and mighty zeal in our ancestors'.[27] Subsequently (1681–2), Wren completed Christ Church, Oxford, in Gothic rather than the 'better forms of architecture', to avoid an 'unhandsome medley': the tower 'ought to be Gothick to agree with the Founders worke'. After Wren's appointment in 1697 as Surveyor to Westminster Abbey, he began a 50-year, £100,000 restoration programme. This focused initially on simple repair, with fenestration renewed as existing, but eventually became more ambitious, including simplified rebuilding of the north transept and rose window, and completion of the western towers in a style that (like the earlier transepts at Orleans) was Gothic in outline and classical in detail. After Wren's death in 1714, his pupil, William Hawksmoor, completed the towers after 1735 and continued repairs. Hawksmoor also consolidated Beverley Minster in 1716 against collapse, using timber framing. But by the early 18th century, as we will see later, attitudes to Gothic architecture were changing, and a new, generalised Gothic detailing became accepted as a legitimate way to obtain a sense of 'period'.[28]

Ancients versus moderns: the distancing of antiquity in Enlightenment Europe

IN parallel with the devolution of 'antiquity' to the new nation-states, the same principles of precise research reached back towards the 'centre', as the new classical mania spread to northern Europe. Along with bold, monumental new classical buildings, there was a new enthusiasm for classical collecting and archaeology, with large collections sold to France or England, and antiquarians travelling to the Mediterranean to measure or research. Although as early as 1575 French antiquarian Etienne du Perac had published a protest tract against destruction in Rome, *Vestigi dell'Antichità di Roma*, the movement of classical antiquarianism was first seriously initiated in the late-17th-century France of Louis XIV. There, a lively debate arose between 'ancients' and 'moderns'. Both believed in classical architectural principles, but the 'ancients' believed that reason was immanent in history, whereas the 'moderns', led by Charles Perrault, believed that they were independent of each other. Charles's architect brother Claude, designer of the new, grand east façade of the Louvre (1667–74), tried to devise a new, 'modern classical Order' to replace the traditional, inherited ones – as did M Ribart de Chamoust a century later, just before the Revolution (1776).[29]

This conscious foregrounding of the 'modern', in contrast to antiquity, is of the greatest importance in our story, presaging the consciously dialectical polarisation between the Old and the New that characterised the mature Conservation Movement. This dialectic would grow gradually in force, until its climax in the mid-20th-century Modernist concept of sharply differentiated, but complementary, old buildings and modern architecture. This new, more detached attitude to antiquity naturally required the best-quality research. The French Academy in Rome, founded in 1666 by Louis XIV, tried to establish a two-way traffic of ideas between France and Italy and to appropriate for France the best of antiquity, and publications like Desgodetz's *Les edifices antiques* of 1682 raised scholarly expectations. The 1670s saw the first drawings of the Parthenon, before its devastation in a 1687 explosion during a Venetian siege: the lack of an impact in Europe of this destruction of a once-cherished monument, owing to the inaccessibility of Ottoman-ruled Greece, illustrates the vital role of location and communication in the 'valorisation' (and 'devalorisation') of monuments. Corresponding to the new scholarship of the antique, a distinctive French perspective on medieval buildings emerged, focusing on their structural character. Around 1700, Félibien the Younger pioneered the subdivision of medieval architecture, into 'Gothique Ancien' (Romanesque) and 'Gothique Moderne' (Gothic). As with the English Gothic Revival, there was a significant overlap of architectural history and conservation history.[30]

In the 17th and early 18th centuries, Rome was at the height of its pride and confidence in the wake of the Council of Trent. The city's great papal families, the Borghese, the Ludovisi, the Barberini, the Pamphili and the Chigi, unleashed spectacular programmes of town-planning, church-building and restoration, employing architects such as Bernini, Borromini, Cortona, or Fontana to reconstruct their palaces – campaigns that were ominous in their implications for surviving built fabric. For example, Paul V (1605–21) demolished the eastern half of St Peter's and the Baths of Constantine, and in 1625 Urban VII (Maffeo Barberini) commissioned Carlo Maderno (succeeded by Borromini and Bernini) to build the new Palazzo Barberini using stones from the Colosseum – leading to the apocryphal popular saying, '*Quod non fecerunt barbari, fecerunt Barberini*' (The Barberini did what the barbarians failed to do).[31]

Developing the patterns initiated in the 16th century, every aspiring artist, architect, antiquarian and collector now came to Rome to discover its past art and latest developments (in the case of France, with the possibility of French Academy support), and diplomats and clerics returned home with private collections of works of art, or, at least, plaster casts of the most famous pieces. Numerous papal decrees of the late 17th and early 18th centuries failed to stem the tide of exported antiquities. By the late 18th century, a vast network of excavators, dealers and entrepreneurs linked the classical sites with private collectors abroad, especially in Britain, and a corresponding bureaucracy developed in the Papal Courts to administer these procedures. Under Pius VI (1775–99), the Vatican expanded its own holdings into a huge 'Vatican Museum', open to the public. Here, the history of conservation overlaps significantly with the history of collecting.[32]

During the 18th century, the balance of authority started to shift, as the contemporary power of Rome faded once again – while the imperialist power of France and Britain grew. Rome was left secure, however, as the focus of international cultural education, and the home of a huge, cosmopolitan community. Subsequently, other sub-centres of the 'antique' would emerge, both in Italy and in Greece.

History, heritage and the Enlightenment

EXPRESSING the new, more equal north-south relationship, Britain and France became the cradles of 'The Enlightenment', a movement that would fundamentally alter concepts of monuments and their conservation. Previously, for all the efforts of antiquarians, most treatment of old buildings had still been a matter of practical expediency or 'tradition'. But, in the wake of the Enlightenment, the past, and its monuments, became intellectualised and valorised.

The Enlightenment transformed the old universalist strand of culture, through new concepts of history and art. Historically, there was a new understanding of chronological relativity, enhancing confidence in human-directed progress through systematic, scientific procedures of understanding. The mushrooming of antiquarian scholarship in Rome, the unearthing of the buried cities of Pompeii and Herculaneum (see below), and the establishment of new sciences of painting restoration and of archaeology (a word first used from the mid-18th century), helped revive the old Renaissance tension between the historic and artistic significance of monuments, while connoisseurs like Winckelmann or Giovan Pietro Bellori further elaborated restoration theory. The late-18th-century explosion in tourist travel to Rome, especially from Germany, generated an extensive supporting infrastructure of hotels and tours. But northern European attention to Italy and classical Rome was beginning to shift from amateurism to Enlightenment scholarship – something that would later support the Conservation Movement's insatiable demands for the guaranteeing of authenticity. The Society of Dilettanti, originally founded in 1732 as a dining society of elite young men who had visited Italy, by the late 18th century had become a central element in the British Enlightenment – a forerunner of institutions such as the Royal Academy and the British Museum. It was also the first European institution to subsidise an archaeological expedition to Greece, and commissioned James Stuart and Nicholas Revett in 1751 to begin publishing their *Antiquities of Athens*.[33] The founding of the Soane Museum in the former house (1813–37) of architect Sir John Soane also spanned the amateur/scholarly division.

The Enlightenment's new understanding of history was a key precondition for the emergence of the modern concept of the historic monument, something that would become an essential ingredient in the mature Conservation Movement. Previously, when people saw everything as mixed together – sacred and secular, past, present and future – 'monuments' could not be perceived as a separate category or concept. Now, with everything more distinct, the stage was set for the modern concept of the monument to emerge: something unambiguously secular and tied to a strong sense of historical progress and authenticity, especially under the authority of the nation-state.

By the late 18th century, even before the downfall of the *ancien régime* throughout Europe in the wake of the French Revolution, the word 'national' was coming into currency, and the normative concept of a national identity or national culture, embracing the whole people (as opposed to feudal elites) was popularised, especially in the writings of J F Herder, and then of Romantic critics like Friedrich Schlegel.[34] Correspondingly, the old structures of trans-national religious authority that had pervaded the urban fabric of 'Christendom' went into decline. In northern Italy, where up to a third of some urban areas was occupied by monasteries and convents, the late 18th century saw a wave of secularisations and suppressions: in the province of Milan, the number of religious houses halved between 1769 and 1781, and the trend continued following the Napoleonic upheaval. In Bologna, for example, religious houses diminished from 79 in 1796 to only three in 1873.[35]

The art of oldness: towards a Romantic aesthetic of conservation

ALONGSIDE the respect for History, another chief support for secularisation was a growing veneration of Art. New philosophical concepts of the autonomy of art as an intellectual faculty distinct from reason – for example in Kant's *Critique of Judgement* – found enthusiastic echoes in literature and the visual arts, in England, France and Germany, where they mingled with powerful moral concepts, and coalesced into the Romantic Movement. Here, the authority of Art was enhanced by the energising power of nature and natural beauty. In France, Voltaire's essay on 'History', published in Diderot's mid-18th-century compendium, *L'Encyclopédie*, and the explorations of Thomas Cook and others in the South Seas, inspired the writings of J J Rousseau on a new ideal of simplicity, primitivism and the 'noble savage'.[36] The Abbé Laugier, too, argued in his *Essai sur l'architecture* in 1753, 'Let's keep to the simple and natural.' Among the elite in France,

Rousseau's lecture stimulated interest in humble buildings – anticipating the architectural concept of the vernacular. This was expressed most famously in the retreat commissioned by Marie-Antoinette from architect Robert Micque in Versailles in the 1780s: a facsimile of a Normandy farmstead. These concepts were echoed in Germany in the emphasis by Goethe, Kant and others on integration with 'Nature'.[37] Eventually, these ideas would all help shape the modern Conservation Movement, with its constant preoccupation with ever more 'ordinary' and 'natural' expressions.

In mid-18th-century England, where Enlightenment ideas were generally absorbed via the filter of empiricism, a wider range of ethical-aesthetic concepts was invented by writers like Edmund Burke, ranging from the Beautiful to the Sublime (with its associations of awe and primitivism) and the Picturesque, which bound together visual irregularity and ideals of individual freedom. Partly, these ideas were a codification of existing practices, especially in landscape gardening. From the early 18th century, the landed classes of England had started laying out landscapes and gardens which integrated aesthetics with moral didacticism. By the late 18th century, landscape designers and patrons across Europe were embracing the emotional power of Picturesque and Sublime aesthetics in reaction against the stiff formality of the old Versailles-like axes.[38]

Crucially for conservation, the landscape gardens of the Picturesque and Sublime were closely linked with an architectural preference for irregular old buildings, especially ruins. This ideology had emerged within the Romantic movement by the late 18th century, partly under the influence of Herder's concept of all-embracing national culture (see Chapter 3). Writers such as Goethe, Hegel and Coleridge drew close connections between Gothic cathedrals and the wild sublimity of forests in particular, or Nature in general. Goethe, although personally sceptical of the excesses of Romanticism, wrote in 1772 that a cathedral's walls should 'ascend like sublime, overspreading leaves of God'. A complex moral discourse of ruins spread during the late 18th century, with contributions from (e.g.) Diderot and Chateaubriand. Soon, this associative quality would become bound up with newly emerging modern values such as nationalism: already in 1772, in *Von deutscher Baukunst*, Goethe hailed Strasbourg Cathedral, in its vastness, as 'German architecture, *our* architecture', 'stemming from the strong, rough German soul' – an argument that also influenced the Schlegel brothers, von Arnim, Görres, Brentano and an almost limitless range of other writers.[39]

In contrast to the Sublime bias of Goethe's aesthetics, English preferences tended towards the Picturesque: the 18th century saw an ever greater preference for irregularity. This evolving ideology of picturesque ruins was most strikingly expressed in the wooded landscape of Fountains Abbey, Yorkshire, as successively remodelled by father and son John and William Aislabie, owners of the adjacent Studley Royal estate. Fountains, a Cistercian complex founded in 1132 and dissolved in 1539, was one of England's richest abbeys; its ruins were among the largest in the country, and began to attract visiting antiquarians from about 1670. John Aislabie, a powerful Whig politician ruined by the South Sea Bubble scandal of the 1720s, had retired to Studley to develop a vast landscaped 'Water Garden' in a formal, Le Notre style. After his death in 1742, his son continued the work, but in the radically different, Picturesque style of Capability Brown. This programme culminated in 1767 in his purchase of the ruins of Fountains for £18,000, for incorporation as a Sublime eye-catcher in his landscape. Although William's father had unsuccessfully attempted to buy the abbey ruins from at least 1720, it was only now, under the new aesthetic order, that they really came into their own: William even accentuated their impact with some selective demolition. Eventually, with the arrival of the railway in nearby Ripon in 1848, Fountains would become one of the most popular tourist destinations in England. The close, yet strained, relationship between the 'discovering' of new

Figure 2.6 The ruin as Picturesque landscape object: Fountains Abbey, Yorkshire, as remodelled by John and William Aislabie in the 1720s–50s

categories of heritage and their commercial/touristic exploitation would be one of the most enduring themes of the Conservation Movement.[40]

The emergence of the new, emotionalist aesthetic and its link to both the romantic appreciation, and the scholarly recording, of the relics of the past, was not a self-contained northern European matter, but was strongly influenced by dramatic developments in knowledge about classical civilisation in Italy. Sensational discoveries of the buried towns of Pompeii and Herculaneum were made after 1750, when Karl Weber, a Swiss military engineer, revolutionised the previously amateurish, treasure-hunting Bourbon excavations (begun in 1709 by the Prince d'Elboeuf) with a new professionalism, and within a few years excavated set pieces like the Villa of the Papyri and the whole of the Theatre of Herculaneum. The drama of the Campanian discoveries was highlighted by the great Lisbon earthquake of 1755, which underlined the risk of sudden

Figure 2.7 The excavation of Pompeii (begun 1709, accelerated after 1750): view of the Forum

destruction at any time; 1788 saw the first archaeological excavations in the Roman Forum, organised by the Swedish ambassador in Rome.[41] The 18th century also witnessed the spread of archaeological investigations outside Europe, for instance to Mexico, where the mid-century saw growing antiquarian interest in the monumental temple complex of Teotihuacán, with survey work by Lorenzo Boturini and others in the 1740s–60s.[42]

Winckelmann: classical scholarship and the art of restoration

THE word 'picturesque' began to be consistently used at this time not just in England but equally in Rome itself, by Johann Joachim Winckelmann, the foremost pioneer in the new appreciation of the past. Arriving in Rome in 1755 as an unknown German scholar and beginning a career of archaeological and artistic researches as Vatican librarian and secretary to Cardinal Albini, Winckelmann brought a potent new combination of scholarship and emotive passion to the study of classical remains, envisioning a site like the Roman Forum as a petrified yet Romantic setting for excavation and research, rather than a desolate quarry for classical fragments. Drawing on the new disciplines of art history (based on documents) and archaeology (based on study of artefacts) he applied Enlightenment historiographical methods to art and archaeology. Classical art, hitherto a static canon of timeless masterpieces, now became something with a history: only Vasari in the 16th century had previously sketched out a historical analysis of art. Winckelmann elaborated this framework into a drama of rise, perfection and decline, with the 'best' art seen as the product of privileged 'high' societies. In keeping with this ideal of perfection went a fashion for plaster casts, hailed by Winckelmann as even better than the originals, since 'the whiter a body, the more beautiful'.[43]

In his *Geschichte der Kunst des Alterthums*, published in Dresden in 1764, and internationally influential after around 1790, Winckelmann brought the discoveries of Pompeii and Herculaneum to international attention, laying the basis for modern archaeology. He declared for the first time that Greek art and civilisation were intrinsically superior to Roman – an idea also taken up by Herder, Goethe and the *Sturm und Drang* movement. At the same time, in Athens, Stuart and Revett were deploying modern survey techniques to produce the first accurate drawings of the Acropolis: the first instalments were published in 1787. The Vesuvian and Athenian monuments were among the pioneering examples of another future stock theme of conservation: the discovery and sudden celebrity of a previously 'unknown' or disregarded monument. In particular, the Parthenon soared within a matter of decades from obscurity to international cult status: increasingly, modern Western civilisation took it as axiomatic that 'the Parthenon is the most important building in the world, the most perfect structure ever to have been erected'.[44] In a textbook demonstration of the interrelationship of destruction and conservation, this reversal was provoked especially by the removal of the 'Elgin Marbles' from the Parthenon to London in 1802. More generally, Winckelmann's historical concepts of classical art inspired Herder to argue that each type of art is the product of a unique stage in history. In this way, paradoxically for one schooled in a sleepy, pre-industrial German state, Winckelmann helped fuel the forceful 'historical' interpretation of cultural development that underpinned the dynamic capitalist societies of modern Europe.[45]

True to his concept of historical specificity, Winckelmann insisted that restoration of works of art should always aim to recover their innate ideal neo-Platonic essence, and should be clearly distinguished, or 'legible', from the originals, to respect the latter's authenticity – a concept widely adopted subsequently in architectural conservation. Also increasingly vital now was the modern distinction, made possible by scholarship, between authentic works and 'forgeries'. The romantic, Sublime appreciation of Roman ruins, in parallel with the scholarly investigations of Winckelmann, was also whipped up by the work of Giovanni Battista Piranesi, a multifaceted artist/antiquarian whose series of 135 views of Roman monuments, the *Vedute di Roma*, issued in the 1740s, established him as the 'Rembrandt of ruins' and profoundly influenced the architecture and design arts of the day, both in Italy and north of the Alps. Piranesi gave the Renaissance concept of *deploratio urbis* a new vividness, and linked it with the craggy patina of age. This would inspire later artists, such as the specialist ruin-painter Hubert Robert, and admirers of other periods or styles, such as Ruskin and medieval Gothic. Piranesi's drawings also inspired a curious Romantic tradition of drawing new buildings as if reduced to ruins – most famously in Hubert Robert's 1796 view of the Louvre and Joseph Gandy's 1798 ruin perspective of Soane's Bank of England.[46]

Antiquarianism and modern 'Improvement' in 18th-century England: from Vanbrugh to Wyatt

CENTRAL to the dynamic growth of the mature Conservation Movement in the 19th century was the principle of the ever advancing chronological frontier of heritage – a process that was already underway in 18th-century Britain, in the diverse enthusiasms for the (mostly medieval) heritage in both England and Scotland. In England, the early-18th-century precedent of the architect/polymath Sir John Vanbrugh decisively shaped those responses. In parallel to his grandiose new classical architecture, he championed the preservation of medieval and Tudor ruins on both associative and picturesque grounds. In 1709, Vanbrugh unsuccessfully tried to persuade his patron at Blenheim Palace, the Duchess of Marlborough, to preserve the surviving section of the originally 12th-century royal manor house at Woodstock as an eye-catcher in views from the new palace, arguing that it was 'tenderly regarded' as a haunt of past kings, and would help relieve a monotonous vista, as 'One of the Most Agreable Objects that the best of the Lanskip Painters can invent'. In general, Vanbrugh contended, old buildings 'move more lively and pleasing Reflections (than History without their Aid can do) on the Persons who have inhabited them; on the Remarkable things which have been translated in them, or the extraordinary occasions of Erecting them'. Ten years later, in 1719, he was more successful in helping preserve, as 'Oene of the greatest Curiositys there is in London', the 'Holbein Gateway' in Whitehall, a 1530s' turreted Tudor gate threatened with demolition as being too narrow for carriages. Vanbrugh suggested diverting the street round the eastern side of the gateway, winning it a 40-year stay of execution.[47] At this

Figure 2.8 Early-18th-century view of the 1530s' 'Holbein Gateway', Whitehall, London

stage, urban conservation was mostly confined to specific structures, within a regularised aesthetic of classical 'improvement'. In the city of Chester, the medieval city walls were recast in 1701–8 as a genteel outlook-promenade, ringing the regular, brick-refaced main streets: the creation of the 'black-and-white Tudor' Chester that we know today, lay over a century in the future.[48]

Following the 1714 accession of the Hanoverian dynasty, with its forceful ideology of Progress and modernisation, perspectives on the great monuments of the past underwent a range of shifts across Britain. We will see shortly that in Scotland, with its more radical approach to 'Improvement', old castles were often modernised in a harshly interventive manner. In England, there were several parallel approaches. Alongside attempts at more archaeologically 'correct' engagement with originals, following Wren's precedent at Christ Church, after the 1720s a new practice emerged, led by decorator/architect William Kent, of adding detailing in a generalised Gothic style – for example, in new facings for the pulpit at Gloucester Cathedral (subsequently destroyed by fire in 1829). By the mid-18th century, a complex and morally charged attitude to the medieval past was emerging, fusing architecture and landscape design in a development of Vanbrugh's ideas. This was given specific architectural expression above all by the historian and pioneering antique-collector Horace Walpole, in his design of the world's first comprehensively neo-Gothic house at Strawberry Hill, near London. Here, an existing modest house was reconstructed from 1753 and carefully set in an 'enclosed enchanted landscape' that would subject visitors to a balance of moods, from medieval 'gloomth' to floral gaiety. Indeed, the early history of architectural conservation in England was largely identical with the early history of 'Gothic Revival' design.[49]

The following generation of English architect-surveyors tried hard to ensure that additions conformed to accurate, period-specific styles. James Essex of Cambridge, for example, did major repairs at Ely Cathedral (1757–62), where he stabilised the unsound roof and east end, and shifted the choir eastwards from under the octagon. At Lincoln Cathedral (from 1761), Essex added tall pinnacles to the central tower, 'as near as I could agreeable to the Ideas of the architect who built the tower', and reconstructed an internal western screen (in 1775) from a classical to a neo-13th-century style. Essex wrote that 'in order to correct the disagreeable appearance of this wall, I was desirous of tracing the original state of this part of the church, and if possible restoring it to the state which the

(a)

(b)

Figure 2.9 Strawberry Hill (reconstructed after 1753 by Horace Walpole)

(a) Exterior view; (b) The staircase hall

Figure 2.10 James Essex's work at Lincoln Cathedral (from 1761)

(a) The tower pinnacles; (b) The nave west screen (1775); (c) The pulpitum; (d) The bishop's throne

builders intended it'. Similarly, in new secular architecture by the mid-century, archaeological accuracy was often the aim. At Raby Castle, for example, the 18th-century interiors, including work by James Paine (1746–9) and Thomas Carr (from 1771), were serious, symmetrical and orderly.[50]

By the 1780s and '90s, the associative concepts of the Picturesque and Sublime and the new English antiquarianism were, together, exerting a pervasive effect on the treatment of English cathedrals.[51] At first, this was in cooperation with the prevalent spirit of rational utilitarianism, stressing convenience alongside aesthetics, and also influenced by a somewhat classical interpretation of the transcendental Sublime, which preferred cathedral interiors to be opened up into vast, uncluttered spaces. The language of the advocates of these projects was profoundly conditioned by the Whig 18th-century ethos of rational progress. Work was described either as 'necessary repairs' (to correct decay or dilapidation) or 'improvements'. 'Improvement' was the leitmotiv of the age: it encompassed agricultural reform, industrial development, urban extension and much else.

This movement of practical, progressive restoration was led in England by the neo-classical architect James Wyatt, who made radical interventions at several major cathedrals in the late 1780s and early/mid-1790s, notably at Salisbury, Lichfield and Durham. Salisbury Cathedral was, before 1789, hemmed in by a disorderly agglomeration of structures. A hundred yards to the south stood a tall, detached, 13th-century belfry, massively buttressed with a timber upper floor and leaden spire: this had previously served as a tavern, as had some of the low, ramshackle buildings around it. At the cathedral east end, the Trinity Chapel was abutted by two late-15th-century chantry chapels, the Hungerford and Beauchamp Chapels, whose elaborate Perpendicular Gothic style contrasted sharply with the main building's early Gothic. Limited repairs and rearrangements, to increase accommodation for services, were undertaken in the mid-18th century by surveyor Francis Price and the Earl of Radnor but, in 1789–92, advised by Wyatt, the newly appointed Bishop Shute Barrington launched a much more radical programme of Improvement. Externally, the dilapidated belfry complex and chantry chapels were all demolished, with the aim of creating a more Sublime, awesome simplicity, ringing the cathedral by open lawns. Inside, to create a vast through-vista, the 13th-century choir screen was removed, along with the remaining medieval glass, and the medieval monuments were rearranged in orderly lines along the plinths connecting the nave pillars; displaced grave slabs were cut up into paving stones, and a new choir screen (topped by an organ gifted by George III) was faced with decoration salvaged from the demolished chapels. The masonry was also overpainted in 'stone' colouring, covering up medieval murals. This was an important early example of what would become a recurrent approach to key historic monuments in the 19th and 20th centuries: to emphasise and enhance them through orderly isolation, clearing away clutter inside and out, and creating a vast surrounding space.[52]

In their practical effect, Wyatt's Improvements were not greatly different to the works undertaken by Wren or Essex in previous generations. What was very different now, though, was their 'reception', in that they provoked a storm of criticism, mostly from antiquarians. Preservationist commentators condemned schemes such as Salisbury as utilitarian in ethos and visually monotonous. Influenced by the informal, bric-a-brac character of Walpole's Strawberry Hill, people began to feel the Sublime should be mixed with an element of picturesque irregularity. Walpole himself intervened against the Salisbury project, chiefly attacking the bishop, but other critics, such as Rev. John Milner, concentrated their fire on Wyatt. This was the first appearance of one of the most enlivening traditions of the Conservation Movement: the polemical exaggeration of changes in preservationist taste, through condemnations of the supposed mistreatment of

(a)

(b)

(c)

(d)

Figure 2.11 James Wyatt at Salisbury and Durham

(a) Late-17th-century view of Salisbury Cathedral complex by Wenceslas Hollar, showing the detached belfry; (b) 1798 view of Salisbury Cathedral from east, showing the Hungerford and Beauchamp chantry chapels; (c) Interior view of Salisbury from west after Wyatt's works of 1789–92; (d) The Galilee Chapel at the west end of Durham Cathedral, originally proposed for demolition in Wyatt's 1795 proposals

monuments by demolition or more insidious restorations. Eventually, in the conflicts of the 20th century, the scope of such attacks would expand to the scale of entire nations, but for now they were confined to a precocious elite of antiquarians and architects. Successive generations of architects and restorers began vehemently attacking their predecessors, beginning invariably with criticism of their alleged inauthenticity and philistinism. There was a gradually amplifying oscillation between advocates and opponents of restoration. The advocates tended to be practical building patrons and their architects, and the opponents to be connoisseurs and writer-critics. With each successive fluctuation, the anti-restoration argument gained ground. As early as *c*.1765, for example, Thomas Gray had written to James Bentham that 'the rage of repairing . . . will be little less fatal to our magnificent edifices than the Reformation and the Civil Wars'.[53]

At Durham, Wyatt's proposals of 1795 expanded an earlier remodelling scheme by George Nicholson (from 1778, including a new, regular, east façade and refaced stonework throughout the building). Wyatt proposed to unify the interior by removing all partitions to provide an east–west view; to demolish the western Galilee Chapel; and to add a tall spire on the central tower. The entire package stirred up fierce opposition. The onslaught on Wyatt was orchestrated by antiquarian Richard Gough, Director of the Society of Antiquaries, who had in 1786 proposed a national campaign of repair of dilapidated medieval buildings.[54] Writing in the *Gentleman's Magazine* in 1789–90, he blasted Wyatt's Salisbury work, especially the demolition of the chantry chapels and the painting of the walls, as examples of 'the scalping-knife of modern taste' . . . 'Improvement, like Reformation, is a big-sounding word and oftentimes alike mischievous in its consequences'. Stung by Gough's attacks, the Salisbury cathedral verger, William Dodsworth, vigorously defended the improvements as a paradigm of classical good taste: writing with Wyatt's blessing, he contended that the homogeneity and regularity of his scheme gave a 'harmony, propriety and effect' to the interior. Gough was backed by the architect John Carter, also a stalwart of the Antiquaries. In a series of 212 articles attacking Wyatt's work in the *Gentleman's Magazine* from 1798, he established the chief conventions of the new genre of polemical architectural journalism, attacking Wyatt's schemes in an emotional and at times violent language that equated them to iconoclasm – an accusation with very negative resonances in England in the wake of the French Revolution. Carter's underlying ideas,

Figure 2.12 Henry VII's Chapel, Westminster Abbey, as restored by James Wyatt from 1809 (completed 1827 by B D Wyatt) (see also p. 104)

though, differed little from the German romantic writers': a combination of aesthetic preference and pride in Gothic as a national heritage.[55]

As we will see later, mid-19th-century English architects also took it as axiomatic that Wyatt's work was the height of bad taste: A W N Pugin lambasted him as 'this monster of architectural depravity – this pest of cathedral architecture'. But where the later-19th-century opposition focused on the 'falsity' of restoration, the target of the contemporary attacks was Wyatt's Improving modernity. When, in 1797, Gough's faction tried to block Wyatt from joining the Society of Antiquaries, they were crushingly defeated, on the grounds that his cathedral projects, as new building works, lay outside the Society's competence. However, the Durham controversy, and the linked furore in the Antiquaries, proved a turning point for Wyatt's Improvement schemes. Although his local executant architect in Durham, William Morpeth, proceeded from 1795 with the east facade remodelling and the external refacing in regular, smooth stone, the most radical element – the demolition of the Galilee Chapel – was blocked by Dean John Cornwallis and, by 1798, Wyatt's activities at Durham were confined solely to fabric repair. It would be another generation before a fresh wave of restorers would stir up renewed, and even more fanatical, opposition.[56]

The same spirit as Wyatt's also informed contemporary church Improvers in other countries, some of whose schemes have survived better than the English ones. In Austria, for example, Ferdinand von Hohenberg proposed a *stilrein* (stylistically pure) rebuilding of the Augustinerkirche (1784–5) and Minoritenkirche (1785–7) in Vienna. His approach, still visible today in strikingly intact condition in the Minoritenkirche, involved the imposition of a 'correct' but rather sparse neo-Gothic style, emphasising a symmetrical, architectonic, Sublime roominess similar to Wyatt's.[57]

Castle Romanticism and the 'historical landscape': 18th-century Scotland

OWING to the strong sense of *genius loci* among early antiquarians, the growing engagement with old buildings took varying forms in different countries. In Scotland, which combined an early ideology of 'national history' (exemplified by the pioneering 1582 book by George Buchanan) with enthusiastic participation in the modernising enterprise of Great Britain, the focus of the Romantic Movement was almost exclusively on the Sublime. Its energies were focused not on great cathedrals, largely lacking in Presbyterian Scotland, but on castles in rugged landscaping. The Scottish Improvement movement was even more radical than in England: the entire pre-Improvement rural building stock and landscape was relentlessly liquidated in the late 18th and early 19th centuries, and 500 new classical planned villages were founded between 1720 and 1860. The change from Stuart to Hanoverian Whig rule was reflected in a shift from respect for castellated architecture, as in Sir William Bruce's rebuilding of Holyrood Palace in 1671–9, to a preference for utilitarian regularisation, changing 'castles' into 'houses'.[58] In 1747–58, the 2nd Duke of Atholl commissioned architects John Douglas and James Winter to reshape his Perthshire seat, Blair Castle, into a roughly classical appearance by chopping off turrets and upper floors (a scheme which he dubbed the 'clipping of the Castle'); on completion of the work, the building was renamed 'Atholl House'. The

facing page
Figure 2.13 The Minoritenkirche, Vienna, as restored by Ferdinand von Hohenberg in 1785–7

(a)

(b)

The North East prospect of old and New House

Figure 2.14 Banishing and reviving the castle

(a) Blair Castle, 1736 view of east front by C Frederick; (b) The east front of the 1748 scheme by John Douglas for regularising the castle into 'Atholl House'; (c) The revived Blair Castle, as rebaronialised in 1869–76 by David Bryce; (d) Inveraray Castle, by Roger Morris, 1744–9

(c)

(d)

INVERARAY CASTLE.

ideal of 'purifying' unfashionable older buildings by purging them of supposedly tasteless features, would recur in later centuries – as in the mid-20th-century German fashion of '*Entstückung*' (removal of 19th-century plaster ornament from tenement facades). John Adam's 1765 remodelling of Castle Grant was a variation on the same theme: a massively austere new classical block was planted in front of an unloved, truncated castle.[59]

Although the mid-18th century was also an age when similar castles, such as Corgarff and Braemar, were being refortified by the government in a utilitarian manner against possible Jacobite insurrection, there were also the first glimmerings of a reversal in taste, with castles, new and old, returning to fashion. In Inveraray, the 3rd Duke of Argyll, a bold Improver with a strong sense of feudal obligation, in the 1740s began a vast programme of Improvement, including a planned classical new-town and replacement of his dilapidated tower house with a 'new castle' (1744–9, by Roger Morris), castellated/Gothic outside and classical inside. The last quarter of the 18th century saw the development of a fully fledged castle style by Robert Adam. Similarly, in Kassel, in Germany, 1793–5 saw the building of a new 'ruined castle', the Löwenburg, in the landscaped park of Wilhelmshöhe, by designer H C Jussow.[60] Old castles were also increasingly valued: by the 1780s and '90s, antiquarian-motivated repair schemes were getting underway across Scotland. In the 1790s, the hammerbeam roof of Darnaway Castle was propped up to allow new replacement walls to be built underneath it and, in 1789, a new Society of Antiquaries for Scotland, dedicated to protection as well as recording, was founded.[61]

But Scotland's chief international influence at that time lay not in those antiquarian efforts, but in the overtly Sublime cult of the ruin in wild landscape. Although Scotland, like England, had been largely free of the ruinous conflicts of 17th-century Europe, the march of Improvement left the landscape dotted with ruined castles. These, by the mid-18th century, were increasingly venerated as objects of Sublime wonder, within a distinctive concept of 'Scottish historical landscape'. This had been presaged in Sir William Bruce's design of his new classical house, Kinross (1679–93), its axial vista aligned on the ruined tower of Lochleven Castle, in which Mary Queen of Scots was imprisoned a century before. During the 18th century, this sensibility developed into a more complex world-outlook. This included both 'negative' elements, celebrating the uneasy sense of *deploratio*, of discontinuity and tragedy in Scottish history, as well as positive elements, reflecting the new, romantic worship of the Scottish landscape and the myth of *Ossian* (authored by James Macpherson).[62] The cult of Ossian helped shape the emergent world-view of northern European Romantic nationalism as something innocent and untainted by 'French' or antique high culture: Goethe's Werner even said, 'Ossian has supplanted Homer'. In 1786, the 11th Earl of Buchan acquired Dryburgh Abbey, one of several Borders religious houses left in ruins by Anglo-Scottish conflict, and made it the centrepiece of a historic landscape focused on Scottish history and classical texts: he later erected a classical statue of William Wallace at Dryburgh in 1814. From here to the early-19th-century writings of Sir Walter Scott (see Chapter 3) was a short, easy step.[63]

By the late 18th century, a diverse array of ideas and practices concerning old buildings had emerged across Europe. Many were connected, positively or negatively, with the emergent world-outlook of modernisation and progress. But this loose network of elite enthusiasts and concepts did not yet amount to a 'Conservation Movement'. What was required was a cathartic spark to fuse these ideas violently together – a traumatic stimulus that would convert the burgeoning concern and enthusiasm for the architectural past into a dynamic mass movement of the modern age.

Growth of the Movement

First modern ideologies of conservation

International revolutions
and national heritages

1789–1850

*An appreciation of our national treasures introduced and implemented . . .
throughout the whole fatherland would perhaps be the finest monument the
present age could erect to itself.*

Karl Friedrich Schinkel, 1815

*Precisely in such periods of revolutionary crisis, they anxiously conjure up the
spirits of the past to their service and borrow further names, battle cries and
costumes in order to present the new scene of world history in this
time-honoured disguise and its borrowed language.*

Karl Marx, 1852[1]

History's new driving force: conservation becomes a 'movement'

IN Chapter 2, we saw how the emergence of 'Modern Europe', with its competing nations and ideas, and its driving ethos of Progress, had stimulated the emergence of a new sensitivity towards old 'monuments'. That trend, however, had remained rather amateurish and sentimental, the business of gentlemen antiquarians, architects or Romantic poets. From the late 18th century, all that changed utterly. Political and economic modernisation in Europe and America accelerated to a furious pace, leaving behind the leisurely tinkering with Improvement and the gentlemanly nostalgia for the things of the past. Now, with everything increasingly in chaos, and all previous norms thrown upside down, the cause of old monuments suddenly took on a new sharpness and urgency. It became an 'ideology', a dynamic 'movement' of a typically modern kind, a movement with clear political significance and its own historical trajectory. It became – in the words of Austrian writer Alois Riegl at the end of the 19th century – a 'modern cult of monuments'. The new ascendancy of Progress was founded on a new sensitivity to history. Increasingly, 'historic monuments' served as a 'mirror of modernity' in the built environment, as a cultural anchor in turbulent times. The Conservation Movement was unambiguously a child of modernity. This modernity, however, took not one, but several forms.

The new respect for history originated in 18th-century rationalism, in the conviction that history could be 'a teacher of private virtue and correct public policy'. By the early 19th century, both the moral and scientific strands of history were being vigorously developed, especially in Germany, where insistence on scholarly use of original sources was combined with a messianic sense of purpose, harnessed to the burgeoning cause of nationalism. History, asserted Böhmer, served 'the love of the Fatherland, the conviction that the knowledge of the past could be instructive for the present'. Leopold von Ranke argued that 'History has had assigned to it the task of instructing the present for the benefit of ages to come.' A century or so later, entire townscapes would be attributed a similar didactic power; for the moment, those ideas were chiefly expressed through the written word.[2]

The new urgency of History for nations and their 'destiny' was first highlighted not by scholarly historians but by the dramatic ruptures of the French Revolution and its Bonapartist offspring. These violently ended the old, religion-legitimised absolutist states and multinational empires and put in their place, as objects of veneration and piety, secular concepts such as national identity or class solidarity. Within the new Enlightenment outlook of human-determined historical Progress, secular monuments assumed an almost sacred character, as a stabilising force that could offset the iconoclastic excesses of revolution.

As we will see shortly, the French Revolution's emphasis on equality and emancipation had conflicting effects in the development of the modern Conservation Movement. On the one hand, the concept of universal fraternity freed people from the old, staid international cultures of antiquity and 'Christendom', dominated by popes and Roman emperors, and substituted an ideal of general common humanity. The long-term effect of this on monuments was to foster the idea of collective responsibility for historic relics, transcending national or social boundaries. As early as 1758, Swiss jurist Emmerich de Vattel had argued that works of art were a common heritage of mankind, which must be protected in war as in peace.[3] Now, Napoleon's overthrow of old regimes across Europe, accompanied by an unprecedented international process of plunder of cultural objects, heightened the sense of common threat and patrimony.

But alongside this internationalism flourished a world-view that would become more important to conservation's growth in the late 19th and early 20th centuries – that of nationalism. At first, its main thrust was emancipatory and egalitarian, but during the 19th century it shifted towards a more assertive formula. Each nation tried to be internally cohesive and, externally, to outdo its competitors – with 'historic monuments' playing a key legitimising role. At first, France took the lead in developing this idea, through Napoleon's conquests. But others, notably Germany, soon caught up, in reaction against French domination.

Nationalism became the most powerful driving motor of the Conservation Movement, although its effects 'on the ground' were expressed not through homogenisation and cooperation but through competition between rival national strands. The emphasis within the early Conservation Movement was on stability and order, yet it became an eager participant in the late-19th- and 20th-century pursuit of national 'destiny' and 'expansion' – a restlessly active 'movement' in its own right rather than just a reactive counterweight to others' modernising initiatives. Left behind for the moment in this nationalistic frenzy was Rome, the old seat of 'universal' culture – not least because Italy also now had to share its prestige with newly independent Greece, highly active from the start in classical archaeology and restoration.

The upheaval of modernity was not simply a by-product of the French Revolution. Equally important was the destabilisation of 'traditional' culture by the iconoclastic social and economic changes of early capitalism. These originated in France's great geopolitical rival, Britain. Fuelled by both Enlightenment logic and commercial greed, agricultural Improvement broadened into a general boom in industrial urbanisation, which would transform Britain in 50 years from a rural landed to urban capitalist society. The alienating effects of this upheaval stimulated in England and Scotland, earlier than anywhere else, a conservation response to modernity, building on the Romantic Movement while answering a 'rational' need for the re-embedding of what had been violently torn apart.[4]

France and Britain, around 1800, developed parallel modern revolutions, both exploiting monuments as agents of stabilisation from within. Existing elite monument types, such as castles or cathedrals, were appropriated for the new purposes. Politically, monuments fuelled concepts of heritage that could help dampen down revolutionary chaos. And economically they provided a reassuring presence, whether in collaboration with the new mobile capitalism, or in critical dialogue with it through anti-modern utopianism. Whether one believed that Progress was alienating and bad, or improving and good, the emergent Conservation Movement had a potential role to play. In his *Eighteenth Brumaire of Louis Napoleon* (1852), Marx wrote of the paradoxically 'progressive' character of the past, arguing that love of the past was an 'ideology' that helped promote but also disguise the underlying forces of history and the interests of the ruling classes. He claimed of the bourgeoisie that, 'precisely in such periods of revolutionary crisis, they anxiously conjure up the spirits of the past to their service and borrow further names, battle cries and costumes in order to present the new scene of world history in this time-honoured disguise and its borrowed language'.[5]

Reflecting their integration with modernity and progress, these two strands of the emerging Conservation Movement, the 'political' and 'socio-ethical', in turn pointed forward to the future, helping structure the path of modern conservation over the following century. They pointed, for example, to the way in which 19th-century Germany would harness monuments to a struggle for national 'rebirth', or the way that mechanisms of state regulation invented in France would proliferate internationally in the later 19th and the 20th centuries. And they showed how the robust debates of Victorian England could create an ethical foundation-narrative of modern conservation, by setting 'restoration' and 'repair' in confrontation with each other – a tension between love of decay, and resistance to it, which the Conservation Movement has never since been able to resolve.

Terror to treasure: the French Revolution, crucible of 'national heritage'

THE complex succession of emotions and attitudes towards the 'national heritage' in France, following the Revolution, hugely intensified the long-standing play between destructive and protective forces, as the same agencies – the Revolutionary Committees – decreed measures of preservation and dissolution almost simultaneously. This activity fell into two main phases.[6] First came the transfer to the nation of confiscated royal and church assets in 1789–92. Then followed a campaign of ideologically motivated violence in the Terror of 1792. Finally, there ensued a protective counter-reaction, far more passionate and far-reaching than the old antiquarianism, owing to the scale of destruction just sustained.

The first phase, of nationalisation of cultural assets, was crucial in establishing the idea that old relics constituted a collective 'heritage' that could anchor the future of a nation in its past. In the writings of revolutionary commentators, such as Armand de Kersaint's 1791/2 *Discours sur les monuments publics*, virtually the entire modern vocabulary of conservation was invented (or borrowed from property law): heritage, patrimony, save, vandalism, conservation, historic monument, etc. The new terms 'national heritage' and 'historic monument' were far broader in scope than the old 'antiquities': almost anything, whatever its provenance or its date, could potentially be included. This principle of inclusiveness and 'semantic homogenisation of values' was of central importance to the growing Conservation Movement, giving it its sense of ever-expanding scope and general self-importance. The same values could also be applied to buildings not yet completed, such as the modern classical Madeleine or Ste-Geneviève, which become a 'French Pantheon'.[7]

The chief practical distinction in post-Revolutionary heritage definitions was between 'moveable' and 'immoveable' assets. From this sprang the two modern disciplines of 'museum conservation' and 'architectural conservation'. Protection of both aspects of 'heritage' began in 1790. At the suggestion of Mirabeau and Talleyrand, a *Commission des monuments* was created to carry out a nationwide inventory. Moveable assets were to be gathered together in depots – a process that received a spectacular boost in 1790–1 when the archaeologist Alexandre Lenoir, concerned at the threat to confiscated religious objects, secured (with support from Jean Sylvain Bailly) the mandate of the National Constituent Assembly for a radical law that authorised the gathering together of displaced art objects in a *Musée des monuments français*, initially housed in the Couvent des Petits-Augustins in Paris.[8]

This process of mass displacement and regrouping of antiquities, under a national banner, was typical of modern civilisation in its radical uprooting of existing patterns. The first stages in this process were not deliberately destructive, although the confiscations, by passing many key monuments suddenly to neglectful private hands, indirectly encouraged large-scale dereliction or dilapidation. Within a year, though, came the flight of the king and the onset of the Terror, with its state-sanctioned, 'patriotic' iconoclasm of *ancien-régime* symbols. In August 1792, the national assembly decreed the 'suppression' of all monumental vestiges of feudalism, unleashing a fury of destruction across France. This campaign was little different from post-Reformation iconoclasm or the purging of 'pagan' monuments in late antiquity. Arguably, it was more pervasive than that of the Huguenots, even if it did not involve the destruction of entire cathedrals. In fact – in the old tradition of appropriating valuables to finance wars – the main emphasis was on melting down statues, preferably for conversion to revolutionary armaments: within days of the August decree, statues of kings were felled across Paris and elsewhere. The destruction inexorably spread to all corners of France: autumn 1793, for instance, saw the smashing of 235 statues within three days at Strasbourg Cathedral. But the most sweeping condemnations were largely rhetorical, as in the case of Versailles: 'In the name of the sovereignty of the people . . . we strike to death this abode of crime, whose royal magnificence was an insult to the poverty of the people and the simplicity of republican morals.'[9]

But even during the Terror – demonstrating the interrelationship of destruction and conservation – articulate counter-arguments were voiced within revolutionary circles. These made for the first time a clear distinction between (positive) artistic and (negative) political-associative values. In August 1792, one speaker pleaded in the Convention that 'monuments of despotism are falling throughout the kingdom, but monuments that are precious to the arts must be spared, conserved'. Of the threatened Porte St Denis (François Blondel, 1672), it was argued that, 'for its dedication to Louis XIV, it deserves to be hated by free men, but this portal is a masterpiece . . . it could be converted to a national monument that connoisseurs would come from all over Europe to admire'. Kersaint argued of the Louvre that 'respect' for its artistic merit should outweigh 'hatred'

Figure 3.1 Heritage vicissitudes of the French Revolution

(a) François Blondel's Porte St Denis (1672); (b) Changing street signs (pre- and post-revolutionary, and modern) in Bordeaux: 2011 view; (c) Hubert Robert's painting of the 'violation' of the St Denis royal tombs in October 1793

for the kings that had built it, and condemned calls for the destruction of Paris, 'the City of Cities, the pride of the empire', as 'counter revolutionary'. Other historic buildings were protected by revolutionary use – for example the Conciergerie, seat of the 1793 show-trials.[10]

Epitomising the confusion of those years, in 1793 the Committee of Public Instruction and the Arts Commissioners simultaneously issued one decree ordering the destruction of 'all signs of royalty and feudalism' and another forbidding citizens:

> to remove, destroy, mutilate or alter in any way – on the pretext of effacing signs of
> feudalism or royalty – books, drawings, paintings, statues, bas reliefs, antiquities . . .
> and other objects of interest to the arts, history or teaching.

Under the *ancien régime*, a concern for regularisation of urban development had been emerging – from the 17th century, Parisian facades were strictly regulated and a 1783 royal ordinance stipulated a maximum 20-metre height for new stone buildings. Continuing this trend, in 1793, a 'Commission of Artists' proposed a grander Parisian replanning scheme, including a vast east–west axis.[11]

Even in this time of chaos, the foundations were being laid for France's pioneering system of centralised state control of conservation. In 1793, a *Commission temporaire des arts* issued an 'Instruction on the methodology of inventorisation', edited by polymath Felix Vicq d'Azyr, defining essential values of 'national monuments' ('cognitive', economic, artistic, etc.), with a comprehensiveness and rigour unequalled until the work of Alois Riegl a century later.[12] Potentially, this concept of French national heritage, and its inventorisation, formed part of a wider scheme to classify intellectual life and shape national identity through educational propaganda – a rationalist, scientific agenda that would help underpin the modern Conservation Movement, alongside the passions of nationalism and the love of Art.

For the moment, though, chaos reigned in France, and the advances in conservation theory were divorced from the reality of events 'on the ground'. To highlight this reality, the word 'vandalism' (a key protest-term ever since) was coined in 1793 by Joseph Lakanal, a member of the Committee of Public Instruction, in a report on heritage to the National Convention; it was then popularised by an antiquarian, the Abbé Grégoire. Now, Lenoir's museum project became energised with a new urgency. In 1793, he intervened following a National Convention decree ordering destruction of the tombs of former kings in the royal abbey of St Denis (where most French monarchs since the 10th century were buried). Although the tombs were opened up and the bones were thrown into nearby pits – following which Lenoir argued that 'this time, St Denis has been purged of the royal "race"' – he successfully argued for the rescuing and transfer of many of the tombs to his '*Musée des monuments français*', which opened as a public museum in 1795.[13]

The following year, France's first 'architect of monuments and conserved buildings', Pierre Giraud, was appointed. But the fate of individual monuments remained, in practice, a private affair, with official intervention haphazard and difficult. For example, the present-day 'Tour-St-Jacques' was the surviving remnant of the medieval church of St Jacques de la Boucherie, sold in 1797 to an entrepreneur. The tower was saved by Giraud, who persuaded the *Conseil des batiments civils* to indemnify the owner for the material value of its stonework and then proposed its inclusion in a shortlist of eleven 'historic monuments' being prepared by the council: the list encompassed buildings of all ages, including even the Madeleine. It was only after 1836, when the tower was bought by the municipality and incorporated as a landmark in its civic-improvements programme, that its future was finally secured.[14]

Napoleon's empire: Europe's first 'international heritage movement'

THE French Revolution, with its claims to universal applicability, unleashed a wave of 'creative instability' in the international aspects of conservation. As France shifted from anarchy to

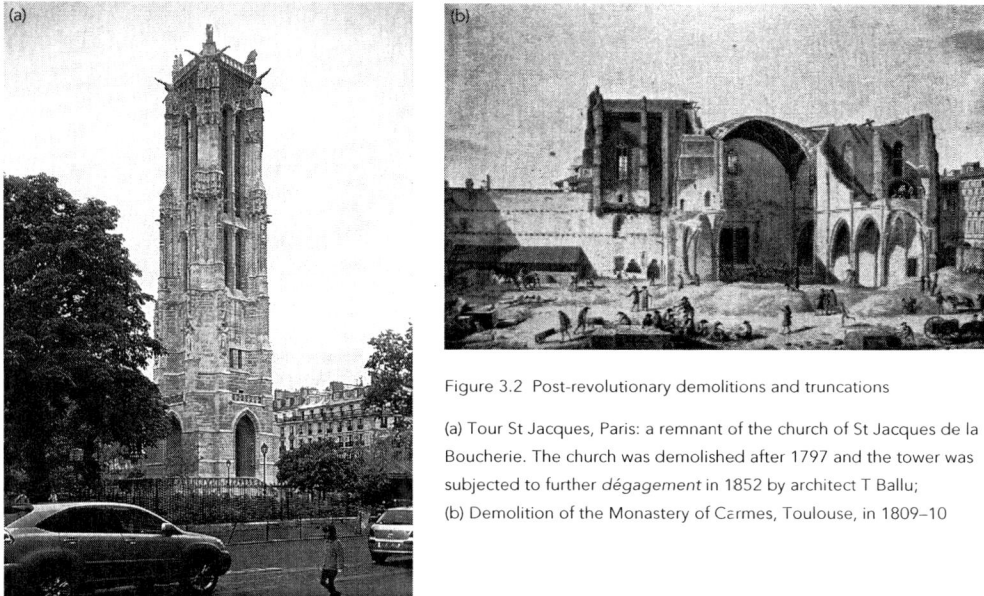

Figure 3.2 Post-revolutionary demolitions and truncations

(a) Tour St Jacques, Paris: a remnant of the church of St Jacques de la Boucherie. The church was demolished after 1797 and the tower was subjected to further *dégagement* in 1852 by architect T Ballu;
(b) Demolition of the Monastery of Carmes, Toulouse, in 1809–10

expansionism under Napoleon, another novel aspect of the new French monument concept emerged. The scope of the 'national' expanded beyond mere right to exist in equality with other nations. France laid claim to the 'universal' status of antiquity, appropriating Rome's 'moveable heritage' in the same way that Augustus had shipped obelisks from conquered Egypt. The 'international' and 'national' became muddled together. Already, in 1793, the Louvre had publicly opened for the first time, under the name 'The Republican Museum', as a permanent exhibition of the royal collections and of art confiscated from aristocrats and from occupied Italy and Holland. In a report that year to the *Commission temporaire des Arts*, the Abbé Grégoire pointed to the international prestige and visitors, attracted by great monuments, and advocated a policy of overt cultural imperialism vis-à-vis the old Universal City:

> If our victorious armies cross into Italy, the removal of the Apollo of Belvedere and of the Farnese Hercules would be the most brilliant conquest. . . . Should the masterpieces of the Greek Republics decorate the city of slaves? The French Republic should be their ultimate home.

In 1794, the National Committee on Public Education suggested that art experts should tour the occupied territories to identify works of art suitable for transportation to Paris. This prompted fierce criticism from some French commentators, such as Quatremère de Quincy, whose 1796 letters to General Miranda criticised France for 'ruining Rome', and argued that Rome itself should instead become a vast, open-air 'museum'.[15]

Within a handful of years, Napoleon's conquests had vastly expanded the scale of this artistic imperialism. Under the terms of the Treaty of Tolentino, 1798 saw the departure from Italy of vast processions of wagons loaded with art booty: on one single day, over 500 carriages laden with artworks rumbled out of Rome under military guard, despite vociferous protests by population and clergy. In 1800, British tourists in France reported Paris to be still in chaos, arguing that 'the French are horrified by everything reminding them of the past'. But Napoleon was now enthusiastically reviving the grand old tradition of French urban order and striving to make the Louvre (now renamed the 'Napoleon Museum') the world's foremost museum – an international as well as national repository of the finest works of Italian art. Simple plunder was buttressed by the new scholarly research: anticipating 1930s'/'40s' interventions by German historians in Poland, Napoleon's Egyptian expeditionary army was accompanied by an illustrator, Dominique-Vivant Denon, who explored the antiquities of the upper and lower Nile Valley and published in 1802 a two-volume book containing the first scholarly record of ancient monuments in Egypt; Denon persuaded Napoleon to commission a comprehensive survey of Egyptian remains.[16]

Following Napoleon's example, the British snatched the Rosetta Stone (discovered in 1799) away from the French, while, from 1800, British antiquarian Lord Elgin exploited Britain's Turkish alliance to secure the removal to London of the Parthenon sculptures. In one of the earliest examples of the interrelationship of rediscovery and protest, these sculptures' new, high-profile status in London duly provoked a polemical counter-reaction. Their removal was denounced as an imperialistic crime by Lord Byron in an 1809 poem, 'The Curse of Minerva', beginning a two-centuries' philhellene protest discourse that continues today.[17] More generally, the ubiquitous activity of English art-collectors across southern Europe provoked growing resentment and appreciation of local art-treasures. The northern European proto-nationalist attempts to appropriate the classical legacy were not confined to France and Britain: 1812, for example, saw the purchase from the Turkish authorities by the Bavarian Crown Prince, Ludwig, of pediment sculptures unearthed the previous year at the Temple of Aphaia at Aegina (first measured in 1765 by Stuart and Revett) by an expedition led by architects C R Cockerell and Carl Haller von Hallenstein. The sculptures were a centrepiece of the new Glyptothek in Munich, opened in 1830, and the Aegina temple became, for over a century, a special focus for German archaeological scholarship; after German unification in 1871, however, a far more ambitious excavation programme was begun at the Olympia complex, heavily financed by the Berlin government and continued into the 20th century.[18]

After Napoleon's defeat in 1814/15, the French plunder of moveable art objects was reversed: after the allies entered Rome in 1815 following Waterloo, the Pope despatched a commissioner for antiquities, the artist and president of the Accademia di San Luca, Antonio Canova, to Paris to organise the return of the treasures. Eventually, a British warship brought them back to Rome and the Vatican Museum reopened in 1822: as a victor, Britain held on to the Elgin Marbles.[19]

The pride of Napoleonic expansionism was also stamped on the occupied territories by 'restorative' interventions in their 'immoveable assets'. In the annexed Rhineland, Napoleon became determined, following an 1804 visit to Trier, to showcase the city as a centre not of Roman colonialism but of Franco-Gallic culture. Accordingly, he authorised a radical restoration of the arched Roman 'Porta Nigra' as a grandly isolated 'Gallo-Belgic' ruin, largely shorn of the medieval church structures and houses that had hemmed it in; ironically, the reconstruction was only completed in 1876 under Prussian rule. Elsewhere in the occupied territories, heritage dereliction was often a by-product of the Napoleonic traumas. In Speyer, the rebuilding of the cathedral from its 17th-century destruction by the forces of Louis XIV was no sooner completed than, within 15 years, revolutionary iconoclasts were busy desecrating the interior (in 1793); in 1806, following the

Figure 3.3

(a) and (b) The Temple of Aphaia at Aegina, first surveyed 1811 by Cockerell and Haller von Hallerstein and later restored largely by German archaeologists

Figure 3.4 The Porta Nigra, Trier, restored after 1804 at Napoleon's instigation

demolition of the Cathedral Palace, the destruction of the cathedral itself was only narrowly averted through a personal appeal to Napoleon by a local bishop. In Italy, too, older patterns of property ownership were radically destabilised, with large-scale redundancy of institutions such as monasteries continuing into the 19th century.[20]

In Rome itself, unlike the looting of 'moveable' works of art, a more careful policy was applied to the 'immoveable assets'. After the 1798 occupation and 1809 annexation of the papal states, Napoleon began a fresh drive of civic enhancements and restoration of classical monuments. Solutions included proposals by Canova as well as by French architects such as Louis-Martin Berthault. A further wave of excavations of the *Forum Romanum* began in 1801 and continued intermittently until 1827. From 1803, Giuseppe Valadier cleared the area round the Arch of Constantine. The Colosseum was the natural focus of attention, with successive efforts of repair and consolidation by eminent members of the *Accademia* after 1806. In that year, Raffaele Stern buttressed the eastern section with reinforcement and infill walling; in 1812 the French began clearing away the luxuriant vegetation from the interior (not completed until 1871). After restoration of Pius VII's rule, in 1824–6 Valadier made further repairs, with the new work sharply distinguished in brick. In 1817–23, Stern and Valadier also 'restored' the remains of the Arch of Titus, dismantling them and remounting them on a brick core, with new parts made 'legible' in travertine.[21] Elsewhere in Italy, the Napoleonic period also transformed the excavations in Herculaneum and Pompeii, focusing on the latter and bringing a new, professional attitude to bear on the work. By the 1820s, when the *Forum Romanum* was still largely an overgrown field, not just typical Roman houses but the entire forum of Pompeii were already excavated and visible to visitors from all over Europe, and in 1834 they were celebrated in Edward Bulwer-Lytton's famous novel, *The Last Days of Pompeii.*[22]

Dynamic nationalism and the anchoring role of heritage

THE most enduring heritage influence of Napoleon's imperialist internationalism would stem not from those direct interventions, but from their indirect effect in provoking a powerful counter-movement of nationalism. The 1814–15 Congress of Vienna and the 'Concert of Europe' attempted to turn back the clock across Europe, reinstalling pre-revolutionary regimes everywhere and instituting international cooperation to 'restore' the looted treasures of the 'Bouquet de Napoleon'. This restoration found powerful resonances in the emergent Conservation Movement. However, this was not a simple return to the *status quo ante*, but a 'return to Ithaca', a 'revival' not of the old dynastic conglomerates but of nation-states of a new assertiveness and coherence. Politically, nationalism across Europe was shaped by successive waves of influence from France. Initially, the Revolution had undermined religious and absolutist regimes and legitimised the ideal of 'national community'. Soon, though, the influence reversed itself, as Napoleon's domination provoked growing nationalist resentment, especially in Germany, where the liberation wars of 1813–14 now involved mass civilian armies and architects such as Karl Friedrich Schinkel outspokenly backed anti-Napoleonic patriotism: Schinkel designed the Iron Cross medal in 1813. The following decades saw mounting unrest against the 'restored' regimes, culminating in the continent-wide revolutions of 1848–9 and widespread destruction of historic buildings: for example, Dresden's Rococo Zwinger, later renowned for its lavish post-1945 restoration by East Germany (see Chapter 10), was also ravaged in the 1849 popular uprising in the city.[23]

These political resentments fell on fertile ground, prepared by wider developments in cultural theory and the philosophy of history in the later 18th and early 19th centuries. Hegel's writings invested the ideal of historical Progress with inescapable driving force: 'progressive History'.[24] And in the work of Johann Gottfried Herder, this driving force became bound up with the concept of

Figure 3.5 Forum and Colosseum

(a) Giuseppe Valadier's repairs to the Colosseum, 1824–6; (b) Raffaele Stern's 1806 work on the Colosseum; (c) J M W Turner, 'Modern Rome', 1839: view from the Capitoline looking across the overgrown and overbuilt Forum zone towards the Colosseum

Figure 3.6 Restoring the Arch of Titus

(a), (b) Piranesi's views of the ruined arch embedded in later urban fabric, 1750/71;

(c) The arch as reconstructed by Stern and Valadier in 1817–23

intrinsic national identity. Before the mid-18th century, nations had been regarded as accidental creations of dynastic convenience, but now they became seen as the results of an intrinsic consciousness (*Volksgeist*) that demanded cultural self-determination. During the early-18th century, writers ranging from the 3rd Earl of Shaftesbury to J J Rousseau had claimed the pre-eminence of feeling over reason – a concept of natural goodness and natural theology, possibly traceable back to the Pietists, which fuelled the late-18th-century *Sturm und Drang* movement in Germany. In opposition to the traditional political history of kings and generals, and elite concepts of culture and art, Herder's philosophy of history was concerned with broad socio-cultural characteristics, including customs, folklore and popular literary traditions and, above all, language. For Herder, language was the key to the formation of a *Volk*, and its enduring 'treasure-trove', giving it natural force and, potentially, political legitimacy. Taking the Jews as a paradigm, he argued in his *Ideen* that cultural forces provided the *Volk* with an invisible continuity through the centuries:

> This is the invisible, hidden medium that links minds through ideas, hearts through inclinations and impulses, the senses through impressions and forms, civil society through laws and institutions, generations through examples, modes of living and education. It is through this medium that we actively establish a continuum between ourselves and those that follow upon us.[25]

Unlike the stock Enlightenment belief in simple linear history, Herder saw progress as something interdependent with tradition: earlier periods were not inferior to later ones. Without tradition, progress was like a plant without roots, while tradition without progress was like a plant without water. Although Herder himself argued that *Volksgeist* should fuel a pan-global ethos of 'Humanity', in the heady atmosphere fomented by Napoleon's conquests, these ideas became narrowed and radicalised into cruder stereotypes, including those of the nation as organism, intrinsic national character, and national 'destiny'. In the works of German Romantics like Fichte, the identification of *Volksgeist* with language continued, elaborated by the philological researches and folk-tale collections of the Brothers Grimm. Later nationalist campaigns were fuelled by sometimes self-contradictory ideas of 'national universalism' – for example in the 19th-century-Italian nationalism of Giuseppe Mazzini, who argued that 'every people has its special passion, which will contribute towards the fulfilment of the general mission to humanity. That mission constitutes its nationality. Nationality is sacred'.[26]

The impact of the French Revolution on this evolving nationalist world-view was complex. Its emphasis on equality and fraternity fuelled the idea of the national community, while the stress on 'liberty' empowered ideas of local identity and diversity, often based on supposed medieval precedents: local 'tradition' could offset disruptive modernity at the national level. Each nation saw itself as a microcosm, as a miniature universe: externally, these microcosms were in egalitarian competition. This was almost a secular religion, a modern equivalent of the ancient polytheistic diversity of places and objects – although religion itself remained a central force in modern nationalism. Unlike the more fluid modernity that developed in North America, grounded in waves of mass migration, the new European nationalisms were structured by a burning sense of place, with obvious implications for the newly defined 'historic monuments', which became vital anchors, narratives in stone that counterbalanced the dislocating force of modernity.

In keeping with this intense sense of *genius loci*, the Conservation Movement itself took widely differing national forms. In this chapter, we focus on three in particular: that of Germany, which forcibly developed the 'political' strand of heritage, as an integral arm of the resurgent 'Fatherland'; that of France, which exploited its strong administrative tradition to create the world's first fully fledged conservation bureaucracy; and that of Britain, which pioneered a private, laissez-faire model of conservation, divergently focused in England on medieval churches and in Scotland on Baronial castles.

As the 19th century advanced, the new Conservation Movement increasingly did its 'anchoring' work through the specific process and ideology of restoration. This was itself a highly modern phenomenon, a doctrine that, although purportedly dedicated to tradition, and chiming with the concerns of political 'Restoration', was just as disruptive of 'authenticity' as any factory or steam railway. This paradox was summed up in an often-quoted definition by France's premier restorer, Eugène-Emmanuel Viollet-le-Duc, whose specific contributions we will trace in detail shortly. In 1866, in Volume 8 of his *Dictionary of Architecture*, he argued that 'the term Restoration and the thing itself are both modern. To restore a building is not to preserve it, to repair, or rebuild it; it is to re-instate it in a condition of completeness which may never have existed at any given time'.[27] In other words, a 'monument' might exist not as an object or as material substance but as an abstract ideal at one remove from historic 'reality', whose modern 'recreation' (or, rather, creation for the first time, in its new form) might, in its greater purity, answer the demands of the nation more effectively than 'authentic' old substance.

During the 19th century, the vast campaign of restoration played a vital role in sharpening the new, historical sense of the national future grounded in heritage. It was also closely interconnected with the system of eclecticism of historical styles within *contemporary* architecture, to convey different meanings and purposes. We last seriously met restoration (rather than, as with Wyatt, Improvement) in the Renaissance in Chapter 1, with the already contentious practice of restoring truncated statues, or the 17th-century scientific restoration of paintings. In the early-19th-century form that concerns us here, restoration aimed at returning the entire monument to a hypothetical original artistic integrity, an ideal essence, which must be deduced and recovered from the present state. But now there were two new developments.

First, with growing national competitiveness, the didactic purpose of restoration became potentially far more strident: to show the superiority of the art of the nation concerned. These feelings of superiority were usually anchored in the supposed medieval golden age and the special Romantic status of Gothic architecture: several countries (France, Germany and England) at first claimed to have 'invented' Gothic. But, as would become clearer in the late 19th and 20th centuries, for nationalists the historic fabric itself was always, ultimately, of secondary importance: what mattered was the intangible underlying reality of the national soul, Herder's 'invisible, hidden medium'. Second, and slightly in conflict with this, there was a new confidence in the diagnostic accuracy of restoration, based on scientific positivist methodology, which made the architect sovereign over old substance: one restored art through the application of science, to reveal a monument's true authenticity.

The nationalistic process of restoration was not confined to reconstruction of already venerated buildings. It also included the plucking of forgotten buildings out of obscurity, or the more literal uncovering of potential heritage set pieces by demolishing accretions that had hemmed them in – a process, previously pioneered in Renaissance Rome, that now went under various names, such as *isolement* in France or *Freilegung* in Germany. Here, the restorers echoed the Wyatt-era Improvers in their concern to clear vast spaces – but the underlying motive, with the rise of nationalism, was new and far more urgent. In the modern Conservation Movement, heritage became steadily more 'instrumentalised' by nationalism: harnessed to urgent objectives of an essentially 'political' character.

'Ein Fenster zum All'

('A window to the universe': Friedrich Gilly, 1794)[28]

THIS process first emerged in a fully fledged form in Germany, where the post-revolutionary and Napoleonic years saw an inexorable shift from the old fragmented provincialism (*Kleinstädterei*) towards more unified concepts of the nation. There was a veritable explosion of nationalistic heritage fervour, within which enthusiasm for secular monuments paralleled the concern for cathedrals and churches. This post-Napoleonic heritage nationalism was accentuated by the precocious emergence, in the Göttingen School, of the formal discipline of history, mixing national fervour with scientific accuracy.[29] A highly emotional, subjective nationalist rhetoric, a modern equivalent of *pietas*, would become dominant within German conservation over the following century. These ideas were fuelled by the explosion of cultural research that began in the 1800s, including Wilhelm and Jakob Grimm's collecting of folk tales around 1807 and the 1805–8 publication of Achim von Arnim and Clemens

Brentano's folk-poetry collection, *Des Knaben Wunderhorn*. In 1838 work began on a 33-volume *Deutsches Wörterbuch*, by brothers Wilhelm and Jakob Grimm (published from 1854). And the achievements of German science and research, and the rise of the *Bildungsbürgertum* (educated middle class), laid the foundations for the future spread of sophisticated conservation concepts.[30]

The epicentre of the new fervour was Prussia, whose growing ambitions had collided with Napoleonic power with particular violence, especially at the humiliating 1807 peace of Tilsit. Within Protestant Prussia, the hearth-stone of the new heritage nationalism was not a cathedral but a castle – the fortress of Marienburg in East Prussia, seat of the Teutonic Knights during their medieval ascendancy, thereafter controlled by Poland for three centuries (and known as Malbork) but, since 1772, with the partitioning of Poland, once more under German hegemony. The Marienburg was *the* archetype of the Conservation Movement phenomenon of an 'unknown' building suddenly catapulted from obscurity to national celebrity. Here, just as in the slightly later case of the Parthenon sculptures, 'rediscovery', 'threat' and 'outrage', followed in quick succession, but the shift was exaggerated by the very low status of the castle during the first quarter-century of Prussian rule. Used as a barracks and as weavers' tenement housing, it was very remote from Berlin. The Teutonic Knights had not been direct ancestors of Prussia itself, and the castle had long belonged to Poland. So when government architect David Gilly visited the grimy old building in 1794 to assess its condition, he was not impressed at all, and recommended it should mostly be pulled down. But Gilly was accompanied by his 22-year-old son Friedrich, a Romantic-influenced designer who later became one of Prussia's most noted architects before his early death. He, like Goethe previously at Strasbourg, was exhilarated by the Marienburg's 'sensual power' and 'painterly form': the ruined refectory vaults, he enthused, were '*ein Fenster zum All*' (a window to the universe). Intoxicated by his 'discovery', like many nationalist preservationists since, he prepared a wealth of superbly finished drawings, some of the exteriors in ruins but others showing the interior restored. When published

Figure 3.7 The Marienburg

(a) Friedrich Gilly, sketch of the refectory, 1794; (b) 1910 view of the complex, as restored from 1882 by Conrad Steinbrecht

on his return to Berlin in 1795, these caused a public sensation and provoked outrage at the neglect of such an emblematic 'German monument'.[31]

After 1800, with German nationalism rapidly growing in reaction to Napoleonic aggression, the Marienburg was targeted by one initiative after another: 1802 saw revised drawings by Friedrich Frick, and 1804 orders by King Friedrich Wilhelm III for preservation of the castle, which in effect became the first German national monument and a symbol of the sacralisation of the Fatherland. Although occupied by French troops between the 1807 defeat and the 1813 victory, restoration work was finally begun after 1815 by Theodor von Schön; only from the later 19th century would the Marienburg become the emblem of an increasingly aggressive, anti-Polish *Ostpolitik*.[32]

After 1815, the lead in German conservation was initially taken by Prussia's leading architect, Schinkel. In his work on medieval churches, such as the Apostelkirche in Münster, his approach was little different from Wyatt's Improvement ethos. But alongside this ran a torrent of passionate romanticism, expressed most directly in fantasy paintings of patriotic medieval subjects, cathedrals or castles, in wild landscape settings.[33] Schinkel was originally, after all, a topographic painter, and his pictures laid a spiritual foundation for the nationalist appropriation and institutionalisation of conservation, including restoration and completion of unfinished structures. He was the first significant architect, rather than antiquarian, to espouse the cause of national architectural heritage and propose a full-blown programme of preservation. This was a significant step in the instrumentalising of conservation, as an object of systematic, government-backed concern.

As one of a new generation of patriotic reformists, Schinkel was deeply affected by Fichte's doctrine of the nationalistic state (*Volk und Staat*) and tried to combine romantic passion with the painstaking efficiency of the civil servant. He was appointed general director of building operations in Prussian territories (Oberbaurat, later Oberbaudirektor/Oberlandesbaudirektor) from 1815 until 1838. In a remarkable memorandum of 1815 to the king, during a campaign to restore the war-damaged Schlosskirche at Wittenberg, he responded to the spread of secularisation and dereliction with the first modern manifesto of 'conservation outrage', arguing that, without urgent state coordination of preservation, 'we will soon be left bleakly naked, like a colony in an uninhabited territory'. He advocated establishment of an official corps of technical government officers charged with inventorisation as the first step towards preservation. During the late 1810s and early 1820s, Schinkel oversaw a systematic inventory by civil servants of monuments (mostly medieval churches) throughout Prussia; 1830 saw initial attempts at bureaucratic protection for Prussian secular monuments, with a cabinet decree against breaking-down of redundant city walls.[34]

Echoing Schinkel's ideas, royal architectural adviser Georg Moller in Hessen devised the first statutory conservation ordinance, issued by Grand Duke Ludwig X in 1818. Ludwig argued that it was a 'patriotic duty' to preserve monuments as a testimony of the 'early morals, spiritual roots and public culture of the nation'.[35] By that time, another Prussian conservation architect, Theodor von Schön, was making his influence felt, leading the first restorations of the Marienburg; in a familiar leapfrog pattern, he and Schinkel were later criticised (by Ferdinand von Quast, first Prussian state monuments conservator from 1843) for their designs' unarchaeological exuberance.[36]

Legitimised by the cult of the Marienburg, castle restoration erupted across Germany, linked (as always in the Conservation Movement) with commercial exploitation of that heritage through tourism. The epicentre of this activity was the Rhineland, recently acquired by Prussia, and already for decades a standard destination for British tourists on Grand Tours. Hitherto a mecca for the elite international cult of the Picturesque and Sublime, the area now became an exemplary symbol of

(a)

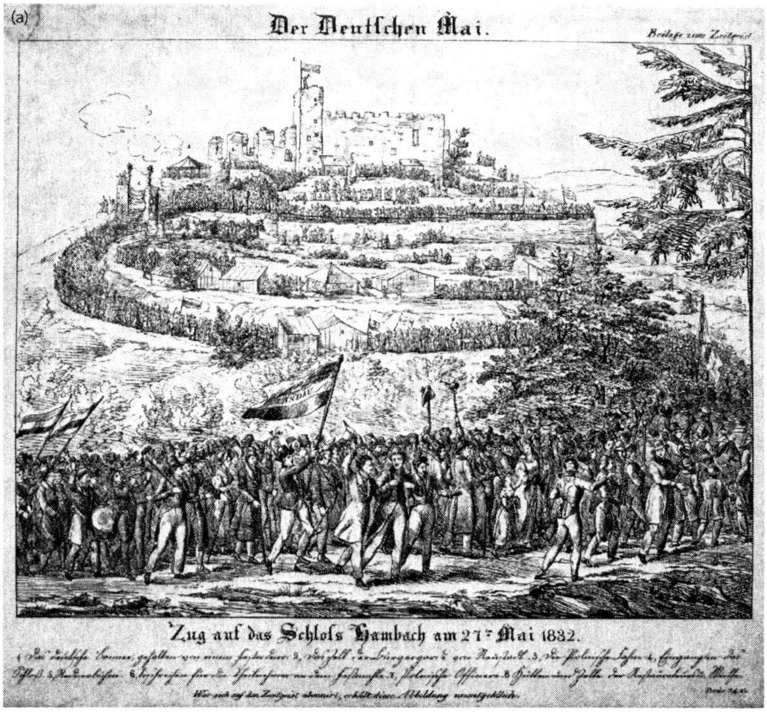
Der Deutschen Mai.
Zug auf das Schloss Hambach am 27ᵗ Mai 1832.

(b)

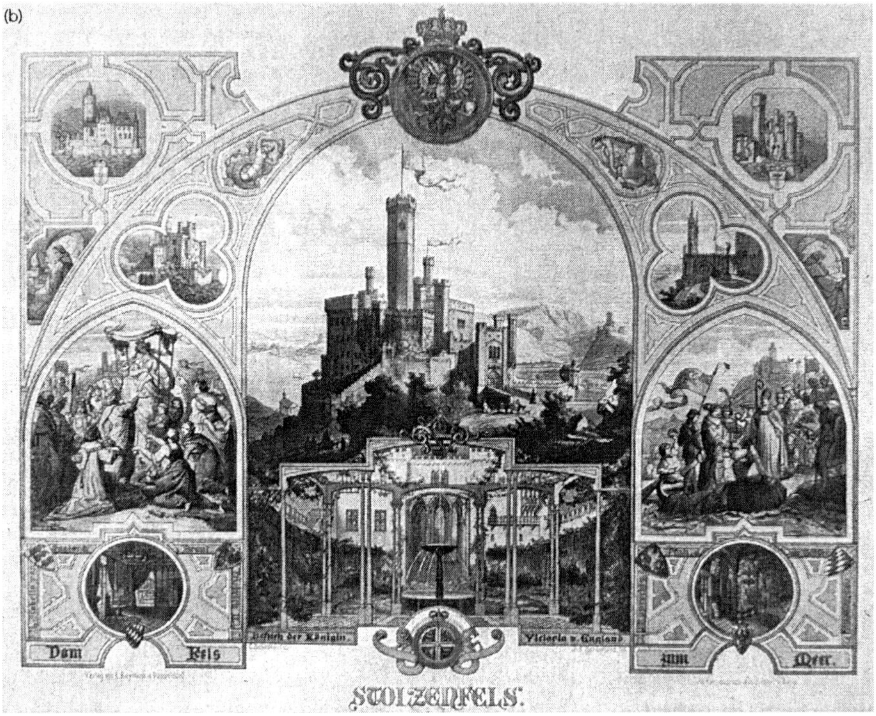
Vom Fels
zum Meer
STOLZENFELS.

Figure 3.8 Castle revivals in Germany

(a) The 1832 'Hambacher Fest' patriotic mass rally; (b) Stolzenfels Castle, restored 1836–47 by Schinkel and Stüler: commemorative views of the 1845 visit by Queen Victoria and Prince Albert

German nationhood, initially in elite circles immediately after the 1813/14 victory, but rapidly embracing mass tourism after the 1827 introduction of regular steamships. An upsurge of castle restoration ensued. Schinkel substantially restored Rheinstein Castle in 1825–9 and Stolzenfels Castle in 1836–47 (with F A Stüler) for the medieval-obsessed Prussian Crown Prince/King Friedrich Wilhelm (IV): Stolzenfels, destroyed by French troops in the 17th century, had been gifted in 1823 to the Crown Prince and rebuilt as the Prussian monarchy's summer residence. Schinkel's neo-Gothic plan incorporated the medieval remains and included a stately, vaulted 'Knights' Hall' inspired by the Marienburg. Similar projects began throughout Germany, as with the 1833–7 restoration of the war-destroyed Hohenschwangau Castle by Crown Prince Max of Bavaria. The patriotic castle mania transcended political differences: liberals opposed to the post-1815 political restoration tried to appropriate its prestige, and May 1832 saw a famous demonstration rally by 20,000 patriotic students and opposition supporters at the ruined castle of Hambach. The castle mania was an international one, but in Germany it chimed with the popular liking for houses in detached villa form, and the early 19th century saw an explosion of tourist guidebook publications on *Ritterburgen*.[37]

Cologne Cathedral: completion or restoration?

EQUALLY international was the fervour for restoration of medieval cathedrals and churches, initially fuelled by the Romantic conviction that Gothic was invented in one's own country. In Germany, though, the architectural focus was not a 'restoration' or a 'ruin' but a 'completion' project, for the 13th-century Cologne Cathedral, left an uncompleted stump in the 16th century, including a medieval crane that had survived on the site for three centuries.[38] Here, too, following Goethe's late-18th-century paeans to the Gothic as 'German architecture', the French Revolution decisively accelerated developments. In 1791, the cathedral was first praised as an aesthetic rather than religious monument by a radical Jacobin, Johann Georg Forster, and in 1814, following the Leipzig victory, explicitly political analogies were drawn between its completion and the reunification of Germany. Medieval enthusiast Joseph von Görres argued that this would be 'a symbol of the new Reich we intend to build!'

Architectural opinion duly followed in the wake of all this. The great 14th-century drawing for the twin-towered main west front was rediscovered in two pieces (in Darmstadt and Paris). Since 1808, architect Sulpiz Boisserée had been pressing with increasing passion for the completion of Cologne Cathedral as a monument to 'German history', with the backing of Goethe and Crown Prince Friedrich Wilhelm, arguing that it constituted a 'spatial representation of the collective history of the German Fatherland'. This exactly fitted Schinkel's conception of Gothic as a symbol of German national unity (in 1814–15, for instance, he proposed a project for a great Gothic cathedral in Berlin). After interventions by Schinkel, a 57-year programme of work started at Cologne in 1823, initially focusing on repair of the existing choir, but then, in a grand royal ceremony (*Dombaufest*) of 1842, moving on to the completion work of this 'eternal memorial of piety, concord and faith of the united families of the German nation on the holy site', latterly under the direction of Ernst Friedrich Zwirner. The first stone was lifted by the surviving medieval crane.[39]

The Cologne project, eventually completed in 1880, was a microcosm of the German neo-Gothic architectural world, with figures such as August Reichensperger (a native Rhinelander and

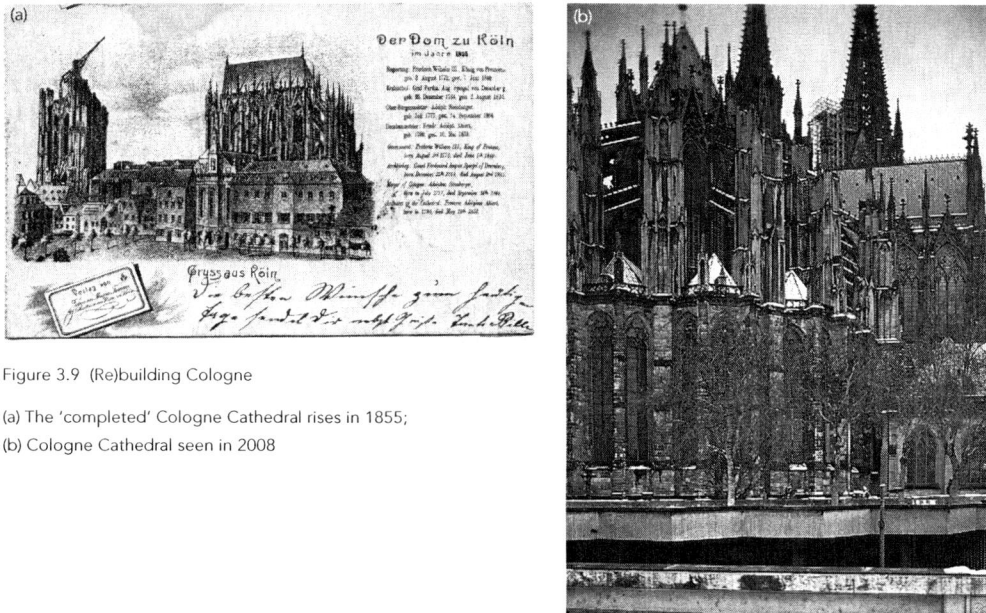

Figure 3.9 (Re)building Cologne

(a) The 'completed' Cologne Cathedral rises in 1855;
(b) Cologne Cathedral seen in 2008

Germany's foremost neo-Gothic theorist) playing a key role in the *Dombauverein*. It legitimised a wave of 'church completions', as we will see in Chapters 4 and 5. It also overlapped significantly with ideals of conservation and monument authenticity, just as, in contemporary England, restoration and 'Gothic Revival' were inextricably linked. Schinkel wrote of Cologne in 1816 as both an 'artistic undertaking' and as an attempt 'to conserve with all care and respect what the efforts of past generations have left to us'.[40] In its juxtaposition of the 'real old' and the 'old-style new', Cologne anticipated both the 19th-century movement of cathedral restoration but also, arguably, the 20th-century facsimile rebuilding of monuments destroyed in war. More immediately, it stimulated an extensive programme of earlier 19th-century cathedral restoration in German-speaking countries and northern Italy, mostly involving stripping out Baroque fittings and plasterwork: for example, the restoration from 1826 of Magdeburg Cathedral, the earliest Gothic building in Germany, with extensive input from Schinkel.[41]

After 1830, many restorations started to shift away from earlier ideas of classically inspired, monochrome stylistic purity to a properly researched conception of the medieval monument as something 'other', including elements of polychromy. These more rigorous concepts of authenticity led, paradoxically, to a sharp increase in both the diversity of solutions and conflicts between them. Both were exemplified in a range of projects stemming from the patronage of King Ludwig I, beginning with an ambitious, multi-phased restoration of Bamberg Cathedral. Here the excesses of the first stage, from 1828, where artist Friedrich Karl Rupprecht tried to expunge all Baroque decoration, led three years later to a more sensitive restoration strategy devised by architect Karl Alexander von Heideloff, including the uncovering of richly painted decoration. This programme, in turn, was reversed in the late 1830s by royal architect Friedrich von Gärtner, and resumed again, in 1846, by the Bavarian crown prince.[42] The architect-patron relationship of Gärtner and Ludwig also stimulated, in the town of Aschaffenburg in 1840–8, a different kind of restoration, the

'Römisches Haus', a facsimile of the recently excavated Castor and Pollux House in Pompeii, stocked with original Roman works of art. In another of Ludwig's projects, at Speyer Cathedral, Neumann's 'archaeological' late-18th-century restoration no longer seemed sufficient for such a symbolically loaded building, so an 1846–53 restoration scheme by artist Johann Schraudolph filled the church with rich new polychromatic decoration in the 19th-century 'Nazarene' style (itself, in turn, expunged by 'purifying' restoration in the 20th century).[43]

Nuremberg and the beginnings of 'Old Town' ensemble conservation

So far, we have traced the nationalist intensification of monumental building-types – cathedrals, castles – already established as objects of antiquarian and archaeological concern. But the Conservation Movement was from the beginning a movement dedicated to territorial expansion, and now its scope began expanding from single monuments to entire old towns. Here, again, it was Germany that drove forward the frontiers of the Movement, proselytising the love of the medieval city in its entirety.

Reflecting the highly dispersed character of the German heritage, the focus of this trend was not a cathedral city but the old imperial capital of Nuremberg. Like the Marienburg, Nuremberg had reached its reputational nadir immediately prior to its 'rediscovery': following a three-century decline, in 1796 the imperial insignia were spirited out, the city was occupied by French troops and in 1806 was annexed by the newly created Kingdom of Bavaria. For much of the late 18th century, travellers had disparaged it as shabby and backward. Mozart in 1790 simply wrote, 'We had breakfast in Nuremberg – an ugly city'. Another traveller in 1796 complained that 'everything is narrow, petty, oppressive', and redolent of 'emptiness and decline . . . nothing great, nothing sublime'; entering St Sebald's Church was like 'coming into a den of bats, it stank so much of vermin and looked so malevolent'.[44]

But in 1796, a radically different view, a manifesto for a Romantic, nationalist rediscovery of Nuremberg, was published anonymously by Wilhelm Heinrich Wackenroder and his friend, Ludwig Tieck: the exuberantly titled *Herzensergiessungen eines kunstliebenden Klosterbruders* (Impassioned Effusions of an Art-loving Monk) – a tract which became one of the foundational texts of German Romanticism, awakening the educated classes to the fantasy of a past golden age. Visiting Nuremberg for the first time three years after Mozart, Wackenroder wrote excitedly that 'there is not a single new building' in the city – it had an 'antique, exciting appearance'. And this unique character was (as always, with such 'rediscoveries') under threat, not from deliberate destruction but from wanton neglect and decay: one chapter of Wackenroder and Tieck's little book contains an impassioned plea to safeguard the city that was Dürer's home, and whose 'quaint' streets bore the 'permanent trace' of his art: 'How deeply I love the structures of that age, which have such a robust, powerful and true language.' In their combination of nostalgic patriotism and yearning for vernacular community, celebrating Nuremberg's irregularity while bemoaning its decay, Wackenroder's text anticipates the

facing page
Figure 3.10 The Dürer House, Nuremberg

(a) Exterior view in 1816 (engraving by J C Erhard); (b) View from the house towards the castle, in 1838 (engraving by G C Wilder)

Romantic apotheosis of Nuremberg throughout the 19th and early 20th centuries, and the cult of the Old Town in general. In tandem with Goethe's sentimental story of the following year (1797), *Hermann and Dorothea*, it began to establish the idea that the seat of German identity was the historic or small town.[45]

As with all the 'advances' of conservation, this new love of irregular medieval urban ensembles was as much aesthetic as ethical in character, rooted in the already established liking for the Picturesque and Sublime irregularity of landscape. It ran in parallel to the work of Romantic artists such as David Roberts, 1840s' painter of Egyptian ruins, or the slightly earlier John Martin, painter of sublime biblical fantasy-canvases. More relevant still, as we will see in Chapter 4, were the polemical 1830s' drawings of A W N Pugin, contrasting an idealised medieval town and its debased modern equivalent. The rapidity of the diffusion of the Old Town 'discourse', outwards from Nuremberg, was reflected, for example, in the nearby fortified town of Rothenburg ob der Tauber, which shifted during the early 19th century from a state of depression and dilapidation, threatened with demolition of walls and towers (despite the preservationist efforts of Ludwig I and von Gärtner), to a celebrated symbol of Romantic German identity and a mecca for artists and tourists. And in the nearby Czech lands, a similar Old Town movement, focused on Prague, sprang into life in 1837–41, in reaction to the threatened demolition of the Town Hall: one protest memorandum to the Emperor warned of the townscape damage to 'one of the most remarkable and picturesque squares of the capital'.[46]

Restoration France: the first government heritage service

IN each country, the burgeoning Conservation Movement took its own specific form, influenced by geopolitical factors and pressures specific to architecture. In France, the period following the 1815 defeat – unlike the simple explosion of nationalistic pride in Prussia – was marked by a complex balance of cultural and political forces, with correspondingly complex repercussions in the heritage field. France shared in the spirit of restored 'tradition' sweeping the continent following the Congress of Vienna. Equally prominent, however, was a spirit of rational, 'scientific' logic, as expressed both in administrative procedures and in the Beaux-Arts tradition of architectural education.

At first, most of the revolutionary initiatives were allowed to lapse. After 1817, Lenoir had to return most of the contents of his museum in the Petits-Augustins to their former owners. St Denis reverted to its role as a royal mausoleum, a protracted process already begun by Napoleon in 1805. At the turn of century, Chateaubriand had written that 'St-Denis is a desert: birds fly through it and grass grows on the broken altars'. But, from 1813, Napoleon put architect François Debret in charge of restoration, and work began on assembling dummy monuments from fragments of buildings destroyed during the Terror. From 1816, the royal bones (and Louis XVI's ashes from the Madeleine) were reverently reinterred in an ossuary, and the monuments recovered from the Petits-Augustins were reinstalled in the crypt; in 1833, the exterior was radically restored, a centimetre of stone surface being scraped off throughout. After further interventions precipitated the collapse of the north-west tower in 1846, from 1847 Viollet-le-Duc took over as architect. Typically of the Conservation Movement, he immediately set about 'de-restoring' Debret's work, transferring the tombs from the crypt in 1859 to an approximation of their original places in the transepts (along with tombs from other churches destroyed at the Revolution) and repairing the exterior.[47]

Figure 3.11 Viollet's church restorations

(a) Perspective of Viollet's 1859 scheme for rearrangement of the royal tombs in the St Denis transepts, showing the architect himself on the right; (b) Present-day view of the north transept of St Denis; (c) Notre Dame, Paris, restored 1844–64 by Viollet; (d) St Sernin, Toulouse, present-day view following 'de-restoration' of Viollet's 1860s' work in the 1970s by architect Yves Boiret

With the abandonment of the revolutionary claim that France was the universal centre of world revolution, a new, patriotic canon of French history emerged, influenced by the romantic medievalism sweeping Europe – for example in the writings of Sir Walter Scott. Displays in the Louvre and elsewhere stressed the diversity of the French nation, combining the tradition of the monarchy with the social consciousness of the Revolution. Influential cultural critics such as Montalembert, Victor Hugo, Guizot and Chateaubriand refocused the role of historic monuments to entrench this new patriotism. Writing from a wide range of political standpoints, they adumbrated, between them, the modern discourse of 'collective memory': from the 'right', Montalembert argued that 'long memories make great peoples', while from the 'left', Victor Hugo claimed, the year after his 1831 publication of *Notre-Dame de Paris*, that medieval monuments evoked both 'the memory of the kings and the tradition of the people'.[48]

Significantly outstripping the earlier German developments at Nuremberg, this new French patriotism spread its scope more widely than the old history of kings and great men: Victor Hugo argued in 1831 that the entire historic 'fabric', rather than isolated buildings, constituted the true national heritage. It was 'a deposit left by a whole people, the accumulation of ages. Great edifices, like great mountains, are the work of ages'. For the passionately Catholic Montalembert, the Middle Ages were crucial as the high-point of Catholic faith and art. And Balzac opposed the old and new France, the France of mechanical, practical progress and the fast-disappearing 'portraits of former eras'.[49] As to conservation and restoration, the repudiation of revolutionary chaos by Napoleon and the restorationists did not automatically lead to better care of historic monuments. One of the most serious demolitions of all, that of the Romanesque set piece of Cluny Abbey, began in 1798 (because of road works) and spanned a quarter-century, leaving only fragmentary remains.[50]

Under the Restoration, in the 1820s and '30s, the pace of destruction significantly quickened, prompting fresh protests against 'vandalism'. These began in 1825 with Victor Hugo's '*Guerre aux demolisseurs*', which evoked a 'universal cry that summons the new France to the rescue of the old', and Montalembert's tract '*Du vandalisme en France*' (in the *Revue des Deux Mondes* in 1833). Even government officials, presenting the 1833 budget, agreed that France must save the Gothic monuments and churches of the Middle Ages, 'those precious survivals of old France to which time brings some new destruction every day'.[51] Following the chaos of the 1830 Revolution, growing public and antiquarian nostalgia found itself confronted with ambitious agendas of planned modernisation, promoted by the state. The response, typically of French public administration, was a comprehensive conservation strategy entrusted to a central state bureaucracy. After 1830, the first steps were taken towards a permanent structure of government-organised architectural conservation in France, rapidly developing into a comprehensive structure unmatched in any other country, and outstripping even the earlier measures in Prussia. This was a crucial building-block in the evolution of the modern Conservation Movement, whose restless expansion in the 20th century would require significant state coordination.

Following an 1830 proposal to the king, Francois-Pierre Guizot, Minister of the Interior and Sorbonne history professor, created a post of *Inspecteur general des monuments historiques de la France*, intended to chart out a permanent strategy of government conservation. The first post-holder was Ludovic Vitet. He, however, resigned in 1834 in order to pursue the conservation agenda through political means. His successor, the 31-year-old Prosper Merimée, presided over a methodical new survey programme of '*classement*' (classification) of buildings deserving the status of a historic monument. For this system, finance and administrative back-up were essential, to allow authoritative designation processes and a proper long-term system of repair and maintenance. Accordingly, in 1837, a *Commission des monuments historiques* was set up (the commissioners initially including Victor Hugo and Montalembert). Merimée remained dominant in the historic monuments service even after his eventual resignation in 1853.[52]

The concerns of this emergent heritage bureaucracy differed significantly from those of its descendants in innumerable countries today. In these years, the role of the French state was not seen as one of imposing sweeping restrictions on old buildings as a whole. The government's concern was solely with classified monuments, and it saw its task as being to supply financial aid for approved repair work, a role reinforced, from 1841, by provisions for compulsory purchase. By the publication of the first government list, in 1840, there were 934 classified monuments, and nine years later 3,000 – mostly medieval churches and Roman remains. Grant allocations rapidly soared as well, from 120,000 francs in 1836 to 800,000 by 1848.[53] Although circumscribed in practice, the process of inventorisation and upkeep was still potentially a vast one and, typically of the French royal and Napoleonic tradition, it was entirely centralised in Paris. Ironically, within the capital itself, the Commission's powers were more restricted: only key masterpieces were classified. In the upkeep of cathedrals, the Commission's role was before 1907 purely advisory; executive authority was exercised by an autonomous body, the *Bureau des edifices diocesans*.[54]

The development of France's formidable system of centralised heritage management had one less desirable consequence. The potential contribution of the country's burgeoning network of local antiquarian societies was largely disregarded – something that partly stemmed from the deep political divisions since the Revolution. Initial attempts to develop a lively voluntary sector had been led by the Normandy antiquarian, Arcisse de Caumont, who founded in 1824 a Norman regional archaeological society, published influential antiquarian overviews from 1830 and established the *Bulletin Monumental*. In 1834, Caumont set up a national *Société Française d'Archéologie*, based in Caen, with a watchdog remit: local correspondents were charged with monitoring threats to monuments in their areas. But although Caumont exercised considerable influence in the national debates before the formation of the Commission in 1837, thereafter his watchdog network was marginalised, largely for political reasons. Its ranks were pervaded by traditionalist aristocrats, who looked with hostility on the new Paris-based empire of government monument administration and its elite of national executant architects (including Viollet, Charles-Auguste Questel and Émile Boeswillwald).[55]

In the Commission's early deliberations, the convoluted issues of prioritisation that still beset conservation organisations today were tackled for the first time. In the early 1840s, for instance, there were earnest debates about whether Laon Cathedral merited repair grant-aid: Merimée dismissed its architecture as 'barbarian', but Baron Taylor countered that few medieval monuments had survived the Revolutionary traumas intact. Anticipating modern 'rescue recording', demolition of the dilapidated abbey church of St Germer (Oise) was reluctantly conceded, on condition that Boeswillwald first prepared measured drawings. The polarisation between government 'archaeological' and 'architectural' approaches emerged for the first time, with archaeologists vainly fighting the architects' radical restoration schemes – such as a proposal by Questel to remove a 17th-century portal at Tournus. The Commission was also preoccupied with the fate of fragments from demolished buildings, such as the Hotel de la Tremoille, and the issue of whether church fittings should be conserved *in situ* or transformed to museum settings. Its post-1848 director, Charles Lenormant, bemoaned the dispersion of Lenoir's museum in the Petits-Augustins as 'a deplorable loss to the study of the history of art'.[56]

The partnership of modernisation and restoration in France

MID-19th-century France witnessed increasingly ambitious efforts to reconcile modernity and heritage. M Martin, keeper of the Seal, argued in the Chamber of Deputies in 1841 that 'public utility was not a purely natural thing; national traditions, history, art itself, are they not in truth matters of public utility, just as much as buildings and arsenals and roads?' And in 1905, the eminent British archaeologist, J Baldwin Brown, concluded in his comparative study of international heritage protection that, 'if France be reckoned a classic land in all that concerns the care of the artistic treasures of the past, her reputation stands equally high as the country of good taste in the outward apparatus of modern life'. This French collaboration of modernity and heritage was associated above all with Baron Georges Haussmann's bold programme of surgical replanning of Paris after his appointment as Prefect of the Seine from 1853, under the Second Empire.[57] In those years, informed by the logical clarity and Beaux-Arts love of classical order and grand spaces, French architects developed a philosophy that allowed the nascent Conservation Movement to act as a constructive counterpart to radical modernisation. This outlook was not unlike the maligned efforts of Wyatt, half a century earlier, to bring Improvement to England's cathedrals; but it sharply conflicted with the romantic love for old city quarters, by now being preached in France by Victor Hugo.

This policy of modernising conservation, 'instrumentalised' in the service of the state, was active both at a wider urban scale and at the level of the individual monument. As with Wyatt, the key urban-planning principle under Haussmann was openness: to open up new, regular streets and spaces for enhancement of traffic, the economy and public security, and demolition of 'slums'. Most older areas of Paris were affected: some, such as the Île de la Cité, were largely cleared and redeveloped in the early 1860s, with new apartment blocks controlled by strict visual guidelines. In the process, churches and other key historic monuments, previously swamped in higgledy-piggledy accretions, were set apart by '*isolement*' or '*dégagement*', a process then seen as ennobling, by liberating them from degrading squalor. For example, the 16th-century Tour St Jacques, in its truncated post-Revolution form, was envisaged as the future linchpin of Haussmann's planned '*grande croisée*' of boulevards at the Châtelet. From 1852 it was subjected to comprehensive *dégagement* by architect Théodore Ballu, who tore down dense housing from all around it; the cleared area was planted with dignified civic greenery by Alphonse Alphand in 1854, and the following year the tower was officially reclassified from a church to a 'civic monument'. In other towns, similar policies were followed: in Orange, 1845 saw a rare use of government expropriation powers to clear over 100 old houses from inside the Roman amphitheatre. The chief large area in Paris unaffected by this surgery was the Marais, where in 1866 the City Council bought the 16th/17th-century Hotel Carnavalet for use as a museum of the city's history: the mansion was restored by Victor Parmentier.[58] But after the chaos and destruction of the Commune of 1871, which left many public buildings in ruins, general calls began for a more wide-ranging policy of urban conservation in the French capital.

Complementing the policy of radical openness in the urban fabric was an equally radical policy concerning the individual building: interventive restoration. Here France reaped the benefits of her precociously developed school of official conservation architects, able and eager to develop a consistent ethos and body of work over decades, aided by lavish government repair funds and the scientific rationalism of French architectural education.

Viollet-le-Duc: Europe's premier restorer

THE new philosophy of 'scientific' restoration was developed most systematically by two architects closely bound up with the state heritage machinery: Eugène Emmanuel Viollet-le-Duc, who became, internationally, 'practically a symbol of the restoration movement', and the slightly older Jean-Baptiste Antoine Lassus.[59] Just as elsewhere, this restoration saga was dominated by the incessant, cyclical battle between cautious and radical-interventive approaches. But French restoration, with its special intensity of state support, developed a unique trajectory, contrary to the standard narrative: more conservative repair, in the 1840s, was followed by more confident and radical restoration, in the 1860s. This 1860s' position, highlighted in Viollet's 1866 *Dictionary* definition, was one of open modernity, aiming to define a building's artistic 'essence' and recover an ideal, unified authenticity for the benefit of the present.

In the 1840s, Viollet and Lassus had foregrounded restraint and objectivity, insisting in 1843 that restoration should be done with 'absolute discretion, completely avoiding any personal opinion'. But after Lassus's death in 1857, Viollet gave fuller rein to his creative urges, in increasingly radical schemes of intervention, linked with clearance and repair. Inspired by his own restoration schemes, as well as the tradition of architectural classification going back to Durand, Viollet became convinced that he had successfully transcended the academic conventions of Beaux-Arts design and reached a new 'universal' expression – a somewhat 'modernist' position. The scientific passion was, however, coherently linked to practical proselytising. Viollet published in 1849, with Merimée, a guidebook to the care and maintenance of historic buildings – the prototype of one of the Conservation Movement's most important categories of publication, the official maintenance manual.[60]

At this point, we should briefly review the main points of Viollet's career. As one of the Commission's preferred consultants, he was one of the first specialist 'conservation architects' – although he also built some new work – and rose to the top of both the state and religious-diocesan conservation systems. In his most important early restoration projects – in 1840–59 for the church of La Madelaine at Vezelay, a set piece of the transition from Romanesque to Gothic (a distinction newly identified in France), and in 1844–64 for Notre-Dame in Paris, with Lassus – he navigated his way through complex public debates about two revered monuments, in which every detail had to be argued through and justified. Merimée's advocacy of the Vezelay scheme emphasised the nationalist urgency of keeping up with Germany and England: this was a 'great work, that is so much in the interest of our national glory'.[61] The first task was simply to repair the dilapidated structure, but soon issues of restoration and authenticity arose: repair of the west front removed some features dating from an uncompleted 13th-century scheme. At Notre-Dame there was a similar story. An initial careful report by Viollet and Lassus warned against cavalier restoration but demanded that the authentic stylistic essence of each building phase could, and should, be revealed by scholarly intervention. Then followed a rolling, multi-phase programme of work, partly based on what now seems conjectural evidence. This included stripping away 17th-century internal alterations, replacing the nave windows in a 12th-century style, remodelling the west portal designed by Soufflot in 1771, and reinstating 28 statues of kings on the west front, mostly destroyed in the Revolution; a *flèche* was built above the crossing, but proposed spires remained unrealised. In subsequent church restorations, Viollet became increasingly confident in restoring a hypothetical original unity: for example, in his Romanesque-oriented restoration (from 1860) of St Sernin in Toulouse.[62]

The restoration of Notre-Dame formed an integral part of the reconstruction of the Île de la Cité, spiritual centrepiece of the Second Empire's reshaping of Paris, and arguably mid-19th-century Europe's foremost example of integrated modernisation and restoration – although much of the new work was seriously damaged during the 1871 Commune. Alongside clearance of congested slum houses, the rambling Palais de Justice complex was comprehensively rebuilt in 1847–71, including extensive new building as well as surgical restoration of the venerable Conciergerie complex: the Tour de l'Horloge was rebuilt, and the Tour Bonbec at the north-west corner was renewed (in 1870) by Joseph-Louis Duc and Etienne Theodore Dommey. Internally, the medieval vaulted halls of the Conciergerie, including the magnificent, double-naved Salle des Gardes and the even grander Salle des Gens d'Armes, were freed of accretions. In parallel with this restoration, Felix Duban and Jean-Baptiste Lassus reconstructed the Sainte-Chapelle in a painstaking archaeological project of 1840–63, completed by Boeswillwald (with advice from Viollet). The Conciergerie was designated a classified monument in 1862.[63]

As the new, spiky tower profiles of the Conciergerie showed, in secular restorations there could be a more pictorial freedom. This freedom was exploited by Viollet from 1846 in his restoration scheme for the fortified city of Carcassonne. This was another in the succession of 'rediscovered treasures' that dotted the history of the Conservation Movement. Like the Marienburg, this 13th/14th-century complex, with its two rings of fortifications, had previously been a neglected ruin, invisible in the accounts of late-18th-century travellers, for whom the only interest in Carcassonne was the later medieval 'new town', the continuously inhabited *ville basse*. Decommissioned as a military fortress in 1804, and with the gap between outer and inner walls (the *lices*) filled with squalid weavers' hovels, the old city began to be used as a quarry. It was only 'saved' through a concerted campaign by local antiquarian J P Cros Mayrevieille, working since the late 1830s in a rare local/central alliance, with Merimée, to secure classification of the complex in 1839–40, and the commissioning of Viollet as consultant architect in 1846. Viollet's restorations began with the Porte Narbonnaise in 1853, and continued until his death in 1879. They were continued by Boeswillwald, who removed the infill houses from the *lices* in the 1890s and finally, in 1911, completed the transformation of the ravaged ruin into an idealised medieval citadel, with fairytale pointed roofs supposedly in the style of the 13th century. In effect, this was a grand operation of *isolement* or *dégagement*, not just of one building but of an entire old town.[64]

A similar approach was applied in a succession of set-piece projects personally commissioned by Napoleon III. In 1857, Viollet was instructed by the emperor to transform the castle of Pierrefonds, ruined since 1617 and classified since 1848, into a royal summer residence, complete with high roofs, sculpture, painted decoration and furniture. The huge project, although curtailed by the defeat in the 1870/1 Franco-Prussian war, was completed in 1884 after Viollet's death by his son-in-law, Maurice Ouradou. The commission was entrusted to Viollet on the advice of Merimée, initially with the idea of rebuilding the keep and leaving the remainder as a ruin, but later expanded into complete restoration, after Napoleon III decided to make it a showpiece for the 1867 Universal Exposition, with grandiose interiors such as the monumental *Salle des Preuses* (completed in 1866, and housing Napoleon's collection of antique weapons). Reflecting Viollet's 1866 affirmation of the modernity of restoration, the result looked very 'new', lacking any patina of age. Viollet's efforts spawned an extensive school of imitators.[65] For example, at the royal chateau of St Germain en Laye, a 16th-century complex expanded by Jules Hardouin-Mansart as a grace-and-favour mansion in 1680 with large angle pavilions, but reduced to a prison and barracks since the Revolution, in 1862

(a)

(b)

Figure 3.12
Viollet's Carcassonne

(a) The Ville-Haute at
Carcassonne, engraving of
c.1850; (b) Late-19th-century
view of the *lices* at
Carcassonne prior to
clearance of infilled slum
dwellings; (c) Present-day
view of the *lices*, following
the Viollet/Boeswillwald
reconstruction of 1853–1911

(c)

(a)

Figure 3.13 Viollet's Pierrefonds

(a) External view of the castle, rebuilt in 1857–84 by Viollet and Maurice Ouradou; (b) Statue of Viollet-le-Duc at Pierrefonds

(b)

Napoleon III decided to convert the building into a museum of Celtic and Gallo-Roman antiquities, and commissioned a restoration by Viollet pupil Eugène Millet. He recreated its 16th-century state, with drastic demolitions (including Mansart's corner pavilions) and rebuilding of 'missing' elements. The programme continued after the fall of the Second Empire: progress photographs of 1877 show the chapel, with its idiosyncratic square window apertures, as a fragmentary shell, and the main chateau block awaiting its surgical reduction.[66]

We should not view these prodigious efforts of reshaping and reappropriation through the narrow spectacles of later anti-restoration rhetoric (see Chapter 4). Seen in their own light, the clear modern light of Second Empire France, they can be recognised as a creative and activist engagement with historic built fabric, on an unprecedented scale. France may have lost her position as world leader of political revolution, but the modernisation-restoration drive of the mid-19th century was no less revolutionary and iconoclastic in character. Nor was this a matter of mere utilitarianism. High ideals of rescue and rejuvenation were involved, as Vitet explained in 1833:

> History, like a clever sculptor, gives life and youth back to monuments by reviving the memories decorating them: it reveals their lost memory, renders them clear and precious to the towns of which they are witness of the past, and provokes public revenge and indignation against the vandals who would plan their ruin.[67]

As time went on, however, the radical restorations by Viollet and others encountered increasing resistance: for example at Bayeux Cathedral, where a proposal of 1855–9, endorsed by the architect, to demolish the crossing tower – a mixture of late Gothic and Baroque – was circumvented by opponents through an ingenious foundation supervised by railway engineer Eugène Flachat. And in Paris, the 1838–48 restoration of the church of St Germain l'Auxerrois by architect E-H Godde, in a homogeneous 15th-century Flamboyant style, complete with surgical *dégagement* from the

Figure 3.14 'Fortress' restorations of the Viollet era

(a) External view of the Paris Conciergerie (restored by J-L Duc and E-T Dommey); (b) Interior of the Salle des Gens d'Armes, Conciergerie (restored by Duc from 1868); (c) The reconstruction of the chateau of St Germain en Laye (architect, Eugène Millet) seen in progress in 1877

Figure 3.15 The church of St Germain l'Auxerrois, restored by E-H Godde in 1838–48 following successful pleas by Chateaubriand against its proposed demolition for road works

urban fabric, stirred up even more opposition; it was violently denounced by Victor Hugo and Montalembert, who argued that 'Our cathedrals were not built for the desert or the pyramids of Egypt, but to soar over the dense habitations and narrow streets of our city'.[68]

French and German influences in 19th-century European restoration

THE national cohesiveness of the French conservation system ensured it was widely imitated abroad, both in the immediate post-1815 period and during the subsequent history of the Conservation Movement – especially in countries with a strong 'statist' governmental tradition. French-style historic monument commissions were founded in Belgium (1835), Spain (1844) and the Netherlands (1874). In Austria, 1853 saw the establishment of a Central Commission for Research and Preservation of Historic Monuments. Initially proposed during the 1848 uprising, the new Austrian commission was partly intended to bolster 'artistically' the legitimacy of ethnic German hegemony within the empire: its first head, Karl von Czoernig, initially steered it along Viollet restorationist lines. But the local reflections of the French system, far from being homogeneous, were highly diverse. In Russia, where Peter the Great had issued the first antiquities law in 1704 and Alexander I had demanded protection for classical archaeological sites on the Black Sea coast as early as 1805, Nicholas I in 1826–9 began a systematic nationwide register of architectural 'antiquities', with prohibitions on their demolition (exempting churches). This was organised regionally, including a pioneering 1830 register in Novgorod. However, from the mid-century the initiative passed to successive national bodies: the Imperial Archaeological Commission (from 1859), the voluntary Moscow Archaeological Society (founded in 1869), and an expanded Imperial Commission (in 1874). Hungary of the 1860s, in the decade of 'national rebirth' and establishment of the Dual Monarchy, witnessed successful efforts by architect Imre Henszlmann, trained in France in the 1850s, to build up a state-preservation organisation. In 1881, Henszlmann secured passage of Hungary's first, precocious monument-protection law (which placed effective restrictions on private property) and the establishment of a national historic monuments committee – for which prominent architects such as Ferenc Schulcz and Imre Steindl acted as French-style nominated consultants.[69]

Compared with the vigour of nationalist conservation struggle north of the Alps, the old 'universal' centre of Rome remained, during the early/mid-19th century, partly still an 'object' whose role was to stimulate the aspirations of others. However, the Vatican passed pioneering conservation laws in 1802 (Lex Doria Pamphili) and 1820 (Lex Pacca), which entrusted antiquities to the Permanent Council of the Apostolic Chamber; and an 1823 fire at the Basilica of S. Paolo prompted a massive reconstruction scheme, partly by Pasquale Belli, completed in 1854. The English poet Shelley visited Rome in 1819 in a state of despair for European liberty, with reactionary regimes reinstalled everywhere, but rediscovered hope in the ruins of the Baths of Caracalla, with the conviction that Nature would ultimately dissolve the bonds of tyranny, and proceeded to write *Prometheus Unbound*.[70] Only with Italian unification in 1870, however, would a more activist stance be taken towards the heritage.

The situation was very different in independent Greece. Here, the creation of the new state in 1830 and selection of Prince Otto of Bavaria as first king in 1833 was followed by the introduction, under the aegis of Bavarian royal architect Leo von Klenze, of seemingly the most precocious conservation laws in Europe. However, their object of attention was exclusively the classical past.

Unlike northern Europe, medieval buildings were looked on with contempt, as relics of a shameful, benighted period of national history. The year 1834 saw the passing of the first Greek law on antiquities, far earlier than elsewhere and enshrining the strict ethno-nationalist ideal that 'all objects of antiquity in Greece, as the productions of the ancestors of the Hellenic people, are regarded as the common national possession of all Hellenes'; 1837 witnessed the founding of a national Archaeological Society. This ensured active protection of the main classical sites, along with the first steps in a two-century-long programme of forcible reshaping of the Athenian Acropolis from a fortified town to an archaeological monument, stripping away all remains of later phases of life. This programme, proposed in 1834 by Klenze and commenced in 1835–6, promised that 'the traces of a barbaric age, its rubble and formless debris, will disappear here as everywhere in Hellas, and the remnants of a glorious past will arise in new splendour'. The plan included *dégagement* through demolition of Turkish infill houses, and reassembly and re-erection of fallen columns (*anastelosis*) 'in a picturesque manner' under Generalkonservator Dr Ludwig Ross. The 1835–9 reconstruction

Figure 3.16
Early restorations on the Athenian Acropolis

(a) Uncovering the fragments of the temple of Athena Nike in 1835; (b) The temple pictured in a commemorative German publication of 1839, following its initial reconstruction

of the temple of Athena Nike on the Acropolis, destroyed in 1686–7, was the first of several radical restorations of this diminutive structure; its fragments were disinterred and reassembled by Ross (later restorations in 1935–40 and 2000–8 would also require complete dismantlement).[71]

Private preservationism: conservation in early Victorian Britain

WHILE France and Germany, between them, explored the implications of contemporary political upheavals for the emergent Conservation Movement, it was above all in Great Britain that the forces of social and economic modernity were most comprehensively projected into the heritage field. Despite the increasingly friendly cultural relations between Britain and France, especially at the time of the Crimean War and the Great Exhibitions of London (1851) and Paris (1855), a gulf still separated the two countries concerning conservation.[72] Britain in the 19th century pursued a course radically different from France and the rest of Europe, in its reliance on private organisations and forceful individuals rather than state bureaucracy or national mobilisation.

Like a 'Russian doll', this difference overlaid a further internal split, reflecting the cultural polarisation between England and Scotland. Within Scotland, the 19th century's assertive Unionist nationalism focused proudly and straightforwardly on the secular heritage of Baronial castles. In England, a more complex situation prevailed. Its free-enterprise system of conservation was divorced from explicit political nationalism and focused instead on burning social and religious controversies. England also saw an exceptionally fierce and complex debate for and against church restoration, spanning several decades. All this would ultimately allow English writers to achieve critical insights, unmatched elsewhere in the Conservation Movement, into the meaning of heritage for turbulently modernising societies and to develop the polemical concept of architectural heritage as a golden-age bulwark against disruptive change – a passionately utopian idea that would later, in turn, energise nationalist conservation elsewhere.[73] Indeed, despite the impact of Viollet, Schinkel and others, theoretical reflection on conservation in England was significantly 'in advance' of elsewhere until the late 19th century. And the English interpretation of heritage through a socio-cultural polarisation between traditional community and modern alienation anticipated the more pointed contrast drawn by Tönnies in the 1880s between *Gemeinschaft* and *Gesellschaft*.

All these more emotive developments of conservation values, including the violent rejection of restoration, stemmed from the mid-19th century – a story that we will follow in Chapter 4. In this chapter, we trace earlier developments in reaction to the France of Merimée and Viollet. For, paradoxically, Britain was also the place where a positive conception of heritage as an active ally of progress became most fully developed. This special trajectory of the Conservation Movement stemmed from the particular context of Britain as a maritime, imperial power where the spirit of capitalist Improvement had established early and deep roots. Just as the French Revolution had up-ended European politics, so the revolution of Improvement had turned upside down the social structure of Britain in the same years, first by expunging the entire pre-modern agricultural landscape in a mania of rectangularisation, and then by funnelling the displaced population into a fury of urban industrialisation at home and imperial expansion overseas.[74]

By the mid-19th century, Britain, uniquely, had almost a century of experience of such transformations. It was a largely urbanised society, from which, unlike continental Europe, 'vernacular' or 'traditional' culture had been largely purged, although elements of elite feudal culture had survived. The precocity of Improvement and industrialisation had created a tangible divide between past and present that was sharper and more 'real' than the political somersaults in France, yet more ideologically permeable, reflecting the political consensus against 'continental extremism'. The relationship of heritage and modernity in early-19th-century Britain was generally peaceable, with strenuous efforts to exploit 'invented traditions' as an anchor for modern society. As W Bagehot argued in his *English Constitution* of 1867, the British political system's 'essence is strong with the strength of modern simplicity; its exterior is august with the Gothic grandeur of a more imposing age'.[75] Where, in statist France, the fragments of antiquity rescued from the Revolution were consigned to museums, in Britain there was a widespread feeling that they were, and should remain, part of everyday life – something that was both good and bad, given that much of the population still lived in shabby old houses.

Much of Victorian culture was steeped in a new and complex awareness of modernity, as something separately appreciable for the first time, yet actively grounded in the past: John Stuart Mill's 1831 articles on 'the Spirit of the Age' argued that the systematic comparison of the present with former ages was unprecedented. Contemporary new architecture, in Britain as in other countries, reflected this vigorous yet ambiguous modernity through its 'historicist eclecticism' – its use of a range of archaeologically accurate 'historic styles' in novel combinations. Following the 1851 Great Exhibition, there was an upsurge of collecting of exemplars and plaster casts of historic precedents and details, especially in the Architectural Museum, a private society founded in 1851 with support from key Gothic Revival architects; by 1855, it already housed 3,500 casts, and thereafter moved into the South Kensington Museum.[76]

In early-19th-century London, there was still much pride in the progress of Improvement and the opening up of the city – including isolation of key monuments in Wyatt style. J Elmes's *Metropolitan Improvements* of 1827 praised John Nash's programme for increasing London's 'magnificence and its comforts, by forming healthy streets and elegant buildings, instead of pestilential alleys and squalid hovels'. Elmes bemoaned the post-Fire failure to implement Wren's plan, perpetuating the intricate street plan of a city 'whose streets are lanes, and whose lanes are alleys': the Corporation of London should 'emancipate their fine Cathedral from its monstrous thraldom' by radical *dégagement*.[77] The same determination to modernise applied across Britain: in 1799, echoing contemporary denunciations of Nuremberg (see below), Robert Southey simply said that 'Exeter is ancient and stinks!'[78]

By the 1820s and '30s, especially after the 1835 Municipal Reform Act, ambitious city improvement schemes were in full swing everywhere, reflecting the growth of a new civic pride. In Edinburgh, for example, new boulevards and viaducts were being rammed through the medieval Old Town to connect it to the classical New Town, and in Newcastle-upon-Tyne, a multi-level classical redevelopment scheme for the congested medieval core, hatched by Richard Grainger and designed by architect John Dobson, was rampantly underway by the mid-1820s – a programme both welcomed and lamented in an 1812 local account. By the mid-century, the pace and scale of Improvement had only accelerated further, at times out of sight, as with the Metropolitan Board of Works's prodigious emergency programme to construct a sewer network across London in only seven years (1858–65; directed by Sir Joseph Bazalgette). And a Haussmann-style ethos of *isolement* had begun to spread to the colonies, for example in British-ruled India, where, especially following the 1850s' insurgency, cities were 'opened up' and rationalised with boulevards driven through the old town fabric by Improvement Trusts.[79]

Strange bedfellows: romantic medievalism, popular archaeology and tourism

THIS frenzy of Improvement stayed, throughout, in dialogue with the ideologists of the Romantic Movement, and their concept of the golden-age integrity of medieval Gothic culture. Building on the nostalgic primitivism of Macpherson's *Ossian*, the early-19th-century novels of Sir Walter Scott lamented the loss of pre-modern society, whether in medieval England or Jacobite Scotland. Yet Scott's own work exemplified the paradoxical and often positive engagement between past and present in early-19th-century Britain. Although he himself was trenchantly opposed to extreme laissez-faire economics and utilitarianism, his books were a runaway global success (beginning with his 1810 poem, *The Lady of the Lake*, which sold 25,000 copies across the world within eight months) and energised the growing mass tourist industry. His own Borders villa, Abbotsford, enlarged in 1812–23, itself became a significant tourist attraction: in 1844, Fox Talbot recorded it for a book, *Sun Pictures in Scotland* (only the second photographic book ever published), devoted largely to places featuring in Scott's novels. By the mid-1850s, Abbotsford welcomed over 4,000 visitors a year, and by the mid-'70s over 2,000 American visitors were recorded annually.[80]

The forms that this 'constructive medievalism' took in England and Scotland were markedly divergent. In England, ideas of romantic heritage were loosely bound up with the movement of 'Tory medievalism'. This held that England could be saved from modern vices by revival of courtly, medieval-style Christianity, grounded in traditional church stability: in the antiquarian Kenelm Henry Digby's *The Broad Stone of Honour* (1822), the medieval knight seemed to presage the 19th-century gentleman. In Cobbett's *Rural Rides* (1830) and Southey's *Sir Thomas More* (1829), the polarisation between present alienation and past wholeness became yet more explicit.[81]

The English Tory idealisation of the Middle Ages continued into the 1830s and '40s, with the 1839 Eglinton Tournament, and Disraeli's novels *Coningsby* (1844) and *Sybil* (1845) stressing the restorative effect of the Middle Ages on the present, and Charlotte M Yonge's 1853 *The Heir of Redcliffe* reviving the language of medieval chivalry.[82] But political liberals were equally addicted to medievalism and, in the 'Gothic Revival' in mid-19th-century England, it went hand in hand with love of progress. The design work of A W N Pugin combined intense golden-age medievalism with a fevered addiction to railways and all forms of modernity. We will return later to Pugin's passionately utopian *Contrasts* of past and present and his profound effect on the later development of the Conservation Movement.[83]

Although English antiquarianism in the 17th and 18th centuries had often been associated with closet Catholicism, the mid-19th-century boom in archaeology, following the 1843 foundation of the British Archaeological Association and the establishment of prehistory as an object of scholarly enquiry following Darwin's *Origin of Species* (1859), was also bound up with modernisation, especially as voluntary amenity societies evolved into mass movements linked to tourism. The year 1824 saw the foundation of the first local conservation society (the Association for the Protection of Ancient Footpaths), dedicated to preserving York's city walls, and 1846 the establishment of the first local amenity society, the Sidmouth Improvement Committee, dedicated to safeguarding the town's picturesque walks and promoting 'the general improvement of the place'.[84] The 1840s, as in France, saw early instances of the modern phenomenon of the 'rescue survey' prompted by new development. In 1845, for example, the unearthing of the remains of the Priory of St Pancras at Lewes, Sussex, by railway excavations – a discovery welcomed by the railway directors as a spur to tourism – provoked the foundation of the Sussex Archaeological Society (one of the first in England). The

ARMOURY AT ABBOTSFORD.

(a)

Figure 3.17 Abbotsford

(a) Contemporary view of the entrance hall; (b) Present-day view of the exterior

(b)

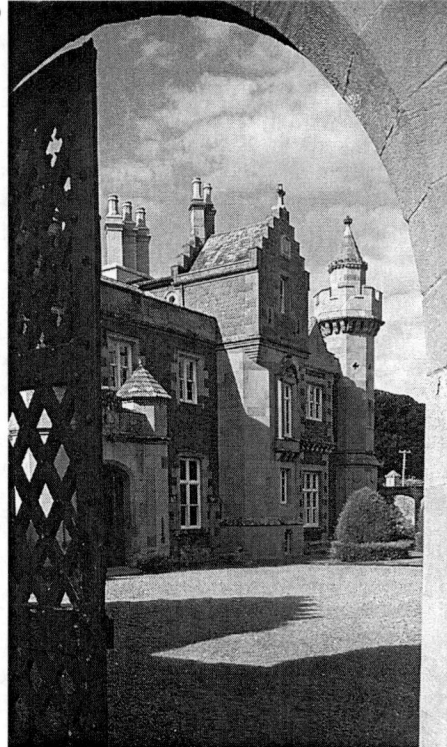

Illustrated London News highlighted the paradoxical interrelationship here between railway and archaeology: it was 'strange indeed [that] these relics have been uprooted in a work peculiar to our own time'.[85] The coming of the railway also brought mass tourism to heritage sites already identified by elite cognoscenti: Fountains Abbey became one of England's major tourist attractions following the 1848 opening of a railway to Ripon.

In the long run, the 'French' administrative-led system would provide the template for the 20th-century state-supported systematisation of the Conservation Movement. Initially, however, the voluntary efforts in Britain were more effective 'on the ground' than the early state bureaucracies, especially in producing significant quantities of scholarly or popular publications on the heritage.

Figure 3.18 The remains of the Priory of St Pancras in Lewes (dismantled 1537) were unearthed in 1845 by the construction works for the Brighton, Lewes and Hastings Railway. This present-day view from the north-east shows the close juxtaposition of the two: the railway cutting runs directly across the site of the main priory church building

Sir George Gilbert Scott and the Victorian restoration movement

MID-19th-century England and France seemed especially close in their mutual zeal for church restoration. However, where France and Germany focused on high-prestige, state-backed restorations of great cathedrals and abbeys, here too the English approach was one of decentralised voluntarism – contrasting starkly with Merimée's careful strategy in France in the unstructured, non-hierarchical way that it dispersed its passionate efforts and controversies into the humblest parish churches across the country. Prominent here were specialist pressure groups such as the Cambridge Camden Society (CCS, or 'Ecclesiologists'), founded in 1839 by Cambridge undergraduates John M Neale and Benjamin Webb, and closely aligned with mid-19th-century 'high church' Anglicanism. Influential, too, were the architectural journals (including the CCS's own *Ecclesiologist*, founded in 1841) – a specialist arm of Britain's outspoken free press. Vast sums of money were involved in this haphazard yet cumulative activity: for instance, no less than £530,000 was expended in the diocese of Peterborough on parish church restorations between 1844 and 1874. The state-subsidised Incorporated Church-Building Society (founded in 1818) modernised large numbers of churches in a more practical manner. But in the long-term development of the international Conservation Movement, mid-19th-century England's foremost contribution was in converting heritage into an effective vehicle for mass private or recreational activities – an approach that would later become especially characteristic of 'historic preservation' in North America.[86]

Through the restoration mania, the Conservation Movement also fed into the broader currents of architectural debate in early Victorian England. Under the influence of Pugin and subsequent Gothic Revivalists, the regular, symmetrical architecture of Regency England was condemned as sterile and monotonous and a new, irregular medievalism was demanded, linked to the growing architectural emphasis on individualism and personality, as well as the moral concepts of honesty and simplicity. Here, restoration projects formed a more sober counterpoint to the ferociously original High-Victorian new architecture designed by the same architects, such as Butterfield or Street. Restoration fashions also closely reflected shifts in the period of medieval architecture most favoured within the Gothic Revival: the English late-Gothic 'Perpendicular' style, previously liked for its regularity and its large, square windows, was supplanted by the earlier 'Decorated' Gothic, which now also became the favoured style for restoration schemes. At a meeting of the Ecclesiologists in 1847, co-founder J M Neale rhetorically claimed he would not mind the demolition of the mainly Romanesque/Perpendicular Peterborough Cathedral, provided it were replaced by a Decorated building that was 'as good of its sort'.[87] Through the efforts of the Ecclesiologists and others, restoration was closely integrated with contemporary religious ideals; it served as a metaphor for the anticipated restoration of a religious faith undermined by modernity and encouraged a more sacramental, ritual atmosphere.[88]

One of the stock-in-trades of this English system of architectural debate was the rapid evolution of fashions, with each movement violently denouncing its predecessor, and conservation was no exception. The early 19th century saw an upsurge in condemnations of Wyatt-style restorations as utilitarian and unscholarly. Paradoxically, this was coupled with growing aggressiveness towards the fabric of old buildings, especially any classical elements. Wyatt himself had in 1809 commenced a remarkably conservative restoration of Henry VII's Chapel at Westminster Abbey (completed in 1813–27 by B D Wyatt, with mason Thomas Gayfere Jr.), cautiously refacing it and salvaging around a third of the decayed medieval surface as well as stabilising the eastern flying buttresses (see Figure 2.12); but in the 1827 restoration of Temple Church by Sydney Smirke and Decimus Burton, 'ecclesiology engulfed antiquarianism', with 17th- and 18th-century furnishings and classical tombs ruthlessly scrapped. This scheme, in turn, fell victim to the rapid march of fashion, and was attacked by *The Ecclesiologist* in 1842 as insufficiently advanced. What that meant was demonstrated the previous year, when architect Anthony Salvin (assisted by J L Pearson) began a showpiece Ecclesiological restoration of the circular Holy Sepulchre Church, Cambridge, shaving off later alterations and crenellations and leaving a 'pure, Norman' structure with a neat, pointed roof, intended internally to evoke the Jerusalem original of the Holy Sepulchre: the project's partisan associations with religious Tractarianism made it the centre of national controversy, even in the mainstream press.[89]

Within the fevered world of mid-19th-century English architecture, crazes followed one another in bewildering succession. The growing demand for brutal 'honesty' in structural expression also prompted a sudden mania around 1840 for the stripping, or 'scraping', of internal plaster and external stucco to reveal the 'real' stonework. Over several decades after 1844, for example, 103 out of 390 churches in Leicestershire had their internal plaster removed. The plaster-removal fashion, in turn, was severely attacked in 1873 as 'a nakedness compared with which puritan whitewash is decency' by a later Gothicist, G G Scott Junior.[90]

Precisely because of their zealous commitment to their partisan 'causes', English restorers lacked the dispassionate self-awareness of their French contemporaries. Whereas Viollet acknowledged the hypothetical character of much of his work, *The Ecclesiologist* in 1842 argued simplistically that 'to restore is to revive the original appearance lost by decay, accident or ill-judged alteration'. Although respect for later alterations was not ruled out, what was paramount was 'to recover the original scheme of the edifice as conceived by the first builder', avoiding both destructive intervention and excessive conservatism.[91] The historian and Oxford Movement polemicist Edward Augustus Freeman hailed as a paradigm the restoration of St Mary Stafford, of 1841, by George Gilbert Scott, in which Georgian features were removed and 'degenerate' Perpendicular windows replaced by smaller windows in an 'earlier' Gothic style. With Scott, the modern entanglement between restoration and eclectic new architecture reached its most extreme point, as he himself recorded in an 1848 lecture, published later as 'A Plea for the Faithful Restoration of our Ancient Churches'. Here he differentiated between the absolute respect for all phases required in classical archaeological sites and the flexibility permissible in Christian churches 'erected for the glory of God and the use of Man'. Old-style Improvement-era restorations, in which 'the restorer too often preserves only just what he fancies, and alters even that if it does not quite suit his taste', was absolutely to be condemned. Yet one might venture 'to replace features which have been actually destroyed by modern mutilation, where they can be indisputably traced', with the aim of 'raising the architectural character of the building'. This balancing-act would require archaeological detective work of some complexity.[92]

A crucial factor was the extreme dereliction and decay of many churches at this time: Scott brought to bear great practical competence in structural repair. He also actively enhanced many medieval churches with sumptuous fittings of characteristic 19th-century pattern, such as the Skidmore iron screens at Hereford, or Salisbury (where Scott reversed much of Wyatt's work, removing his screen and reredos in 1859). At Chichester Cathedral, he rebuilt the central tower in facsimile (1865–6) following its collapse in 1861 during botched restoration works. Similarly, at St Albans, where a public right of way cut across the cathedral, Scott (working with clerk of works John Chapple) saved the central tower from collapse in 1870, his workmen labouring in 24-hour shifts to shore up the crumbling masonry, and straightened the five western bays of the nave through prodigious efforts of scaffolding and hydraulic props – a task completed after his death, in 1878. At Westminster Abbey, where Scott succeeded Blore as Surveyor in 1849, he reconstructed the chapter house from 1866 from its use as a public record office, complete with galleries inserted in 1703 by 'some barbarian': 'Seldom do we see a noble work of art reduced to such a wreck!' More controversially, at Westminster, he began the recasting of the north transept front into its 'ideal' state (beginning in 1875; work was completed in 1892 under J L Pearson), although admitting that it was 'nearly impossible to form anything like a correct idea' of the original arrangement. And at Chester Cathedral, he drastically rebuilt the exterior, adding a welter of eclectic elements, replacing some Perpendicular windows with Early Pointed-style lancets, and others with a 'late Norman' wheel window. In his work at Ely, by contrast (from 1848 to 1864), Scott pursued a cautiously archaeological approach in restoring the octagon, where he removed Essex's 1757 work, and installed a magnificent set of neo-Norman ceilings in the nave and west tower, by Le Strange and Gambier Parry.[93] But the furious rapidity of the march of fashion in English architecture had, by 1850, left even Scott behind, as a new, even more intransigent position was proclaimed by John Ruskin – as we will see in the next chapter.

Figure 3.19 Scott's cathedral interventions

(a) Rebuilding prospectus for the tower of Chichester Cathedral following its 1861 collapse; (b) The facsimile tower and spire as rebuilt by Scott in 1865–6; (c) The Westminster Abbey Chapter House, seen in its condition as a galleried public record office, prior to Scott's 1866–73 restoration; (d) Chester Cathedral prior to its 1868–76 restoration by Scott; (e) Present-day view of the east end of Chester Cathedral

Secular heritage in mid-19th-century England

IN England, churches remained the focus of conservation attention during the 19th century. However, the shift of new architectural fashion towards 'later' styles, including Tudor, was reflected in growing respect for secular medieval architecture, inspiring both new buildings and preservation – the difference between the two sometimes being fudged. Architect George Devey applied his extensive research on 'minor' domestic architecture both to repair and extension schemes (e.g. the Leicester Square infill cottages, Penshurst, 1848–51) as well as radical rebuilding projects, such as Betteshanger House (from 1856), where his design imitated a centuries-long accretion of elements. Wordsworth's writings had already focused attention on rural 'vernacular' cottages, naturalising a theme pioneered in *ancien-régime* France in Marie-Antoinette's '*hameau*' in the 1780s. In 1810, Wordsworth's *Guide to the Lakes* argued that the Lake District's indigenous architecture should inspire new houses in the area, a line frequently echoed over the following decades: Ruskin argued in the late 1830s that new rural architecture should respect local environments and materials.[94]

As yet, however, England witnessed no concerted admiration for old towns or 'everyday' urban medieval architecture as at Nuremberg. Pre-1666 London now only existed in two-dimensional street-layout form, and the prestige of Oxford and Cambridge eclipsed everything else: to many Victorians, Oxford was *the* ideal city, with its rich mixture of public formality and collegiate Picturesque.[95] By the mid-19th century, many in England were rejecting the neo-classical Sublime as tainted by industrial utilitarian patterns such as workers' terraces and factories, and were instead favouring individualised suburbs of villas – although the neo-classical Sublime still reigned in Scottish cities such as Glasgow and Edinburgh. Only later in the century would England see anything like a cult of the collective Old Town, with the publication of the *Historic Towns* series from 1889 (edited by Edward Freeman) and the 1875 foundation of a 'Society for Photography of Relics of Old London'.[96]

New Town versus Old Town in Romantic Edinburgh

THE early-19th-century Conservation Movement followed a subtly different path in Scotland. Although the addiction to laissez-faire privatism was as great as in England, the general outlook was shaped by a 'small-nation' concept of heritage as a bulwark against external domination. As we saw above, Walter Scott's writings powerfully bolstered this sense of modernising, yet rooted, identity.[97] Early Scottish conservation initiatives by private individuals focused on ecclesiastical monuments, including the 1809 purchase and restoration of Melrose Abbey by the Duke of Buccleuch. Paralleling these, though, was a pioneering government preservation policy, unmatched elsewhere in Britain, which exploited the fact that since James VI's 'annexation' in 1587, the greater abbeys and cathedrals had been under state control. At Dunkeld Cathedral in 1762, the government granted £300 towards repair work on the tower, and in 1789 an Exchequer grant helped consolidate the early Christian St Rule's Church at St Andrews. From 1809, architect Robert Reid was appointed 'Architect and Surveyor to His Majesty in Scotland'; in 1827, Reid secured establishment of a separate Office of Works, charged with an ongoing programme of repair (rather than active 'restoration or embellishment') of ruined or intact cathedrals and abbeys, including Elgin (from 1823), Fortrose (from 1828), Dunfermline (1829) and Glasgow, St Andrews and Arbroath (1834). This was not state intervention on the scale of Merimée's operation in France. Here the government was acting rather like an enlightened landowner: Reid's powers were strictly limited.[98]

But, by then, the emphasis in Scottish heritage debates had shifted to secular architecture. As with the interrelationship between restoration and new architectural eclecticism in the English Gothic Revival, Scott's highly publicised work at Abbotsford spawned a new architectural style linked with national 'heritage' – only, here, with castellated domestic architecture rather than churches. This 'Scotch Baronial' style, developed by architects William Burn and David Bryce on the foundations of scholarly antiquarianism, above all the lavish 1847–52 *Baronial and Ecclesiastical Antiquities of Scotland* by R W Billings, in turn stimulated a re-Baronialising of previously Improved castles. At Blair Atholl, from 1869 Bryce added a forest of turrets and castellations, 'restoring' Atholl House to Blair Castle, but in a form very different to its original state before its 1740s' classical 'clipping' by Winter.[99]

The conflicts between modernity and heritage, between the pressures of Improvement and the concept of the Scottish historical landscape, came into especially violent collision in Edinburgh, a city which was an outstanding exception to the lack of a concept of the 'historic town' in early-19th-century Britain. During the late 18th century, as in Nuremberg, the medieval city had been an object of escape and disgust, only here confronted across a valley by the regular, modern classical 'New Town'. This pattern was decisively broken by the visit of George IV in 1822, choreographed by Walter Scott: in 'modern' Highland dress, the king banqueted in Edinburgh Castle and processed down the High Street. The surgical inroads of Improvement into what was now called the 'Old Town' were still gathering pace, and John Britton's *Modern Athens* of 1829 argued that recording would sufficiently compensate for demolition of a medieval tenement in the Lawnmarket: such relics 'have a picturesque appearance, when delineated on paper, but few persons will regret their removal, to make way for modern improvements'.[100] But by then, a swing in opinion against the New Town was in full swing, with aesthetic and ethical criticisms in parallel. In 1818, author Susan Ferrier branded it a 'desert', and the 1826 *Provincial Antiquities and Picturesque Survey of Scotland* lambasted the 'tameness' of the 'commonplace . . . monotonous' New Town plan.[101]

During the second quarter of the 19th century, the growing cult of the Edinburgh Old Town became the basis of Europe's first fully fledged urban amenity and conservation pressure group – a forerunner of a key element of the modern Conservation Movement. It was led by one of the country's foremost judges, Lord Henry Cockburn, a friend of Walter Scott who combined support for the ethos of Improvement with demands to temper its excesses, arguing in 1847 that 'a wise man would like to have seen the past age, but live in this one' and pioneering a more strident type of civic protest against 'philistine' commercial Improvement, or utilitarian railway building. Cockburn, along with the Society of Antiquaries of Scotland, orchestrated the first fully fledged 'civic demolition protest', against the demolition of the 16th-century Trinity College Church from the Old Town for railway works: the proposal was 'an outrage by sordid traders [who] would remove Pompeii for a railway, and tell us they had applied it to a better purpose in Dundee'.[102] Other protestors highlighted the potential loss of tourist visitors to 'our Metropolis, to which a thousand interesting associations are connected', and claimed that 'our English neighbours are instituting committees to protect the antiquities of their ancient churches'. The protests proved futile, and the church was dismantled into numbered pieces in 1848: in 1870, it was partly reconstructed nearby. In reaction to the march of Improvement in the city, in 1841 an 'Edinburgh Association for Illustrating Local Antiquities' was formed, to compile 'accurate sketches and authentic historical records'.[103]

Figure 3.20
Rediscovering
'Old Edinburgh'

(a) Trinity College
Church seen in
1753, a century
prior to its
dismantlement
and removal in
1848 and
subsequent part-
rebuilding in 1870
in Chalmers Close;
(b) Perspective of
Peddie and
Kinnear's Scotch
Baronial scheme of
1853 for the new
Cockburn Street

(a)

(b)

Increasingly, Improvement schemes within Edinburgh echoed the new sense of Old Town heritage through a modern but context-sensitive Scotch Baronial style, beginning with the cutting of an access route to the railway station from 1853 (Cockburn Street) and proceeding in 1867 to a City Improvement Act and further road incisions (overseen by official architects David Cousins and John Lessels), whose 'elevations shall be of plain but marked character; in harmony with those fine specimens of national architecture in many of the neglected and overcrowded areas'.[104]

The beginnings of historic preservation in the USA

THE privately led conservation framework pioneered in Britain – although destined to be superseded there by strong state interventionism in the mid-20th century – spread with more enduring effect to other anglophone countries, especially the newly independent USA, which made its distinctive contribution to the story of the Conservation Movement from this point. Here, by the 1810s, the first notions of heritage and historic preservation were being developed, reflecting the expansive, nation-building character of American national identity. Chief objects of concern were revolutionary sites – buildings, by definition, far more recent than the medieval structures that preoccupied contemporary European conservationists. But the overall process was the same, beginning with the 'rediscovery' of a previously forgotten monument: the Old State House in Philadelphia, a classical 18th-century building that had been the site of the Declaration of Independence a quarter-century earlier. Truncated in 1781 by the loss of its tower, it lay neglected and forgotten since the state government had moved elsewhere in 1800. In 1812, a proposal was exhibited for 'restoration of its ancient steeple' but, the following year, the State of Pennsylvania tried to sell the building for redevelopment, and in 1815–16 its Greek Revival interior woodwork was modernised, a project widely condemned as a 'sacrilegious outrage'. Eventually, in 1828, after the building had been sold to Philadelphia City Council, the city invited William Strickland and John Haviland to submit proposals to rebuild the steeple: Strickland's design was executed, while Haviland supervised the restoration of the interior in 1831–2. The name, 'Independence Hall', was first applied to the building in the 1830s, and the 1752 Old State House Bell was rebranded the 'Liberty Bell', to symbolise the anti-slavery movement.[105]

The restoration of 'Independence Hall' was a joint government/private effort. Likewise, the first preservation acquisition of a building associated with George Washington, the Hasbrouck House in Newburgh, New York, in 1850, resulted not from private effort but from an initiative, and expenditure of nearly $2,000, by the State of New York: it was argued that,

> if our love of country is excited when we read the biographies of our revolutionary heroes, or the history of revolutionary events, how much more still the fires of patriotism burn in our bosoms when we tread the ground where was shed the blood of our fathers – when we move among the stones where were conceived and consummated their noble achievements.

However, in most other pioneering schemes, private initiative reigned – a pattern that would dominate the US preservation scene. Thomas Jefferson's house, Monticello, was sold as a farm in 1831, following Jefferson's death in debt in 1826, but subsequently was bought by an admirer of Jefferson's, Lt. Uriah P Levy, USN, and preserved privately until the 1920s. Likewise, Mount Vernon, Washington's house, was preserved after the deaths of Washington and his wife, respectively in 1799 and 1802, initially by relatives, then by the initiative of the Mount Vernon Ladies Association of the Union – a grouping founded by Ann Pamela Cunningham from 1853, after proposals for federal government acquisition were rejected three times by Congress (in 1846, 1848 and 1850). Mount Vernon was repeatedly remodelled to restore its 1799 appearance, beginning with replacement of the entire front portico in 1860.[106] From the 1840s and '50s, more humble 'heroes' houses', including log cabins, also claimed heritage attention. But only after the 1860s' Civil War would a more broadly based concept of national identity and heritage emerge, along with a new concern for protection of untamed natural wildernesses.

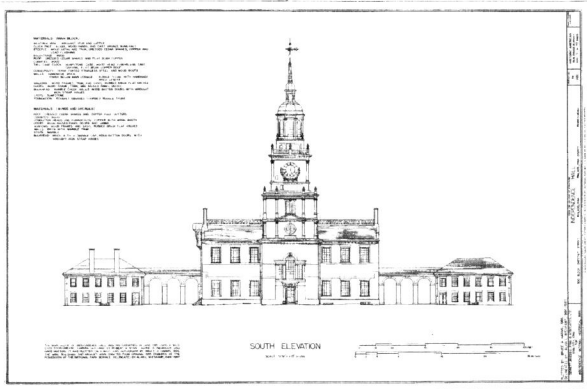

(a)

(b)

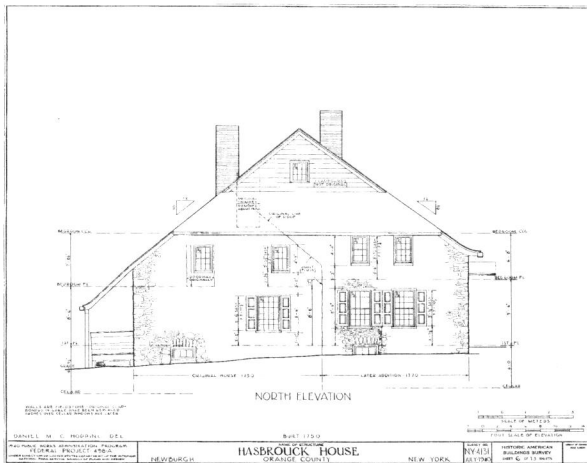

(c)

Figure 3.21 Hero sites of the American Revolution

(a) Old State House (Independence Hall), Philadelphia, PA (restored 1828–32), elevation drawing by HABS; (b) Old State House, HABS staff seen at work after World War II; (c) The Hasbrouck House, Newburgh, NY (acquired for preservation in 1850), HABS drawing; (d) The Hasbrouck House, HABS photograph; (e) Mount Vernon, 2009 photograph, showing the front portico renewed in 1860

(d)

(e)

111

The cult of 'great men's homes' in Europe

IN Europe, as in the USA, a movement to preserve architecturally 'humble' buildings associated with 'great names' flourished from the early 19th century. This was not a new phenomenon: what was novel was its systematisation as a mass movement, linked to the wider proselytising agendas of the Conservation Movement, agendas often initially dominated by political/ethical ideals, but later infiltrated by touristic/commercial interests.

This change is traceable, for example, in the long-standing cult of Martin Luther, whose 15th-century birthplace house in Eisleben was rebuilt after a 1689 fire following an appeal throughout Protestant Germany; from 1815, this semi-private shrine was elevated to a Prussian national monument, its upkeep subsidised by King Friedrich Wilhelm III.[107] Likewise, the Wartburg castle at Eisenach, scene of Luther's 1521 translation of the New Testament into German (and thus the 'birthplace' of the modern German language) was already an object of 'pilgrimage' during the 16th century and was visited in subsequent centuries by numerous eminent personalities, including Goethe (several times between 1771 and 1801). In the early 19th century, this stream suddenly became a flood, as the Wartburg became the focus of mass nationalist-cum-Protestant fervour, including a huge student celebration (the Wartburgfest) in 1817 at the tricentenary of the Reformation. In 1849–60, the local ruler, Grand-Duke Carl Alexander von Sachsen-Weimar und Eisenach, and the castle's 'Kommandant', Bernhard von Arnswald, commissioned architect Hugo von Ritgen to reconstruct it as a museum-cum-summer residence, enhancing it with a dramatic new tower and transverse courtyard block (Torhalle) and recasting the 12th-century 'Palas' in elaborate neo-Romanesque style.[108]

Figure 3.22 Luther and the Wartburg

(a) 1845 painting of the Wartburg by F Preller, prior to its 1849–60 restoration; (b) Present-day view of the castle; (c) Luther's room in the Wartburg (the Lutherstube) as depicted on the front cover of an East German commemorative guidebook to Luther heritage sites (Institut für Denkmalpflege der DDR, *Martin Luther, Stätten seines Lebens und Wirkens*, East Berlin, 1983); (d) Present-day view of the Lutherstube; (e) Detail of the banqueting-hall ceiling in the restored Palas, depicting architect Hugo von Ritgen; (f) 1890s' watercolour by Emil Büchner of the banqueting hall; (g) Interior of the Elisabeth Boudoir, restored by Ritgen and recast in the 1900s with mosaic decoration

Martin Luther
STÄTTEN SEINES LEBENS UND WIRKENS

Urban 'great men's homes' were especially suitable as a focus for the early-19th-century boom in mass commemoration. In Nuremberg, for example, the house owned by Albrecht Dürer at the end of his life had attracted civic interest over subsequent centuries, but only became the focus of concerted private and municipal action in 1826. Then, it was acquired by the Town Council and rented to a new 'Albrecht Dürer-Verein', which commissioned architect Carl Alexander Heideloff to carry out a neo-historic conversion into a Dürer museum. The failure of the 1848/9 Revolution stimulated a fresh burst of cultural nationalism and conservation of 'great men's homes', including the establishment in 1859 of a Freies Deutsches Hochstift (Free German Foundation) to restore the birthplace house of Goethe in the Grosser Hirschgraben in Frankfurt: it was bought by the Foundation in 1863 and converted into a museum and headquarters.[109]

The growing mid-19th-century trend to integrate house museums with commercial tourism was most highly developed in Great Britain, with its pioneering advances in capitalism and mass railway travel. Here Carlyle's writings, especially his 1841 essay, 'Heroes and Hero-Worship', stressed the edifying effect of great men's relics and helped popularise house-museums: Abbotsford, as an antiquarian tourist site created by the author himself, was a hybrid and especially 'modern' case (see above).

Of the tourist-orientated 'birthplace museums' that began to spring up across Britain, the most prominent was the timber-framed house in Henley Street, Stratford-upon-Avon, where the poet Shakespeare was born in 1564. Owned by family descendants until 1806, and partly occupied by a public house, the Swan and Maidenhead, the building became a focus for pilgrims in the mid-18th century, as the poet's national reputation soared: the first Shakespeare Festival, organised by actor David Garrick, was held in Stratford in 1769, with the house as a focus. During the early 19th century, it was protected only informally by tenants and hostel-keepers, and visitors frequently hacked off keepsake lumps from the furniture. Scratched names on the window record visits from literary giants, including Keats, Scott, Carlyle, Dickens and Tennyson; in the first visitors' book, instituted in 1812, the initial signatures are those of American visitors.[110]

All that changed in September 1847, when the house was put up for auction, and a national Shakespeare Birthplace Trust was formed overnight. It raised £3,000 by public subscription (from donors including Dickens, Tennyson and Prince Albert) to purchase the house, and in 1857 began a five-year museological restoration programme. In Haussmannesque style, the terraces of houses on either side were demolished as being a fire risk, the public house was expelled, gables were restored, bric-a-brac relics were swept out, and the building became a 'Shakespeare Library and Museum', fit to receive the flood of tourists who arrived with the railway (which reached Stratford in 1859). The annual number of visitors rose from 700 in the early 19th century to 2,500 in 1850 and 30,000 in 1900.[111]

The great men's house movement exemplified the way in which, during the early and mid-19th century, the growing Conservation Movement played an integrated part in the wider processes of social, economic and cultural modernisation of the age. Private enthusiasts and government administrators worked together to ensure that the historic heritage acted as a constructive counterbalance to the furious excesses of political and industrial change. So far, so good: but the new process of discovery and defence of threatened monuments, and the increasing organisation of that heritage into autonomous, potentially rival national blocs, had unleashed passions that would begin

Figure 3.23 The cult of Shakespeare

Henley Street, Stratford-upon-Avon, 2010 view showing the Shakespeare Birthplace House (restored from 1857) and in the background the 1962–4 Shakespeare Centre (architect Laurence Williams)

to develop their own momentum and gather force during the later 19th century, to the point where the Conservation Movement was ready to play a full role in the violent national and ideological confrontations of the next century. In the next two chapters, we trace that fateful process of intensification, beginning with the development of a new and angry rhetoric of conservation protest in mid-19th-century England.

– 4 –

The life-force of age

Restoration and Anti-Scrape, 1850–90

How cold is all this history, how lifeless all imagery, compared to that which the living nation writes, and the uncorrupted marble bears! – how many pages of doubtful record might we not often spare, for a few stones left upon another!

John Ruskin, 1849

The real principle of our action is this: let dead things go, let living things be kept.

Leonard Courtney (SPAB), 1878[1]

Pugin and Ruskin: preservation as polemic

So far, we have traced how the concept of the historic monument, as something decisively 'other' than modern society, emerged following the great political and social ruptures of the late 18th and early 19th centuries – and how the growing Enlightenment ideals of History and Progress correspondingly stimulated the formation of a dynamic 'Conservation Movement', spanning Europe, yet fragmented between competing nationalisms. Historic monuments, long-established or newly 'discovered', seemed ideal supports for growing nation-states, giving their competitive self-assertions a solid, apparently enduring foundation. Yet, before the mid-19th century, this apparatus of monument nationalism remained incompletely mobilised to the cause of the Conservation Movement. True, the enthusiastic researches of amateurs were now augmented by the agitations of cultural nationalists against 'vandalism' or neglect and, in some countries, by the first government preservation efforts. What still lay in the future, though, was the passionate, even fanatical, zeal that would mobilise monuments in the life-or-death struggle of the nations. Ironically, a key pointer towards that future emerged in the 'private' heritage culture of England, semi-detached from the growing continental national rivalries, and lacking any explicit nationalist discourse until around 1900. There, a new, ethically charged formula of conservation emerged during the 19th century: the so-called 'Anti-Scrape' movement.

Anti-Scrape was concerned less with the traditional conservationist fight against demolition, than with the supposedly more insidious evil of restoration. Venerating the time-worn fabric of old buildings with almost religious passion and, eventually, under the influence of socialist utopianism, extending that passion from individual buildings to entire old towns, the Anti-Scrape faction branded any restoration (or 'scrape') of that 'living' fabric as intrinsically immoral, and the restorers as wicked men. Anti-Scrape was important in its own right, as the beginning of a new, international phase of debate over the pros and cons of restoration. But, stemming from a more general English predisposition to debate architecture in exaggerated terms of good and evil, it was even more important internationally in the broad way that it legitimised extreme, polemical positions within the Conservation Movement: the 'opposition', whether individual architect or entire nation, from now on must be ruthlessly pilloried. Henceforth, conservation was always, at least potentially, fighting a kind of 'war', whether against external enemies ('the demolishers', 'the planners'), or a civil war against rival factions or competing national heritages. Like the Church Militant, this was the Heritage Militant – an ethos that culminated in the next century, with the involvement of conservation in the brutal reality of 'total war', as a bastion of 'national community'.[2] And even once the fury of world war was finally exhausted, conservation's battling spirit burst out again in the late 1960s and '70s, in a final furious campaign against modernity in all its guises – a campaign that left it, for a few years of delirious triumph, as the only surviving combatant on the field.

For the moment, all that was far in the future. Initially, as we saw above, the attitudes of English Gothic Revival architects towards new architecture and old buildings differed little from the calm modernity of the Viollet circle in France, despite their mainly private, rather than state, patronage. But a great shift was underway, and first emerged in the fevered writings of Pugin. He, alongside others such as Walter Scott, had reflected the general romantic critique of utilitarian progress in his celebrated *Contrasts*, which compared the medieval golden age with the harsh utilitarianism of the present. But *Contrasts* had a more general impact on architectural debate in Britain, establishing the convention that architectural change should be advocated through fierce oppositions between absolute good and absolute bad. And it popularised the still influential concept that the architecture of an evil social system must itself be intrinsically evil. For Pugin, that present-day evil included not just factories or workhouses but also the cathedral-rebuilding schemes of Wyatt and his contemporaries, who had crassly mixed 'Improvement' and 'restoration', and violated the authenticity that allowed old buildings properly to represent the medieval golden age. Outbidding Carter's polemics, Pugin condemned Wyatt in the most vituperative terms: he later recalled an 1833 visit to Hereford Cathedral where, 'horror! dismay! the villain Wyatt had been here, the west front was his. Need I say more? All that is vile, cunning and rascally is included in the term Wyatt.'[3]

Although Pugin's own work, ironically, legitimised the next phase of Gothic restorations, that inconsistency was almost incidental: 'authenticity' now rested not in any specific qualities, but in maintaining a sharp separation between old and new. In an epoch that relied on historical styles to communicate architectural meaning, it became vital whether a building was 'original' or a 'fake'.[4] The immediate impact of Pugin's writings within English architectural and antiquarian culture was impeded by his personal eccentricity and his passion for Catholicism, but it was not long before the architectural rhetoric of good versus evil, and the rabid hatred of restoration, was taken up by a critic anchored securely within the mainstream of modern Protestant culture: John Ruskin.

In contrast to Pugin's religious conservatism, Ruskin's work formed part of an alternative strand of medievalism, initially concentrated in Britain, which condemned modern utilitarianism not just for its modernity but for what it saw as its oppression of ordinary people. Along with Thomas Carlyle, Ruskin pioneered a new, utopian reaction against industrialisation or modernity in general. For these writers, the Middle Ages were a golden age not of chivalry and hierarchy but of egalitarian community. Carlyle's writings were of seminal importance in this tradition, combining Calvinist emphasis on the role of the 'hero', and medieval nostalgia, with the conviction that, ultimately,

117

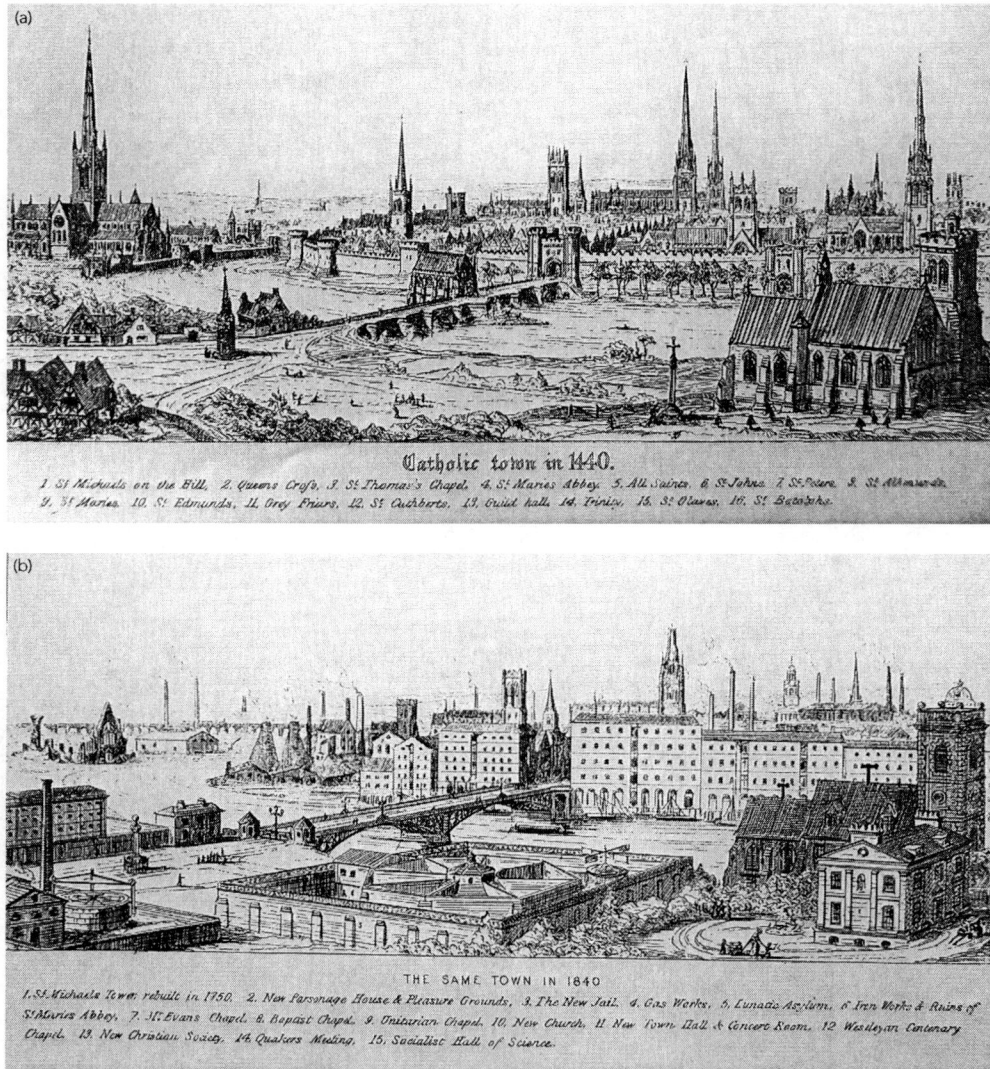

Figure 4.1

(a), (b) Pugin's *Contrasts*: medieval and modern views of an imaginary English town

industrialism and imperialism can be refined and ennobled. In Carlyle's *Past and Present* (1843), the degradation of the modern workhouse is contrasted, Pugin-style, with the harmony of Bury St Edmunds Abbey, led by 'Abbot Samson'.[5]

Ruskin applied that world-outlook to architecture. He was a medievalist only in a general sense, arguing in the *Seven Lamps* that 'we have built like frogs and mice since the thirteenth century'. Although his rhetoric was constantly underpinned by strong aesthetic preferences, by a love of old, irregular texture (in other words, by the Picturesque and Sublime), his argumentation was equally concerned with the way the built environment, in his view, expressed the social ills of modern

Figure 4.2 John Ruskin

(a) View of Ruskin in 1843; (b) Illustration of weathered carving on Rouen Cathedral, from Ruskin's *The Seven Lamps of Architecture* (1883 edition); (c) Extract from *The Seven Lamps*, denouncing restoration

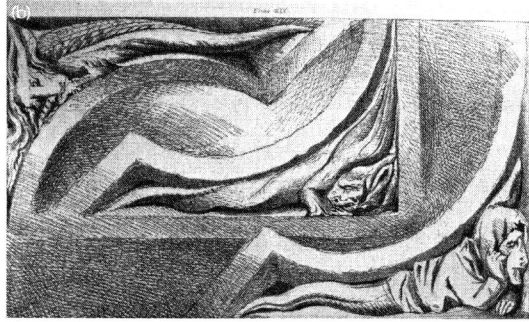

the preservation of the architecture we possess: but a few words may be forgiven, as especially necessary in modern times.

APHORISM 31. Neither by the public, nor by those who have the care of public monuments, is the true meaning of the word *restoration* understood. It means the most total destruction which a building can suffer: a destruction out of which no remnants can be gathered: a destruction accompanied with false description of the thing destroyed.[54] Do not let us deceive ourselves in this important matter; it is *impossible*, as impossible as to raise the dead, to restore anything that has ever been great or beautiful in architecture. That which I have above insisted upon as the life of the whole, that spirit which is given only by the hand and eye of the workman, never can be recalled. Another spirit may be given by another time, and it is then a new building; but the spirit of the dead workman cannot be summoned up, and commanded to direct other hands, and

Restoration, so called, is the worst manner of Destruction.

[54] False, also, in the manner of parody,—the most loathsome manner of falsehood.

capitalism. His admiration for the rich Gothic architecture of medieval Venice was partly based on its beauty as 'art' in decay, but partly also on its character as a lost golden age, exemplifying the processes of decline and fall. This posed an urgent warning to contemporary England, whose social cohesion was, he believed, mortally threatened by the alienation of modern capitalism. Ruskin hailed the physical substance of buildings, as revealed in their surface appearance and texture, as a supreme expression of the joy of human labour. For him, historic buildings were significant not for their artistic design, or their evocation of some ideal state, or their external association (patriotic or religious), but for their cumulatively accumulated substance, ennobled spiritually through participation in the processes of 'life', whose essence could be grasped physically through the picturesque patina of age.[6]

Ruskin and the 'living' monument

THE ideal of authenticity, or 'truth' as Ruskin usually put it, now rested not in the form but in the material, and was attributed vastly higher moral importance. In Ruskin's *Lamp of Memory*, the Enlightenment emphases on historic and artistic value were still present. But they were now bound into the built fabric, whose picturesque beauty, he argued, stemmed from its participation in nature's sacred evolution processes, and whose historical value stemmed from its unique ability, having stood for centuries, to physically embody the collective social and natural memory. This was a status almost like the *pietas* value of the Roman *lares* – as Ruskin himself suggested, arguing that 'Our God is a household God, as well as a heavenly one; He has an altar in every man's dwelling; let men look to it when they rend it lightly and pour out its ashes.' Like holy places in polytheistic religion, the old stones seemed almost animate. Henceforth the word 'life' became a key rhetorical prop of conservation. From 'life', in turn, stemmed the fabric's value as 'art'. For Ruskin, like previous Romantics such as Goethe or Wordsworth, art was not a 'formal' quality but an expression of character, or 'another nature' (Goethe). For Kant (*Critique of Judgement*, 1790) art linked the here and now and the eternal, and for Hegel, in the 1820s, works of art were a 'spiritual' phenomenon.[7]

This meant that aesthetic preferences were no longer a matter of mere stylistic labels like 'Picturesque' or 'Sublime' – as with earlier Romantic admiration of ruins – but could be expressed in profound-sounding existential and ethical language. It was for fundamental moral reasons, Ruskin argued, that no-one had the right to intervene in the fabric of an architectural monument, other than make basic repairs: 'That which they laboured for . . . in these buildings they intended to be permanent, we have no right to obliterate. . . . Architecture is always destroyed causelessly.' For Ruskin, restoration of any kind was an active destruction not just of dry history but of truth and life. Its evil went far beyond its utilitarian modernity. He condemned it in the most thundering, sermon-like invective, eclipsing Carter and even Pugin:

> Neither by the public, nor by those who have the care of public monuments, is the true meaning of the word *restoration* understood. It means the most total destruction which a building can suffer: a destruction out of which no remnants can be gathered: a destruction accompanied with false description of the thing destroyed. . . . There was yet in the old some life, some mysterious suggestion of what it had been, and of what it had lost; some sweetness in the gentle lines which rain and sun had wrought. There can be none in the brute hardness of new carving. . . . Do not let us talk, then, of restoration. The thing is a Lie from beginning to end. You may make a model of a building as you may of a corpse, and your model may have the shell of the old walls within it as your cast might have the skeleton . . . but the old building is destroyed, and that more totally and mercilessly than if it had sunk into a heap of dust, or melted into a mass of clay.[8]

This opposition between 'living', 'artistic' truth and 'dead', 'mechanistic' falsity would condition much conservation rhetoric over the following century, even after the specifics of the anti-restoration debate had faded from prominence. The theme of restoration's corruption of authenticity had already been sketched out by others – for example Stendhal (M-H Beyle) complained of the 1817–23 rebuilding of the Arch of Titus by 'this lamentable figure' (Giuseppe Valadier), that 'What remains is a *copy* of the arch of Titus, although one has to concede that the site is still the same and that the reliefs on the inside of the arch have been preserved.'[9] And, of course, the earlier French attacks against 'vandalism' had pioneered much conservation rhetoric – in opposition to demolitions, not restorations. But it was Ruskin who gave anti-restorationism its new moral force, through the emphasis on 'life' and the extremism of the rhetoric.

This, in turn, made possible a new, more emotionalist interpretation of national history and heritage, in terms of an organic community open to all, linking past generations with the future. Ruskin argued,

> Every human action gains in honour, in grace, in all true magnificence, by its regard to things that are to come. . . . Therefore, when we build, let us think that we build for ever. . . . For indeed the greatest glory of a building is not in its stones, nor in its gold. Its glory is in its Age, and in that deep sense of voicefulness, of stern watching, of mysterious sympathy, nay, even, of approval or condemnation, which we feel in walls that have long been washed by the passing waves of humanity.

As mere custodians of this living force, 'we have no right whatever to touch them. They are not ours. They belong partly to those who built them, and partly to all the generations of mankind who are to follow us.' Ruskin himself was not a nationalist, but in this passionate argumentation the Conservation Movement had almost arrived at the 20th-century rhetoric of nationalist conservation, mobilising old buildings as soldiers in the struggle to defend the 'total' community, for the sake of present and future.[10]

Ruskin's defence of the 'living' monument, in the Romantic tradition of the worship of ruins, relied not upon logic but upon passion – in forcible contrast to the measured argumentation that underpinned Viollet's interventionism.[11] Not surprisingly, in terms of strict logic, Ruskin's polemic was riddled with anomalies. For example, it was a fairly bold sophistry to argue that a building could suffer any worse destruction than being demolished. And the veneration of decay posed the obvious question of whether, and at what stage, it was legitimate to arrest that decay. In a further inconsistency, both the original form and its degeneration seemed to share in the precious life-force: Ruskin argued that 'there was yet in the old some life, some mysterious suggestion of what it had been, and of what it has lost'. Was the 'life' accrued through decay a different type of life from that conferred by the workman's hand in the original building – or was the entire sequence of argumentation just rhetoric? It was also unclear whether, in a *reductio ad absurdum*, the logical consequence of Ruskin's doctrine was to allow buildings ultimately to collapse. Charles Dickens spotted this particular self-contradiction within Ruskin's theory and mischievously proposed that buildings should be allowed to decay by one inch a year, until they disappeared altogether. Yet all these inconsistencies seemed to be reconciled and gain further impact by the very boldness of Ruskin's metaphoric leaps, hailing inanimate stones as having a living force and societies as having a collective 'memory' which could be nurtured by that living force, to combat the forces of modern alienation.

Demonising the restorers: William Morris and Anti-Scrape

THE inconsistencies of Ruskin's doctrine were reconciled by having a substantial target: the rapidly booming craze for restoration, which was increasingly dominating the Gothic Revival in England. This was a movement that, like many architectural crazes on the verge of overthrow, was becoming increasingly vulnerable, and sensitive, to principled criticism. It was above all the work of Sir G G Scott, combining gargantuan output and open embrace of modernity with an increasingly reflective and self-critical outlook, which epitomised this self-contradictory attitude. The first stage of the anti-restoration 'battle' focused on a proxy combat between Scott and Ruskin. In 1850, the year after Ruskin's *Seven Lamps* appeared, Scott published a self-exculpatory book (written in 1848), *A Plea for the Faithful Restoration of our Ancient Churches.* It tried to tread a middle line between the 'destructiveness' of indiscriminate restoration and the positive possibility of recreating 'features which have been actually destroyed by modern mutilations, where they can be indisputably traced'.[12]

Typically of architectural rhetoric in the modern age, the vast practical benefit of Scott's structural interventions was simply disregarded by his critics, who branded him a coarse barbarian. In contrast to the confident modernity of the Viollet philosophy that reigned in France, in the English debates the field was conceded to Ruskin's emotionalistic world-outlook: Scott-style 'reluctant restorers' were forced ever more on the defensive. Gradually, opposition within England crumbled, and Ruskinian values infiltrated even relatively staid heritage organisations: in 1855, the Society of Antiquaries agreed to help finance a 'conservation fund' (the first use of the term 'conservation' in this context) to help oppose the 'pernicious influence' and 'evil' of restoration, which (in Ruskin's words) resulted in 'falsified buildings, untrue in art, unjustifiable in taste, destructive in practice, and wholly opposed to the judgement of the best Archaeologists'.[13]

Increasingly, Scott was forced onto the battleground chosen by his opponents, agreeing in 1864 that a church untouched by any restoration was the best. Defending his restrained works of the early 1870s at St Albans Cathedral, where he had managed to curb the interventive zeal of the local Restoration Committee, he pleaded that

> I am in this, as in other works, obliged to face right and left, to combat at once two enemies from either hand, one wanting me to do too much, and the other finding fault with me for doing anything at all.

It was an irony typical of the wilfulness of modern architectural polemic that the arch-bogeyman of the Ruskin faction should have been someone unusually sensitive to conservation issues: for example, Scott's winning entry in the 1872 competition for a new Episcopal Cathedral in Edinburgh carefully planned the new complex so as to preserve a small, 17th-century castle, Easter Coates, on the site. As for Ruskin, however influential his world-outlook was in the long term, his own immediate contribution to this campaign was impeded by his unworldliness: rather than trying to win over adversaries, in 1874 he refused the gold medal of the Royal Institute of British Architects (RIBA) because of 'the destruction under the name of restoration brought about by architects.'[14]

In the late 1870s, the balance of educated opinion in England shifted even more decisively against the restorers. This resulted not from the initiatives of Ruskin himself, who was no longer driving the 'struggle'. That role was played by a new, dynamic private pressure group, the Society for the Protection of Ancient Buildings (SPAB), led by the designer-propagandist William Morris. The foundation of SPAB, in 1877, was of immense importance to the development of the Conservation Movement, as it was the ancestor of all modern conservation campaigning societies. SPAB was a new kind of 'private' heritage group, dedicated not to antiquarian researches or restoring individual monuments but to confrontational public agitation on a national scale against 'threats' to old buildings – something previously confined to gentlemanly protestors in individual towns, such as Lord Cockburn in Edinburgh. SPAB's protest campaigns were more aggressively focused than these predecessors, thanks to its voracious exploitation of the Ruskin ideology, which it gave the more populist name 'Anti-Scrape'; yet it was all the more effective for being pragmatic and outcome-oriented where Ruskin had been idealistic and unworldly.

Its populist appeal gained, too, from its links to the egalitarian political doctrine of socialism, espoused fervently by Morris himself. He powerfully developed Ruskin's anti-modern utopianism, combining his romantic concept of emancipated artistic labour with the historical materialism of Marx. Rather than demanding a radical rupture between future and past, Morris argued for an organic continuity of community, linking the Greek polis and the self-governing medieval guilds to the 'Vision Splendid' of the socialist future: 'Fellowship is heaven and lack of it is hell; fellowship is life and lack of fellowship is death.' Morris's stance was equally an aesthetic one. One of England's foremost decorative artists and furniture makers, he became a leader of the late-19th-century movement – known as 'Arts and Crafts', 'Aesthetic Movement', or 'Queen Anne'

Figure 4.3 Cartoon cf William Morris by Frederick Waddy, 1873

– to reject as tastelessly commercial the 'High Victorian' phase of architecture and design, with its boldly geometrical use of historical motifs, and advocate instead a more 'artistic' and 'informal' approach, inspired by the 'vernacular'.[15]

Morris argued that reformed art and social revolution could progress hand in hand, especially in architecture. The Middle Ages, he felt, provided an ideal exemplar. Before the alienating onrush of modernity, art had been 'a part of everyday life': in 1884, he claimed, 'Anyone who wants beauty to be produced in the present day in any branch of the fine arts . . . must be always crying out, "Look back! Look back!"' In his *News from Nowhere* (1890) a contemporary Englishman awakes after a 100-year sleep to find industrial London replaced by a green, harmonious landscape of small houses, 'alive and sympathetic with the life of the dwellers in them'. The time traveller delightedly discovers that Morris and Rossetti's Oxfordshire Arts and Crafts rural retreat (from 1871), Kelmscott Manor, had survived into the new utopian age, its continuing 'life' directly embodying the values of the old times: 'It seems to me as if it had waited for these happy days, and held in it the gathered crumbs of happiness of the confused and turbulent past.'[16] This elaborated the animistic stream of thought developed by Ruskin, sacralising and attributing a quasi-eternal life to the monument. From a strictly Christian perspective, this was arguably a kind of idolatry, but the socialist Morris was no longer concerned with conventional religion. For him, the sacred lay not in heaven, but in the environment around us, which was available to everyone. And though Morris's political outlook was firmly socialist rather than nationalist, the new concept of old buildings as seats of life would powerfully shape nationalist conservation, by allowing almost any kind of built environment to assume a quasi-sacred aura.

As part of that ethos of living heritage, Morris's attitude towards restoration became ever more intransigent and, by the mid-1870s, he agreed with Ruskin that not just poor-quality

restoration, but *all* restoration was to be condemned on principle – albeit for the socialist reason that it was destroying one of the most powerful defences against modern alienation. Like Pugin, Morris cultivated an exaggeratedly polemical public persona, on one occasion spluttering as a restoring clergyman passed him, 'Beasts! Pigs! Damn their souls!' Like Pugin, and unlike Ruskin's unworldly critiques, Morris set out to put his principles into practical action – through the foundation of SPAB.[17]

The spark that provoked this step, was, inevitably, a project by Scott. In March 1877, Morris wrote to the *Athenaeum*, protesting against the 'barbarism' of Scott's proposed restoration of Tewkesbury Abbey and suggesting the foundation of 'an association for the purpose of watching over and protecting' old buildings. He argued in forcibly secular terms that this society should strive to 'awaken a feeling that our ancient buildings are not mere ecclesiastical toys, but sacred monuments of the nation's growth and hope'. Heritage could also potentially serve as spurs to mass national mobilisation, given that now, 'the newly invented study of living history is the chief joy of so many of our lives'. Morris's new society reflected his eclectic outlook in its diverse membership of antiquarians, architects, artists and art-collectors. SPAB's *Manifesto* joined Ruskin's *Lamp of Memory* as a founding text of modern conservation. Developing Ruskin's identification of 'life' with the ordinary houses of the people, the manifesto laid claim to almost the whole 'cultural landscape': a monument was not just a cathedral but 'anything which can be looked on as artistic, picturesque, historical, antique, or substantial; any work, . . . over which educated, artistic people would think it worth while to argue at all'. In 1878 a Liberal MP, Leonard Courtney, told a SPAB meeting that 'the real principle of our action is this: let dead things go, let living things be kept'.[18]

The founding of SPAB formed part of the late-19th-century explosion of social-reformist voluntary environmental groups within Britain: the 1867 foundation of the first nature-protection society, the Commons, Open Spaces and Footpaths Preservation Society, followed the celebrated 1866 'liberation' of 430 acres of Berkhamsted Common from enclosure, by 120 workmen who travelled from Euston by special train. In the words of one of its key members, housing reformist Octavia Hill, more small parks were urgently needed, as 'open air sitting rooms for the poor': her 1877 essay, 'Our Common Land', linked together advocacy of public ownership of land and open space preservation. This social agenda was also prominent in groups such as Ruskin's 'Guild of St George' (1871), or the preservation societies for Dartmoor (1883) and the Lake District (1885). Later volunteer associations included the Society for the Protection of Birds (1889), the Coal Smoke Abatement Society (1899), the Garden Cities Association (1899) – and the National Trust (NT, 1895), whose foundation we will trace in detail in Chapter 5.[19]

The SPAB immediately launched vigorous struggles against both restorations and demolitions of historic buildings, including post-medieval churches – reflecting the all-embracing scope of its manifesto. In 1878 Morris began a passionate campaign against demolition of several of Wren's classical City of London churches: this 'outrageous and monstrous barbarity' would leave England

> liable to be looked upon by foreign visitors and by our own posterity as the only people who have ever lived who, possessing no architecture of our own, have made themselves remarkable for the destruction of the buildings of their forefathers.

Morris also vigorously opposed intrusive advertising hoardings: the expunging of commercial clutter would become a key concern of urban conservation across Europe at the turn of the century.[20]

But the SPAB's central preoccupation remained the fight against restoration, and here its work was shaped by the distinctively English tradition, stemming from Pugin, of violent architectural polemic and journalism. Immediately, SPAB members became involved in the most blistering attacks on Scott, continuing even after his death: in 1878, Morris referred to him as 'a happily dead dog'. To replace him, a fresh bogeyman luckily appeared, in the person of amateur architect Lord Grimthorpe, who carried out extensive restoration works costing £100,000 at St Albans Cathedral from 1879, including the recasting of the Perpendicular west front in neo-13th-century style and a new rose window in the north transept, also replacing a Perpendicular one. Here the attacks could be more vituperative still, not being restrained by residual respect for an elder statesman of the profession. The *Builder* lambasted his new transept as 'railway station Gothic' and branded Grimthorpe an 'architectural pretender who can pay for the privilege of amusing himself . . . with a building which is the property of the nation'. Another critic described his rose window as a 'stone colander'. On behalf of SPAB, Stevenson became involved in a furious exchange over the issue, and Morris condemned the work in 1884:

> As it goes on, a terrible dullness settles over this once romantic and deeply interesting building. The new roof is dull, the west front is more dull, the changes of style on the long south side now fail to raise the slightest emotion, since they are so obviously the work of one time . . . the whole is . . . an architectural freak.

As this rhetoric reveals only too clearly, strong visual preferences still underpinned all these moral tirades. Restorations by other respected Victorian architects, such as T G Jackson, J L Pearson and even G F Bodley, were similarly attacked in their turn.[21]

Conservation campaigning: SPAB's early activism

Spab's influence was immeasurably increased by the way it combined this moral zeal with a practical organisational system of monitoring and proselytising, including a national network of subscription-paying members, executive and casework committees (dominated by Arts and Crafts sympathisers) and local correspondents. As honorary secretary from 1877 until his death in 1896, Morris shaped this structure through his furious energy and force of personality, maintaining a punishing pace of personal casework visits, helped by co-founder Philip Webb and, from 1883, by a part-time paid secretary, Hugh Thackeray Turner. During these two decades, nearly 2,000 individual buildings and 2,862 cases were dealt with.[22]

Just as important as the sheer quantity of its work was the way the society popularised and institutionalised a confrontational style of public representation, both in its regular casework letters and propaganda, and in Morris's own intemperate language on site visits. W R Lethaby, for example, recalled that, 'at one cathedral, at having some commercial work of stalls shown to him, in pride he burst out, "Why, I could carve them better with my teeth!" A canon of —, on pointing out the "improvement" he wanted in the cathedral, was answered, "[Is this place] not good enough for you?" After a painful visit to — Church, which being skinned alive, he rushed to the window of the inn shaking his fist as the parson passed by.' As we will see later, the metaphor of the 'living' heritage under threat from murderous opponents would be an enduring one, and would assume even more exaggerated forms in the 20th century.[23]

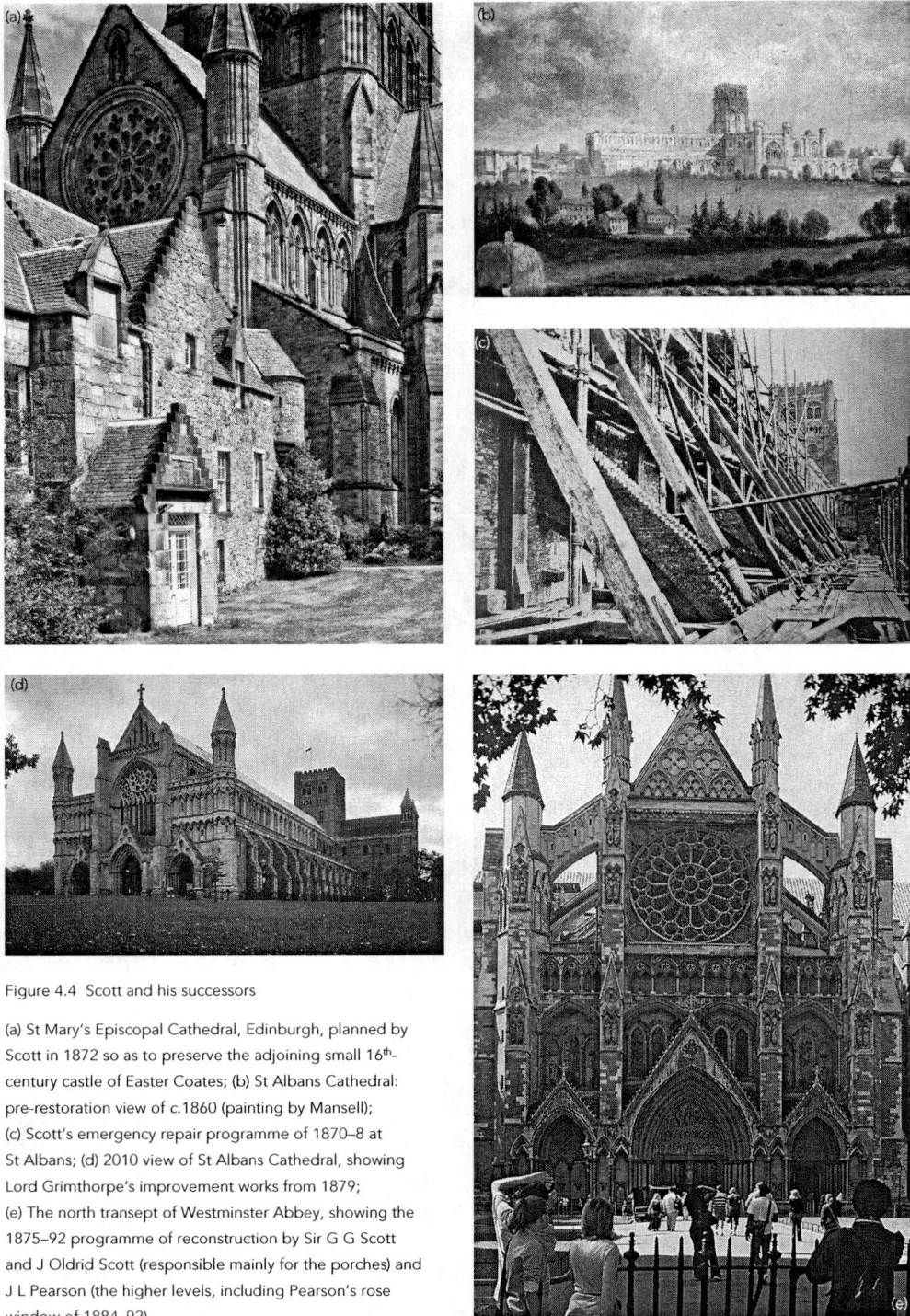

Figure 4.4 Scott and his successors

(a) St Mary's Episcopal Cathedral, Edinburgh, planned by
Scott in 1872 so as to preserve the adjoining small 16th-
century castle of Easter Coates; (b) St Albans Cathedral:
pre-restoration view of c.1860 (painting by Mansell);
(c) Scott's emergency repair programme of 1870–8 at
St Albans; (d) 2010 view of St Albans Cathedral, showing
Lord Grimthorpe's improvement works from 1879;
(e) The north transept of Westminster Abbey, showing the
1875–92 programme of reconstruction by Sir G G Scott
and J Oldrid Scott (responsible mainly for the porches) and
J L Pearson (the higher levels, including Pearson's rose
window of 1884–92)

The enduring international influence of Anti-Scrape stemmed not just from this 'campaigning' combination of moral extremism and practical organisation, but also from SPAB's all-embracing heritage definition – presaging the modern Conservation Movement's existential drive to push out its boundaries in all directions, both thematically, to embrace ever more types of built environment, and chronologically, to come ever nearer the present day. With each nation seeing itself as a miniature universe, it was only logical that the entire cultural landscape of one's own nation should potentially count as heritage. We saw the beginnings of that concept in the writings of Victor Hugo or Ruskin on the 'homes of the people' as foundation-stones of national heritage, or (earlier still) in the 18th-century Romantic 'discovery' of vernacular culture by Wordsworth and others. On his 1840–1 visit to Rome, Ruskin scoffed at the city's great monuments, while enthusing about its everyday streets. And in England he attempted to revive 'folk' practices, such as 'pagan' maypole dancing in 1881.[24] Increasingly, the emphasis of the 'living monument' concept was shifting from elite structures to wider environments.

Among English architects, the late-19th-century move towards more informal styles (such as Queen Anne) was also linked to a shift in historic architectural fashion from the ecclesiastical Middle Ages to the secular architecture of the Tudor age, the 17th century, or even the 18th century. Novelists had pioneered the same path decades before: Charles Kingsley's *Westward Ho!* (1855) presented the Tudor age as a redoubt of 'burly triumphant Protestantism and nationality'. By 1857, Trollope's loving description of the old mansion of Ullathorne in *Barchester Towers*, in words that could have applied to any of Devey's new or recast houses, emphasised its combination of picturesqueness with 'quaint English homely comfort', enhanced by the patina of age: 'No colourist that ever yet worked from a palette has been able to come up to this rich colouring of years crowding themselves on years!' And by 1855, Thackeray's novels were already popularising the idea that the previously despised 18th century also embodied these values and was a 'merrier England', full of both 'quaintness' and 'vigour'. Soon, heritage fashion would advance again, and the early-19th-century 'Regency' period would be hailed as the last spontaneous, pre-modern age – an English equivalent to the German *Biedermeier*.[25] Architecture and interior design followed where these authors led. The Queen Anne and Aesthetic movements of the 1870s and '80s began to value more informal historic precedents, as part of the new concept of the artistic 'everyday home', populated with judiciously chosen 'antiques'.[26] In architecture, the Vernacular Revival and the cult of the objet-trouvé appropriated elements of the Romantic cult of ruins. The cottages first 'discovered' by Wordsworth were increasingly venerated; as we will see in Chapter 5, when the National Trust was founded in 1895, the first building it acquired was a timber-framed, thatched-roof vicarage at Alfriston, Sussex.[27]

Concern for the future of dense old towns was less widespread in England, with its lower-density residential patterns than the Continent (or Scotland). A transitional case was Stratford-upon-Avon, whose status derived chiefly from the older type of individual associative values, but whose picturesque, timber-framed street architecture soon began to attract tourist attention as an ensemble in its own right. The only major exception to this rule within the UK was still the Old Town of Edinburgh, where, in parallel with the early City Improvement schemes from the 1860s, antiquarian efforts were stepped up, with the restoration of the 16th-century John Knox House (from 1855; also associated with a 'great man').[28]

In its first ten years of life, SPAB's pragmatic dissemination campaigns, deploying a range of articulate people to agitate in the press and lobby in parliament, were extraordinarily effective. They succeeded through purely voluntary means in building up a wide middle-class enthusiasm for historic buildings whose pervasiveness, like that of the cathedral restorers before them, exceeded the centralist, bureaucracy-led movements of France or Germany – although Britain lagged far behind in legislative protection, and the first measure for voluntary state 'guardianship' of (largely prehistoric) 'scheduled monuments', the Ancient Monuments Act, was only passed in 1882. At early SPAB meetings, central issues of heritage philosophy were repeatedly debated – for example, the relative importance of the artistic and social values of monuments, or their relationship to religious beliefs. And in these debates, the Pugin-Ruskin intermixing of social-ethical rhetoric with aesthetic preferences continued. In an 1887 keynote lecture on 'The Sacredness of Ancient Buildings', barrister Frederic Harrison argued that

> The multiplicity of pasts in a great edifice, the vast range of its power over an infinite number of human souls, the concentration of efforts by which it was built up and the countless generations of men who have contributed to its beauty, or have been touched by its majesty, give it a collective glory which no statue or picture ever had.

Morris agreed with Harrison that monuments had a 'sacred' character, 'what you may honestly call a religious feeling'. And Thomas Hardy argued in 1904 that

> the protection of an ancient edifice against renewal in fresh materials is, in fact, even more of a social – I may say a humane – duty than an aesthetic one. It is the preservation of memories, history, fellowship, humanity, life. Life, after all, is more than art.[29]

By the 1880s, these protest campaigns often met with success. For example, in the city of York, an 1885 threat by the Archbishop to shut six medieval parish churches inspired a fierce resistance campaign run by SPAB and a local clergyman: it culminated in a great public meeting at which Morris pleaded that the proposal would discredit York's cultural reputation. After months of acrimonious debate, the Archbishop withdrew his plan. Likewise, in the case of the 12th-century Kirkstall Abbey, symbolic hearthstone of industrial Leeds, an 1888 threat to put the ruins up for public auction provoked an outraged campaign, here fronted by the *Birmingham Daily Post*, and eventual purchase by a Yorkshire-born businessman, Colonel John T North, who presented the ruins to Leeds Corporation: the Abbey House was eventually converted into a municipal museum in 1925–7. But articulate voices were raised on the other side, too. For example, in 1890, Havelock Ellis complained that England was no longer a centre of industrial vitality but 'a museum of antiquities'. And in Peterborough, in 1897, SPAB protests at the proposed replacement of the medieval Guildhall by a new town hall were vociferously opposed by local lobbyists, who blustered that removal of the 'sad anachronism' was imperative in a 'go-ahead' town like Peterborough, whose cathedral in any case 'beat it [the Guildhall] in antiquity'. By that time, a compromise had been generally agreed across the country: to preserve a limited number of symbolic ancient buildings with great care, but to allow alterations or demolition elsewhere. These intensive schemes included 'lesser' buildings as well as major monuments: for example, the small parish church of Inglesham, near Lechlade, near Morris's Kelmscott Manor, targeted in 1885–1900 for a complex conservative-repair programme overseen by Westminster Abbey surveyor J T Micklethwaite and partly financed by Morris himself.[30] In Scotland, where the implicitly English nationalist element in SPAB's propaganda sometimes struck a negative chord, the society was of relatively little influence, and radical restorations of ruined castles continued apace.

Figure 4.5 Kirkstall Abbey, Leeds, as preserved following the 1888 controversy

England versus the rest: international restoration contrasts and conflicts

By now, from a position of relative closeness earlier in the century, the heritage world-outlooks of England and France had diverged sharply; in 1881, for instance, J J Stevenson patronisingly described Viollet's restoration philosophy as 'unintentionally amusing'. Morris explained this divergence partly in nationalist terms, as due to the widening gulf since the 17th century between England's 'solid and settled' society and the supposedly meretricious absolutism of France, or the 'corrupt' civilisation of Catholic Rome. The stage was set for several decades in which English and continental preservationists talked past each other. During the first two decades of SPAB, the restoration movement was in full flood on the Continent, in a wide range of projects that intermingled 'completion', 'restoration' and 'isolation'.[31] Ongoing church completion projects in France included St Ouen in Rouen, realised in bristling, twin-spired form by H Gregoire in 1838–52. And in Germany, energised by the Cologne project, there was furious activity, including completion of Ulm Cathedral in 1844–90 with a 520-ft-high tower and of Regensburg Cathedral with twin spires in 1859–69 (as part of a Gothicising and Baroque-expunging campaign started by Ludwig I in 1828). Restoration and 'enhancement' projects included Matthias Berger's 1858–68 restoration of the late Gothic Frauenkirche in Munich, sweeping away Baroque side altars and other decoration, the refacing of the Lübeck Rathaus with a more 'artistic' Gothic facade in 1872–91, and the elaborate schemes of historicist decoration inspired by August von Essenwein, director of the German National Museum in Nuremberg, that were executed in Gross St Martin (1883–91) and St Gereon's churches in Cologne, and in Braunschweig Cathedral (1876–7).[32]

The reconstruction of fantasy castles in the Viollet tradition combined elements of both restoration and completion. Within Germany, a leading figure was King Ludwig II of Bavaria. His fanatical enthusiasm for Wagner's opera, *Tannhäuser*, fuelled by an incognito 1867 visit to the newly restored Wartburg, prompted him to launch a vast project to remodel a ruined stump near Prince Max's reconstructed Hohenschwangau into the vertiginous showpiece castle of Neuschwanstein.

Figure 4.6 19th-century cathedral completions

(a) Ulm Cathedral, 1974 view showing main tower on left (completed 1844–90); (b) Regensburg Cathedral, completed in 1859–69 with twin spires

Conceived as a royal country residence 'in the true style of the old German Ritterburgen', the first designs for Neuschwanstein (1869) were appropriately not by an architect but by a theatre designer, Christian Jank; his concept was implemented, up to Ludwig's death in 1886, by successive architects including Eduard Riedel and Georg Dollmann.[33] Elsewhere in Europe, castle-reconstruction projects sprouted haphazardly, their distribution reflecting the enthusiasms of the super-rich rather than nationalist zeal. Some took an idiosyncratic form, as in the two lavish projects in south Wales by

Figure 4.7 Post-Viollet castle restorations

(a) Neuschwanstein Castle, radically reconstructed/expanded from 1869 by Christian Jank and others: this view shows the castle under construction in 1885; (b) Hungary's equivalent to Pierrefonds was the post-1867 reconstruction of the fire-destroyed 15th-century Vajda Hunyad, by Ferenc Schulcz and Imre Steindl. This reduced-scale copy of the castle was built as a pavilion in the 1896 Millennium National Exhibition in Budapest; (c) Holland's Pierrefonds: P J H Cuypers's reconstruction of Kasteel de Haar, 1891–1912; (d) The metal roof of the staircase hall at Kasteel de Haar, inspired by Cuypers's Amsterdam Centraal-Station, being installed

131

English Gothic Revival architect William Burges for one of Britain's richest aristocrats, the 3rd Marquess of Bute, at Cardiff Castle (from 1868) and Castell Coch (from 1871) – both crammed with flamboyantly individualistic decoration. In resurgent Hungary, the years from 1867 saw a spectacularly spiky reconstruction of a fire-destroyed 15th-century castle, Vajda Hunyad, by Ferenc Schulcz and Imre Steindl – a project indebted to Viollet yet exuberantly 'Hungarian' in expression. A similar approach was followed at Kasteel de Haar, a later project (1891–1912) by the eminent neo-Gothicist, P J H Cuypers. Ruinous since its destruction by troops of Louis XIV, the 15th-century rubble stump was reconstructed in a straightforward Dutch translation of Viollet, including brick facing, stepped gables, corbelling and window shutters, and a fantastically ornate central stair-hall with arched iron roof reminiscent of Cuypers's Amsterdam Centraal Station (opened in 1889). The project was commissioned by an aristocrat, Baron Etienne van Zuylen, who had married into the Rothschild family and had been put in touch with Cuypers by Victor de Stuers, one of the Netherlands' foremost conservation campaigners. Zuylen envisaged his house as 'a kind of museum like Pierrefonds, but still comfortable to live in'.[34]

In Italy and Greece, the strength of the classical tradition posed Viollet-style restorers a more complex challenge than in northern Europe, looking towards several ideals at once. In 'classical Athens', in an unambiguous attempt to prioritise the monuments of antiquity, there unfolded probably the most extreme example of *isolement* anywhere. What now became known as the 'sacred rock' of the Athenian Acropolis was literally scourged of millennia of accretions – a campaign dwarfing the puny restorations against which Morris raged in England, as buildings that would have been treasured in any northern European country were relentlessly destroyed. This began in earnest in 1876 with the demolition of the early medieval 'Frankish Tower' in the Propylaea, and accelerated in a turn-of-century programme of combined *dégagement* and *anastelosis* directed by government engineer Nikolaos Balanos, including not only the stripping of the surface of the rock but also the part-reconstruction of the Erechtheion and caryatid porch in 1902–9.[35]

The ambiguities of the Italian position were exemplified in the city of Florence, where a burgeoning cult of Dante looked both backwards to the Middle Ages and forward to the Renaissance: a group of old houses in the historic centre was labelled 'Dante's House' and acquired by enthusiasts in the 1860s. Medievalism was satisfied by a campaign of Viollet-style restoration of key monuments, including Santa Maria Novella church ('liberated' from accretions in the 1860s), the Gothic facade of S. Croce (heavily restored by N Matas in 1857–62, with a grand new 1865 statue of Dante in front) and the cathedral (completed with new facade by Emilio de Fabris in 1868–87); the Bargello was drastically restored as an arts centre by Francisco Mazzei, complete with reopened courtyard arcading (finished in 1865). But Florentine restorers were equally concerned with the Renaissance, and the Corridoio Vasariano, by the Uffizi, was made 'more classical' by opening its arcade to the river and removing additions (by Giuseppe del Moro, from 1885).[36]

Following the French, German and Italian experiences, similar Viollet-style restoration campaigns fanned out across Europe. In Spain, where the heritage was complicated by the double focus on Christian and Islamic sites, there were extensive restorations, generally with a 10/20-year time lag from France, Italy and England. The semi-ruinous Leon Cathedral, Spain's first designated national monument (1844), became a Spanish paradigm of the Viollet approach: the relatively conservative work of architect Juan de Madrazo (1869–76), including vault repairs and restored north facade, was followed in 1880–92 by the more adventurous interventions of Demetrio de los Ríos, aiming at creating an 'original Gothic purity' inside and out, and finally by the consolidatory

Figure 4.8 Castellated and classical restorations in Florence

(a) The Bargello, restored by Francisco Mazzei (completed 1865); (b) the Corridoio Vasariano, opened up to the river in 1885 by G del Moro

work of Juan Bautista Lázaro. Sometimes, successive stages of restoration reflected reversals of values. At the Ridderzaal in the Hague parliament complex, for example, the late 19th century saw a tangled sequence of restorations, beginning in 1861 with installation of a neo-Gothic iron roof (by government architect and medieval enthusiast Willem Rose) instead of the medieval timber original; this in turn was demolished and rebuilt in timber by C H Peters as part of a Viollet-style restoration of the complex in 1896–1905, including a castellated rather than an informal classical exterior.[37]

Soon, however, the tide of public opinion on the Continent began to shift against the modernising isolation/restoration policies associated with Viollet and Haussmann. In France itself, where Boeswillwald continued restoring Carcassonne beyond 1900, the first cries of opposition to Haussmannian 'isolation' of historic monuments, and calls for preservation, began in response to the demolition of the Cité in the early 1860s. After middle-class houses on the island started being acquired for demolition, *Le Figaro* in 1868 argued that the area should have been preserved. After the devastations of the 1870–1 Commune and Prussian invasion, people questioned the Viollet restoration ethos more generally. However, as late as 1888, opinion was still sharply divided: Charles Delon's book, *Notre Capitale*, questioned the appropriateness of 'artistic' appreciation in decayed old areas: 'The word Picturesque means something like, "makes a beautiful picture". That is excellent in painting! But in real life? The streets of old Paris may be interesting to visit, but that is no reason to live in them!'[38]

Figure 4.9 Leon Cathedral, Spain, restored in 1869–76 by Juan de Madrazo and in 1880–92 by Demetrio de los Ríos

Figure 4.10 The Ridderzaal, The Hague, following the castellation and reconstruction scheme of 1896–1905 by C H Peters

Elsewhere, too, opposition increasingly surfaced. In Sweden, architect Fredrik Lilljekvist's Viollet-style early-1890s' restoration of Gripsholm Castle provoked fierce debate. And in Austria, where Honorary Conservators (established in the post-1850s' system) had undertaken major restorations, a similar pattern prevailed. At the St Stephansdom in Vienna, for instance, although Friedrich von Schmidt carried out a lengthy restoration from 1863, his ambitious 'completion' proposal for a second spire remained unrealised, and 1880 proposals by the *Dombauverein* for radical restoration of the Riesentor were attacked by the influential art history professor, Moritz Thausing; the proposal was only finally abandoned after 1902. And at Diocletian's Palace in Split, an aggressive programme of *Freilegung* initiated by cathedral architect Alois Hauser in 1876 was blocked after 1903 by pressure from Vienna-based anti-restorationists. A typical sequence of ideas is traceable in the Czech Lands, where Viollet-like stylistic-unity restoration was introduced in the 1860s by architects such as Bernhard Grueber, only to be condemned in the '80s and '90s. Eventually, when the country's chief neo-Gothic purist, Josef Mocker, put forward proposals for completion of St Vitus's Cathedral in Prague and for restoration of Karlštejn Castle (from 1887 onwards, with von Schmidt), these met a crescendo of attacks, for example from conservator Max Dvořák, or in 1897 by Jaroslav Kamper, who asked of Mocker's restoration of St Barbara's Church, Kutná Hora, 'How can we renew something that never existed?'[39]

By the turn of century, as we will see in Chapter 5, a general vogue for the ideas of Anti-Scrape and Ruskin would pervade the Continent. These included the concept of the 'living' building, which was internalised into the debates of individual countries, and accentuated the shift from individual monuments to ensembles.[40] Ruskin's influence spread after his death in 1900, which removed a long-standing ban on translations. In France, he was hailed in the 1890s as the inspiration of a movement rejecting rationalism for primitivism and symbolism, while in imperial Germany, where a new enthusiasm for 'simple', 'massive', 'monumental' architecture was building up, his ideas were welcomed as a way of reconciling the new industrial society with German Romanticism: Hermann Muthesius hailed him as 'the prophet of a new artistic culture'.[41] More generally, Ruskin's impassioned ideas, such as the importance of 'life' in monuments, began to fuel the great ferment of heritage nationalism, and the shift in attention from individual monuments to ensembles. If the entire heritage of one's own nation was a vital fount of life, then the heritages of rival nations could now seem as worthy of condemnation as had the interventions of Grimthorpe at St Albans to the English SPAB activists.

But while the gradual, long-term assimilation of English ideas on the statist Continent would prove uncontentious, it was a different matter in the 1870s and '80s when initial attempts were made to export SPAB's ethos of polemical privatism to 'defend' threatened monuments. On the Continent, where the English tradition of knockabout architectural polemic and journalism was unfamiliar, the society's aggressive approach seemed like an offensive and 'imperialistic' intrusion.

The first SPAB overseas campaign began in 1877, when Anti-Scrapists, including Ruskin, learnt of a scheme to reface and regularise St Mark's in Venice, under Giovan Battista Meduna, surveyor to the fabric from 1853. Meduna had already refaced the north facade in 1860–4 and the south facade in 1865–75. A storm of protest arose in England, and Ruskin wrote to a sympathetic Italian contact, Count A P Zorzi, of his 'horror and contempt' at the restoration. Zorzi issued a polemical pamphlet condemning the 'destruction' of the external walling, followed in 1879 by an outbreak of outrage in the British press, orchestrated by Morris and SPAB. Presciently, one newspaper, *The Globe*, warned of a likely reaction in newly reunified Italy, arguing that Italians 'may find it awkward to be told they are vandals'. Sure enough, counter-attacks began immediately: Pietro Saccardo, assistant director of the building scheme, retaliated with a pamphlet, *San Marco, gl'inglesi e noi*. While not denying the substance of Zorzi's critiques, Saccardo thundered at the SPAB campaign as 'a comedy' that insulted 'the honour of Italy': self-respecting Italians should, on principle, ignore the 'indignation of John Bull'. In reply, the *Builder and Architect* published three detailed critiques of Meduna's scheme by heavyweight neo-Gothic architects – although it was unkindly pointed out at an 1880 SPAB meeting that one of them, G E Street, had overseen an equally invasive restoration in England, at Oswestry parish church.[42]

Eventually, in 1882, faced with a crescendo of criticism, the Venetian authorities unobtrusively climbed down, following new government regulations favouring preservation as opposed to

Figure 4.11 SPAB on the attack in Italy

(a) St Mark's, Venice: the north facade, showing the refacing carried out by G B Meduna in 1860–4; (b) The Ca d'Oro, Venice, following the 'de-restoration' of Meduna's work after 1894; (c) The Loggia del Bigallo, Florence, originally built in the 1350s; in 1881, SPAB unsuccessfully opposed a radical reconstruction scheme by Giuseppe Castellazzi (1881), including the recreation of its arcading (previously closed for nearly 200 years)

restoration; the 1880s saw efforts to remedy some of the effects of Meduna's work, although during Ruskin's final visit to Venice in 1888, he still lamented the injury inflicted on the church by the 'march of Progress'. Ruskin's disapproval also helped undermine a Meduna restoration scheme at Venice's 15th-century Ca d'Oro and inspired a corrective rebuilding after 1894 by Baron Giorgio Franchetti. By contrast, the SPAB met with unambiguous failure in 1881 when it agitated against a radical restoration of the Bigallo loggia in Florence, opposite the cathedral, by Giuseppe Castellazzi, including boldly conjectural reconstitution of its arcading.[43] Ottoman Turkey also attracted blistering attacks in 1876–8 from Morris, who was obsessed with the 'Eastern Question' and the need to recover St Sophia, Constantinople, from its alleged neglect by 'the infidel'.[44] As we will see later, a 'Ruskinian' slant became increasingly prevalent at the turn of the century among Italian restorers, especially in the growing trend to differentiate sharply between old and new – whereas in Greece the practice was to merge the two. From now on, the position of Italy and Greece within the international world of conservation was established as one of double-edged prestige, attracting the admiration and attention of all, but also compelled to suffer a more invasive level of outside scrutiny of domestic conservation policies than other, less prestigious countries.

– 5 –

Militant monuments

Nationalist conservation rivalries, 1890–1914

For like the Jews of old, most of us work with two hands: in one we bear a sword and fight for the works of our forefathers; and in the other we bear a trowel, and build up, as best we may, the works that are to express our age to future generations.

Heywood Sumner, 1895[1]

IN previous chapters, we saw how admiration and restoration of ruined castles became a vehicle for the appropriation of heritage by 19th-century nationalism, culminating in Viollet's grand projects in Second-Empire France. The greatest Viollet-style castle-restoration in France, however, dates from the turn of the century: the 1901–8 reconstruction of the hilltop castle of Haut-Koenigsbourg above the town of Sélestat. A mound of rubble for 270 years since its destruction by Swedish troops in the Thirty Years War, Haut-Koenigsbourg was transformed into an ensemble of awesome monumentality, crowned by a 250m-long, 50m-high keep. This late flowering of Viollet's restoration doctrine was inspired by Pierrefonds, but differed from its predecessors in one significant respect, hinted at in its subtly different monumental style, composed of blocky shapes faced in rough, rock-faced rubble. For the restoration of Haut-Koenigsbourg was a project not by a French architect for a location in France, but by a German architect, Bodo Ebhardt, for a location then in Germany, in Alsace: Höhkönigsburg im Elsass, an area that reverted to French control in 1918 and again in 1945. And its restoration, by a Berlin architect who had in 1899 founded a stridently nationalistic 'Association for the Preservation of German Castles', formed part of a militantly imperialist agenda, backed by the Emperor Wilhelm II, of bolstering German, Prussian and Protestant control over 'recovered' Alsace-Lorraine.

Presiding over the grand inauguration of the restored castle in 1908, the Emperor triumphally declaimed, 'Long may the Höhkönigsburg, here on the west of the Reich, like the Marienburg on the east, serve as a symbol of German culture and power into the farthest future': this was 'a mighty fortress for a mighty nation'. The Viollet-like, scenographic restoration, however, did not please everybody in Germany, but stirred up a storm of opposition from anti-restorationists (such as Otto Piper) indirectly influenced by the ideas of Ruskin. These critiques were transformed into biting anti-German propaganda by Alsatian-French artist Jean-Jacques Waltz ('Hansi'), whose satirical drawings in the style of children's cartoons mercilessly pilloried the 'Prussian' culture of the rulers of the 'Reichsland Elsass-Lothringen', and made the pomp of the Höhkönigsburg inauguration the subject of a special 1908 collection of cartoons.[2] During the second German reoccupation of Alsace

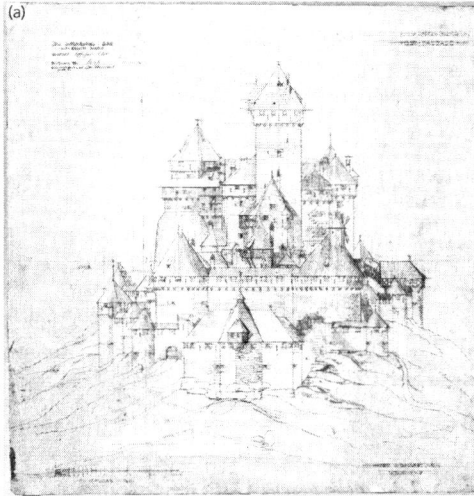

Figure 5.1 Contested heritage in the 'Reichsland Elsass-Lothringen'

(a), (b) Höhkönigsburg im Elsass: drawings by architect Bodo Ebhardt; (c) Satirical cartoon of the 1908 inauguration of the Höhkönigsburg by artist Jean-Jacques Waltz ('Hansi'); (d), (e) Hansi cartoons of 'German tourists in Alsace', c.1913

Touristes allemands dans un village d'Alsace.

139

in World War II, the Höhkönigsburg was surprisingly little emphasised, as being out of keeping with the collectivist nationalism of Germany's new rulers, and after 1945 it disappeared altogether from German public consciousness, becoming a popular French (and Alsatian) monument – and, of course, tourist attraction.

This bewildering tangle of overlapping national affiliations and discourses, all in relation to a single building, highlights the paradoxical condition of the Conservation Movement around the turn of century: riven by growing national conflicts, yet sharing a single set of underlying values as strongly as ever. In the last chapter, we traced how the mid-19th-century world of historic monuments, with its increasingly stale recipe of Romantic nationalism, medievalising restoration and Picturesque-Sublime aesthetics, was radically revitalised by the polemical English doctrine of Anti-Scrape, which attributed the old substance of buildings with a living force. Now, this passion for the 'living monument' was reinforced by urgent new currents in architecture, culture and politics, and the intensification of movements such as nationalism and tourism.

In geopolitical terms, the years round 1900 saw a parallel upsurge in both nationalism and internationalism, the first becoming vastly more important to conservation, with the ever more strident competition among the imperialist powers. Greater discipline and cohesion was called for – a spirit which, despite strenuous counter-efforts towards an international, universal outlook, would draw the Conservation Movement into its influence and foster a subtle convergence between the various national traditions of heritage – especially in the growing concern for protection of old towns. The new discipline and order were partly a reaction against everything the 19th century had stood for – including the idea that national identity could be evoked architecturally through literal historical styles. G K Chesterton wrote in 1904 of the sudden overthrow of earlier Victorian culture, of the way in which 'a whole generation of great men and great achievement suddenly looks mildewed and unmeaning'. Now, instead, identity would come from more overarching ethical values, ideologies, and aesthetic/formal preferences. German designer Richard Riemerschmid simply dismissed everything built between 1840 and 1897 as rubbish, and declared, 'We simply had to start again from the beginning.' Within new architecture, the call was for buildings that were more 'artistic', more 'rational', and certainly more 'simple' – all at once. For conservation, the implication was a further demotion of the post-French Revolution stress on the *historical* value of the monument, in favour of more unambiguously present-day concerns and motives. The powerful yet sometimes incongruous way in which all these fitted together was exemplified in the writings of German 'Heimatschutz' propagandists like Paul Schultze-Naumburg, combining visual and ethical praise of 'plain', 'quiet' old houses with an escalating nationalist sense of dynamism, of a Conservation Movement ceaselessly 'on the move'.[3]

The challenge of Darwinism and science to the Christian account of history prompted a furious search for religion-substitutes. In Germany, where writers such as Nietszche and Julius Langbehn (author of the famous 1890 book, *Rembrandt als Erzieher*) had condemned the rational and called for humans to generate their own values (the 'will to power'), art seemed a new kind of hyper-reality, unlocking deeper truths: in a 1902 exhibition in the Vienna Secession building, visitors worshipped Beethoven in a special sacred enclosure.[4] Some German critics and art historians of the late 1880s and '90s, such as Wölfflin and Burckhardt, declared that 'artistic feeling' should be conveyed, psychologically and empathetically, through more individual interpretation of mass, 'form' and 'space' – characteristics they identified not only in the revered monuments of the Middle Ages but also in the classicism of the Renaissance and in the massive plasticity of the Baroque era.[5]

In France and the Anglo-Saxon countries, interpretations of classicism remained more literal, but everywhere there was now a Ruskin-like reverence for the vernacular simplicity of 'minor architecture', as a special expression of national or regional culture. Everywhere, too, the new artistic and scientific emphases in architecture were paralleled by a new emphasis on its social tasks and on collective mass movements and building types, such as social housing. The built environment became increasingly politicised and the pursuit of community became both an object of scientific sociological research and a potential religion-substitute: modernising and anti-modernising movements marched in parallel. Morris wrote in 1887 that 'Christianity will be absorbed in Socialism', and three years earlier, biologist Patrick Geddes, a young disciple of Le Play, argued that Ruskin had heralded 'a new Utopia' that would rectify the spiritual and material impoverishment of the modern city.[6]

Alois Riegl: towards an international heritage?

FACED with all these impassioned new world-outlooks, the Anti-Scrape obsession with the restoration issue seemed increasingly narrow and tame, although it continued to develop new offshoots and interpretations across Europe. Instead, especially in the German-speaking countries, a new controversy arose, not over restoration techniques, but over the broadest meaning of monuments, linking them to the great political concerns of the age. The opening shot in this debate, and one of the most significant of all conservation texts, was a pamphlet entitled *Der Moderne Denkmalskultus* (the Modern Cult of Monuments), authored in 1903 by Alois Riegl, Professor of Art History at Vienna University and newly appointed government General-Conservator: the pamphlet was originally envisaged as a theoretical appendix to a policy report, the *Outline of a Legal Structure for Conservation in Austria*. As an academic, Riegl participated eagerly in the contemporary pursuit of 'art' and 'form' through the application of subjective feeling (*Stimmung*). But he was equally aware of the breadth of other potential criteria of monument value and set out to systematically define all the key values of conservation.[7] The wording of his title emphasised that the monument was very much an idea of the modern age and indeed amounted to nothing less than a religion-substitute – one of many within modern secular civilisation. Indeed, by *Denkmalskultus* he meant something similar to 'Conservation Movement' as used in this book. Developing the Ruskinian concept of monuments as 'living' rather than 'dead', his central insight was that whether and why something was a monument, and thus any concept of authenticity of the monument, derived not from its origin, or from eternal values, but from its present-day *reception*.

Riegl subdivided monument values as follows. On the one hand, there was a group of 'present-day values', *Gegenwartswerte*. These corresponded to Vitruvius's criteria of architecture (i.e. practical use value, *Gebrauchswert*, and artistic value, *Kunstwert*) along with another, more complex concept, 'newness-value' (*Neuheitswert*): the highly finished perfection of a monument restored in the Viollet style. For Riegl, unlike Ruskin, artistic value had to be kept in its place, not attributed a transcendent or timeless status. On the other hand, there was a grouping of 'recollection values' (*Erinnerungswerte*) concerned solely with the past. This division reinforced the old tension between artistic and historic motives for conservation, identifying the former firmly as a present-day concern.[8]

The 'past values' *also* reflected modernity, however. They had undergone their own process of evolution during the modern age, from the 'intentional commemorative value' (*gewollte*

Figure 5.2 Alois Riegl

(a) Portrait of Riegl; (b) Front page of Riegl's *Moderne Denkmalskultus* (1903)

Erinnerungswert) of the statues and inscriptions of the Renaissance, to the 'historical value' (*historische Wert*) of the restoration age, and finally the 'age-value' (*Alterswert*) of his own day. *Alterswert*, which valued any and all old buildings solely for their atmosphere of age, was an all-embracing value for an egalitarian era, democratically levelling the old hierarchy of elite monuments *versus* the remainder. Riegl explained that *Alterswert* 'completely leaves out of consideration, as a matter of principle, the local single phenomenon, and . . . merely cherishes the subjective atmosphere – before age-value, all monuments are equal'.[9]

This sequence of past values – 'intentional-commemorative', 'historical', 'age' – was not a cut-and-dried chronological sequence; there was a vast overlap. The 19th century had actually seen a boom in 'intentional-commemorative' political monuments, with hundreds of tower-like memorials erected to Bismarck throughout Germany, and in Britain a more literal commemoration of innumerable imperial and civic dignitaries through statues. And the South African War pointed to a bigger and even more 'modern', commemorative explosion – in war memorials – around the corner. Following 1914–18, war memorials would become the linchpins of vast memory landscapes of suffering and redemption: memorials that would warn as well as commemorate. The sequential relation between intellectual historical value and feeling-based age-value corresponded roughly to that between Viollet restorationism and Ruskin-Morris conservation.[10]

The chief gap in Riegl's value-system was any explicit reference to monuments' political-ideological significance – perhaps surprisingly, given that he himself had a distinct political agenda, within which monuments would reinforce the universal collective values of socialism and Christianity. Like Morris, Riegl was a humanistic socialist, who believed that socialism amounted to universal love. But whereas Morris did not exploit fully the potential contribution of conservation to the cause, for Riegl its chief value was the way it could, he felt, help stimulate the spread of fellowship. Developing Herder's concept of the equality of national artistic expressions, Riegl believed that state sponsorship of conservation should avoid nationalist propaganda and promote collective understanding. Through the universality of *Alterswert*, people would appreciate the value of monuments in other countries as easily as in their own. Whereas 'artistic and historical' monuments were still tainted by elite, aristocratic overtones, the socialist character of *Alterswert* lay in its simple, tangible oldness.[11]

Alterswert was especially suggestive of religious values because of an inherent inconsistency: the fact that eventually 'nature's unhampered processes will lead to the complete destruction of a monument' and that thus 'the cult of *Alterswert* . . . stands in ultimate opposition to the preservation of monuments'. The pragmatic Morris had not worried about this hypothetical conflict between love of decay and conservation, but Riegl made it the basis of his key insight, stressing the ultimate transience, rather than (as with Ruskin) the quasi-eternal quality, of the built environment. For Riegl, every monument was a pantheistic work of nature; natural monuments and cultural monuments were part of the same phenomenon. Not only was *Alterswert* not concerned with any individual monument; it was not really concerned with the built substance or the facts of heritage at all. Instead of specific History, generalised Time was substituted.[12]

Around this recognition, he argued, one could build a new, universal, pantheistic conservation philosophy of *Menschheitsgefühl* (human sentiment). The monument itself, and its authenticity, had become almost irrelevant,

> a mere perceptible substratum whose task is to stimulate that feeling which is evoked in modern people by the inexorable ritual of being and passing away, of the arising of the particular out of the general and its inevitable reabsorption back in the general.

For Riegl, *Stimmung* (feeling) was not just an expression of religious-substitute empathy, but a way of emancipating the individual, overcoming the Darwinian framework of ruthless struggle by providing 'a suspicion of order above chaos, of harmony above dissonance'. Riegl argued that

> from the standpoint of *Alterswert* one need not worry about the eternal preservation of monuments, but rather one should be concerned with the constant representation of the cycle of creation, and this purpose is fulfilled even when future monuments have supplanted those of today.

Thus monuments 'should not be shielded from the loosening effects of the forces of nature, so far as these are realised through peaceful, regular continuity, and not in sudden violent destruction'. At the end, there beckoned religious faith: 'Conservation is, however, based on a true Christian principle, that of humble resignation to the will of the Almighty, which cuts down to size the presumption of impotent humanity.' For Riegl, the appropriation of monuments by nations was a damaging distraction from the primacy of the individual and the universal. Thus he criticised the pioneering 1902 decree by Archduke Ernst Ludwig of Hesse, restricting private rights over monuments.[13]

Riegl's conception of the power of monuments to convey universal fellowship anticipated strikingly the late-20th-century explosion in international conservation and 'charters'. But it also formed part of a thriving architectural and urbanist strand of the internationalist movement of around 1900–10, a movement especially prominent within town-planning, with its incessant international congresses and exhibitions (e.g. at the 1913 Ghent World's Fair), industriously driven by pioneering planners such as Patrick Geddes.[14]

Suspended uneasily between the national and international was another prominent element of the nascent Conservation Movement – the commercial exploitation of heritage for tourism. International cultural tourism still remained a largely elite matter, focusing on a limited range of well-trodden sites but steadily increasing in scope. In Rome, for example, a growing Anglo-American community of artists and intellectuals congregated around the Piazza di Spagna area ('il ghetto degli inglesi'), and in 1903 founded the Keats-Shelley Museum in a house beside the Spanish Steps. Turn-of-the-century Florence saw a similar clustering of cultivated travellers from Europe and America, attracted by a cultural centre on the cusp of modernity. But tourism at this stage was bound up with nationalism as much as internationalism: a pre-1914 cartoon by Hansi, for example, showed 'German' tourists in Alsace being teased by 'French' local inhabitants in 'traditional dress', even though the reality was more likely the other way round.[15] We will trace shortly the especially strong interconnections between heritage tourism and the cult of the 'Old Town'.

The inflation of *Heimat*: Paul Schultze-Naumburg

IN the turn-of-century Conservation Movement, the burgeoning force of nationalism far outweighed any efforts at international cooperation. We will examine the specific effects of this split on conservation policy later: here we are concerned exclusively with general ideological aspects. Of all the mass ideologies of the 20th century, it was above all nationalism that enlisted the Conservation Movement in its service, a service that involved complicity in warlike antagonisms as well as benign small-country emancipation and pride. As an 1899 conference put it, conservation had now become a matter of 'life or death'. The boldest developments in the 'mobilisation' of the Movement took place in Germany. There, the Ruskin tradition of sweepingly emotional argumentation, praising old buildings as redoubts of living national identity, was intensified into a foundation for the growing sense of *Gemeinschaft* (community) and *Heimat* (homeland, or home environment).[16] Ruskin and Morris had identified old buildings' authenticity exclusively with their fabric, which must be handed on inviolate to future generations. Now, by a further transference, that built legacy, through its reality and authenticity, was to support the destiny of the national community, providing that intense local intimacy that, within modern nationalism, complemented the abstract world of the nation and the state. The very vagueness of *Heimat*, and its concern with the whole environment, emphasised that it was there for everybody, not just an elite few.

This ideology appropriated and politicised the polarisation famously defined by Ferdinand Tönnies in 1889, between *Gemeinschaft* (community) and *Gesellschaft* (modern society). 'Living' old buildings and towns became physical embodiments of *Gemeinschaft*, able to bring everyone organically together, in opposition to the mechanical and alienating character of *Gesellschaft*. The main bête-noire of all this was utilitarian liberal capitalism, now spreading fast across Europe and North America – the first international 'globalisation' movement. Modern conservation was, and is, always most effectively mobilised by *threat*. Here, the threat was that of capitalist modernisation. In reaction to this menacing cosmopolitanism, each nation not only tried to outdo all others in its cultural excellence (as always since the French Revolution) but also trumpeted the fact through

144

maximal assertion of its own values. To help in this, nationalists appropriated the language of universal significance, combining claims of the all-embracing scope of the national heritage with forceful 'special' arguments, sometimes under the aegis of *Kunstgeographie*, or 'art-geography'. These could take on a distinctly circular character. For example, Wilhelm Pinder, art-historian and early mentor of Nikolaus Pevsner, claimed in his 1944 book, *Special Achievements of German Art*, that 'In the end, the whole of German art is one single special achievement.'[17] For people like Pinder, Art potentially included the entire cultural landscape of the *Heimat*, complementing the natural landscape in its didactic reinforcement of national rootedness (for example, through outdoors youth movements). Architecturally, as we will see, it was not individual buildings, such as cathedrals, that now most powerfully embodied the national community, but the secular, collective fabric of the nation's old towns.

The vehemence and modern radicalism of this purportedly anti-modern reaction in turn-of-century Germany stemmed partly from the fact that, here, all three explosive elements of Western modernisation – industrialisation, class society and nation-building – were telescoped together at once. Historic towns suddenly became industrial cities – Frankfurt-am-Main, for example, doubled in size from 100,000 to 200,000 between 1874 and 1895, and doubled again by 1910. The regional diversity of the country was challenged by efforts at cultural unification. As early as the 1850s, commentators such as Wilhelm Riehl had begun to codify a discipline of German folk studies and ethnography, to counterbalance the chaos of the failed 1848 revolution and the threats of urbanisation. By the turn of the century, the scope of the *Heimat* seemed all-embracing. Where the late-19th-century concern was to build statues of great leaders, now the focus was the national community as a whole. Within monumental design, this was epitomised in the shift, within 20 years, from the 1897 Kyffhäuser Monument to Wilhelm I in Thuringia to Bruno Schmitz's gigantic Völkerschlachtdenkmal of 1913.[18]

The concept of a culture of mass collective memory had been first developed in the very different context of post-Civil War America. But in turn-of-century Germany, the vastly extended scope of national culture became bound up with a wider pessimism among the bourgeoisie, sickened with crass materialism and growing state interventionism. Architecturally, the main targets of dislike were 'jerry-building', eclectic historicism and restoration. Against these, collective memory and *Heimat* protection provided a wholesome protection, stimulated indirectly by Ruskin's emotionalism and the Arts and Crafts ideal of 'simplicity' and the 'homely' interior.[19] The anti-Enlightenment yearnings of romantic German nationalists around 1800 had now finally culminated in a fully fledged *Heimat* Movement, led by the *Bund Heimatschutz* (founded in 1904), and shading into an impassioned ideology of 'Blood and Soil'. Like the Arts and Crafts, Pugin and others, this movement was shaped by the ideal of a lost golden age – in this case, the pre-industrial '*Biedermeier*' of around 1800. Even more than the Arts and Crafts, the *Heimatschutz* movement venerated the rural and hated the urban: writers such as Prof. Ernst Rudorff or Hugo Conwentz argued passionately for both natural and cultural conservation, and J Reimers proclaimed in 1911 that 'the soil on which love of the fatherland could grow, is the same soil in which the *Denkmalpflege* movement has grown'. Heady, mystical rhetoric ascribed life-or-death significance to the nation.[20]

The central figure in this fevered activity was the first chairman of the *Bund Heimatschutz*, the architect Paul Schultze-Naumburg. In 1908, echoing Schinkel's rhetoric a century earlier, he denounced the threats to the *Heimat* in apocalyptic terms, laced with a touch of Ruskinian anti-restorationism:

> The threat confronts us that Germany may lose its character as our trusty *Heimat*, and be reduced to a barren wilderness. If things carry on as they are, our cities and villages will degenerate into working-class colonies and poorhouses. Our rich cultural inheritance, handed down by our forefathers, will be destroyed or 'restored' into unrecognisability. And in the process, all its original natural beauty and distinctiveness will vanish.

Like Pugin, Schultze-Naumburg published a polemical text of good and bad pictures, *Kulturarbeiten* (1906), intended to 'combat the hideous devastation of our land in all areas of visual culture'. Its photographs contrasted the simple classicism of the *Biedermeier* with fussy late-19th-century eclecticism. But we should not overemphasise the society-wide impact of this movement, which was largely confined to an artistic and intellectual elite, and their houses and interiors.[21]

Dehio versus Riegl: *Heimat* versus *Menschheit*

A different dynamic operated in Austria, a culturally diverse society where the work of the Central Commission for Research and Preservation aimed to bolster the heritage legitimacy of the multinational Austro-Hungarian Empire, by fostering local diversity. Riegl's argument, with its self-negating overtones and its suspicion of strong, nationally specific concepts of authenticity, fell foul of German conservationists, whose foremost theorist, the Strasbourg-based (and Tallinn/Reval-born) art historian Prof. Georg Dehio, engaged Riegl in a significant debate in 1905. In his famous '*Kaiserrede*' rectoral speech delivered on the 56th birthday of Wilhelm II (and in the presence of the Emperor himself) at Strasbourg University, Dehio contested Riegl's dreamy vision of conservation,

Figure 5.3 Schultze-Naumburg's *Kulturarbeiten* (1906): defending the *Heimat*

(a) *Gegenbeispiel*: 'bad new' house; (b) *Beispiel*: 'good old' house

and demanded a more pragmatic, realistic, socially and nationally committed approach, combining *Menschheit* with *Heimat*. While openly admitting the modernity of this concept – indeed, he claimed, the very word *Denkmalpflege* 'can hardly be older than twenty-five years' – he was confident in its enduring power. Unlike the subjectivity of romantic antiquarianism,

> Its ultimate motivation is the respect for historical being as such. We do not conserve a monument because we consider it beautiful, but because it is a piece of our national existence. To protect monuments is not a matter of enjoyment but of piety. Aesthetic and even art-historical judgements fluctuate; here, we have found an unchanging mark of worth.

Bound up with this 'worth' was the primacy of the nation, or race (*Volk*): Dehio later declared in 1930 that, in all his heritage activity, 'my real hero is the German *Volk*'. Typically of the Bismarck conservative tradition, Dehio advocated 'socialist' measures, such as restriction of property rights, in the paramount interest of the nation. His faith in the continuity of objective historical values was firm, although he admitted the hypothetical possibility of 'monstrous cultural catastrophes'.[22]

Coming from the borderland of the new *Reich*, on territory newly recovered from the French – whom he labelled in 1914 as 'destroyers' of European art – Dehio's overtly nationalist rhetoric contradicted the urbane cosmopolitanism of Riegl, who duly riposted that Dehio was trapped under 'the spell' of the 19th-century notion that looked for the significance of the monument in the 'historical moment'. Uncannily anticipating the violent border changes of the 20th century, including the return of Alsace, and Höhkönigsburg, to France, and the occupation of Tallinn by the USSR, Riegl argued that monuments could not be kept as 'one's own', and that national 'egoism' must eventually give way to the wider picture: 'Monuments [will] attract us from now on as testimonies to the fact that the great context, of which we ourselves are part, has existed and was created long before us.' Everything could now potentially become a monument, but the experience of *Alterswert* in every monument was always identical.[23] Inexorably, humanity would be drawn towards this contemplative cult of cultural heritage by 'the existence and the general diffusion of a feeling, akin to religious feelings, independent of special aesthetic and historical education, inaccessible to reasoning, a feeling that would simply be unbearable to have unsatisfied'. Riegl died in that same year, preventing further development of this argument, some of whose aspects have resonated in subsequent conservation practice. For example (see Chapter 12), the upsurge in the concept of 'intangible heritage' since around 1990 poses the issue of whether the contemplation of transience in the built environment requires old buildings, or physical oldness, at all.

The revolt from restoration in turn-of-century Germany

THIS theoretical clash for the soul of conservation within the German-speaking world was only indirectly mirrored in the world of day-to-day practice, where the turn of the century saw numerous advances on a broad front. The Ruskin/Morris polemic against restoration began to have serious effect in Germany only from the 1890s: the first German edition of the *Seven Lamps* appeared in 1900. Following in the wake of Cologne Cathedral, the assertive wealth of the post-1871 unification era (*Gründerzeit*) stimulated an upsurge in 'completion' projects, such as the Lambertikirche

147

spire(s) in Münster (1898), or the addition of twin towers to Meissen Cathedral in 1904–8 by Carl Schäfer. There were significant restorations and facsimile projects, too, such as the rebuilding of St Pantaleon Church in Cologne from 1890, originally dating from the 10th century but latterly rather Baroque in form: the scheme conjecturally rebuilt the 10th-century towered *Westwerk*. Probably the most extreme of these projects was the restoration of Dankwarderode Castle in Braunschweig, whose *c.*1160 palace block, previously embedded in a Romanesque/Baroque building, had been left exposed as a ruin in an 1873 fire. In 1887–1906, this ruin was 'restored' in a Romanesque style as the local residence of Prince Albrecht of Prussia, including a fabulously sumptuous two-aisled banqueting hall (designed by City Architect Ludwig Winter); the project involved removal of all the remaining fragments of the 12th-century ruins, owing to their alleged structural instability![24]

The Marienburg, naturally, became a prime target for this Indian summer of Viollet-style restoration schemes in Germany, in a campaign (in 1882–1922) overseen by architect Conrad Steinbrecht. Far from a forgotten backwater, the castle was now the spearhead of an aggressive *Ostpolitik*: 1884 saw the foundation of an Association for the Restoration and Embellishment of the Marienburg, and in an 1896 drawing, Steinbrecht proposed the total rebuilding of the castle. In 1902, exactly a century after Frick's visionary sketch, the Emperor Wilhelm II, on one of his over 50 visits to the complex, summoned the German *Volk* to 'marshal its national resources to resist the Polish

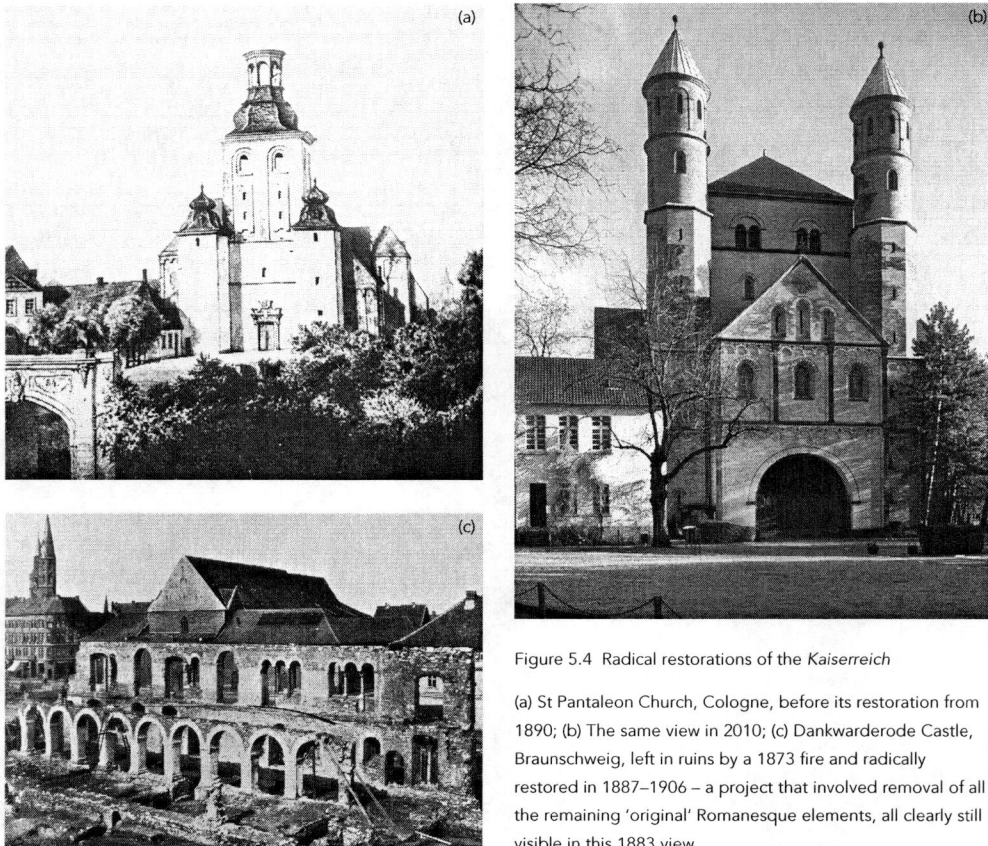

Figure 5.4 Radical restorations of the *Kaiserreich*

(a) St Pantaleon Church, Cologne, before its restoration from 1890; (b) The same view in 2010; (c) Dankwarderode Castle, Braunschweig, left in ruins by a 1873 fire and radically restored in 1887–1906 – a project that involved removal of all the remaining 'original' Romanesque elements, all clearly still visible in this 1883 view

hordes', and in World War I the rebuilt castle assumed an almost mythic status as Germany's eastern redoubt.[25]

By the turn of the century, though, under the influence of Ruskin's translated works, a full-blooded reaction by German conservationists against this belated restoration frenzy was underway. Partly, that reaction stemmed from Ruskin's demand for 'living' monuments, as opposed to 'dead', museum-like structures. Henceforth the word '*lebendig*' (living) became the single most important term in the German conservation vocabulary, denoting the most vital kind of authenticity, whether applied to a castle or to a complete old town. Initially, this living/dead dichotomy was associated mainly with the new opposition to restoration schemes: in 1901, in the debate over restoration of

Figure 5.5 Inventorisers and preservers of the *Kaiserreich*

(a) Portrait of Dehio; (b) Front page of the fourth of Dehio's *Guidebooks*; (c) 1889 area inventory of part of Württemberg – part of the national strategy begun in the 1880s; (d) Paul Clemen, pictured in 1936

149

Heidelberg Castle (see below), Dehio criticised rebuilding of 'living' monuments, which could only result in 'a dead, academic abstraction'. And it was, appropriately, Dehio who led the way out of the old, grandiloquent phase of heritage nationalism towards a new anti-restorationism, responding to the demands for 'simplicity' and 'honesty', as well as the growing sense of external threat, and arguing that restoration was an 'illegitimate child' of 19th-century architectural historicism. Dehio's summary of this ideology in his famous 1905 Strasbourg speech – '*konservieren, nicht restaurieren*' ('Don't restore, but conserve') – passed into the mythology of modern German conservation as its founding definition, neatly concealing its partly English origins.[26]

But Dehio was only one of a brilliant constellation of turn-of-century conservators and architects in Germany, whose attitudes were coloured by the tensions between Germany's old cultural diversity and the new, nationally unifying pressures. Many of the most internationally admired aspects of German conservation were bitterly opposed by regional representatives. These national initiatives included the 1878 proposal for a *Reich*-wide inventorisation strategy, promoted by the Emperor despite regional boycotts (the first inventory being begun in Baden in 1887); the 1899 foundation of the world's first professional conservation journal, *Die Denkmalpflege*, attacked by some as a mouthpiece of Prussian hegemony; the instigation in the following year of an annual Conservation Convention (*Tag für Denkmalpflege*); and Dehio's initiation at the inaugural convention, in Dresden, of a series of area handbooks of historic monuments, the *Handbuch der deutschen Kunstdenkmäler* (the famous 'Dehio guides', whose first volume appeared in 1905). Prior to World War I, as German foreign policy became more assertive, the centralising trends strengthened. In 1910–13, the *Bund Heimatschutz* and the Conservation Conventions were amalgamated; in 1923 their respective journals were merged.[27]

Many of the most 'progressive' German conservation initiatives combined regional and national tendencies. Stimulated by the Conventions, a series of conservation laws were passed by regional governments anxious at Prussian domination, beginning with the Hessen-Darmstadt law initiated by Freiherr von Biegeleben in October 1902. This restricted private rights over 'classified' monuments and inspired others (Sachsen-Altenburg in 1909, Oldenburg in 1910, Württemberg in 1914). The same years saw tough Prussian laws against 'defacement' of the urban and rural *Heimat*, with local authorities given strong powers of planning regulation in 1896 (strengthened in 1902 and 1907).[28] The same regional-national ambiguity applied to many conservation leaders besides Dehio, such as Paul Clemen, first provincial conservator (from 1893) in Prussia's Rhineland province, who famously claimed that 'the Rhine has become my destiny', or Cornelius Gurlitt, eminent Dresden historian of the Baroque; Gurlitt was an early critic of restorations, in 1895, and in 1897 supported a modern Jugendstil scheme (by Rudolph Schilling and J W Graebner) for internal rebuilding of Dresden's Kreuzkirche after a fire. The turn of the century saw other significant victories of conservation over restoration: the Berlin-based architect and design reformist, Hermann Muthesius, reporting in 1897 on his reconnaissance of advanced English practice, advocated a combination of SPAB Anti-Scrape principles with a generally 'scientific' approach. In conservation, the lead-up to World War I was associated not with architectural bombast but with demands for restraint.[29]

The growing anti-restoration consensus among conservation leaders culminated in a celebrated debate of 1901–5 over the proposed rebuilding of one of the Renaissance palace buildings in Heidelberg Castle, the Ottheinrichsbau. Much of the castle had been burned by French troops in 1689 and 1693, and (unlike Speyer Cathedral) not restored, but left as a romantic ruin. The late 19th century saw nationalistic calls for rebuilding, as a 'duty of the entire German race', of 'this jewel of

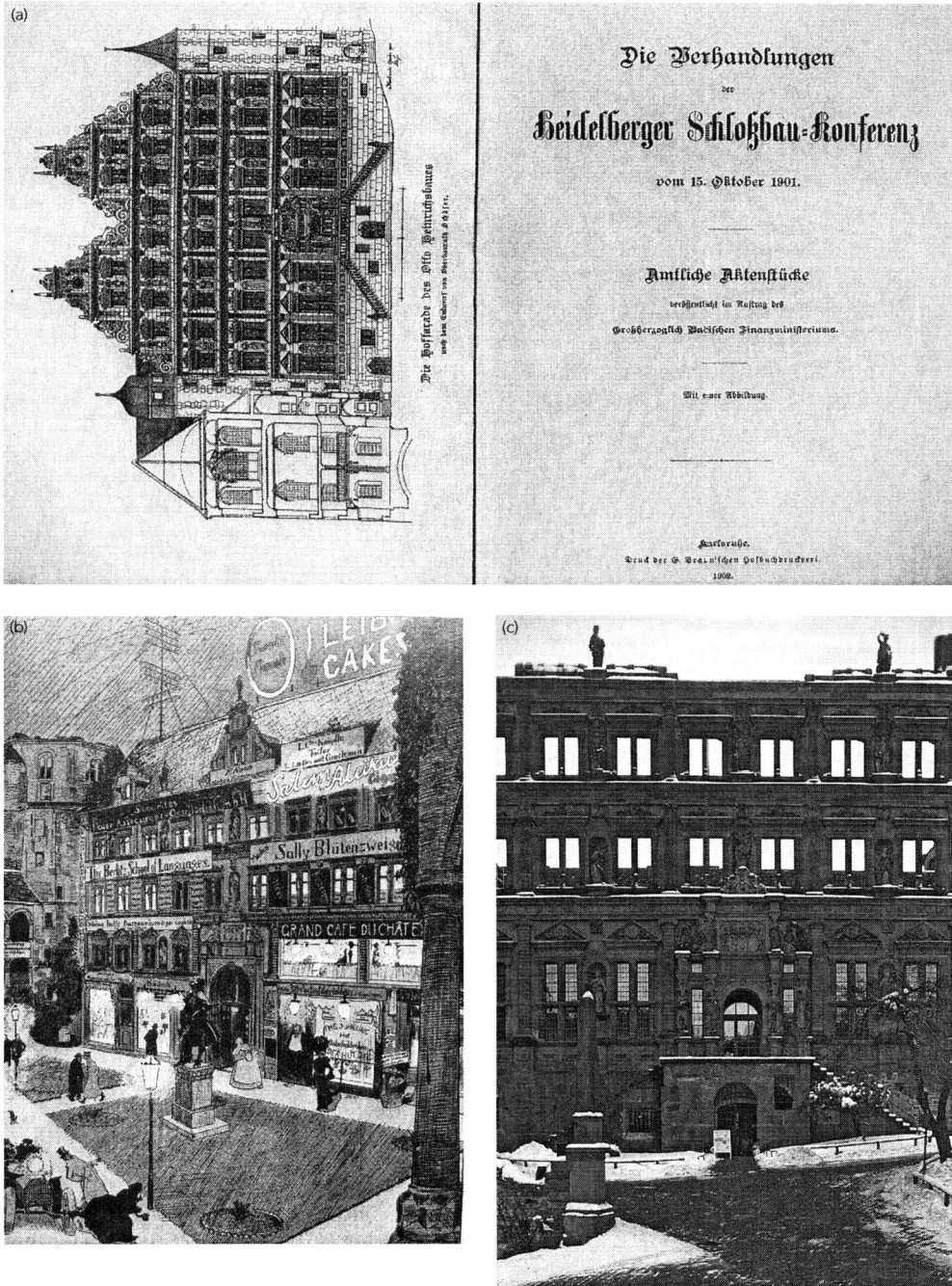

Figure 5.6 Heidelberg's Ottheinrichsbau: restoration frustrated

(a) Programme of the 1901 conference on the Ottheinrichsbau rebuilding controversy; (b) Contemporary cartoon by opponents of the rebuilding, equating the rebuilding scheme with commercialised desecration of the site; (c) The Ottheinrichsbau ruin in 2008

German architecture, as a monument to the regained power and splendour of the fatherland'. After one section was restored and reroofed from 1895, 1901 proposals for reconstruction of the Ottheinrichsbau, justified by claims that the surviving façade would become unstable if left unroofed, stirred up vigorous controversy, including lengthy exchanges in the *Deutsche Bauzeitung* on the issues of structure and artistic significance versus historic integrity. A special conference was held about the controversy in 1901 and, in 1905, the entire proceedings of the sixth Conservation Convention, at Bamberg, were devoted to the debate; no less than 227 scholars picked over it from every conceivable viewpoint, some impassioned and others highly technical. Some (including, predictably, Bodo Ebhardt) firmly backed the scheme, but the climate of opinion was shifting quickly. Eventually, Georg Dehio proposed that the ruin should simply be left as it was and carefully preserved for another century, 'and then let us open a new debate on the subject of Heidelberg'. Instead, conservationists should refocus on national heritage challenges, as 'threatened Heidelbergs are everywhere'.[30]

National-romantic conservation in Austria and elsewhere

ACROSS 'statist' Europe, equivalents to *Heimatschutz* formed a central element in the homogenisation of heritage thinking around an agenda of nationalist assertion and heightened consciousness of 'borderlands'. In each country, the precise formula varied according to its geo-political situation. The Austro-Hungarian Empire witnessed especially diverse outcomes. The turn of the century saw a sharp split between cosmopolitan and regional/national-romantic ideas, and between pan-German and Slavic brands of *Heimat*. In the Czech lands, the 1890s saw growing public enthusiasm for protection of medieval, Renaissance and Baroque monuments, and the foundation in 1900 of a Club for Old Prague. But subsequent years became more dominated by modern architectural influences, especially Cubism.[31] Riegl himself was, of course, uncompromisingly opposed to German heritage nationalism: he even condemned the 1902 Hesse monument law as a step towards 'unitary nationality and culture'. His own official casework, however, took a careful middle position on the restoration question. The revived proposal of 1902 by the *Dombauverein* for restoration of the Riesentor on the Stephansdom – fiercely opposed by Secession members – was cautiously criticised in the Central Commission's newsletter (edited by Riegl), and the following year he became the Commission's first salaried chief conservator. In his arbitration on the plan for *Freilegung* of the palace and cathedral of Split by demolishing old houses (as was happening at the Athens Acropolis) Riegl recommended refusal of the proposal, arguing that the 'picturesque street views' and 'atmosphere' of the old town should have priority. In a pioneering Austrian law of 1903, he proposed that all publicly owned buildings over 60 years old should be automatically protected – a bold measure eventually enacted in 1923 – and that other buildings should be designated according to a range of pragmatic criteria, including artistic, historical and 'national' significance.[32]

In conformity with the generally consolidatory course of conservation theory from now on, Riegl's successor as General Conservator and Art History Professor at Vienna University, Czech-born Max Dvořák, elaborated his ideas into a flexible system of state protection, beginning with a Dehio-style inventory. The first volume was published in 1907: Dvořák described this as a survey of *Denkmallandschaft* (monument landscape). He followed a compromise line between Dehio and Riegl in his official conservation manual, the *Katechismus der Denkmalpflege* (Catechism of

Conservation), published in wartime Vienna in 1916 and 1918. This combined Schultze-Naumburg-style good-versus-bad presentation with an urbane variant of *Heimatschutz*, attuned to a dissolving empire. This aimed to integrate architecture, art and nature conservation into a wider concept of heritage landscape. Upholding the entrenched Ruskin anti-restoration ethos, he incorporated into his value system a special concept of the *Gewachsenes Kunstwerk* (evolved art-work), which stressed the special characteristics of multi-phase, organic development: unlike late-19th-century restorers like Mocker (whose work he bitterly criticised), Dvořák especially cherished hybrid ensembles such as Gothic churches with Baroque fittings. In Hungary, too, the Baroque was actively cherished: the new Dual Monarchy, in the 1870s and '80s, saw a proliferation of national monument committees and commissions, instigated by figures such as Imre Henszlmann, but the focus was not on 'humble' old or new timber cottages but on the Baroque era's everyday classical architecture.[33]

In the empire's smaller constituent countries, emancipatory offshoots of the national-romantic movement drew on old houses as inspiration for new vernacular-style architecture. In 1890s' Slovakia, responding to perceived Hungarian oppression, traditional timber Carpathian buildings spurred a new Arts and Crafts style, proselytised especially by Dušan Jurkovič. In the Austrian-controlled section of Poland, the relatively liberal cultural regime allowed development of an indigenous Arts and Crafts style inspired by wooden cabins in the Tatra Mountains, the *Styl Zakopiański*, by Stanisław Witkiewicz and other Arts and Crafts architects, and its dissemination via new middle-class villas.[34] At the same time, a grandiose and most un-'vernacular' palace-restoration project, equalling the Marienburg in ambition, was underway at the Wawel Castle in Krakow, former seat of the Polish kings. There, reacting against a lengthy process of militarisation and fortification by the Austrian army during the 19th century, a campaign for restoration had already emerged by 1830–3. It was only in the 1880s and '90s that a proposal to convert the Wawel to a Galician imperial residence finally took shape. It was transferred to royal use from 1905. By then, a coalition of scholarly Polish architects, including chief conservator Zygmunt Hendel as well as Tomasz Prylinski and Slawomir Odrzywolski, had deflected Austrian anti-restorationist attacks and won approval for a steep-roofed, towered reconstruction. The scheme's centrepiece, executed externally in 1907–14 (with interiors 'recreated' in the 1920s–30s under chief conservator Adolf Szyszko-Bohusz), was the opening out of the slender arcades around the main courtyard – a bold concept that required the dismantling of the roof supports and the substitution of a light steel structure.[35]

Elsewhere in Europe, similar patterns applied to other emergent or small-country national-isms. In Finland, for example, the early-19th-century work of Elias Lönnrot in collecting and transcribing the *Kalevala*, the national epic poem, foreshadowed a wave of emancipatory efforts by artists and architects in the 1880s and '90s, all seeking an 'authentic' Finnish visual culture: key participants included composer Sibelius, artist Akseli Gallen-Kallela and architects Lindgren, Gesellius, Sonck and Saarinen. In 1890 Gallen-Kallela and Louis Sparre toured Karelia, collecting notes on indigenous architecture, and Sonck drew on the inspiration of Karelian farmsteads for his rural houses. In Gesellius, Lindgren and Saarinen's National Museum in Helsinki (1905–10), the wing devoted to national art was shaped like a rubble medieval church. In 1890s' Catalonia, likewise, there was pressure, led by writer Antoni Aulestia Pijoan, for private and public conservation initiatives, linked to the rising interest in Catalan vernacular buildings: eventually, in 1914, the newly founded devolved Mancomunitat (Commonwealth) administration of Catalonia established a service of cataloguing and conservation of monuments, headed by architect Jeroni Martorell i Terrats.[36]

153

Figure 5.7 Krakow's Wawel Castle: restoration realised

(a) View of the castle in the 1860s, still used as an Austrian military fortress; (b) Watercolour painting by Tadeusz Ajdukiewicz of the 1880 visit of the Emperor Francis Joseph I, during which the castle's conversion to a royal palace was first proposed; (c) Zygmunt Hendel's 1906–9 plans for the castellated reconstruction of the castle; (d) 2010 view of the castle courtyard, from the same viewpoint as (b) above

Regional and central systems in Italy and France

Turn-of-century Italy resembled Germany in its combination of intense regionalism and unifying nationalist ideologies, such as the cult of Garibaldi, sacred symbol of the new secular nation. However, its main conservation innovations were, if anything, rather 'in advance' of those in Germany and Austria. In its post-*Risorgimento* eagerness to shake off its old status as a playground of northern European antiquarians, and appropriate its own heritage, it began synthesising relevant architectural conservation theory from France and Britain from the 1870s and adapting it to its own characteristic conditions of urban multi-layering. Traditional classical archaeological work also continued apace, with ongoing excavations at Pompeii and Herculaneum. In Rome itself, the whole Forum was now uncovered, during excavations that began in 1870 and carried on until World War I, led by archaeologists such as Rodolfo Lanciani and Giacomo Boni; a complete 1:1,000 archaeological map was published in 1893–1901.[37]

Unifying pressures in Italy included initial attempts at centralised national legislation. A 1902 Monument Act (and 1913 regulations) established a single procedure for identifying and designating monuments, later subdivided in 1939 to differentiate between publicly and privately owned buildings. But not everyone in Italian architectural circles backed the consensus of support for a generally conservationist position: the Futurists rhetorically demanded that old, ornate buildings should be burnt and destroyed.[38]

As in Germany, the Italian conservation movement was organised on fundamentally regional lines. The country's most important theorist, whose writing significantly pre-dated that of Riegl, Dehio and Dvořák, was Milan-based: Camillo Boito, professor of architecture and designer of new buildings in eclectic styles. He cautiously adapted the Ruskinian and Anti-Scrape ideology to Italian conditions, opposing any uncritical embrace of either, and claiming that while Viollet's restorations had negated history, Ruskin had mistakenly assumed history had ended. In a 1893 paper, 'I restauri in architettura', Boito invented a comically exaggerated Socratic dialogue between pro-Viollet and pro-Ruskin extremists, highlighting the self-contradictions of both sides: when the latter attacked skilful, deceptive restorations and argued that 'pompous' plaster Baroque reconstructions were preferable because easily removed (a harbinger of the doctrine of 'reversibility'), the latter riposted, 'Oh! That's a good one! Better, thus, an ass of a restorer than a learned restorer!'[39]

What was instead needed, Boito contended, was a theory of 'critical restoration' (*restauro critico*) that recognised that any interaction must be frankly grounded in contemporary values and prejudices. Such a philosophy could integrate monuments into the architectural and social reality of the 'living' city, applying an ordered 'hierarchy of intervention' to allow various appropriate levels of conservation work. Buildings, he argued (echoing Ruskin), were not just dead objects but also 'living' documents of human achievement that one should be able to 'read', if each phase in their history was shown proper respect through a critical philological approach distinguishing different layers of intervention. The old technique of highlighting monuments by clearing vast spaces around them was rejected. But, Boito contended, there should be no confusion about the contribution of the restorer. The 1883 *Carta italiano del restauro* (Italian Restoration Charter), which he co-authored, and which transmitted his principles into official conservation practice, argued that, where restoration was admissible, a distinctive 'restoration style' was essential, to accentuate 'stylistic difference' between old and new work. Boito's ideas shaped the later Italian legislation of 1902.[40]

155

However, the implementation of these precocious ideas on the ground, by a variety of regionally based architects, firmly kept one foot in the world of Viollet. Following Boito's hierarchy of approaches, the Turin-based Alfredo d'Andrade adopted a very conservative line with Roman monuments, but treated medieval or later buildings in a much more diverse, creative way. D'Andrade's restoration of the Palazzo Madama in Turin (from 1884) juxtaposed conservatism and scenographic freedom on opposite sides of the same building, with a preserved Baroque facade on one side and an imaginatively restored, turreted castle on the other side. A more straightforward throwback to the Viollet outlook, against Boito's principles, was the work of Luca Beltrami of Milan. His famous 1903–12 project to rebuild in facsimile the Campanile in Venice, after its sudden collapse, was based on substantial previous restoration experience. This included his bold reconstruction of Milan's Castello Sforzesco (as a museum/art gallery) from 1893, with a Viollet-style recreation of the so-called Filarete tower, the leitmotiv of the restored complex, in 1903–5. The original 1452 tower had only stood for 69 years before being destroyed in a munitions explosion, and the flamboyantly spiky reconstruction was based on sketchy sources. The highly differentiated local Italian permutations of the Conservation Movement's international palette of theories were seen at their most exaggerated in Bologna, where artist and architect Alfonso Rubbiani combined elements of Pugin, Morris and Viollet in an idiosyncratic but highly effective way. On the one hand, Rubbiani campaigned passionately against heritage demolitions and 'alienating' new classical apartment blocks, and praised nature and the crafts. On the other, he pursued a programme of Viollet-like, battlemented, medieval-style restorations of key buildings in the historic centre, including the Palazzo di Re Enzo (1905) and the Palazzo dei Notai and Palazzo Comunale (1908), unifying and scenographically enhancing these sometimes nondescript old complexes into articulate participants in a modern urban theatre. Rubbiani's scenographic, yet conservative reshaping of historic Bologna paralleled the work of foreign contemporaries, such as Patrick Geddes in Edinburgh (see below). Within Italy, it also anticipated the interwar Fascist attempt to revive the sturdy medieval civic spirit through creative conservation.[41]

In Rome, where concern was naturally most intense, 1890 saw the foundation of a SPAB-style 'Artistic Association', chaired by Giovanni Battista Giovenale, for the protection of monuments and enhancement of the old city. Its early interventions fluctuated between restoration and repair, no quarter being shown to Baroque features: the early-18th-century façade of Sta. Maria Cosmedin was removed by Giovenale in an Early Christianising restoration of 1893–9. From around 1910 until World War II, the central conservation figure in Rome was Gustavo Giovannoni, a planner and academic who vigorously advocated scientific teaching and research in conservation, as well as emphasising (as we will see shortly) sensitive, study-based conservation of historic areas, including humble as well as grand buildings.[42]

In France, too, there was a strong dichotomy between centralising national tendencies and an increasingly sharp feeling of regional, rural authenticity. Growing support for the ideas of Ruskin was tempered by the continuing resilence of Haussmann-style stately urban replanning. More generally, the overwhelming prestige of the French classical tradition restrained the diverse neo-medieval enthusiasms that were bubbling up elsewhere in northern Europe. The already formidable apparatus of state power in France was further reinforced around the turn of the century by complex administrative and legal measures. The 1887 conservation law was the first text to propose the present-day system of formal *classement* (listing) of both buildings and moveable works of art (a special concern in France), supported by exhaustive research and preparation of comprehensive

Figure 5.8 Turn-of-century Italian restorations

(a) The Castello Sforzesco in Milan was restored by Luca Beltrami as a museum/art gallery from 1893, including a dramatic recreation of the 1452 'Filarete Tower' in 1903–5; (b), (c) Sta Maria Cosmedin, Rome, reconstructed by G. Giovenale in Early-Christian style in 1893–9

dossiers de recensement. In practice, this law proved unexpectedly weak, but the 1905 disestablishment of the church bolstered the state administrative apparatus by removing the preservation responsibilities of the Ministry of Religious Worship, leaving the churches in the care of commissioners. A threefold increase in requests for state intervention followed, and a rise in *classements* from about 20 to 200 annually. All this was regularised in the famous French law of 31 December 1913. This radically extended the scope of classification, allowing provisional designation without compensation, on the basis of the concept of 'public interest', greatly extending the criteria of eligibility for classification and allowing the possibility of a protected perimeter area around classified monuments – a potential only realised in a later law of 1943. Where the list of 1840 had designated 880 monuments (increased to 1,700 by 1900), the first list of the 1913 law proposed 4,800 – mostly religious buildings. Even after 1900, with the translation of Ruskin into French, regional discourses of architecture in centralised France were never as prominent as in Germany: *Heimat* was a stronger word than *terroir*. Nevertheless, regional voluntary groups burgeoned, as in the outspoken campaigns of the Société des Toulousains de Toulouse from *c.*1905 to protect the brick architectural heritage of *la ville rose.*[43]

The French twin-track model of protection of moveable and immoveable heritage was also adopted in Japan, the only country outside the European/Western orbit to develop a fully fledged state protection system. This was launched following the Meiji Restoration by an 1871 government proclamation, an 1880 grant-aid fund and 1888 national survey, and culminated in the 1897 Ancient Temples and Shrines Preservation Law. Architectural professors and graduates from the Imperial University were charged with developing an indigenous conservation profession, and over 1,000 buildings were designated by 1929 and 500 repaired, aided by French-style 70–90 per cent government grants. Pioneering temple restoration projects (e.g. Shin-Yakusuji Hondo in 1897–8) provoked authenticity controversies that echoed the Ruskin-Viollet opposition: an authoritative 1901 paper by Prof. Zennosuke Tsuji of the Imperial University championed material authenticity. The outstanding exception to this rule was the Ise shrine, the only complex to have perpetuated the Shinto principle of periodic material reconstruction. All similar temples abandoned the practice in the mid-19th century, and were subsequently preserved under conventional criteria of material authenticity. The precocity of Japanese policy can be judged by comparison with Republican China, where a government Antiquities Preservation Section was established in 1912, but the first legislation (the Ancient Artefacts Law) was not passed until 1930.[44]

National Trust: private initiative in turn-of-century Britain

IN the anglophone countries, the position was significantly different, reflecting the ingrained laissez-faire ethos, the deep British involvement with overseas imperialism, the precocious and comprehensive 'improvement' of both town and country throughout Britain, and the emergence of a distinctive preservation movement in the USA.

Despite the importance of English movements such as Arts and Crafts in the growth of the continental *Heimat* ideas, and although a more assertive English/British nationalism grew around 1900 (marked in architecture by the 'Wrenaissance' neo-Baroque style), Britain saw nothing like a *Heimatschutz* movement. Here the definition of 'historic' was narrower, focused on famous people, the reigns of monarchs, military victories and the empire: there was no equivalent to the intense, all-embracing character of the *Heimat*, or to the continental outpouring of speeches, pamphlets or books on conservation. In Scotland, the 'Unionist-nationalism' of the years of Walter Scott was focused on the archaic field of heroic kings and leaders, with conservation efforts concentrated in the thriving castle-restoration movement.[45] Key heritage discourses, far from emphasising local *Heimat*, were broadcast throughout the empire and the anglophone world, especially through tourism. The cult of Shakespeare proliferated globally following the mid-19th-century recasting of the Birthplace House. A new Memorial Theatre opened in 1879, and in the Queen's Golden Jubilee year of 1887, a US philanthropist paid for the erection of a memorial fountain: numerous lavish photo-books documented the 'Home and Haunts of Shakespeare' for an American audience. The year 1891 saw a special Act of Parliament to empower restoration of other 'Shakespeare sites' around Stratford, including the 16th-century 'Anne Hathaway's Cottage', established as a tourist site from the 1820s and bought in 1892 by the Birthplace Trust for preservation.[46]

Overall, the emphasis in Britain was still on voluntary efforts rather than state intervention – as highlighted in the impact of SPAB. The private emphasis was strongly reflected in legislation, where, before 1931–2, the only effective measures were 'ancient monuments' acts, empowering state 'guardianship' of archaeological remains. Although limited protective action had already been attempted in Ireland following the 1869 Irish Church Act, the first UK-wide monuments preservation bill, introduced by Sir John Lubbock in 1873 under pressure from archaeological societies, was finally passed in 1882. It was largely ineffective, partly because ecclesiastical buildings were excluded. General Pitt-Rivers, first inspector of ancient monuments, had to work without any supporting staff. After his death in 1900 the office lapsed for a decade, until the appointment of Sir Charles Peers in 1910. He committed the inspectorate to an extreme Morrisian repair-only approach to guardianship. Although a further Ancient Monuments Act of 1910 allowed proper protection of medieval monuments, it still excluded churches and occupied buildings.[47] Only from 1913, with a new, more comprehensive law (the Ancient Monuments Consolidation and Amendment Act), was it possible to protect later buildings, through advisory lists of noteworthy monuments and imposition of *ad hoc* 'building preservation orders' without owners' consent. Provoked by the threatened dismantling of Tattershall Castle in 1911 for shipping to the USA, the 1913 Act tried to give the Ancient Monuments system some teeth, while avoiding the public opposition stirred up by the more draconian French system. But even its powers were shown as inadequate in 1913–14, when a preservation notice on an early Georgian house at 75 Dean Street, Soho, was overturned by court order; the interior was subsequently dismantled and shipped to America.[48]

Building on the 19th-century proliferation of local amenity societies, the turn of the century saw a further upsurge in local efforts. In 1911, residents of the picturesque timber-framed East Anglian village of Lavenham mounted a protest campaign against demolition of a key building, including posters and public meetings. Voluntary efforts also dominated contemporary archaeology. One extreme example was in 1898–1905, when excavations of the Dunbuck Crannog on the Firth of Clyde (previously investigated in the 19th century by amateur groups), undertaken by artist-archaeologist William Donnelly, became mired in bitter controversy, with rival archaeologist Dr Robert Munro arguing that its chief finds were fakes – a claim later largely substantiated.[49]

The 1880s and '90s saw a boom in systematic photography by private societies of local areas and geological features of Britain, including heritage – the so-called 'survey movement'. Prominent here was the Warwickshire Photographic Survey, begun in the 1890s by W Jerome Harrison, who championed a 'science of truth' based on a 'taxonomy' of photographs. In 1895, Sir Benjamin Stone founded a National Photographic Record Association, with ambitious plans for publication and archives, but the public remained chiefly interested in views of picturesque tourist sights. In the field of architectural drawings, the eminent Scottish architect, Sir R Rowand Anderson, inaugurated (in 1886) a long-running recording initiative, the National Art Survey of Scotland, whose first lavish volume of drawings was published in 1921.[50]

Much the most important of Britain's turn-of-century voluntary heritage initiatives was the establishment of the National Trust for Places of Historic Interest or Natural Beauty in 1894–5 – an organisation intended both to preserve buildings and landscapes and to help revive rural life. The

National Trust was a belated realisation of Ruskin's call for foundation of a society to buy threatened buildings and land. It also sprang logically from the 1867 establishment of the Commons Preservation Society. In 1884, one of the CPS's main organisers, Robert Hunter, called for the foundation of a body empowered to hold land permanently for the people. In 1894 he joined philanthropist Octavia Hill and Rev. Canon Hardwicke Rawnsley, defender of the Lake District against railway intrusion, in convening an inaugural meeting at Grosvenor House. The following year, the new trust was registered, with Hunter as chairman, Rawnsley as secretary and the Duke of Westminster (a friend of Hill) as president; of its 45 council members, 15 were Liberal MPs. The NT's foundation was also prompted by modernisation of local government in the 1880s, creating powerful county councils whose programmes threatened old rural buildings and landscapes. And it formed part of a range of new cultural-amenity groupings, such as the Society for Protection of Birds (founded in 1889), or the Coal Smoke Abatement Society (1899). In 1907, the Trust was incorporated by Act of Parliament (the National Trust Act), allowing it to declare its property as 'inalienable'.[51]

The physical objects of the Trust's efforts – landscapes and vernacular buildings – differed little from those of *Heimatschutz* in Germany; its first property, donated in 1895, was Dinas Oleu, a stretch of clifftop in Wales, followed in 1902 by its first Lake District acquisition; its first building, bought for £10 the following year, was a timber-framed Sussex cottage, Alfriston Clergy House. Only in 1937, following a further Act of Parliament, could the Trust acquire large country houses – an elite building type that later became its focal concern. But the wider cultural context of the Trust's work was remote from the blood-and-soil nationalism of *Heimatschutz*. Rather, it exemplified how heritage helped cement global anglophone culture – broadening from mementoes of 'great men' such as Shakespeare to more generalised themes. The dual architectural/natural landscape orientation of the NT was reinforced by its links with the USA, which had already (see below) moved decisively towards national-park nature conservation. The Trust's constitution was itself modelled on the recent precedent of the Trustees of Public Reservations in Massachusetts (TPRM), established in 1891 at the instigation of Charles Elliot, president of Harvard University, in alarm at modern development pressures around Boston. The NT encouraged moral and financial links with American supporters by focusing on the shared heritage of pre-revolutionary years. Around the turn of the century, Arts and Crafts architect Charles Ashbee became closely involved in promotional tours of America and the colonies; ambassador James Bryce also played a key role.[52] In its first decades, the Trust was a small pressure group rather than a mass organisation: as late as 1928, when its publicity committee was established, membership was still below 2,000.

Heritage and empire

As on the Continent, conservation in the anglophone world was certainly affected by a conflation of national and 'racial' values. But the position, as also partly with France and the Netherlands, was complicated by the relatively mobile culture of imperialism, emigration and maritime trade. Paradoxically, the concepts of mutual respect between different cultures, that would dominate the later 20th century's boom in peaceful heritage internationalism (Chapter 11), first appeared in embryonic form within these overseas empires, especially in dealings with cultures which had their own imperial, monumental heritage. Here, Napoleon-style cultural plunder was increasingly seen as unacceptable, and recording often took its place. Following Cambodia's annexation as a French protectorate in 1863, for example, the 12th-century temple complex of Angkor Wat was carefully protected, and its epigraphic texts and relief sculptures were recorded using an ingenious paper-casting technique developed by Pierre-Victorien Lottin de Laval. His 1857 manual boasted that explorers could thus 'bring back, from the furthest reaches of the earth, in a trunk . . ., an immense series of monuments [. . .] to ensure for France all these precious remains'. By 1900, photography had replaced these ingenious techniques as the recording medium of choice. Within the anglophone world, there was airy talk of a global English-speaking 'family': leading archaeologist G Baldwin Brown trumpeted race and empire, arguing that 'the mother country must remain the soil in which are [rooted the virtues of] the race, and the sense of the continuity of the stock'.[53]

At the same time, the British Empire in the 1890s and 1900s also saw an upsurge in devolutionary attitudes towards heritage, especially concerning the formidable cultural and architectural legacy of India. Here, the 19th century had witnessed early but sporadic initiatives for preservation (by the Bengal government in 1810) or national archaeological survey (1861/71 onwards), combined with utilitarian, aggressive 'City-Improvement' clearances. A decisive turning point came in 1898, with the appointment as viceroy of the 39-year-old Lord Curzon. His controversial seven-year 'benevolent despotism' revitalised the conservation of India's architectural heritage.[54] Finding on his arrival in India that the government office of Director-General of Archaeology had lapsed, Curzon, an eager amateur archaeologist, assumed the duties of the post himself for four years, increasing the conservation budget sixfold, and declaring that a governing power was morally obliged to protect the ancient monuments of the governed, especially those of non-Christian faiths. He argued that 'Imperialism will only win its way in this country if it wears a secular and not an ecclesiastical garb.' Key set pieces were lavishly restored, including a £50,000 repair programme for the Mughal monuments of Agra – a programme which exposed him to SPAB-style anti-restorationist criticisms. Curzon's wider conservation efforts, systematised by a 1904 Ancient Monuments Preservation Act, spanned the whole country: Sir Thomas Raleigh, a member of the Viceroy's Council, wrote that 'I never visited an ancient building in India without finding that the Viceroy had been there before me, measuring, verifying, planning out the details of repair and reconstruction . . . with his own passionate reverence for the historic past'.[55] In Nehru's opinion, 'after every other Viceroy has been forgotten, Curzon will be remembered because he restored all that was beautiful in India'. In activities like these, or the plans of Patrick Geddes (see below) for non-invasive city regenerations, commissioned by princely rulers, centre and periphery moved together, as Indian elites were drawn into heritage protection.[56]

America: heroes' homes and colonial houses

BEFORE World War I, North America also formed part of this loose heritage alliance, contrary to today's assumption of American 'exceptionalism'. As we saw earlier, previous American preservation discourses were concerned less with broad 'national qualities' like the British Picturesque or German noble austerity than with relics associated with national heroes. Nor was there a specific concept of regional *Heimat* – perhaps partly owing to the fragility of the American state in the mid-19th century. It was safer to celebrate the unifying work of revolutionary leaders, efforts that increased in vigour once the trauma of the Civil War was safely behind, using the 'Expo' as a key vehicle, as well as preserved artefacts and restored historic houses. Frank M Etting, restorer of Independence Hall for the 1876 centennial exposition in Philadelphia, argued that 'So long . . . as we can preserve the material objects . . . which these great men saw, used, or even touched, the thrill of vitality may still be transmitted unbroken'. And a reproduction of Boston's John Hancock House, demolished in 1863 after accurate drawings had been made, was rebuilt as the Massachusetts Pavilion at the 1893 World's Columbian Exposition: Mount Vernon and Independence Hall were also copied in many other places.[57]

Although the celebration of heroes was an equally prevalent practice in Britain, it arguably continued to flourish in America for longer in the 20th century, for example in Thomas Tallmadge's interwar architectural history studies. The period 1913–14 saw an Anglo-American initiative to preserve Sulgrave Manor, the Northamptonshire home of Washington's ancestors.[58] American reliance on private initiative was at this stage no greater than elsewhere in the anglophone world: after all, the pioneering acquisition of Hasbrouck House in 1850 had been carried through by the New York state government; the 1860s saw War Department preservation of Ford's Theatre in Washington (scene of Lincoln's 1864 assassination); and in 1872 the federal government intervened decisively in nature conservation, by establishing the first national park, Yellowstone. This started a vast, governmental movement in which the USA led the world, especially after the founding of the National Park Service (NPS) in 1916 – a veritable government empire that encompassed not only wilderness holdings but also designed and even urban landscapes. The 1906 Antiquities Act – equivalent of the 1882, 1907 and 1913 Ancient Monuments Acts in Britain – authorised active protection of archaeological sites on federal land, mainly in the south-west. However, the most spectacular American archaeological recovery initiatives were those in Mexico under the Porfirio Díaz regime, which established a National Monuments Inspectorate in 1885 and pushed through radical, Viollet-like restorations of complexes such as Mitla and Teotihuacán (especially in 1905–11, under chief archaeologist Leopoldo Batres).[59]

Architectural conservation in British North America (Canada) was slightly slower to emerge than in the USA, despite the tradition of greater state interventionism in property rights, and the eager public appetite for heritage: Manitobans, for example, had founded a historical society in 1870, only nine years after creation of their province. A prototype nature-protection park was established by the federal government near Banff in 1885, a national Commission of Conservation, concerned mainly with urban town-planning and nature conservation, was instituted in 1909 by the Laurier Administration, and a 'Dominion Parks Branch' was founded in 1911, five years before its US counterpart. A successful Canadian/American joint campaign of the 1880s to preserve the Niagara Falls was arguably the first international 'conservation initiative'. The first advisory government panel on the built heritage, the Historical Sites and Monuments Board of Canada, was first established after World War I, in 1919, together with a National Parks organisation; the first Canadian 'national historic park' (i.e., incorporating built or cultural heritage) had been established two years previously at Fort Anne, Annapolis Royal, Nova Scotia, and the earliest provincial heritage legislation was passed in 1922 by Quebec.[60]

The diversity of US architectural preservation increased further around the turn of the century. The first state-wide preservation body, the Association for the Preservation of Virginia Antiquities (founded in 1889), continued the long-standing stress on American Revolution nationalism and heroes' homes, while also celebrating southern and even Confederate values. The prominent role of women in voluntary preservation continued in the 1880s' work of the Ladies' Hermitage Association (to preserve the Nashville home of Andrew Jackson) and in that of the

Figure 5.9 Early-20th-century landmarks in America

(a) The Paul Revere House, Boston, restored in 1905–8 (HABS survey photo); (b) The interior of the Abraham Lincoln Birthplace Historic National Site at Hodgenville, KY (HABS photo): the first purpose-built Lincoln memorial, constructed 1909–11 and inaugurated by President William Howard Taft; (c) The 'Spanish Governor's Palace', San Antonio, Texas, preserved and restored between 1915 and 1930 (HABS drawing)

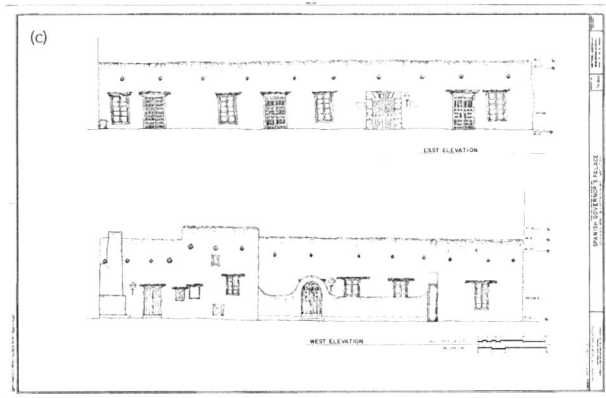

Daughters of the American Revolution (founded in 1890). But there was also growing stress on the intrinsic architectural qualities of historic buildings, with schemes sometimes involving radical, Viollet-style restoration. Boston was a particular hotbed of early preservation initiatives. The successful efforts to safeguard the Old South Meetinghouse in the four years following the 1872 Great Fire, and the similarly effective 1881 campaign by a newly founded Bostonian Society to preserve the Old State House, were followed in 1891 by the foundation of the TPRM. In 1905, the Paul Revere House was bought by a voluntary group, led by antiquary William Sumner Appleton Jr, because of its revolutionary associations, and also as the city's oldest surviving timber building. It was radically restored to its supposedly original 17th-century timber-framed condition and opened in 1908. Appleton's pioneering group, named in 1910 the Society for the Preservation of New England Antiquities, went on to restore over 50 others and shape the course of US preservation, not least by inventing the so-called 'revolving fund', in 1912. Later, Thomas Jefferson's Monticello, acquired in 1923 by the Thomas Jefferson Memorial Foundation, was appreciated as much for its classical beauty as for its associations.[61]

For obvious reasons, the USA led the movement to preserve 18th-century classical architecture. This paralleled the 'Colonial Revival' in new architecture, a style first seriously advocated in Arthur Little's 1877 book, *New England Interiors*, and which reached its climax in the 1880s and '90s.

163

Ingenious argumentation, increasingly focusing on 'national character' as in Europe, turned the 'provincial Georgian' character of the originals to advantage: architect Horace B Mann argued in 1915 that 'there is between Americans and Europeans an actual difference of habit of mind. We are less complex and more direct, less formal and simpler, and all this our Colonial work expresses'. Even more 'simple' were the log cabins featured in world's fairs, such as the Louisiana Purchase Exposition in 1904 at St Louis. The quasi-deification of Abraham Lincoln inspired a more extreme juxtaposition of 'historic' and 'memorial' at the Lincoln Birthplace Historic National Site at Hodgenville, Kentucky, which encased a reconstructed birthplace log cabin within a temple-like Beaux-Arts classical pavilion (1909–11, by architect John Russell Pope). Clashes sometimes occurred between promoters of reproduction ensembles, such as the period rooms in the Metropolitan Museum's American wing, and scholars such as Appleton.[62]

By the 1910s, antiquarian recording had become a widespread, popularised activity in America. Albert G Robinson's *Old New England Doorways* of 1919 presented a survey of surviving 17th-century doorways in the north-east, and argued that 'hunting for old doorways is a harmless and interesting amusement, much like . . . collecting postage stamps'.[63] In America, too, the emphasis within recording was shifting to photography: Frank Cousins's work on the East Coast included an influential survey of 50 historic buildings in New York City in 1913. In New York City, contrary to the popular misconception that preservation only began with the demolition of Penn Station in 1963, the turn of the century saw a plethora of initiatives, often championed by developers as well as groups such as the American Scenic and Historic Preservation Society (ASHPS), founded in 1895. Some were successful but others ended in failure, such as the decade-long ASHPS struggle in 1908–18 to save the early-19th-century St John's Episcopal Chapel in Varick Street. The city also saw pioneering efforts at landscape preservation, including the prolonged fight of *c.*1900 to preserve the City Hall Park and an innovative attempt to regenerate a degraded area through landscaped highway construction, in the Bronx River Parkway project of 1906–25.[64]

The chief exception to the relative lack of European-style *Heimat* heritage in the USA was the Spanish-American architecture of the south-western states. Here, too, although later rhetoric emphasised its anonymous, 'organic' vernacular character, the pioneering examples were associated with elite figures or heroic historical events. In Santa Fe, New Mexico, the 1909–13 restoration of the Spanish Colonial Governor's House by the Museum of New Mexico, evoking regional vernacular styles, informed subsequent preservation initiatives. For example, in San Antonio, Texas, the Chapel of the Alamo was rescued in 1904 by a wealthy heiress and donated to the state of Texas as a shrine to its independence from Mexico (1835–6). Subsequently, an 18th-century single-storey house of the *presidio* in Military Plaza was identified in 1915 as the 'Governor's House' by pioneering Texas preservationist Adina de Zavala, who planned to convert it into a Texan hall of fame; it was grandiosely restored in a free Spanish-Colonial manner in 1929–30 by local architect Harvey P Smith as the 'Spanish Governor's Palace'.[65]

Old Town-*Altstadt*: the 'living home' of community nationalism

THESE national variants of ethos and organisation, whose uniqueness was trumpeted by the competing national powers, overlaid a more fundamental trend within the turn-of-century Conservation Movement towards convergence, even uniformity. The issue which formed the lightning-rod for this paradoxical process of homogenisation was the new, intense love of the 'Old Town' as opposed to isolated elite monuments. This movement was signalled by the late-19th-century fashion for coupling the word 'Old' to the names of individual cities – Alt-Nürnberg, Vieux-Paris, etc. – something that also distantly anticipated today's obsession with city branding (see Chapter 12). Here was a new kind of authenticity, no longer concerned, as with Ruskin, with substance and detail, but with 'life' at an urban, landscape scale. The cult of the Old Town became a microcosm of the wider dynamic of the times, featuring violently fluctuating ideologies and discoveries of neglected jewels, alongside underlying continuities. Like the wider dialectic relationship between conservation and destruction, the love of old towns was intimately bound up with urban modernisation, which acted on them negatively, through destructive provocation, and positively, as a stimulus. For example, the time-honoured and continuing policy of isolating key monuments and creating 'space' by driving new roads through dense old cities, was beginning to be vilified as destructive, yet it also heightened the sense of authenticity of the remaining fabric. And the Old Town also formed a stage on which the uneasy relationship of those two parallel modernities, nationalism and commercial tourism, was played out. With its mainly enclosed spatial structure, the stereotypical Old Town was 'designed' and 'staged' at the same time.[66]

The Old Town started becoming a special concern in the intermittent early and mid-19th-century pronouncements of figures such as Ruskin or Victor Hugo. Ruskin and Pugin appreciated its irregularity both aesthetically and ethically. Within the polarised framework of Pugin's *Contrasts*, the Old Town was a focus of the lost golden age, while Ruskin saw it as a vast collective memorial, silently analogous to the old-style commemorative monument. His drawings of decay (and those of Roberts and others) stimulated the growing enthusiasm for the crumbling fabric of old towns – an enthusiasm that had to be reconciled with the more urgent demands for health and sanitation.[67] Construction of planned suburbs and 'new towns', especially in Britain and Ireland during the Improvement era, as in Edinburgh from the 1760s, or in post-1757 Dublin under the Wide Streets Commissioners – an urban equivalent to Wyatt's opening-out work on cathedrals – had already created a new, stronger awareness of old quarters as an 'other', in both aesthetic and ethical-cultural terms.[68] But it was only in the mid- and later 19th century, with the simultaneous international advance of a number of interlocking phenomena – the Haussmann-inspired movement of urban systematisation by carving grand boulevards and spaces through existing fabric and planning new city extensions; the drive for sanitary housing improvement; the general destruction of obsolete city fortifications across continental Europe; and the general upsurge in tourism and urban commercialism – that a critical mass was reached in staging the 'Old Town' as an international phenomenon.

These new influences were synthesised especially in the German-speaking countries, which devised a special word for this modern concept: *Altstadt*. Although the word appeared even in the first edition of Grimm's *Deutsches Wörterbuch* in 1854, it only assumed its modern, town-planning sense from the 1890s – for example, in an 1896 article on Vienna by planner Joseph Stübben – spreading to popular publications such as Baedeker guidebooks by 1914.[59] It attributed a curiously schizophrenic character to the 'Old Town': on the one hand, a constituent element of the modern zoned city; on the other, a unique, 'living', cultural and visual entity, almost like an individual person, with every part a vital element in the whole. The consequences for the Conservation Movement were clear: from now on, 'background' buildings now occupied the foreground of attention.

As with Viollet and his restorations of individual monuments, it was Second-Empire France that provided an international bogeyman for Old Town enthusiasts. 'Haussmann' became a by-word for crassly brutal redevelopment: in the 1910s, for example, Patrick Geddes branded the 'death-dealing Haussmannising' approach 'one of the most disastrous and pernicious blunders in the history of civilisation'.[70] Yet by the late 19th century this national stereotype was rather misleading, as in France itself there were mounting attacks against the alleged monotony of Haussmann's planning, and against isolating key monuments in vast spaces, as with the Tour-St-Jacques or Notre-Dame. The Paris Opera architect, Charles Garnier, decried 'the odious abuse of the straight line' in 1885, and even Viollet himself pleaded that city extensions of Bern in Switzerland should respect the old town's irregularity. A 1902 planning law attempted to encourage greater flexibility in French urban development.[71]

This growing anti-Haussmann revulsion in France in the late 19th century was not immediately reflected elsewhere in the statist countries of the Continent. In Germany, the practice of radical *Freilegung* around cathedrals continued unabated. Examples included Cologne, from the 1860s, including demolition of three churches and numerous lesser buildings; Ulm, in 1874–9, where a monastery was razed for a new, stately square; and Regensburg, in 1894, where the 13th-century Archbishop's Palace was sacrificed to the demands of openness.[72] And in Italy, Giuseppe Poggi's 'amplification' of Florence from 1864 onwards, during its brief period as capital of unified Italy, 'systematised' the medieval fabric with new right-angled streets lined with neo-Renaissance palazzi, allowing *isolamento* of all key monuments. In Rome, the 1870s and '80s saw the piercing of new boulevards through the old *rioni*, and massed building of apartment blocks outside the Aurelian walls. The 1883 general plan for Rome envisaged a Haussmann-like layout, while ambitious proposals of 1909–10 envisaged *isolamento* of key monuments in both Rome (including the Markets of Trajan and the Torre delle Milizie) and Bologna (in a proposal to isolate the city's two remaining towers, vigorously opposed by conservation architects Alfonso Rubbiani and G Pantini). Likewise, in colonial India, the end of the century saw vigorous clearances and driving through of new 150-foot-wide streets in old cores, for example by the Bombay City Improvement Trust (established in 1898, following a plague panic) or in Delhi and Lucknow. Here, the Haussmann-like planning principles were also intended to segregate different ethnic groups.[73]

The new concept of the Old Town as a living collective heritage, a 'city monument', was not just a piece of architectural polemic or a visual fashion, but was also shaped by major scientific and artistic developments in the understanding of the city, with the German-speaking countries taking the lead in both areas. Correspondingly, the old, utilitarian process of 'improvement' was now widely condemned as brutally philistine. Charles Darwin's evolutionary theories, since the 1860s, had popularised ideas of the influence of environment on human welfare, undermining the old individual-moral concepts of urban improvement, and encouraging advocates of a total vision of transformation of the *whole* city, such as sociologist Georg Simmel. In *Soziologie der Geselligkeit* (1910) Simmel argued that the highest potential aspiration of the modern city was not mechanical efficiency, but sociability. Other contemporary urban theorists, such as Geddes, also called for a spiritual collective vision with near-religious fervour. At a more practical, policy-application level, civic-order and 'planning' movements sprang up everywhere, especially in the vigorous German attempts to overcome chaotic urbanisation through land-use zoning of urban extensions. These were exemplified in Josef Stübben's handbook, *Der Städtebau* (1890), with its orderly planning procedures, beginning with traffic flow and proceeding logically to an aesthetic conception of public spaces, and zoning reform.[74] The effectiveness of these initiatives in Germany benefited from the power of its municipal planning officials and the high esteem of professionalism in public life,

Figure 5.10 The 'amplification' of Florence, directed by Giuseppe Poggi from 1864: this view shows the triumphal-arch portico of the Piazza della Repubblica, reconstructed in 1885–95 (Vincenzo Micheli, architect)

allowing Franz Adickes, *Oberbürgermeister* of Frankfurt-am-Main from 1891, to introduce a bold scheme of graded density control across the city.[75]

Camillo Sitte: the Old Town and 'artistic' city-planning

WITHIN this emergent structure of zoned city-planning, the *Altstadt* commanded a central and almost self-explanatory place, a keystone-like role, its 'oldness' logically reinforcing the modernity of the remainder. That role implied a generic, standardised quality, and indeed many *Altstädte*, in Germany and elsewhere, were spatially rather similar. But they could most efficiently help anchor modernity if they were perceived not as standardised but as the opposite – as intensely place-specific jewels of *genius loci*. Here the Ruskinian concept of the 'life' and intrinsically artistic character of old buildings played a vital role, in a theory developed not in Germany itself but by an Austrian architect and art historian, Camillo Sitte. His immensely influential work, *Der Städte-bau nach seinen künstlerischen Grundsätzen*, showed how an individual, idiosyncratic initiative, at a precisely opportune moment, can exert a dramatic effect. It appeared in 1889, and proved such an instant success that its ideas pervaded Stübben's *Städtebau* the following year.[76]

Sitte devised a powerful new variant of the painterly Picturesque for urban contexts, focused not on static views but on kinetic experience and concerning itself not with small-scale antiquarian picturesqueness but with a broader concept of artistic city-scape, stemming partly from the traditions of landscape-painting. His key insight was to apply the new architectural concepts of space (as in Wölfflin's appreciation of Baroque form and mass) to the entire city, and especially to the interrelationships of squares and visually enclosed civic spaces. In this framework of spatial art, the whole city became a three-dimensional *Gesamtkunstwerk*. The challenge now was to plan new classical cities in a non-utilitarian way, infused by tradition, 'to go to school with Nature and the old masters . . . in matters of town planning'.[77] Sitte's book followed the familiar golden-age framework, tracing 'artistic' planning from antiquity through the Italian Renaissance and medieval Germany to the (supposedly) brutally utilitarian dystopias of modern France and the USA. For Sitte, as a hater of French culture, Haussmann's axial boulevards and wide open spaces (including the *dégagement* or *Freilegung* of historic buildings), exemplified the worst sort of engineer-led city development, although James Hobrecht's 1862 General Development Plan for Berlin, with its relentless grid of dense street-blocks, seemed little better.[78]

Sitte argued that town-planning should stress enclosure rather than openness and an artistic rather than engineering ethos. Where the Haussmannian city was socially alienating and visually faceless, the authentic, artistic city would foster community and project a genuinely individual image, just like a living individual person. In practice, the confrontation was milder than this: in the planning of city extensions, Sitte's ideas interacted positively with Stübben's efficiency-led framework, helping distance both from Haussmann. In 1893, Sitte intervened in the competition for the Vienna Structure Plan (*Generalbebauungsplan*), won by Stübben; this proved a key turning-point in the heritage-sensitive, zoned modernisation of historic cities.[79] In due course, this new individuality of the 'living' city would also, of course, help foster tourism.

Out of this fusion of art and municipal efficiency, a new German-Austrian city-planning philosophy rapidly coalesced – for example, in a plan by Sitte for Ljubljana. The most enlightened cities rapidly absorbed and reflected his artistic ideas, as in Theodor Fischer's 1893 proposals for

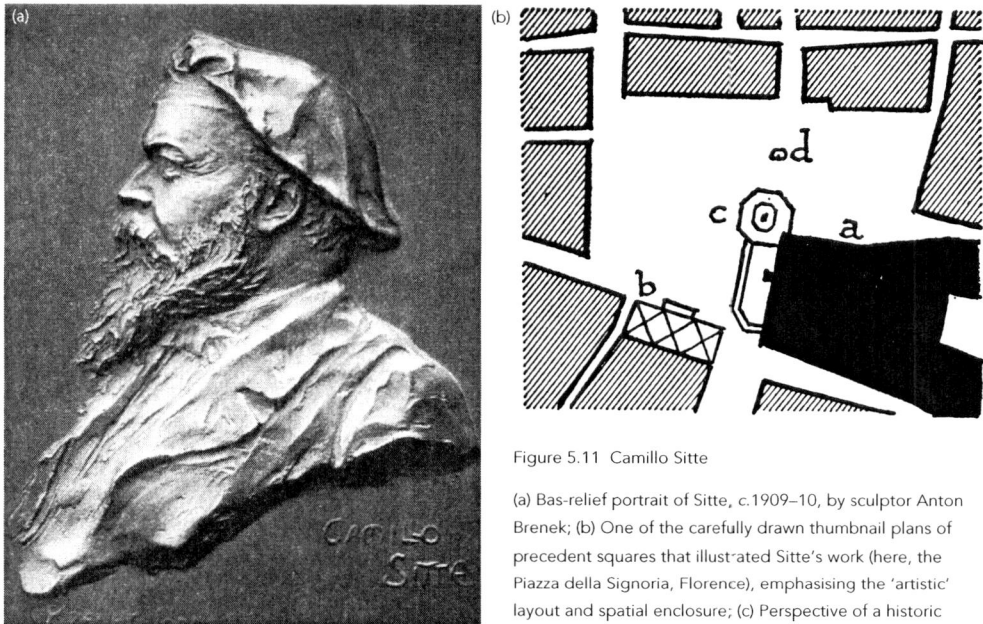

Figure 5.11 Camillo Sitte

(a) Bas-relief portrait of Sitte, c.1909–10, by sculptor Anton Brenek; (b) One of the carefully drawn thumbnail plans of precedent squares that illustrated Sitte's work (here, the Piazza della Signoria, Florence), emphasising the 'artistic' layout and spatial enclosure; (c) Perspective of a historic Germanic street (Breitestrasse, Lübeck) from the 1902 French edition of Sitte's book

part of Munich, with enclosed spaces and picturesque, high roofs, or Karl Henrici's 1890 extension plan for Dessau. Even in go-ahead Hamburg, where most historic areas had been destroyed in the great fire of 1843, city architect Fritz Schumacher (from 1909) attempted, against opposition from the City Engineer, to introduce an artistic and somewhat conservationist planning ordinance in 1912.[80]

An oblique precedent for Sitte's claims of the artistic character of traditional urban ensembles came from the field of house decoration, where the 1870s and '80s saw the emergence of the idea of the poetic home, its designed unity conveying a specific character, often national or ethnic (and again, using the prefix 'Old': 'Old English', 'Alt-Deutsch', etc.). In this mixture of commerce, moral-social utopianism and psychology, 'simple old furniture' or 'antiques' played a key role. Here, there was a similar story to Riegl's sequence of historic-value and age-value. Before the mid-1870s, the authenticity of old furniture was seen as resting in its style. After that, under the influence of the Arts and Crafts, its material oldness became all important: new furniture in accurate historic styles was called 'fake'. This concept of authenticity differed superficially from that of architectural heritage – for example, in its higher esteem of 18th-century work – but the underlying spirit was similar. Just as, for interior designers, the artistic presentation of materially old articles could create a living home, the same was true of the concept of the artistically living *Altstadt*, as brought to maturity by Sitte.[81]

In both cases, a vital animating force in creation of staged ensembles stemmed from commercial enterprise – a force, in architecture, especially channelled through tourism. By the late 19th century, it was firmly established that 'special' old towns could be targeted both for preservationist campaigning and tourist exploitation. Echoing the staging of real old towns was the presentation of *Altstadt* tableaux in national or international expositions, offsetting the threatening mechanistic power of structures such as the 1889 Eiffel Tower. These tableaux began with the displays at Turin (1884) and Edinburgh (1886), and in the Vienna International Music Expo of 1892 (an event which first branded Vienna as a 'City of Music'). Here an 'Alt-Wien' ensemble of stalls and shops was devised by scenery-designer Oskar Marmorek, loosely based on a 17th-century sketch. Anticipating the controversial connotations of post-1945 Disneyland, these hybrid efforts were accused of commercial debasement: Alt-Wien was fiercely criticised by the Wien Alterthums-Verein as 'pictorial puppet-theatre scenery'. The 1889 and 1900 World Expositions in Paris (the first of which included an 'Old Bastille quarter') were the first to embrace mass international middle-class tourism: by 1913, Paris was hosting 300,000 tourists a year.[82]

An alternative form of ensemble-staging, with a rural rather than urban flavour, was pioneered in Scandinavia, where the timber-building tradition encouraged open-air museums populated with demountable pavilions. The first was founded in Kristiania/Oslo, Norway, in 1881 by King Oscar II (of Sweden and Norway) and expanded into a full-scale open-air museum at Bygdøy in 1898–1902. By then, a more ambitious venture was underway in Stockholm: the 75-acre Skansen museum in Djurgården, established by Artur Hazelius (founder of the Nordiska Museet in 1873) to display the pre-industrial life of the regions of Sweden. Having preserved pre-industrial artefacts from 1872 onwards and exhibited his collection in Stockholm from 1873 (and at the 1878 Paris Expo), Hazelius upscaled this moveable heritage from objects to buildings, and brought about 150 dismantled houses from across the country to the Djurgården site (which he acquired in 1882 and opened in 1891–2). There, he staged them in careful ensembles, complete with livestock and costumed staff, inspired by Charles Garnier's historical reconstructions at the 1889 Paris Expo. The ensembles did not just depict rural environments: one represented a 19th-century small town.[83]

In discussing the influence of Sitte on urban conservation, we need to bear in mind that his own concern was new-city-planning; he never advocated direct copying or preservation of the picturesque old quarters themselves. But the connection with conservation was an obvious one, both in his opposition to stranding historic monuments in vast spaces through *Freilegung* and in his artistic animation of the concept of the living old town. And very soon, others did make that explicit connection: his criticism that modern life had turned 'man into a machine', for example, struck an obvious chord with Schultze-Naumburg and the *Heimatschutz*. Diffusion of his ideals outside the German-speaking world was impeded by lack of accurate translations, but the journal he founded in 1904, *Der Städtebau*, inspired Britain's *Town Planning Review* (founded in 1910), and his ideas strongly influenced Raymond Unwin's advocacy of picturesquely planned housing estates and 'civic art' in general.[84]

170

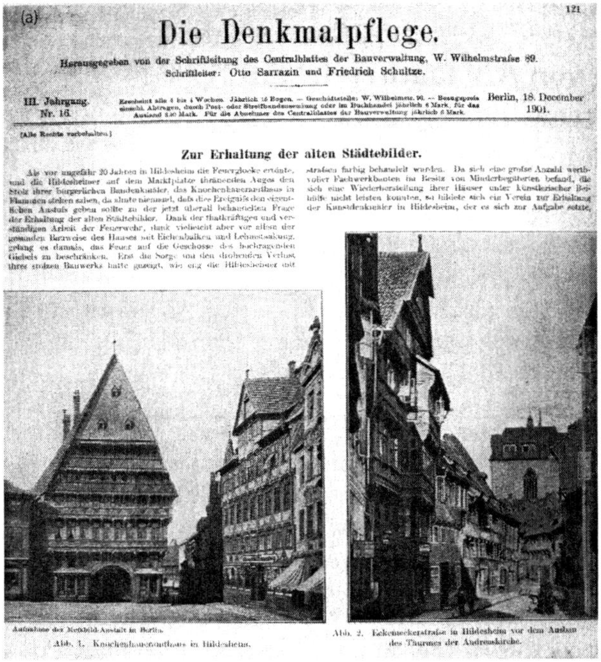

(a)

121

Die Denkmalpflege.

Herausgegeben von der Schriftleitung des Centralblattes der Bauverwaltung, W. Wilhelmstraße 89.
Schriftleiter: Otto Sarrazin und Friedrich Schultze.

III. Jahrgang. Nr. 16.

Erscheint alle 3 bis 4 Wochen. Jährlich 16 Bogen. — Geschäftsstelle: W. Wilhelmstr. 90. — Bezugspreis einschl. Abtragen, durch Post- oder Streifbandzusendung oder im Buchhandel jährlich 6 Mark, für das Ausland 8,00 Mark. Für die Abnehmer des Centralblattes der Bauverwaltung jährlich 5 Mark.

Berlin, 18. December 1901.

[Alle Rechte vorbehalten.]

Zur Erhaltung der alten Städtebilder.

Als vor ungefähr 20 Jahren in Hildesheim die Feuerglocke ertönte, und die Hildesheimer auf dem Marktplatze thränenden Auges den Stolz ihrer bürgerlichen Baudenkmäler, das Knochenhaueramthaus in Flammen stehen sahen, da ahnte niemand, daß dies Ereigniß den eigentlichen Anstoß geben sollte zu der jetzt überall behandelten Frage der Erhaltung der alten Städtebilder. Dank der thatkräftigen und verständigen Arbeit der Feuerwehr, dank vielleicht aber vor allem der gesunden Bauweise des Hauses mit Eichenbalken und Lehmstakung, gelang es damals, das Feuer auf die Geschosse des hochragenden Giebels zu beschränken. Erst die Sorge um den drohenden Verlust ihres stolzen Bauwerks hatte gezeigt, wie eng die Hildesheimer mit ... straßen farbig behandelt wurden. Da sich eine große Anzahl werthvoller Fachwerkbauten im Besitz von Minderbegüterten befand, die sich eine Wiederherstellung ihrer Häuser unter künstlerischer Beihülfe nicht leisten konnten, so bildete sich ein Verein zur Erhaltung der Kunstdenkmäler in Hildesheim, der es sich zur Aufgabe setzte,

Aufnahme der Meßbild-Anstalt in Berlin.

Abb. 1. Knochenhaueramthaus in Hildesheim.

Abb. 2. Eckemeckerstraße in Hildesheim vor dem Ausbau des Thurmes der Andreaskirche.

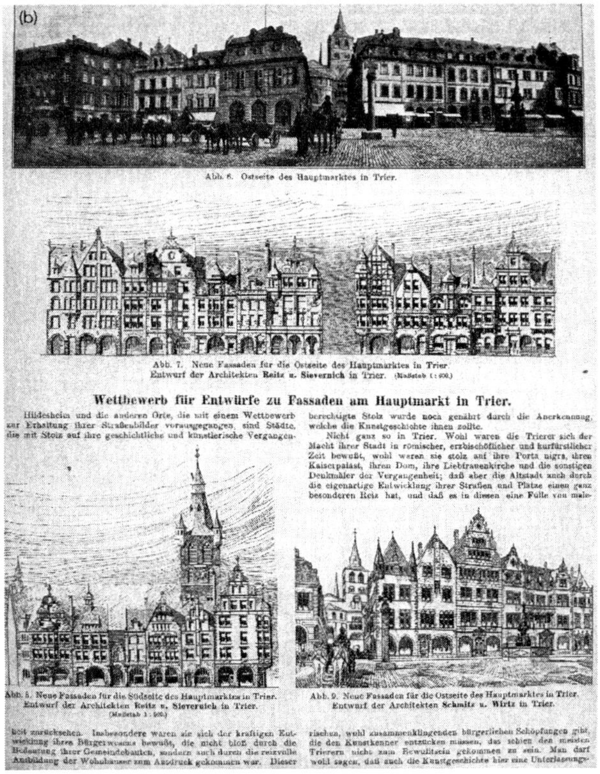

(b)

Abb. 6. Ostseite des Hauptmarktes in Trier.

Abb. 7. Neue Fassaden für die Ostseite des Hauptmarktes in Trier.
Entwurf der Architekten Reitz u. Sievernich in Trier. (Maßstab 1 : 400.)

Wettbewerb für Entwürfe zu Fassaden am Hauptmarkt in Trier.

Hildesheim und die anderen Orte, die mit einem Wettbewerb zur Erhaltung ihrer Straßenbilder vorausgegangen, sind Städte, die mit Stolz auf ihre geschichtliche und künstlerische Vergangenheit zurücksehen. Insbesondere waren sie sich der kraftigen Entwicklung ihres Bürgerwesens bewußt, die nicht bloß durch die Bedeutung ihrer Gemeindebauten, sondern auch durch die reizvolle Ausbildung der Wohnhäuser zum Ausdruck gekommen war. Dieser berechtigte Stolz wurde noch genährt durch die Anerkennung, welche die Kunstgeschichte ihnen zollte.

Nicht ganz so in Trier. Wohl waren die Trierer sich der Macht ihrer Stadt in römischer, erzbischöflicher und kurfürstlicher Zeit bewußt, wohl waren sie stolz auf ihre Porta nigra, ihren Kaiserpalast, ihren Dom, ihre Liebfrauenkirche und die sonstigen Denkmäler der Vergangenheit; daß aber die Altstadt auch durch die eigenartige Entwicklung ihrer Straßen und Plätze einen ganz besonderen Reiz hat, und daß es in diesen eine Fülle von malerischen, wohl zusammenklingenden bürgerlichen Schöpfungen gibt, die den Kunstkenner entzücken müssen, das schien den meisten Trierern nicht zum Bewußtsein gekommen zu sein. Man darf wohl sagen, daß auch die Kunstgeschichte hier eine Unterlassungs-

Abb. 8. Neue Fassaden für die Südseite des Hauptmarktes in Trier.
Entwurf der Architekten Reitz u. Sievernich in Trier.
(Maßstab 1 : 400.)

Abb. 9. Neue Fassaden für die Ostseite des Hauptmarktes in Trier.
Entwurf der Architekten Schmitz u. Wirtz in Trier.

Figure 5.12 Early *Altstadt* conservation

(a) Front page of 1901 issue of *Die Denkmalpflege*, focusing on conservation of urban ensembles and including illustrations of Hildesheim; (b) 1903 competition for *Altstadt* infill in Trier (published in *Die Denkmalpflege)*, showing eclectic historical styling of entries

(a)

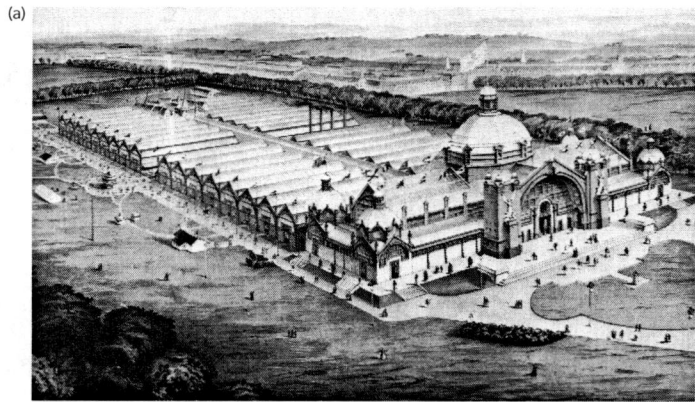

(b)

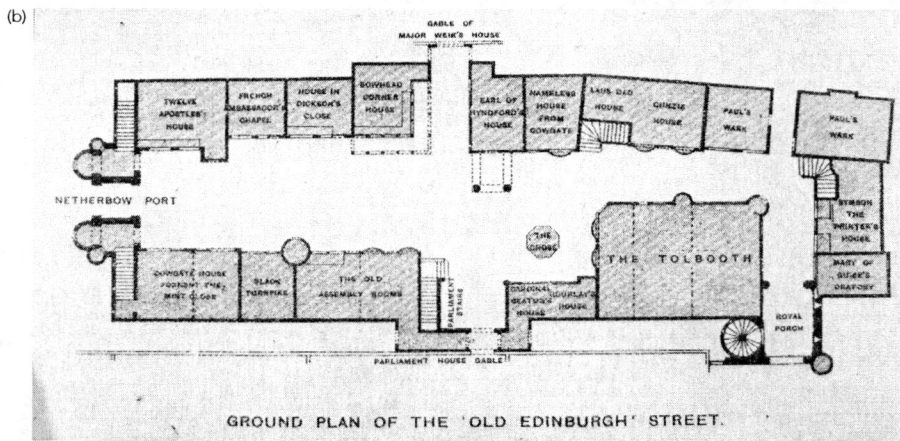

GROUND PLAN OF THE 'OLD EDINBURGH' STREET.

Figure 5.13 Expos and open-air museums

(a), (b), (c) 'Old Edinburgh', 1886, by architect Sydney Mitchell
(the ensemble is seen protruding from the roof at the far end of
the exhibition buidling)

facing page
(d) 'Alt-Wien', 1892, by stage designer Oskar Marmorek;
(e) Oktorpsgården, Skansen's first completed farmstead display,
assembled in 1896

(c)

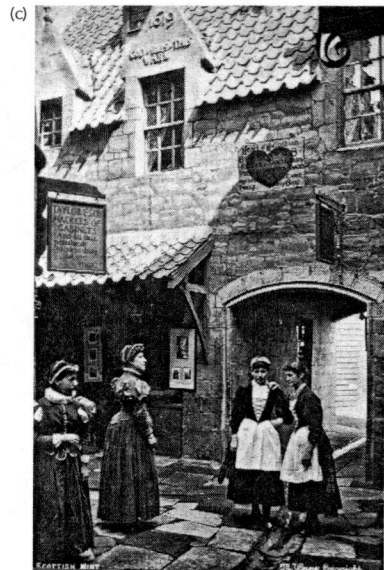

(d)

Alt-Wien (rechte Langseite).

Alt-Wien (linke Langseite).

(e)

In Austria itself, Sitte's concepts were eagerly embraced by conservationists. Riegl, for example, advocated preservation of a 'malerisches Strassenbild' (painterly street-picture) and 'Simmungsreiz' (atmospheric charm) in his report on Split in 1903. In Germany, too, his writings had a profound effect on conservation debates. The 1880s had already seen a swing in opinion against Freilegung, with demolition of city walls approved in Cologne in 1882 but blocked in Nuremberg around 1890. Under Sitte's influence, the process went much further. Joseph Stübben, always a weathervane of planning opinion, shifted from outright support of Haussmann-style axial planning (in his 1881 plan for Cologne's Neustadt area) to fierce opposition: in 1893, he railed against proposed removal of a Roman city gate near Cologne Cathedral (opposed by antiquarians and community leaders but eventually forced through after the personal intervention of the Emperor), protesting that the Cathedral would be left 'surrounded by nothing but flat pavement'.[85] In a keynote speech at the 1903 Conservation Convention, Stübben demanded greater respect for the individuality of historic cities, and at the 1905 convention Prof. P J Meier argued that 'the street layout of a city constitutes the most monumental title deed of its history'. For the leaders of German conservation and Heimatschutz, such as Gurlitt or Schultze-Naumburg, Sitte-like artistic analysis of urban space seemed an organically natural extension of their own ideas. And the years around 1900 witnessed a flurry of public competitions for picturesquely gabled remodelling of Altstadt facades in various cities.[86]

These German debates were anticipated in a decisive initiative undertaken in Belgium, by Charles Buls, a liberal reformist alderman, and subsequently mayor, of Brussels between 1881 and 1899. He opposed the grandiose, Haussmann-like planning schemes of Belgium's imperialist king, Leopold II, by developing a preservation-sensitive planning philosophy, initially on his own, and later drawing on the ideas of both Sitte and Stübben. The first stage of Buls's campaign was an ad hoc response to the growing threat to old Brussels, both from the royal reconstructions and from attempts by individual owners in the Grand' Place – still the most cherished ensemble in the city, two centuries after its postwar rebuilding – to modernise their facades. In 1883, Buls secured a municipal ordinance protecting the facades of the square and facilitating restoration and repair of it and the surrounding streets. Six years later, the publication of Sitte's Städtebau struck a profound chord with Buls. He used it as ammunition against proposals for isolement of historic monuments by demolition of accretions, arguing, 'Isolate a colossus, and you dwarf it.' He fiercely attacked Joseph Poelaert's gargantuan Palais de Justice (1866–83) as both overweening in scale and paltry in its grand isolation. In 1893 Buls published a pamphlet summarising Sitte's concepts, L'esthétique des villes, followed by an 1895 abridgement of his book.[87]

Rapidly, Buls's efforts moved onto the international plane, where a wide range of initiatives was underway, such as the international garden cities movement or the international town-planning congresses. Here, he became coordinator of a 'public art' civic-amenity movement: the first International Congress of Public Art was held in Brussels in 1898. Stimulated by the urban beautification ethos of the 1893 Chicago World's Columbian Exposition, but less aggressive than the Heimatschutz campaigns of Rudorff and others, this movement acted as an international umbrella for the proliferating civic-national campaigns against commercial 'disfigurement'. Examples of such campaigns included those in Edinburgh by the Cockburn Association (founded in 1875), which secured in 1898 the first local legislation for control of advertisements after campaigning against a giant flashing 'Bovril' sign on the Old Town skyline; or the foundation in 1893 of the Society for the Checking of the Abuses of Public Advertising (SCAPA). As we will see later in this chapter, the Sitte/Buls concepts of genius loci conservation and civic enhancement chimed with a particular approach to planned urban renewal in decayed Old Towns: a cautious, piecemeal, place-sensitive formula that went under various names, such as *diradamento* in Italy or 'conservative surgery' in Geddes's Edinburgh work.[88]

Old Town conservation across turn-of-century Europe

MORE important than these growing international efforts, though, were the initiatives at a national and civic level. National differences between patterns of Old Towns were reflected in divergent approaches to their conservation. In Germany, where most great cities had grown from medieval cores, the tightly packed medieval fabric contrasted sharply with the 19th-century modernity around it. Others, such as in Italy or England, were more complex in their stratifications, but dominated by public buildings of a relatively formal, even classical character. Some required little active intervention whereas in others, radical sanitary-led housing improvement and decongestion was demanded. What they had in common, however, was arguably more important. Most were, at the point of 'discovery', threatened by decay or demolitions – indeed, perceived threat was almost intrinsic to the *Altstadt* concept. Most, on the Continent, were sharply defined by the course of old city walls, some surviving but others demolished in the 19th century. And all shared the same public discourse of individuality and intense *genius loci*, the same rhetorical projection as quasi-living individuals – even although their built substance was often so similar. And, in many cases, the Old Town narrative followed a similar path, from a phase of Europe-wide rediscovery at the turn of the century, to a period of implementation and intense political exploitation by nationalist regimes in the interwar years and, finally, a downfall of fiery destruction in World War II.

The most exemplary *Altstadt* was, predictably, Nuremberg, both in its stereotypical binary pattern of walled town and surrounding modern industrial city and in the way it was idolised by nationalists. Here the 19th century built on Wackenroder's precocious 'rediscovery' of the old town in successive cultural initiatives, including the Germanic National Museum (founded in 1852) and Richard Wagner's *Die Meistersinger von Nürnberg* (first performed in 1868), a heroic idealisation of the former imperial capital's 16th-century guild community. Architecturally, these were complemented by the restoration of the Dürer-Haus from 1881 by artist Prof. Friedrich Wanderer (for the Albrecht-Dürer-Haus-Stiftung) as an exemplary *altes deutsches Haus*, stripping away Carl Alexander Heideloff's 1827–9 neo-Gothic scheme and creating a more massive, primitive impression, furnished with antiques and pieces designed by Wanderer himself.[89]

The upgrading of the Dürer-Haus, however, was intended to promote tourism as well as national pride – a double vision of Nuremberg, national-ideological and international-commercial, that climaxed in the 1890s. In 1894, the 400th anniversary of the birth of Hans Sachs inspired an outburst of festivities, including medieval-costumed parades, tableaux, music competitions and *Meistersinger* performances, intended to allow citizens and tourists to 'enter completely into the era of the past'. In parallel to this fairly conventional exercise in domestic staging, Germany's foremost *Altstadt* was also projected dramatically onto the world stage, through an Expo-style 'Old Nuremberg' ensemble in potentially one of the world's most *Heimat*-unfriendly settings: New York City's Madison Square Garden, an extravagantly towered Moorish-Beaux-Arts entertainment complex of 1889–90 by architect Stanford White. From December 1894, visitors paying 25 cents for entry to its 70,000-square-foot exhibition hall, the largest in the world, were confronted not with the usual boxing matches or football games but with an 'artistically' designed miniature *Altstadt*, its 'streets, houses and antique places' lined with actors in 'quaint costumes' and commercial booths selling 'novelties from all parts of the world', as well as an extensive toy fair stocked by Nuremberg's many toy manufacturers.[90] The stalls and stage-scenery for the 1894 New York ensemble – a forerunner of the late-20th-century globalised explosion in 'German Christmas Markets' – had been shipped to America the previous year (by the North German Lloyd Line, in 325 crates), for assembly at the World's Columbian Exposition in Chicago, as an 'Old Nuremberg' display alongside German, Irish and Indian 'villages'.[91]

A vast gulf separated this 'Old Nuremberg' exercise, and subsequent repetitions (e.g. in Buffalo, NY, in 1901), from any conventional concept of architectural authenticity. And its exaggerated rootlessness and commercialism was also suspect from the viewpoint of political nationalism. From now on, the Old Town in Europe and America would be a field of both tension and mutual reinforcement between political, cultural and architectural rhetoric and the forces of commercialism – the latter embraced enthusiastically in the USA, but more cautiously in parts of Europe, including Germany. Here, Nuremberg's cultural-political status as the archetypal *Altstadt* – 'the treasure-chest of the nation' – was foregrounded by conservationists, and the inaugural issue of *Die Denkmalpflege* stressed the need to protect it from modern development: '*Alt-Nürnberg in Gefahr*' ('Old Nuremberg in Danger').[92] Yet the promotion of national and international tourism followed closely behind – as shown even more clearly in the case of Rothenburg ob der Tauber. Seized on by Wilhelm Riehl in 1865 as an exemplar of old German wholeness, and linked by railway in 1873, the town became the focus of a feeding-frenzy of tourism, rising from 400 visitors in 1871 to 10,000 in 1903 and 21,000 in 1911. This cultural/commercial movement, in turn, stimulated various heritage initiatives, including the invented tradition of the *Meistertrunk* (from 1881; a play-cum-festival glamourising Rothenburg's siege in the Thirty Years War) and the 1898 foundation of a '*Verein Alt-Rothenburg*' dedicated to conservation and civic art. The latter helped encourage the uncovering of timber-framing, campaigned successfully in 1900 for a municipal preservation and civic-amenity ordinance (intended to impede proposals for garishly modern or 'fake old' new buildings) and in 1903 appointed architect Theodor Fischer as town conservator.[93]

All in all, *Altstädte* could be politically instrumentalised much more effectively than individual elite buildings and could adapt more flexibly to changing ideological contexts. There were sharp differences between early-19th-century visions of 'Old Nuremberg', the spiky neo-Gothic eclecticism of the 1870s/'80s, and the soberly vernacular post-1900 visions of the *Heimatschutz* writers: eventually, after yet another ideological exploitation in the 1930s, Nuremberg would undergo its most radical restaging of all, in its devastation by bombing and subsequent reconstruction.

Across Europe, the turn of the century saw the simultaneous development of the Old Town as a transnational discourse, albeit with detail variations in its spatial and socio-political aspects – as one would expect in a discourse stressing place-specificity. In Switzerland and Austria, in cities such as Bern or Basel, the spatial framework of Sitte-influenced picturesqueness resembled Germany, but the overarching force of strong nationalism was absent. Legislation of 1911 in Switzerland ensured that conservation was devolved to cantonal level. By contrast, in the impassioned movement for preservation of 'Old Prague', the main driving force was Czech nationalism. Building on an 1873 memorandum by the Bohemian Union of Architects against the uncontrolled suburban expansion of 'one of the most magnificent towns in the world', the 1890s saw a succession of voluntary initiatives (including the 1896 'Manifesto to the Czech Nation!' and the 1900 foundation of the Old

Figure 5.14 *Alt-Nürnberg* at home and abroad

(a) Dürer-Haus, view of rooms restored in '*alt-deutsch*' style by Friedrich Wanderer in 1881–2; (b) Present-day exterior of the Dürer-Haus, as restored after 1945; (c) Postcard of 'Alt-Nürnberg'-themed carnival display in Nuremberg, 1896; (d) Postcard of the 'Alt-Nürnberg' exhibit at the 1901 Pan-American Exposition, Buffalo, NY, created by Jakob Schöllkopf and other leading German-American Buffalo citizens; the entrance fee was 25 cents

Prague Society), but little concrete action until the mid-20th century. In Stockholm, the Gamla Stan (an island Old Town with a medieval layout but 17th/18th-century built fabric) attracted similar initiatives around the turn of the century. The threat of redevelopment provoked the 1901 foundation of a civic heritage trust, the *Samfundet S:t Erik*, by zoology professor Wilhelm Leche and architect I G Clason, to cooperate with the Nordiska and Historiska museums; in 1906, with 1,600 members (more than the National Trust in England) it carried out a government-assisted inventory of the Gamla Stan and in 1916 published this in Ragnar Josephson's *Borgarhus i Gamla Stan*.[94]

Even France soon began to feel the influence of the Old Town ideal. Here, owing to the country's 17th-century prosperity, the archetypal *Altstadt* was less prominent, as the major cities were dominated by formal facades, and 'Vieux-Paris' itself had been split into disconnected fragments by Haussmann. With the swing in public sentiment against him, a strategy to rescue surviving fragments of Vieux-Paris, such as the Marais and the Ile St-Louis, emerged in the late 19th century, led by the city council. Under the Third Republic, historical and antiquarian studies boomed in Paris: the 1880 opening of the Hotel Carnavalet as a museum of civic history unleashed a flood of popular publications and initiatives, including the Society of Friends of Paris Monuments (founded in 1884 by antiquarian Charles Normand) and the Old Montmartre Society (1886): agitation for the preservation of Montmartre climaxed after 1910. Actual preservation initiatives still focused on individual monuments: the city council established an Old Paris Committee to work with the Historic Monuments Commission to accelerate monument classification. However, the 1913 preservation law included early provisions for ensemble preservation, allowing in principle protection of the surroundings of classified monuments; ensembles already given *ad hoc* protection included the classical Place Royale in rapidly industrialising Nancy, designated in 1889.[95]

Reunified Italy, with its mosaic of old city states and regional cultures, had a similar geopolitical situation to Germany, but a very different urban heritage ethos. The simple polarisation between medieval wholeness and alienating modernity was lacking. There was a more complex pattern: in Rome, the presence of antiquity still seemed overwhelming and, in most places, post-medieval classical buildings (Renaissance and even Baroque) were ubiquitous and widely respected. So there was no universally shared reverence for the heritage of medieval *Gemeinschaft*, although Italian commentators also advocated 'living' rather than museum cities: in 1880s'/'90s' Venice, Boito passionately argued for integration of monuments into the 'living city'. Instead, there was a wealth of *ad hoc* debates and positions, ranging from all-inclusive urban conservation concepts more precocious than any other in Europe, to a feeling on other occasions that almost anything, even medieval fabric, was expendable when classical remains were being unearthed. Italian cultural nationalism was complicated by a double orientation, towards the independent *campanilismo* (localism) of traditional Italy *(italianità)* as well as the grandeur of Roman antiquity *(romanità)*. Those years also saw the trenchantly polemical tirades of the Futurists, who violently attacked all aspects of the past.[96]

Although there were vigorous protests against Haussmann-like demolitions in various Italian cities (including Florence, where 1898–1901 saw successful agitations, led by English cognoscenti and a new Society for Defence of Old Florence, against demolition of the Piazza di Parte Guelfa), these pressures came to a head in Rome. There, radical *isolamento*, *sventramento* and *sistemazioni* were imposed in the 1870s and '80s on a city that had just experienced significant population growth, from 135,000 in 1805 to 226,000 in 1870, while lacking any effective middle class able to mount civic-amenity campaigns.[97] The foundation of the Artistic Association in 1890 heralded growing concern, not just for elite monument restoration, but for protection of the collective urban fabric. The extreme complexity of that fabric prompted the development, especially by Giovannoni, of a precocious and flexible doctrine of contextual urban development, or *ambientismo*. This often involved active replacement of decayed existing fabric, through a small-scale redevelopment pattern, or *diradamento edilizio*, that contrasted starkly with the sweeping *sventramenti* of the Haussmann tradition. *Diradamento* (or 'thinning out') presupposed a careful treatment of historic areas, including humble as well as grand buildings. Unhygienically congested areas could be selectively thinned out by demolition, while conserving key elements, and avoiding clearing huge spaces round monuments. In 1913, Giovannoni advanced a Sitte-style proposal for *diradamento* of a specific area of Rome, the so-called 'quartiere Rinascimento' around the Piazza Navona.[98] In the works of Giovannoni (as systematised in his 1913 articles) and in his town-planning schemes and those of younger architects like Marcello Piacentini, old-town conservation and classical new-town development were juxtaposed, without any sense of incongruity, notably in Piacentini's 1907 plan for extension of Bergamo, implemented after World War I.[99]

Urban England: historic towns, industrial cities

THE Old Town conservation concept was at its strongest in statist Central Europe. At the extremities of the continent, its hold was rather weaker. In Russia, for example, the traditional, timber-built urban heritage was more dispersed in character and had been ravaged by many fires over the centuries, as well as by a massive rebuilding drive in the late 18th and 19th centuries: protests against demolitions in Moscow and unplanned industrial development in classical St Petersburg only really began to gather pace in the early 20th century. In Britain, detached from continental tensions over borderlands, the picture was quite complex. Here, in an extension of the 18th/19th-century split between English and Scottish approaches to conservation, the turn of the century saw a polarisation between relatively low-key concepts of country-town heritage in England, and an extreme and idiosyncratic Old Town ideology that developed in the city of Edinburgh – a striking example of the cultural parallelism that has characterised the history of conservation.

In England, old-town conservation efforts were given a distinctive character by the generally more diffuse form of old towns, lacking the sharp division between *Altstadt* and modern city. The English countryside and country towns were already relatively sacrosanct, as the absolute contrast between industrial squalor and historic villages, towns and cathedral cities was seen as a simple given fact. England had less need for a Schulze-Naumburg or a *Heimatschutz*, as most places seen as meriting preservation were already 'safe'. Few of the great industrial cities had a significant medieval core, yet the precocity of Improvement and urban growth had left its mark everywhere; most older towns were dominated not by gabled medieval-style buildings but by the classical horizontality of the 18th and 19th centuries.[100]

In the late 19th century, the 'Domestic Revival' in new architecture conferred a special prestige on the timber-framed architecture of the Tudor period. In the city of Chester, whose streets were still dominated by brick classical facades well into the 19th century, the mid-century saw growing agitation by local antiquarians, led by the Chester and North Wales Architectural, Archaeological and Historical Society, for a restoration of the city's supposedly traditional black-and-white timber-framed architecture. There ensued a succession of flamboyant reconstruction schemes by local

architects, such as John Douglas, T M Lockwood and T M Penson, financed by the super-rich Duke of Westminster, and culminating in Douglas's speculative rebuilding of one side of St Werbergh Street in 1898–9. By 1914, this programme had largely expunged Georgian brickwork from the centre and transformed it into the fabulous black-and-white ensemble we know today.[101]

A final difference between English and continental old towns was the strong hierarchy that characterised the former, the apex being occupied by Oxford and Cambridge, high-prestige urban ensembles free of any suggestion of threat, dominated by the introverted, yet often monumental complexes of the colleges and studded with institutional buildings of the 18th and 19th centuries – in other words, as unlike the stereotypical 'decayed *Altstadt* embedded in industrial *Grossstadt*' as possible.[102]

Leading British conservationists were, however, well aware of the conceptual advances being made in rampantly urbanising Germany. J Baldwin Brown wrote admiringly in 1904 of the German care for the street-picture. The relative dearth of theoretical conservation texts in England contrasted with the Continent's torrent of institutional initiatives, meetings, journals, reports, as well as its sheer density of verbiage, polemic and images (including illustrated books, photographic albums, etc.) And as a result of England's sharp separation between 'tradition' and 'modernity', urban conservation efforts were focused on towns far removed from aggressive development, as with the continued enhancement of Stratford, or the 1911 campaign to 'save' Lavenham. Within the larger towns and cities of England, the lack of medieval built fabric and the dominance of 18th- and 19th-century buildings, led to an overwhelming emphasis on sanitary redevelopment and slum clearance, technologies in which Britain was far ahead of the Continent, but which relied on engineering technologies not dissimilar to the Haussmann tradition. In industrial centres such as Manchester, the East End of London or the Birmingham of Joseph Chamberlain one layer of modernity was piled upon another, leaving only fragments to represent the pre-industrial community that supposedly pervaded places such as Nuremberg: for example the Trinity Hospital almshouses in east London, first subject of the Survey of London in 1896.[103]

Patrick Geddes and the 'conservative surgery' of the Edinburgh Old Town

IN the Old Town of Edinburgh, City Improvement took a strikingly different form. This approach was provoked by the way in which, in Scotland, a more continental pattern of industrial development of old urban centres had prevailed, followed by the projection of radical sanitary City Improvement into historic urban cores. A succession of City Improvement Acts, beginning in the earlier 19th century, applied aggressively corrective approaches to decayed medieval urban fabric, amounting in Glasgow to total clearance of the medieval core. These techniques of radical Improvement were doubtless influenced by the invasive replanning schemes for cities across the British Empire, often directed by Scottish engineers (see above).[104]

In reaction to all this, a more contextual City Improvement policy was developed in Edinburgh from 1867, including infilling of slum pockets with new tenements in neo-historical

facing page
Figure 5.15 St Werbergh Street, Chester, rebuilt in black-and-white style in 1898–9 by architect/developer John Douglas

181

styles. This reaction was shaped by the city's entrenched cult of historical self-veneration since the days of Walter Scott. By the turn of the century, that tradition had generated a largely self-contained Old Town discourse of unusual complexity. Partly, this was shaped by literary influences, by the celebration of romantic Scotland in the novels of Robert Louis Stevenson. Stevenson, although condemning the 'insufferably ugly' gabled villas of contemporary suburbia, was no pessimistic, anti-modern Schultze-Naumburg, and valued the late-18th-/early-19th-century classical New Town as highly as the Old Town: 'Few places, if any, offer a more barbaric display of contrasts to the eye . . . the point is, to see this embellished [old town] planted in the midst of a large, active and fantastic modern city, for there the two react in a picturesque sense, and the one is the making of the other.'[105]

This advocacy of the positive coexistence of picturesque-old and classical-modern was hardly new, and Stevenson's appreciation owed much to the 18th-century Sublime and Picturesque. But in Edinburgh in the 1880s–90s it was developed in a strikingly new direction by an outsider to architecture, the biologist and geographer Patrick Geddes. Geddes's early career exemplified the late-19th-century addiction to all-embracing philosophies linking art and science. He developed a personal world-outlook of planned social progress, grounded in cultural heritage and using Edinburgh's buildings as a didactic foundation. In a whirlwind of initiatives from the 1880s, combining books, exhibitions and theatrical events with social rehabilitation and interventions in the Old Town, Geddes put forward a doctrine of urban regeneration grounded in the historic-cultural landscape of the city, rather like a Greek *polis*. His interventions included building the dramatically sited Ramsay Garden and selectively clearing Lawnmarket back-courts under a Municipal Improvement Scheme – both in 1893. The juxtaposition of Old and New Towns would provide not just a Picturesque and Sublime visual stimulus, but also a symbol of humanistic social evolution.

As a firm believer in scientific progress, Geddes's aim was different from the English Ruskin-Morris tradition: what was needed was not to turn back the clock but to regenerate modernity, replacing the blind materialism of outmoded 'palaeotechnic' society with a 'neotechnic' utopia of progressive planning.[106] Anticipating many mid-20th-century Modernist planners, Geddes argued that city regeneration must be grounded in a comprehensive social and economic civic and regional survey. Then, the regeneration of the city could follow, combining modern suburban development with renewal of the old core as a spiritual urban acropolis. Within the existing fabric, change and renewal was vital, but not through the 'coldly fanatical iconoclasm' of Haussmann or the clinical isolation of old buildings. Influenced by Octavia Hill's struggles for public open space and by admiration for the collegiate formula of Oxford, Geddes envisaged small public courtyards and gardens as the focus of the regenerated Old Town: degraded cities must be transformed into 'Centres of Life'.[107]

The name Geddes gave to this process of selective urban renewal, 'conservative surgery', betrayed his own roots in biology and medicine. Although its practical effect was similar to the *diradamento* developed in Italy two decades later, this was not a conservation philosophy as such. Although he described himself as 'a strong conservative' about demolitions, old buildings were only important to Geddes as an element within a policy of cultural regeneration, driven by 'scientific recognition' of the value of heritage in building a future better than the past. Geddes had no truck with golden ages:

> Conservation of the memorials of the past and the interpretation of its development
> are of encouragement and of service to that opening future which, with our greater

THE TOWN PLANNER.

Figure 5.16 Patrick Geddes

(a) Geddes pictured in a cartoon of c.1915 when working as a town-planning consultant in India; (b), (c), (d) Illustrations from Geddes's 1904 plan for the civic enhancement of Dunfermline (*City Development: a Study of Parks, Gardens and Culture Institutes*). Images (c) and (d) are 'before' and 'after' proposals

(c)

(d)

resources, it should be possible to achieve . . . the past is still here in the present, but the present is already great with the future.

Heritage was not 'mere archaeology nor mere annals' but 'the study of social filiation'.[108]

Inheriting the cultural baggage of Walter Scott and Stevenson, Geddes inevitably saw Edinburgh as an exemplary case: it was 'the chief culture city in the world' and a 'microcosm of geography'. Unlike Haussmann's Paris, with its 'fine' architecture but 'evil' planning, Edinburgh's value was as an ensemble. And Geddes openly welcomed the link between Expo tableaux and urban conservation, arguing of the 1886 'Old Edinburgh' display that it was 'probably the most admirable reconstruction of an ancient city yet effected', meriting erection 'in permanent form', and that its lessons could be applied elsewhere, for instance in his 1904 proposals for civic regeneration of the historic town of Dunfermline.[109]

The role of these renewal and conservation schemes was as much didactic as anything else: Geddes's chief importance was as a propagandist and proselytiser rather than a Riegl-style theorist. Like Riegl, though, Geddes was also a committed internationalist. He and his son-in-law, Frank Mears, prepared a range of displays on the heritage and development of Edinburgh for an international planning exhibition at the Royal Academy in London in 1910 and at the 1913 international Expo in Ghent. Typically of the English-speaking world, Geddes saw the lessons of conservative surgery in imperial and global terms. After 1914, the chief international impact of his ideas was not in Europe but across the British Empire, in drawing up plans for many towns and cities in India, Ireland and Palestine. And in Edinburgh itself, Mears and a range of others such as City Architect E J MacRae perpetuated elements of the conservative-surgery fashion into the interwar period.[110]

All these ideas anticipated Giovannoni's *diradamento* and also paralleled the work of the *Heimat* advocates or of Sitte, with their stress on the role of old buildings as spiritual tokens of a better future. However, although he had earlier admired Buls's work in Brussels, Geddes only became aware of the specifics of German and Austrian planning and conservation on a 1909 tour, interpreting Sitte as a vindication of his ideas of conservative surgery and praising his work in the 1915 edition of his book, *Cities in Evolution*. And there remained fundamental differences, not only between his scientific optimism and the cultural pessimism of the *Heimatschutz*, but also between his internationalism and the intensely politicised nationalism of later *Altstadt* ideology – as we will see in the next chapter.[111]

Crisis of the Movement

Mass heritage, mass destruction

– 6 –

Monument wars

Devastation and rebuilding, 1914–39

The eternal monuments of our history must be allowed to tower in solitary grandeur.
Benito Mussolini, 1934[1]

THE years from 1914 to 1945 saw the climax of the Conservation Movement as a Church Militant. Its chief 'denominations', the national heritages, furiously attacked each other, even as their underlying common values continued to strengthen. The Conservation Movement was swept along by, and in turn fuelled, the new, ferocious currents of the age. These curbed some of its more long-standing passions, such as the debates for and against restoration, and infused others, such as the idea of living heritage, with a new and hyper-politicised urgency. The turn of the century's intense focus on the sacrosanctity of the built fabric, and its elaborate theoretical debates and speculations, faded from view. Never again would there be a Heidelberg-style debate, with thousands of opinions about a single, relatively small building. The new preoccupations were external to the Movement, and above all political: in an age of nationalist extremism, some of the most respected strands of 19th-century conservation thought were simplified into stereotyped, bellicose rhetoric. There was an unprecedented inclusiveness about what could be part of the heritage, yet buildings also took their place in a wider palette of 'memory landscapes'. Mirroring the way in which the 20th century's other favourite 'total' architectural doctrine, the Modern Movement, was co-opted by socialism, it was almost exclusively nationalism that enlisted conservation into its service, a service that involved complicity in warlike antagonisms as well as support for benign forms of small-country emancipation and pride. Nationalist movements mobilised the Conservation Movement; and that driving force would increase until its eventual exhaustion in 1945, allowing a more peaceful renewal of energies to follow.

World War I: from area conservation to area destruction

THE first and foremost task of conservation, for the half-century after 1914, would be with repairing the wounds inflicted by increasingly destructive wars. The late 19th century had seen the expansion of conservation theory to embrace entire landscapes; now, the challenge was the actual devastation of vast swathes of landscape. The urgency and emotional force of reconstruction dwarfed the old nit-picking restoration debates, as it necessitated recreation not just of cherished individual

187

monuments but of entire cities. This was, in a way, a return to Viollet, but not as he would have recognised it.

Of course, the physical scale of these challenges was not altogether unprecedented. Built environments of the past, including modern Western cities, had suffered accidental or natural catastrophe, without consequent cultural trauma. What was crucial – extending the logic of Riegl's argument – was the *reception* of these destructions, and the consequent intensity of moral capital invested in reconstruction. Partly, this was affected by the age and prestige of the destroyed landscapes: the vast expanses reduced to ashes within hours in the devastating Chicago Great Fire of 1871, or flattened a century later in the 1966 Tashkent earthquake, were utilitarian and little valued in character, allowing a simple narrative of optimistic modern reconstruction, with few calls for recreation of what had previously existed.[2]

Nevertheless, it was widely foreseen that the devastation in historic towns during modern warfare was likely to be unprecedented. The destructions of the 16th, 17th and 18th centuries had been forgotten, but Napoleon's depradations in Italy and elsewhere were still within relatively recent memory, and the American Civil War was more recent still. The urban destruction wrought in that wide-ranging conflict mainly concerned utilitarian rather than 'historic' fabric: a German visitor to Richmond, Virginia, in 1868, commented that 'these ruins are not romantic, like our old ivy-covered castles. They are still blackened by fire, and gape glumly at the sky through blank, empty window-sockets.'[3] But the American devastation had prompted a pioneering international debate in the late 19th century about wartime protection. In 1874, in Brussels, Tsar Alexander II convened the world's first conference on codification of the conventions of war; delegates from 15 states agreed a draft Declaration of the Laws and Conventions of War. This argued that cultural treasures were the common heritage of humanity as a whole, and that occupying powers had the rights of administrators, not owners. The implications were fleshed out at an 1889 conference on international heritage protection in Paris (instigated by the French government) and two Hague conferences in

Figure 6.1 The aftermath of the Chicago Great Fire of 1871, seen from the Tribune Building (stereoscopic view)

1899 (convened by Tsar Nicholas II) and 1907. The latter resulted in a convention which stipulated that the trustee duties of occupying nations included safeguarding of historic monuments and buildings dedicated to religion, art and science – provided they were 'not being used . . . for military purposes'. These conferences formed part of the period's often abortive upsurge in international cooperation initiatives: 1910, for example, saw a proposal for establishment of an International Commission for Protection of Nature – a project abandoned on the outbreak of war in 1914.[4]

The Great War soon showed the ineffectiveness of these protection measures, especially given the temptation to exploit the solidity of historic monuments for military purposes. Partly, the conflict continued the 17th–19th-century pattern of towns destroyed by field artillery, and damage was confined to army campaign zones – most notoriously and continuously the 'Western Front' in north-eastern France and Belgium. But enhanced modern firepower vastly intensified the destruction, and the heightened modern sensibility of 'living' heritage elevated the places affected to a martyr-like status. Again, 'reception' was everything. The neglected backwater of East Prussia suddenly became the focus of national concern following its devastation in a 1914 Russian invasion. In the West, the focus of outrage was the town of Ypres, previously disparaged in a Baedeker of 1910 as 'dull and lifeless', despite its 'broad, clean streets' and 'imposing old buildings', including one of Europe's largest secular Gothic medieval buildings, the Cloth Hall.[5] The most severe initial damage in Belgium was actually sustained by the city of Leuven, whose medieval centre was reduced to ruins in September 1914; the renowned university library was gutted by a fire for which responsibility was hotly contested.[6] But it was Ypres that fell victim to the most thoroughgoing destruction of any European historic town, owing to its position in a strategic salient, with the front line less than two miles away during almost the whole war. As early as 22 November 1914, with 10–12 shells falling each minute, the Cloth Hall and St Martin's Church were burnt out. After May 1915, remaining civilians were evacuated, and by 1918 only a handful of buildings were left standing: a man on a horse could look right across the town. Although it is difficult if not invidious to rank cities by degree of wartime damage, contemporary photographs suggest the destruction of Ypres was far more extreme than that of more notorious World War II bombing targets such as Hamburg or Dresden, which were left mainly as fire-gutted ruins rather than (as with Ypres) a sea of rubble.[7]

The bombardments of Leuven and Ypres provoked correspondingly thunderous allied condemnations, including letters to the German government from SPAB and from Sir Edward Poynter, president of the Royal Academy.[8] Equally devastating were the attacks on front-line towns in France, a campaign that reinvigorated the grand old French tradition of criticism of heritage 'vandalism'. Some towns, such as Amiens, St Quentin and Soissons, were bombarded relatively briefly in free-flowing campaigns of 1914 or 1918, while others were close to the front line during the whole war and suffered almost as badly as Ypres: in Arras, the front line sometimes ran through the town centre itself.

Most severely damaged, and elevated to the status of France's '*ville-martyre*', was Reims, briefly occupied in September 1914 and then left just on the French side of the front for the rest of the war. By 1918, three-quarters of all its buildings were destroyed or irreparably damaged, including the archbishop's palace, city hall, basilica of St Remi and most of the city centre. Most controversial, though, was the fate of the cathedral of Notre-Dame, traditional coronation seat of the kings of France. In the first bombardment of 19 September 1914, it was badly damaged, and further shelling, especially in February 1915, brought down the vaulting and apse flying buttresses and severely damaged the ornate pinnacles, stained glass, internal decoration and sculptural figures on the main

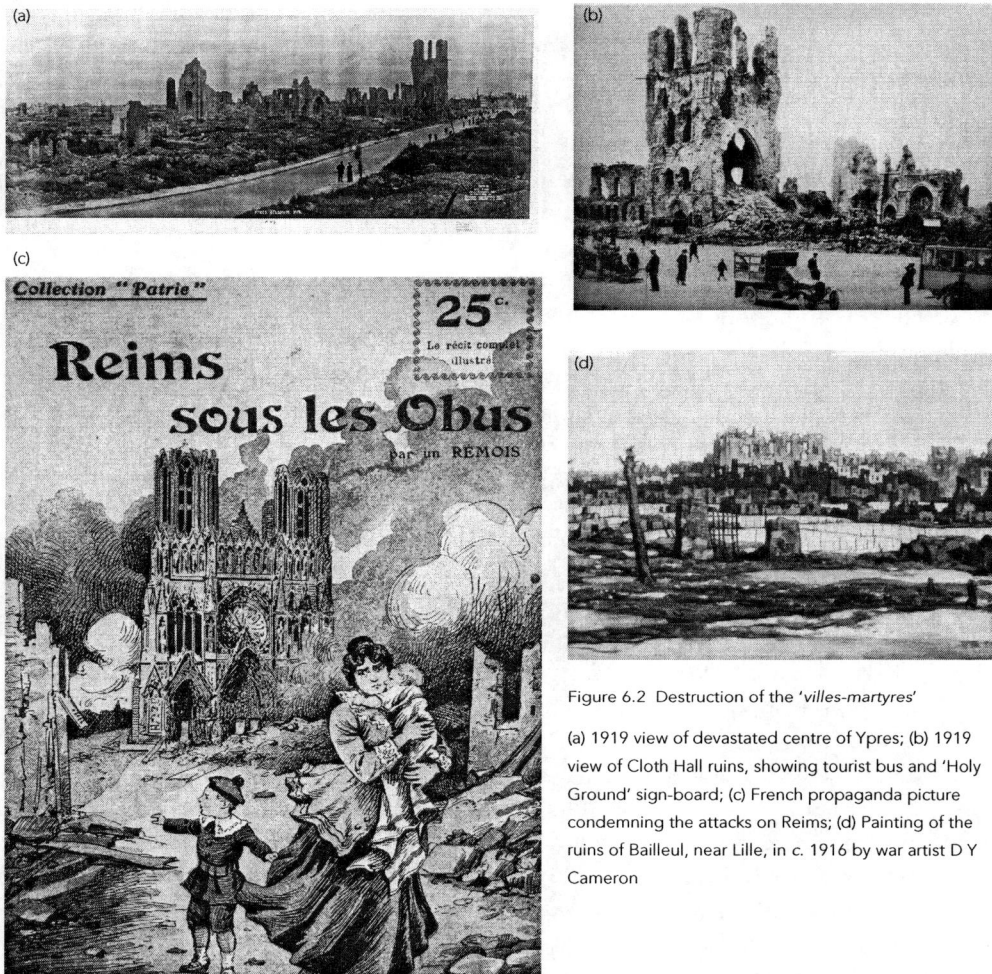

Figure 6.2 Destruction of the 'villes-martyres'

(a) 1919 view of devastated centre of Ypres; (b) 1919 view of Cloth Hall ruins, showing tourist bus and 'Holy Ground' sign-board; (c) French propaganda picture condemning the attacks on Reims; (d) Painting of the ruins of Bailleul, near Lille, in *c.* 1916 by war artist D Y Cameron

portals and frontages. No significant attempt had been made to remove fittings beforehand – a lesson that was subsequently influential in World War II. In northern France, the French military mission attached to the British Army established an emergency *Service de protection des oeuvres d'art* (Art Protection Department) to secure ruins and evacuate moveable pieces, but by then most of the damage was done.[9]

Although the bombardment of Reims accounted for only 2 per cent of all the wartime devastation in north-eastern France, the cathedral's symbolic importance ensured that it became the focus of a bitter propaganda war, based on the hyper-accentuation of existing stereotypes. French writers argued that the Germans were militaristic barbarians, jealously set on destroying superior French culture, who had left the cathedral 'no more than a heap of ruins'. German writers, conceding the awkward point that German shells had undeniably caused the damage, riposted that the French authorities had cynically placed military observation posts on the cathedral to provoke bombardment and that their claims were thus hypocrisy. A more credible victim of intentional 'cultural destruction' was the Chateau de Coucy, a massive, part-ruined 13th-century castle with donjon and corner towers, renowned before the war as one of France's half-dozen most visited tourist sites. Occupied by the German army as a headquarters from 1914, it was even visited by the Emperor himself. In March 1917, however, with the army in retreat from the area, the entire complex was systematically blown up on the orders of General Erich von Ludendorff: the French duly decided to keep the rubble as it was, to serve as a 'monument to barbarity'.[10]

Acutely aware of the negative impact of actions such as this on Germany's image as a *Kulturnation*, the government established an extensive monitoring apparatus in the occupied areas, appointing Paul Clemen in 1915 as 'cultural protection officer' (*Kunstschutzbeauftragter*) in the army general staff and attaching art-historian officers to front-line units to inventorise and protect monuments. In 1915, a Conservation War Convention was held in Brussels, with representatives from Germany, Austro-Hungary and Switzerland. Researchers such as Gurlitt were also sent to Warsaw during its comparatively mild German World War I occupation. The Emperor publicly forbade attacks on culturally prestigious monuments, reputedly threatening with the death penalty any German pilot who bombed Buckingham Palace, Westminster Abbey or the Houses of Parliament in London. In 1919, Clemen published a systematic record of the wartime protection campaign, *Kunstschutz im Kriege*, and continued protesting its legitimacy in later years, for example in his 1933 book, *Die deutsche Kunst und die Denkmalpflege*.[11]

Beyond the main conflict, more episodic acts of heritage destruction flared fitfully. Naval bombardments could bring sudden damage to monuments far from the battlefields, as in the December 1914 bombardment of the English coastal towns of Scarborough and Whitby by German battleships, resulting in the partial collapse of the west facade of the ruined Whitby Abbey.[12] Damage also resulted from conflicts indirectly triggered by the war. For instance, the long-simmering Irish national question eventually escalated into a general civil war in 1919–23, when the grand 18th- and 19th-century architecture of the Protestant 'Ascendancy' was systematically targeted by the anti-British guerrillas of the Irish Republican Army. The Ascendancy had shaped the urban infrastructure of the country, including the brick classical terraces of Dublin, and had dotted the country with hundreds of monumental country houses, classical or neo-medieval in style. These houses now became prime targets for piecemeal republican attack (usually by arson), in an effective campaign of ethno-cultural decapitation of the outgoing ruling class. The war's end and the collapse of Germany and Austro-Hungary also saw a brief resurgence of the Napoleonic phenomenon of cultural looting, for example in the 1919 Italian removal of art treasures from Vienna under armed guard in exchange for food aid, an arrangement fiercely denounced by chief conservator Dvořák.[13]

'Holy ground': the rebuilding of Ypres

So much for the destruction itself – but what was its impact on the now mature Conservation Movement, or the 'modern cult of the monument' as Riegl put it? Paradoxically, one immediate effect was to revitalise and radicalise the intentional-commemorative values that Riegl had pronounced obsolete. The destruction had created for the first time an intensely politicised 'memory landscape' of mass conflict, focused on the Western Front. Here, ruined towns were only one element, alongside the vast campaign of conventional-style war memorials – a new commemorative built environment of community, whose intensity was intensified in the confrontational borderlands shaped by the Versailles land exchanges. Henceforth the concept of the memory landscape would be a key instrument in collective identity-building, but the specifics would constantly and radically fluctuate, including the relative role of historic buildings as opposed to intangible traces. One enduring common thread was the relative unimportance of direct material authenticity: even more than the restoration 'falsehoods' denounced by Ruskin, the modern memory landscape was largely a work of fiction rather than recollection.

Unlike earlier war memorials, the new memory landscape of the 'fallen' would warn as well as celebrate, extending the Christian framework of martyrology to the whole people. In France, state subsidies under a 1919 law facilitated the proliferation of tens of thousands of mass-produced *poilu* figures in villages and towns across the country. The position in Britain was different, with a strong geographical polarisation between overseas and home: the most important Scottish home war memorial, the Scottish National War Memorial of 1924–7, designed by Robert Lorimer, was a conversion of a historic building, a barracks, with an incredible concentration of Arts and Crafts symbolism.[14]

The chief demand now facing the Conservation Movement was how the damaged towns and monuments should be restored, to take their part in these new landscapes of collective 'sacrifice'. In some cases, restoration was ruled out by changes in political and cultural authority: most country houses burnt in the Irish civil war were left to moulder forgotten. In less comprehensively damaged places, some restoration had been possible during the war, as in the post-1916 rebuilding of East Prussia under *Heimatschutz* influence.[15] For the almost annihilated towns of the Western Front, solutions far more radical than rebuilding and repair were necessary, solutions so far-reaching as to completely supersede the old Anti-Scrape debates about restoration, which were concerned with buildings that were still standing. Buildings that had largely disappeared posed different ethical challenges, and it was the achievement of the 20th-century Conservation Movement to develop a rich diversity of solutions to this problem.

The postwar rebuilders did not begin their work altogether without recent precedents. For example, a vigorous controversy had flared briefly in 1902 after the 323-foot-high, largely 16th-century Campanile of St Mark in Venice, following minor alteration works, collapsed without warning on 13 July into a vast pile of bricks. It was ironic that the fallen tower had been the chief landmark of the favourite city of arch-anti-restorationist Ruskin, but an emergency city council meeting the same evening voted decisively to build an exact replica. This provoked a brief but heated international debate over the issue of authenticity. Many English commentators naively assumed that one could apply Anti-Scrape dogma to this unprecedented situation: the London *Daily Telegraph* chided that

> The charm of the Campanile of St Mark's was the stern and imposing beauty of proportion and strength. It has fallen, with all the associations that have clustered around it so long, and it may fervently be hoped that no civic zeal for restoration will ever lead a Venetian prefect to try and build it up again. Its memories are safe upon the canvases of Canaletto; its proportions are enshrined in every architectural text-book in the world; but the grace its builders gave it is lost for ever, with the art they knew.

This unworldly rhetoric pointed to the likely limitations of doctrinaire anti-restorationism in the new century – not least because any replacement would not, of course, be a restoration but a

Figure 6.3 Resurrection of the Venice Campanile

(a) St Mark's Square immediately following the collapse in 1902;
(b) Inauguration of the rebuilt tower in 1912

completely new copy! The building of that facsimile duly began straight away, in 1903, under the direction of restoration specialist Luca Beltrami, and was inaugurated in a grand patriotic ceremony in April 1912.[16]

In the terms set out by Riegl the following year, Beltrami's facsimile Campanile was a logical and 'authentic' expression of several key monument values. It displayed historical value, present-day art-value, newness-value, and even use-value, given the requirement of the tourist industry for a recognisable skyline image – a link that would become even more blatant a century later in the reconstruction of Dresden. The one monument value completely absent from the Campanile rebuilding, of course, was age-value (*Alterswert*), in the sense of respect for old substance. Yet even here, if *Alterswert*, as Riegl suggested, is simply a way of highlighting transience, irrespective of monuments' specific identity, and if transience can also include change by violent means, then might not a copy sometimes be even better at evoking *Alterswert* than a repaired original? A facsimile building solution was also adopted in another turn-of-century example, the St Michaelis Church in Hamburg, whose massive 1751–62 tower collapsed following a 1906 fire and, despite opposition from strict conservationists, was replaced in 1906–12 by a copy – designed by architect Julius Faulwasser with the aid of copious drawings from an 1883–6 survey.[17]

Now, however, conservationists faced not just vanished landmarks but entire vanished towns. And in 1918, as the guns fell silent on the Western Front, fierce debates erupted about the most appropriate form of postwar reconstruction. These were most intense in Ypres, where the issues of *tabula-rasa* rebuilding were first fought through, and authoritative formulae were established. This was largely the achievement of one individual, the Ypres municipal architect Jules Coomans, a self-effacing figure who, despite his decisive intervention in the course of the Conservation Movement in the 20th century, remained internationally unknown – a typical example of the way that conservation, unlike its alter ego, Modern architecture, consistently avoided flamboyant self-projection. Trained as a neo-Gothic designer and engineer, Coomans was appointed town architect in 1895 and embarked immediately, and fortuitously, on a systematic programme of inventorisation and restoration of its heritage. Although this programme was only half completed in 1914, it had already generated a comprehensive archive of plans and photographs, which Coomans promptly moved to safety in his wartime home in Boulogne-sur-Mer.[18]

What were the possibilities and choices available in Ypres? Ruled out from the start was any suggestion of rebuilding the town on 'rational, modern' lines – although several Dutch Modernist architects, led by W M Dudok, initially floated such an approach. The two chief alternative strategies proposed for Ypres were both concerned with conservation of the past, but in very different ways. British public opinion, articulated forcibly by Winston Churchill, wanted the ruins of the entire town preserved as 'Holy Ground', to symbolise the British sacrifice – a concept that combined intentional commemorative value and traditional British Anti-Scrape, citing Coucy in support of its case. Several key Belgian architects backed this position, including Eugene Dhuicque of Brussels, who advocated the safeguarding of a ruined 'zone of silence' in the centre, along with the ruins of St Martin's Church, the Cloth Hall and Belfry; Dhuicque argued that the Middle Ages had no greater right to commemoration than the late war. These speculations by outsiders, though, were completely unacceptable to public opinion in the town itself. As early as 1916, with the support of mayor Colaert, Coomans began drawing up a counter-strategy for rebuilding the town in its entirety on a two-tier pattern, precisely reconstructing the principal monuments against a background of reconstructed street architecture in a simplified, somewhat generic 'Flemish Renaissance' style of stepped gables – later dubbed the 'Ieper Style' by Coomans.[19]

By 1920–1, the population's rapid return had created a fait accompli in favour of Coomans's plan. The British decided to focus commemorative efforts on conventional war memorials, including a stately new classical memorial gateway, the Menin Gate, designed by Reginald Blomfield. Coomans was now free to flesh out and implement his two-tier strategy. With the arrival of swarms of outside architects and builders from across Belgium, a frenzy of building was soon underway: most houses in Ypres were built in the years 1922–4, in Coomans's favoured Ieper Style, and reconstruction of St Martin's Church was begun, in an enhanced form including 'restoration' of a spire that had collapsed as long ago as 1433. That project, begun in 1914 but frustrated by the war, was finally completed in 1930. Rebuilding of the Cloth Hall and Belfry followed from 1928, with the latter completed in 1934, but the rear of the Cloth Hall only finished in 1967, by Coomans's successor, P A Pauwels. Ten years after the armistice, the ruins of Ypres had largely been eradicated, and an international boom in commemoration-tourism was underway, culminating in the inauguration of the Menin Gate in 1927.[20] Most Belgians, and many others too, made a 'pilgrimage' to the restored town. The solution devised by Coomans at Ypres became the foundation for the most influential formula of postwar historic-town reconstruction throughout 20th-century Europe. In practice, the rival 'memorial-ruin preservation' strategy would only be successfully applied to individual buildings, not entire urban ensembles.

'Martyr towns': the reconstruction of France

In France, although 1915 had seen calls for ruin-preservation as a warning to future generations, culminating in 1917 in the decision to retain the remains of Coucy, the main postwar debate was between *conservateurs* and *novateurs* (the latter advocating a measured modernity). True to France's tradition of centralised heritage administration, the Historic Monuments Commission (CMH) threw itself into the task of restoring not just key individual monuments but entire historic towns.

Figure 6.4 Reconstruction in Flanders and Picardy

(a) Ypres Cloth Hall, rebuilt in 1928–67 by J Coomans and P A Pauwels; (b) Ypres, S. de Vosteghe Straat 20, (architect F Verheyen, 1922): a typical rebuilt house in Coomans's generic 'Ieper Style'; (c) The Grand Place in Arras, rebuilt 1919–23

Charles Gennuys, inspector-general of monuments (with chief architect Henri Deneux) took charge of heritage reconstruction in the devastated areas, reporting after a rapid inventory that 600 *monuments classés* (later raised to 808) required total reconstruction *à l'identique* and securing a 1919 law stipulating rebuilding of key monuments in 'the same character' as before. The scope of the classification system was expanded by the decision to list some key ensembles, such as the surviving facades of the Grand Place, Petit Place and surrounding streets in Arras, burned out in October 1914.[21]

Of the 808 monuments selected for reconstruction, half were completed by 1929 and almost all the remainder by 1934. A few were facsimiles of completely destroyed structures, such as the church of Vesseny (Aisne), or were reconstructed in openly Modernist styles: at the church of St Pierre de Roye, Picardy, architects Duval and Gaze attached a Perret-style concrete nave and tall clock tower to a restored medieval choir. Usually, a twin-track policy similar to that of Ypres was followed, intended to promote both cultural regionalism and tourism: by 1919 the city of Arras had already established a tourism agency. In Arras, the main monuments (the cathedral, St Vaast Palace and the town hall) were reconstructed exactly, as were surviving facades in the Grand Place and Petit Place, but other nearby facades were built in a generally 'Flemish' style using local stone: the street-rebuildings and the town hall were complete by 1923 and the cathedral by 1934 (architect, Pierre Paquet).[22] At nearby Bailleul, destroyed almost as comprehensively as Ypres, a similar 'Flemish style' reconstruction was pursued. A 1919 plan by architect Louis-Marie Cordonnier proposed 'restoration' in brick rather than the actual, mainly colour-washed facades of prewar days. And at Béthune, where the Town Belfry in the Grand Place had been abutted by a clutter of later houses, their ruins were cleared (despite local protests), leaving the belfry isolated in a wide square ringed by neo-Flemish facades, many by architect Jacques Allena; the soaring brick church of St Vaast was reconstructed by Cordonnier in a modern style, with prestressed concrete frames and brick cladding. A Haussmannesque proposal for Soissons by municipal architect Paul Devouchelle was modified by omitting a proposed triumphal boulevard, to avoid isolating the restored cathedral. The reconstruction of villages, in Picardy and elsewhere, emphasised creative design of new ensembles, inspired by records of existing traditional architecture. In January 1917, *L'Illustration* showcased an exhibition of peasant architecture in the devastated regions, and André Ventre, chief architect of historic monuments, undertook a campaign of measured surveys (published 1930). The 1925 Paris Expo included a *village moderne*, with regularised cottage-pavilions by Historic Monuments architects – a forerunner of the famous *Centre regional* in the 1937 Expo.[23]

The vast scale of reconstruction preoccupied most French conservation architects in the 1920s and '30s, although a US $1 million grant in 1924 from John D Rockefeller encouraged a last burst of Viollet-style radical restorations in the Versailles marble courtyard and at Fontainebleau. There were also further advances in the French administrative system, including steps towards inventorisation and protection of ensembles. The *Loi Cornudet* of 1919 and a further act of 1924 required all large towns and historic centres to prepare a structure plan agreed with CMH, and a 1930 law extended protection in principle to landscapes and townscapes, establishing an initial inventory. Significant examples included Mont St Michel, put under protection in 1928 by CMH (which then began demolishing illegally built shacks) and the landscape of Versailles, where a system of optical 'cones' was introduced in 1929 to safeguard the main vistas. CMH also strongly promoted technical research and education in conservation, establishing a special 50-lecture conservation course at the École de Chaillot from 1920.[24]

Reconstruction debates within France were dominated by the case of Reims, the 'martyr city'. This status went hand in hand with a boom in tourism, with thousands of people visiting the city daily during the 1920s. Here, in contrast to Ypres, a reconstruction plan of a distinctly modern character was implemented. Earlier in the war, the Germans had planned to reconstruct the city after their eventual victory in a 'Rhineland' rather than 'Prussian' or 'southern German' style, but in 1917–19, aided by the Rockefeller funding, there was a flurry of classical French and American proposals: the sculptor Rodin suggested leaving the city in ruins, but that proposal was ignored. Some citizens resented the US interventions, as with the British in Ypres, but eventually, in 1920, following a visit by US

Figure 6.5 Rebuilding Reims

(a) Reconstruction of the roof of Reims Cathedral underway in 1926: photo by Henri Deneux; (b) Present-day view of the west facade of Reims Cathedral (restoration completed 1938); (c) Present-day view of the cathedral north facade, and adjoining streets rebuilt in a Beaux-Arts classical style in 1921–8

President Wilson and rejection of an unimaginative proposal by the city engineer, a far bolder plan by a Harvard-trained, Beaux-Arts architect-urbanist, George Burdett Ford, was approved. It envisaged that the monumental centrepieces would be faithfully restored, with Deneux supervising work in the cathedral, St Remi Basilica and St Jacques church, and principal CMH architects Bernard Haubold and Max Sainsaulieu overseeing the others. The wider fabric of Reims would be rebuilt on modern, functionally zoned Beaux-Arts lines, with a ring of garden suburbs, parks and industrial zones linked to the centre by broad axial avenues and two new central boulevards cut through to improve views of the cathedral. Funded by two city council reconstruction loans of over 300 million francs in 1921–2, the plan was implemented in 1921–8 by various Parisian architects and left rebuilt Reims richly endowed with Art Deco urban architecture, including both ordinary street facades and public set pieces like the Carnegie Library (designed by Sainsaulieu and completed in 1928); by 1926 the prewar population had almost been reattained. At the cathedral, Deneux installed a temporary roof, with a new reinforced concrete roof structure of his own design below. Financed by Rockefeller, the vaulting was completed by 1936 and the exterior details were reinstated, including the elaborate gabled arcading in front of the nave roof. The cathedral was solemnly rededicated in July 1938 in the presence of French president Albert Lebrun, senior clerics and international dignitaries: a German reporter enthused that 'the towers of the cathedral stand anew in the midst of the province of Champagne, to evoke the past, and to exhort reconciliation and lasting entente between two ancient civilised nations'.[25]

Interwar internationalism: the Charter(s) of Athens

THOSE particular sentiments of reconciliation were soon forgotten the following year. But overall, the interwar period saw important steps towards a more concerted internationalism in architectural heritage.

In some cases, heritage was an accidental beneficiary of postwar geopolitical shifts. In Palestine, for example, the British Mandate – a formalisation of post-1917 British rule following the Treaty of Sèvres in 1920 and League of Nations ratification in 1922 – entrusted British engineers, planners and architects with safeguarding the character of the old city of Jerusalem. In the 19th century, Jerusalem had witnessed bitter rivalry between Greek Orthodox and Catholic churches over control of the Holy Sepulchre Church (and the Basilica of the Nativity at Bethlehem), followed by a competitive building frenzy of churches and hostels by colonial European powers, each set on securing a foothold in the sacred soil – a campaign that culminated in the triumphal horseback visit by the Emperor Wilhelm II in 1898.[26] Now, a succession of British-sponsored plans advanced a conservationist agenda, within a geopolitical framework of international 'custodianship'. Guidelines were set by Alexandria's City Engineer, William McLean, whose town-planning scheme of July 1918 proposed to channel new development to the outskirts and to place skyline-impact and building-material restrictions on the historic core, which would be enveloped in a belt of undeveloped land. Following interventions from 1919 onwards by Patrick Geddes (who called for the old city to be isolated in a green 'Sacred Park') and Charles Ashbee (who founded a 'Pro-Jerusalem Society' dedicated to civic improvement), a policy emerged that combined Geddesian conservative surgery with Haussmann-like *isolement* and clearance of shanty excrescences around key monuments: subsequent British building ordinances stipulated that all new buildings should be stone-faced.[27]

More important in the ultimate evolution of the Conservation Movement were the growing attempts at systematic internationalism, channelled through professional organisations: a strategic vision for definition and protection of the 'common heritage', in the light of the failure of the prewar conventions to protect monuments during World War I. Here, contrary to its later reputation for geopolitical ineffectiveness, the League of Nations took the lead, founding an International Committee on Intellectual Cooperation (ICIC, in 1922) and an International Museums Office (IMO, in 1926), based in Paris; in the Americas, 1935 saw a development of the Hague provisions in the Roerich Pact.[28]

The outcome was the 'Charter of Athens' – forerunner of the international charters that would dominate the Conservation Movement from the 1960s and '70s. In an angry, nationalistic age, this was a mere foretaste of the later explosion of collaborative internationalism and multiculturalism. But this charter only emerged after a protracted, incremental process that would typify the processes of international conservation. First came a 1930 IMO conference on scientific preservation of works of art, in Rome. Then followed the main architectural conservation conference, in October 1931: a 10-day meeting in Athens, attended by 120 representatives from 23 mainly European countries. Chaired by Paul Leon, the French president of CMH, its sessions set out the approaches of various countries. A paper by Giovannoni explained the newly issued *carta italiana del restauro* (1932, updating an 1883 original), and a lecture by Leon conveyed the new anti-restoration ethos of French officialdom. The conference conclusions were forwarded to the ICIC and the League, which approved them in 1932, as the 'Athens Charter'. The Athens conference proceedings (published in Paris in 1933 by the Committee on Intellectual Co-operation, as *Le conservation des monuments d'art et d'histoire*), and the 1932 charter, anticipated the systematic methodology of such events in the future.[29] As its full title suggested – 'The First International Congress of Architects and Technicians of Historic Monuments' – what was valued was thoroughness, consensus, consolidation, homogenisation, as opposed to the bold, dashing innovativeness of individuals such as Goethe, Ruskin, Morris or Riegl.

The Athens conference comprised seven main sessions – 'general principles and doctrines', administration and legislation, aesthetic enhancement, restoration materials, deterioration, conservation techniques and international collaboration. Conspicuously absent was any definition of a 'monument'. The charter based itself squarely on the post-French Revolution concept of historical heritage and the Enlightenment discourse of control and surveillance: monuments (once identified by experts) were 'entails' or 'patrimony', in this case of all humanity rather than of the nation, which must be passed on unaltered, with strict 'custodial protection'.

On the issue of restoration and authenticity, the Athens charter endorsed an institutionalised Ruskinian position, echoing the 1932 Italian charter's repair-only formula, respecting styles of all periods, while exploiting modern technology and construction. Article 1 noted approvingly the general international 'tendency to abandon restorations *in toto* and to avoid the attendant dangers by initiating a system of regular and permanent maintenance', and cautioned that 'when, as the result of decay or destruction, restoration appears to be indispensable, . . . the historic and artistic work of the past should be respected, without excluding the style of any given period'.

An awkward issue of principle was posed by the host city, where an escalating frenzy of *anastelosis* was underway on the Acropolis under the direction of Nikolaos Balanos, with the earlier work on the Erechtheion being outdone in 1923–30 by efforts to re-erect large sections of the Parthenon. Although the invasive character of the programme, involving wholesale dismantling and re-erection, and its refusal to distinguish visually between old and new elements, both defied the new international consensus, the charter politely acquiesced in limited *anastelosis* and sidestepped a proposal that the Acropolis 'should be declared the common property of all humankind'. Only later, in the 1960s and '70s, was Balanos's work conclusively discredited when the damage caused by corrosion of his iron connecting bolts was discovered.[30] Balanos's was not the only initiative of radical intervention in interwar Athens. A vast, externally directed project, echoing the 18th-/ 19th-century excavations in Rome by rich foreigners, commenced in the same year as the Athens charter: the excavation of the Agora, cradle of ancient democracy, by the American School of Classical Studies at Athens. Founded in 1881, this school was re-energised in the 1930s through a large grant from J D Rockefeller. Its showpiece Agora excavation was headed initially from 1931 to 1945 by field director T Leslie Shear, and from 1946 to 1967 by Homer A Thompson. Repeatedly destroyed and rebuilt, the Agora area had been partly excavated in the 19th century and truncated on the north by the Athens–Piraeus Railway. The first, 24-acre portion was acquired in 1925, and the start of its general excavation in 1931 rapidly attracted some of the most illustrious names in Greek archaeology.[31]

In the Athens charter's organisational recommendations, community priority over private property rights was accepted and enhanced international collaboration was demanded, including establishment of documentation and education centres. Practical use-value should be given priority: 'the occupation of buildings, which ensures the continuity of their life, should be maintained, but . . . they should be used for a purpose which respects their historic or artistic character'. The only indirect concession to the idea of area conservation was a call for prohibition of unsightly advertisements and disfigurements near monuments. In the wake of the Athens conference, 1934 saw the foundation of a *Commission internationale des monuments historiques* (again under League auspices) – but by then the worsening geopolitical climate was militating against the success of such initiatives.[32]

Significantly, Athens also hosted, in 1933, the fourth meeting of the key Modernist grouping, CIAM (International Congresses of Modern Architecture). This was one of the landmarks in the evolution of the Conservation Movement's alter ego, the architectural Modern Movement. Held in the School of Architecture in the National Technical University of Athens, its conclusions were published in 1941 by Le Corbusier – also, confusingly, under the title, 'Charter of Athens'. The relationship between the Modern Movement and the Conservation Movement figures ever more prominently in the next three chapters, at least in relation to Western Europe. Most commonly, that relationship is assumed today to have been intrinsically hostile. But in reality it was rather complex in character, with the two Movements having as much in common as in conflict. Both were structured around an internal narrative of progress, one springing organically from the past, the other breaking from it. And they shared a strong belief in the separation and clear expression of new and old architecture, stylistically and spatially.

This concept of a dialectical or complementary relationship of conservation and modernity is visible even in extreme Modernist polemic, such as Le Corbusier's 1925 *Plan Voisin* vision for Paris, which dotted isolated monuments, such as the Tour St Jacques, in a sweeping Modernist landscape of greenery and towers – an approach that curiously resembled the National Socialist 1937 programme of urban reconstruction, which (as we will see below) would have stranded heritage islands in vast, open monumental complexes. Contemporary urban conservation rhetoric, with its emphasis on 'life', was rather closer to the language of the CIAM 1933/41 charter than it was to the

Figure 6.6 Interwar Athens

(a) The Athens Agora, view of the start of excavations in June 1931; (b) Notebook records of the start of site demolition work; (c) 1975 view of the fully excavated and landscaped Agora

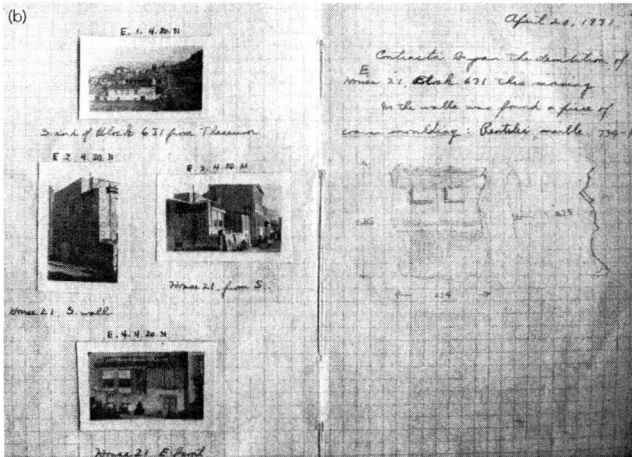

desiccated, museological language of the IMO 1931/2 charter. Le Corbusier's address to the CIAM conference dramatised the new–old dialectic, combining evocations of modern technology with paeans to the Parthenon and arguing that both shared an ethos of purity and light: he hailed the Parthenon, in its newly created isolation, as a timeless symbol of ideal form, as elemental as a modern liner sailing across a wide ocean. The Modernists were well aware, however, of the cultural and aesthetic value of the city as an ensemble. One remarkable passage of the subsequent CIAM charter argued that

> Architectural assets must be protected, whether found in isolated buildings or in urban aggregations. . . . The life of a city is a continuous event that is expressed through the centuries by material works – layouts and building structures – which form the city's personality, and from which its soul gradually emanates. They are precious witnesses of the past which . . . form a part of the human heritage, and whoever owns them, or is entrusted with their protection, has the responsibility and the obligation to do whatever he legitimately can to hand this noble heritage down intact to the centuries to come.[33]

Previously, Ruskin sympathisers had above all tried to differentiate between 'new old' (or 'fake old') and 'old old'. Now, conservationists and Modernists were agreed that the old and new should be clearly different.[34] What was above all condemned, in individual buildings and ensembles, was any 'pastiche' of new and old (a word used in both French and English). This was a more exaggerated version of the old 19th-century bogeyman of fake historic buildings (or fake antique furniture). Among conservationists, there were endless national and international debates about whether new work should be visually distinguishable from old fabric. At the 1931 IMO conference, in contradiction to the *anastelosis* programme underway on the Acropolis, Giovannoni restated the long-standing Italian position that the two must be sharply differentiated. For the moment, though, these debates were mainly confined to a doctrinal and theoretical level, divorced from the fierce controversies on the ground about postwar reconstruction, whether in facsimile or semi-modern form. The conflicts in post-1918 France, with the CMH pitted against modernising civic and commercial forces, gave a foretaste of the countless arguments across Europe after 1945.[35]

Ideological borderlands of interwar heritage

Also of immediate importance within the post-1918 conservation world were the geopolitical schisms that were beginning to permeate the built environment. Countries controlled by totalitarian regimes, left or right, saw new and old architecture in a far more overtly political, instrumental light than did those that retained parliamentary democratic systems. In fact, within new architecture, the entire period from 1918 to 1989 saw incessant change in the political interpretations of the balance between modern and traditional styles: one minute the first might be the preserve of the left and the other of the right, the next minute the distinction might have changed to a democratic-versus-totalitarian split. As far as the relationship between the new and the old in architecture was concerned, the most fundamental differences were over the importance that a regime, or system, placed on historic architecture as opposed to new construction, with innumerable individual variations in the balance between modern and traditional approaches. Here, it was the right-wing dictatorships that exploited their built heritage most forcibly, inflating the prewar rhetoric of *Heimat* and exalting neglected regions into fervently defended borderlands. For communist regimes, old heritage was intrinsically less important, although the use of historic styles in new architecture was a different matter, being highly valued from the 1930s in Stalinist Russia.

In the mid-1940s, as we will see in Chapters 7 and 8, this egregious politicisation of the heritage would fuel its own fiery downfall, along with the general international credibility of nationalist and tradition-based architectural doctrines. The fascist states' political exploitation of heritage was not accompanied by significant theoretical development: what was outstanding was their zeal in putting theory into practice. In the democratic countries, by contrast, the emotional temperature of conservation remained lower, but its boundaries were pushed outwards, especially chronologically: whereas the views of the Nuremberg *Altstadt* in the 1934 propaganda film, *Triumph des Willens*, showed the sacred medieval streets swarming with stormtroopers and stolidly folk-clad peasants, a 1930 advertisement poster of an old-town street in 'York, Walled City of Ancient Days', issued by the London and North-Eastern Railway, depicted it during the early-19th-century Regency period, complete with fluttering ladies and gentleman dandies, and dainty bow-windows.[36] In France, too, the glory of the national classical tradition loomed increasingly large in heritage affairs. Yet there was still much in common between both sides: the pre-industrial golden-age ideal of the 'Biedermeier' in Germany also focused on the late 18th and early 19th centuries, and there was general agreement over such matters as the evils of later-19th-century historicism, the desirability of 'living' rather than 'museum'-like old towns, and the need to purge insanitary and unsightly clutter and commercialism. In the early-20th-century Conservation Movement, strong general parallels were combined with vigorous competition and divergence in detail.

Conservation's most complex ideological relationship in the interwar period was, arguably, that with socialism. As in early 1790s' revolutionary France, socialist attitudes towards the traditional heritage bequeathed by the supposedly obsolete feudal or capitalist systems were deeply ambivalent, with violent fluctuations between an iconoclastic, *tabula-rasa* modernity and a meticulous reverence for key monuments. The initial Soviet regime, under Lenin, generally favoured the heritage cause, appointing writer Anatoly Lunacharsky as Commissar of Education, advocating conversion of old buildings for public use and warning citizens, 'Don't touch one stone – protect the monuments, the old buildings, articles, documents – all this is your history, your pride.' Stalin's rise brought a dramatic reversal, with large-scale closures and demolitions of churches peaking around 1932 and the disbanding of preservation societies such as the Moscow-based OIRU (Society for Study of Russian Mansions, suppressed in 1930). Stalin's increasingly colossal reconstruction plans required removal of any historic fabric in the way, not just in Moscow, whose 16th-century *Kitai-Gorod* wall and 17th-century Sukharev Tower were torn down in 1934, and whose stock of state-preserved buildings had shrunk from 474 to 74 by 1935, but across the whole USSR, under a 1937 urban-beautification strategy. Moscow itself was systematised on a scale outdoing the demolitions of Haussmann or the 19th-century Russian Imperial governments, especially in the 1935 development plan of Lazar Kaganovich and his aide Nikita Khrushchev, with its vast classical axes. Yet that systematisation process featured surprising anomalies, as removal sometimes involved not demolition but shifting of buildings on rollers, exploiting a remarkable Soviet technology developed from 1935 by engineer E M Handel. Some demolitions were undeniably spectacular: in 1931, the Cathedral of Christ the Saviour (1839–83, by Konstantin Ton), a memorial to Russia's victory over Napoleon and a focus for Tsarist Orthodox Christianity, was blown up, with the intention of building a vast, skyscraper Palace of Soviets on the site – a project never implemented. The cathedral was not only a religious building but a late-19th-century one – thus excluded from any interwar definition of valid heritage. Likewise, in Samara, Ernest Gibere's neo-Byzantine Christ the Saviour Cathedral was demolished and replaced in 1936–8 by an opera house and arts museum. In Kiev, however, a far earlier monument, the Monastery of St Michael of the Golden Domes (12th–17th centuries) was demolished in 1935–7, along with the neighbouring Trinity Church. Even St Basil's Cathedral in Moscow's Red Square was only saved from destruction in 1933 when the militant preservationist Petr Baranovsky wired Stalin threatening to commit suicide – after which Stalin rescinded the demolition order (and sent Baranovsky to a Gulag). Reflecting the regime's embrace of grand classicism, Tsarist secular architecture was shown greater respect, especially in the monumental city of Petrograd, sanctified as the cradle of the Revolution and renamed Leningrad in 1924. Already, the Peterhof palace complex had been placed under state protection, and May 1918 saw the first excursion there of 'ordinary Soviet citizens'. In Moscow, too, elaborate precautions were taken to spare monumental structures threatened by the Gorki Street boulevard project: a building as recent as the 23,000-ton Savvinskaya Court, a national-romantic design of 1907 by I S Kuznetsov, was trundled back from the new street-line in 1939 by Handel's building-removers (in this case with sleeping tenants still inside!).[37] Socialist attitudes in the West were very different: left-leaning intellectuals, inspired by the Italian Futurists, indulged in iconoclastic bursts of utopianism, especially in the great variety of proposals for *tabula-rasa* urban replanning.

Mussolini's Italy: *romanità* and *italianità*

FOR the fascist states, by contrast, old buildings were consistently important as a focus of ideological mobilisation. Italy, in particular, experienced a 20-year ferment of diverse ideas about how monuments should be pressed into the service of the dynamic, reborn nation. Ironically, this very diversity, coupled with Italy's continuing status as crossroads of Western culture, would subsequently allow it after 1945 to mutate into the epicentre of the international Conservation Movement – a unique case of a nationalist hotbed regenerating itself into a paradigm of respected internationalism.

During the 1920s and '30s, Mussolini's regime not only enjoyed a prolonged period to mature its architectural policies, but showed great pragmatism towards the built environment, exploiting an ever-changing palette of modern and traditional elements. The two key concerns of any fascist regime were internal mobilisation and external projection, with heritage playing a key aesthetic, legitimising role in both. For these tasks, the Italian Fascists devised two quite distinct cultural world-outlooks, both bound up with the national past and known under the shorthand

facing page

Figure 6.7 Stalinist surprises

(a) The late 1930s' scheme for the widening of Gorki Street (Tverskaya Street) as a grand boulevard: 1981 view of the front facade of Gorki Street 6; (b) Hidden out of public view behind Gorki Street 6, but painstakingly preserved: Savvinskaya Court, by I S Kuznetsov (1907), whose 23,000-ton structure was rolled back from the new street-line in 1939 by E M Handel's building-removers; (c) The General Plan for Moscow, approved 1934 by the Central Committee (and published as *Generalni Plan Rekonstruktsii Moskvi*, 1935): Gorki Street is seen running north-west from Red Square

(a)

(b)

(c)

names of *italianità* (Italian-ness) and *romanità* (Roman-ness). *Italianità* was an inward-looking concept of unity-in-diversity, of a strong Italy grounded in vibrant, regional differences and in the heritage of the sturdily self-sufficient *medioevo* (a term flexibly defined to include both the Middle Ages and the Renaissance). *Romanità* evoked the internationally revered legacy of ancient Rome, especially the age of Augustus, combining external strength and austere self-reliance.[38]

This parallel pursuit of two threads of cultural nationalism was precisely expressed in Fascist conservation policy. In both cases, the old debate for and against restoration was left behind by the new climate of dynamic mobilisation, which radicalised authenticity into a politicised rhetorical concept and redirected restoration away from material fabric towards subjective spiritual essence – a tendency that countered the doctrinal insistence of Giovannoni and others on strong differentiation of new and old, restored and original. Mussolini himself was indifferent to 'antiquarian history': history was only of interest if it could be politicised and instrumentalised. Administratively, the parallel action of national and local forces was vital, with state decrees supporting individual civic struggles. What was excluded from both *italianità* and *romanità* was not only 19th-century historicism but also the Baroque, still widely condemned as decadent. However, there was also a noticeable hierarchy among remains of earlier periods: the archaeological excavations that underpinned *romanità* occasionally required demolition of medieval or Baroque built fabric, as at the Athens Acropolis and Agora.[39]

The pursuit of *romanità* was concerned chiefly with the built fabric of Rome itself, a city whose population was booming (increasing by 75 per cent in the period 1921–6). Here, the regime created a monumental landscape out of a mixture of reconstituted antique sites and stately new classical projects. A 1937 advertisement proclaimed that 'Rome today should appear to the wider world to be as vast, ordered and powerful as it was in the early years of the reign of Augustus.' Paradoxically, the PNF (the fascist movement) originally distrusted modern Rome as provincial and politically hostile. But Mussolini, inspired by poet Gabriele d'Annunzio, set about converting Fascism to the myth of Rome by elevating the city to a new universal capital, a Third Rome, successor to the antique and Christian Romes.[40] In December 1925, Mussolini proposed a five-year plan to make Rome a new wonder of the world, reviving the glory of Augustus through a Haussmann-like building and conservation policy suitably modified for the Eternal City. Surgical demolitions would enable the planning of grand new classical axes and 'liberate' key Roman monuments still cluttered with accretions – the Theatre of Marcellus, the Capitoline, the Pantheon – by clearing vast spaces around them. This approach had been anticipated, before the rise of the PNF, in the planning of the gargantuan Vittoriano monument. As Mussolini memorably declared in a 1934 speech on urbanism, '*I monumenti millenari della nostra storia devono giganteggiare nella necessaria solitudine*' ('The eternal monuments of our history must be allowed to tower in solitary grandeur'). The same years, from 1927, also saw vigorous recommencement of the excavations at Pompeii and Herculaneum, under director Amedeo Maiuri, whose work up to 1961 left the sites in essentially their present condition.[41]

Unlike Haussmann's tightly constrained clearances of corridor-like urban spaces, Mussolini's strategy required more sweeping demolitions, aiming at the archaeological unearthing of a vast swathe of the imperial centre, radiating from the existing excavations of the Forum. Although substantial chunks of picturesque medieval Rome were overlaid on this area, Mussolini decreed that, where conflict arose, the medieval fabric must be sacrificed. There followed a succession of demolition campaigns, starting in 1923–4 with the *liberazione* of the markets of Trajan and the fora of Augustus and Caesar. The year 1926 saw Mussolini give the first pickaxe blow to clear the built accretions round the Theatre of Marcellus; the Campidoglio was 'liberated' by widening the via Tor de' Specchi (with demolitions of Baroque churches and houses) to create a new boulevard, the Via del Mare, inaugurated by the *duce* in October 1930. The period 1928–9 saw the razing of the Piazza Aracoeli to create an uninterrupted open space by the Vittoriano, and in 1931 excavations started for an even grander new boulevard, the Via dell'Impero, linking the Colosseum to the Palazzo Venezia (used from 1929 by Mussolini as his own headquarters).[42]

Individual monuments in previously excavated areas were revisited in a more forcible way. In 1930–8, the Baroque church of S. Andrea nella Curia del Senato, which encased the structural remnants of the Roman Senate (the Curia Iulia) was literally stripped away, inside and out, allowing the bare brick shell to be 'restored' and reroofed. A key player in the PNF's monumental reconstruction strategy was the government's state superintendent of antiquities from 1928, Antonio Muñoz, but Mussolini himself kept a close eye on proceedings: Muñoz recorded in 1934 that 'The *duce* follows day by day, hour by hour, the progress of the vast programme in its minutest details.' Muñoz directed the 1934–6 *sistemazione* of the area around the mausoleum of Augustus: a new Fascist classical square was formed around the stump-like tomb, and Augustus's Ara Pacis was set in a new pavilion of 1938 by architect Morpurgo to its west. The final set piece project, just commenced at the outbreak of war, was the *sistemazione* of the Castel San Angelo and Borghi district, by the opening of the Via della Conciliazione from 1936 (completed in 1950), to provide a grand approach to St Peter's.[43]

There were, however, limits to the reconstruction. In 1928, Mussolini blocked a proposal by Armando Brasini for a more aggressive city-centre demolition strategy, and there was a growing

Figure 6.8 Mussolini – the fascist Haussmann

(a) Contemporary propaganda painting of Mussolini inaugurating demolition work adjacent to the Via dell' Impero in 1935; (b) 1975 view of the 'restored' Curia Iulia

acceptance that large areas of the centre should be preserved. For Giovannoni's doctrine of peaceful coexistence of old and new, the invasiveness of the new *romanità* posed an awkward dilemma of conscience: in 1939, Giovannoni ventured that there should be no more large-scale renewals, only *risanimento*. The demolitions were exhaustively photographed for posterity by Muñoz's department: 7,700 pictures were taken of the city-centre redevelopment zone between 1924 and 1940. The decision to hold in Rome the 1929 congress of the International Federation of Housing and Town Planning stimulated various rival proposals for a Rome masterplan, varying in their degree of confrontation of old and new. A group headed by Giovannoni, despite his advocacy of *diradamento* elsewhere, proposed massive redevelopments similar to Le Corbusier's utopian schemes for Paris, whereas some younger architects, led by Marcello Piacentini, proposed conserving the historic centre and building a completely new administrative centre to the east.[44]

The city's eventual masterplan, drafted in 1930–1 under Piacentini's oversight, combined large-scale preservation with selective new monumental spaces, including the Via dell'Impero. The most grandiose new ensembles would be built away from the historic core, above all in a new city quarter to the south, 'E42' (later EUR), to house a proposed 1942 international Expo – a project that inflated *romanità* to a heroic level. By the late 1930s and early '40s, Mussolini had successfully reshaped the public perception of Ancient Rome in keeping with the values of Fascist modernity and the grandeur of the new Italian empire, following the 1936 occupation of Abyssinia: Edward Schneider argued that year that 'the leader, being a Roman builder, has remoulded the old Roman ideal, stirring up a heartfelt, genuine mystique of citizen, nation and race'. In May 1938 Rome, with Florence, was toured by Adolf Hitler during a much-publicised state visit.

Of all Mediterranean countries, only the Greece of Balanos and the Turkey of Kemal Atatürk witnessed a similar passion for radical archaeological intervention – in the Turkish case, with the aim of establishing national 'roots' in Sumerian and Hittite civilisation. In postwar Israel the same nation-building approach to archaeology would be energetically pursued under the aegis of a succession of soldier-archaeologists, such as the flamboyant Gen. Moshe Dayan: in 1963 archaeologist Yigael Yadin told army recruits at Masada,

> When Napoleon stood among his troops next to the pyramids of Egypt, he declared, 'Four thousand years of history look down upon you.' What would he not have given to be able to say, 'Four thousand years of *your own* history look down upon you'?[45]

The conservation principles that supported the pursuit of *italianità* were very different to those of *romanità*, being derived from Giovannoni's principles of sensitive *diradamento* and the positive coexistence of new and old, as well as Ruskin-like ideas of medieval purity. Yet they also often involved creative remodellings of Italy's historic centres, bringing them closer to Viollet's approach at Carcassonne. In old cities up and down the country, medieval town halls were restored for regime use and piazzas were enhanced for use in revived civic festivals. This was the carefully choreographed face of a new programme of egalitarian cultural tourism for working-class Italians. The prototype which inspired many others was Giovannoni's own 1913 plan for the Rinascimento quarter in Rome itself, a project which he pursued throughout the late 1920s and early '30s.[46]

Although Giovannoni, with his close fascist ties, became marginalised politically after the 1943 division and part-occupation of Italy, his ideas had been well publicised in his 1931 textbook, *Vecchie città ed edilizia nuova* (*Old Cities and New Building*) This inspired a diversity of *diradamento* schemes

by designers in various other cities. These drew extensively on Giovannoni's (and ultimately Sitte's) concept of the need for a designed, artistic unity, as set out in *Vecchie città*. The first step, usually, was preparation of a new development plan, preferably by an architect, integrating the old town into an overall design for city and regional development, to cater for modern traffic and avoid any 'museum'-like atmosphere. Then, the old town itself could be carefully revitalised, with complementary modern interventions, maintaining a dialectical integration of major monuments with vernacular 'minor architecture' (rather than the antagonistic relationship between the two in Mussolini's Rome); sanitary clearances would be concentrated in the centre of street blocks, creating new garden courtyards. The framework echoed both Geddes and Riegl, in the way it exploited the use-value of the old urban fabric.[47]

Figure 6.9 The quest for Augustan glory

(a) New facades of the 1938 'sistemazione' around the mausoleum of Augustus; (b) Museum of Roman Civilisation, 1937 model of Imperial Rome; (c) 1930s' Italian and German stamps commemorating the Augustan Exposition and the Munich Putsch (the latter featuring
the Munich Feldherrnhalle)

209

Although the initiative in these urban regenerations was almost invariably local, the central state helped by targeted interventions, including a pioneering 1928 law passed by the Ministry of Public Instruction for preservation of Siena and San Gimignano. The latter, with its famous towers, was a textbook example of systematic Fascist restoration of a canonical historic town. In 1922, its municipality began promoting itself as a city associated with Dante and started a comprehensive programme of *ripristino* (recovery), inspired by a mildly racist ideology of medieval purification. The main architectural consequence was the replacement of classical features by 'medieval' ones, including the remodelling and crenellation of the town hall, the stripping of a neoclassical facade from the church of the Collegiata and the building of a 'restored' loggia at the corner of the main Piazza del Duomo in 1934–6, picturesquely linking the city's two main spaces into a new and more theatrical unity. Nearby Arezzo, less well endowed with established monuments, underwent a more radical reshaping in the 1930s at the instigation of the local *podestà* (Fascist chief administrator), P L Occhini, with Giuseppe Castellucci as architect. In this case, the campaign promoted an association with Petrarch, including restoration/creation of a 'Petrarch House'. Arezzo's main square, hitherto dominated by 16th-century classical buildings, was medievalised with new towers and castellated, arched rubble facades, becoming the setting for a reinvented civic festival, the 'Joust of the Saracen', to match Siena's *palio* (redesigned in 1928); one palazzo was converted in 1932–3 into a local Fascist headquarters, complete with heightened main tower to symbolise Fascism's medieval roots.[48]

One of the set pieces of the nationwide propagation of Giovannoni-style *diradamento* under Fascism was the town of Bergamo. Its prewar extension plan by Piacentini, juxtaposing a new, stately modern-classical lower town against the backdrop of the hilltop old-town (*città alta*), had, in effect, anticipated the dialectic of *romanità* and *italianità*. Now, one of Piacentini's pupils, engineer Luigi Angelini, an enthusiast for local vernacular architecture, drew up and implemented (in 1930–42) a plan of exceptional sensitivity for *diradamento* of the old town. The main pedestrian east–west axis (via Gombito/Colleoni), focused on the Piazza Vecchia, was reinforced and a new, parallel rear footway was formed from the Piazza Mercato Scarpe to the main square by selective clearance of courtyard slum areas into small, linked garden spaces – one of which was later named Piazza Luigi Angelini in his memory. The resulting project was highlighted for praise by Piacentini himself, as a unique case of 'a sensible theory put effectively into practice'.[49]

In towns where a less harmonious situation prevailed, a conservation programme might be drawn up to counter an aggressively modern municipal development plan. In the southern city of Bari, where a 1926 city plan proposed to drive a massive axial street through the old town, architect Concezio Petrucci produced a counter-proposal in 1931, including careful measures for *risanamento*. Hailed by Giovannoni as the first successfully implemented, concerted old-town revitalisation, Petrucci's plan diverted new streets round the waterfront edge of the old town, whose fabric of minor architecture was reverently preserved. During the '30s, city after city adopted development plans that

facing page

Figure 6.10 *Italianità* in action

(a) 1945 portrait of Luigi Angelini; (b) Progress plan of Angelini's Bergamo scheme as at December 1941; (c) 1942 long-section of the newly created pedestrian route from Piazza Mercato Scarpe to Via Lupo; (d) 2011 view of the new route from north-west; (e) 2009 view of the east side of the Piazza Grande, Arezzo, showing 1930s' towered reconstruction; (f) 2009 view of the Piazza del Duomo, San Gimignano, showing the loggia rebuilt in 1934–6

(a)

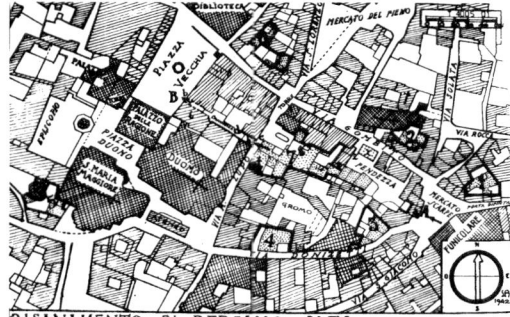

RISANAMENTO DI BERGAMO ALTA (ZONA CENTRALE)
STATO DEI LAVORI DI DIRADAMENTO E SISTEMAZIONE DICEMBRE 1941
IL PERCORSO ABcd è IL NUOVO TRANSITO PEDONALE TRA MERCATO SCARPE E PIAZZA VECCHIA (LAVORI IN CORSO)

(b)

(c)

RISANAMENTO DEI · BERGAMO · ALTA · 1942
SISTEMAZIONE · FACCIATE · SUL · PERCORSO · MERCATO · DELLE · SCARPE · VIA · MARIO · LUPO ·
(LATO DESTRO)

(d)

(e)

(f)

211

balanced planned modernity and heritage: for example in Pavia, in a 1933 competition win by Milan architects Banfi Belgiojoso Peressutti Rogers (BBPR), or Valle d'Aosta in 1936. The year 1938 saw an isolated aberration from this careful norm, in a flamboyant personal proposal by Mussolini for a 164-metre-high neo-Gothic campanile beside Milan Cathedral (not carried out!). But, on the whole, the Fascist reshaping of Italy's historic cities and towns was done with such artistic skill and subtlety that eventually, following the political somersaults and conflicts of 1943–5, the 1930s *diradamento* policies escaped the stigmatisation of Fascism and emerged ready to contribute to the internationalisation of postwar conservation, as well as to the post-1980 architectural doctrine of 'intervention'.[50]

The Third Reich: rhetoric before results

THE situation in Germany under the Third Reich was similar in some ways, but radically different in others, not least in the subsequently taboo character of all its architectural and urban policies, as well as in the wartime destruction of most of what was done. Quantitatively, what was actually achieved was far less than in Italy, as public spending during the regime's six peacetime years was dominated by rearmament (which absorbed half of government expenditure). There was, though, a general equivalent to the Fascist dualism of *romanità* and *italianità*, in the parallel evolution of grandiose classical urban reconstruction plans, almost all unachieved, alongside small-scale *Altstadt* rehabilitation work.

The conservation policies of the regime developed against the background of the chaotic early years of the Weimar Republic, when a wide range of different heritages and histories was thrown into play, and each political and cultural faction devoted intense efforts to shaping its own collective memory. Among the socialist progressives, concern for the future blotted out the past: the old turn-of-century alliance between *Heimat* and modernity, exemplified by figures such as Cornelius Gurlitt, had now disappeared. Social-utopian architects and artists, including the November 1918 Art-Workers Council founded in Berlin, generally rejected the Old and the *Heimat* altogether, as obsolete and discredited: what was needed was a focus on social housing and city-planning. The year 1924 saw Ludwig Hilberseimer's scheme for redevelopment of Berlin's Friedrichstadt area, with relentlessly parallel *Zeilenbau* slabs, and the decade of the 1920s the comprehensive replanning of central Hamburg by city architect Fritz Schumacher (including the cutting through of wide city centre streets, and conservation confined to restricted *Traditionsinseln*, or 'heritage islands'). More generally, the revolutionary upheavals of 1918–19 left many conservationists feeling under threat, not least because of the savage cuts in government subsidies during the economic crises of those years. Many historic buildings, including the properties of the deposed royal family, had been wrecked by revolutionary militias, and forces opposed to the *ancien régime* were pressing forward on all fronts. The year 1922 saw a complex dispute over proposed removal of a 1910–11 equestrian statue of Kaiser Wilhelm II from the Hohenzollernbrücke in Cologne – a proposal opposed by the Prussian Culture Ministry on grounds that bridge and statues were an integrated cultural monument. Some old-fashioned conservationists stayed loyal to the old monarchical system and opposed the new, standardised *Landesämter*: for example the Bavarian *Generalkonservator*, Georg Hager, insisted on still using the royal title of his department.[51]

The heritage community generally welcomed the 1933 accession of the National Socialists (NSDAP) to power, given their emphasis on safeguarding the *Heimat*, and their promised return to generous state subsidies. Paul Clemen enthused about Hitler's 'deep empathy for the mysterious magic of the world of monuments'. The substance of NSDAP conservation policies, however, contained little new. Hitler's own views on architectural heritage, as expounded in *Mein Kampf* (1925/6) or in his 1935 Nuremberg party rally address, were rather conservative and not unlike Haussmann's France. On the one hand, heritage was a key element in 'the eternal foundations of our life and our existence as a *Volk*'; but, for him, the essence of this heritage lay in traditional grand public buildings and in overpowering symbolic monuments like the Pyramids, rather than the banal homes of the people, or even medieval *Altstädte*, with their 'paltry clutter of timber or brick buildings'.[52]

Overall, the heritage policy of the Third Reich followed the same opportunistic pattern as in Mussolini's Italy, emphasising propaganda image rather than substance. This pushed the old Dehio/Ruskin conservation restraint and reverence for built substance very much into the background, in favour of a '*schöpferisch*' (creative) freedom of action. To reflect the regime's rhetoric of radical renewal, conservationists such as art historian Wilhelm Pinder emphasised overarching collective concepts, such as *Ganzheit* (totality). The old heritage nationalism of the Kaisers, as exemplified by Höhkönigsburg, was rejected as pompous and old fashioned in its bourgeois individualism: what was needed was something more modern, yet also rooted in the glorious German past – reconciling tradition and modernity within a total national community.[53]

That modernity, however, did not require the building of Modernist architecture, other than in utility structures. In the forefront, instead, was the Third Reich's equivalent to *romanità*: its programme of monumental building. Having no large-scale Roman remains as a starting point, Germany had to build anew, in a programme of grandiose projects designed both with modern circulation and propaganda in mind. These began with the autobahns, intended to endure 'silent and strong down the centuries'. A 1940 painting showed the medieval town of Limburg an der Lahn with an arched autobahn bridge in the background, like a Roman aqueduct. Hitler's most cherished architectural cause was his programme of monumental urban reconstruction, signalled in 1934 and fully set out in the 1937 *Gesetz über die Neugestaltung deutscher Städte*. This colossal revival of the Haussmann ethos of grand axes and spaces would have culminated in Berlin, which was to be transformed into a 'representative world city' by constructing a wide north–south axis. Unlike Mussolini's work in Rome, this would have involved demolition not of established historic areas, but of the widely despised zone of 19th-century tenements. Elsewhere, a more invasive approach to historic fabric was envisaged: the proposed transformation of Cologne into a *Gaustadt* would have left only a rectangular *Altstadt* zone south of the cathedral, surrounded by vast new axes and buildings and confronting a monumental *Gauforum* on the opposite bank of the Rhine. The only one of these projects significantly implemented before war intervened, in Weimar (from 1936, by Hermann Giesler), would also eventually have destroyed most of the *Altstadt* by driving through massive axial boulevards. In Hamburg, intended by Hitler to become the most openly modern of the *Führerstädte*, the radical urban clearances since the 1890s were continued by the Third Reich, including demolition in 1934–6 of the dense, medieval Gängeviertel as a seditious 'incubator of Marxism': historic buildings would eventually be confined to isolated 'heritage islands', notably the harbourside Cremoninsel.[54]

Within National Socialist heritage policy, the authenticity of old built substance was downplayed. Rather, the built heritage became one element in a wider *Ganzheit* of landscapes of invented cultural memory, as with the sacrificial cult built up around the Feldherrnhalle and

Figure 6.11 The *Neugestaltung* of Weimar: report on the proposed *Gauforum* (designed by Hermann Giesler) in the *Allgemeine Thüringische Landeszeitung*, 26 August 1937

Königsplatz in Munich, to commemorate the 'fallen' of the 1923 Munich Putsch. Rather than building new monuments, the regime intensified the exploitation of existing ones, such as the octagonal, towered 'Reichsehrenmal Tannenberg', built from 1929 on the site of a World War I German victory. The Neue Wache in Berlin, following its rebuilding in 1929–31 to the competition-winning design of Hans Tessenow, was used as a Valhalla of fallen generals ('Reichehrenmal'). The vision was like a fascist transformation of Geddes's vision of Edinburgh the culture city, or the costumed displays of pre-1914 Nuremberg and the turn-of-century Expos, with military parades replacing 'masques of ancient learning'.[55]

In the treatment of the built fabric, highly politicised restorations were the order of the day. Occasionally, these were focused on elite, traditional monuments, aiming for an 'authenticity' that was ideological rather than architectural. Most symbolic was the eight-year restoration of the Abbey Church in Quedlinburg, undertaken from 1936 under the patronage of Reichsführer-SS Heinrich Himmler, to emphasise the graves of kings Heinrich I and Otto I, and ornament them with Nazi symbols. In other medieval buildings, equally radical remodellings were undertaken, expunging earlier restorations from the historicist 19th century and substituting a rough, rubble finish, as anticipated in the 1928–30 remodelling of St Georg in Cologne in a neo-12th-century rubble style by Clemens Holzmeister. Similar in character, as we will see shortly, was the artistically rustic restoration of the Nuremberg Kaiserburg from 1933, under Hitler's direct patronage.[56]

More prevalent was the improvement and restoration of entire old towns, as always with the insistence that the result must be 'living' rather than museum-like. The concept of the *Altstadt*, and the word itself, gained general currency, but its ideological significance subtly shifted, as illustrated in two Conservation Conventions devoted to the subject and separated by a decade: that of 1928 in Würzburg and Nuremberg, and that of 1938 in Hamburg. The 1928 event (suggestively entitled *Altstadt und Neuzeit*) had as keynote speakers the architect/planners Theodor Fischer, Ernst May and Fritz Beblo and analysed the *Altstadt* as an integrated zone of the modern planned city, sharing wider urban problems such as traffic congestion and bad housing. The 1938 Convention, conversely, saw *Altstadt* conservation as an integrated element of cultural and racial nationalism, as highlighted by Pinder in a programmatic speech of 1934 on 'Saving the German *Altstadt*'. Now, the paramount aim was *Gesundung* (cleansing or healing), a concept encompassing not only traditional sanitary slum clearance but also expulsion of political and racial 'undesirables'. And in 1938, just as in 1928, built-fabric authenticity in the old Ruskin-Dehio sense was no longer a significant factor: authenticity was a matter of national-local-racial community.[57]

Yet, in most cases, the practical rehabilitation policies 'on the ground' differed little from the 1920s. Each city concocted its own recipe out of certain common elements, including slum clearance, housing rehabilitation and traffic improvement. In cases of especially poor housing conditions, there was special stress on *Entkernung* (clearance of courtyard infill), as in the Cologne *Altstadt* programme under the direction of Hans Vogts – here with the avowed aim of expunging 'asocial, criminal and indecent persons' who threatened the moral and racial authenticity of the true *Altstadt*-dwellers. In the influential programme in Frankfurt-am-Main, this was also combined with extensive facade restoration and *Fachwerkfreilegung* (uncovering of timber framing – a practice mainly seen in small towns, but also energetically pursued since 1925 in Frankfurt).[58] Often, facade restoration also involved *Entstückung*: removal of late-19th-century ornamental stucco. This practice was prevalent in many northern and central European countries, but was pursued with special fervour in Germany. Although Modernist architects had pioneered elements of *Entstückung*, as in Bruno Taut's multi-coloured 1921 Barasch department store in Magdeburg, the technique spread to both the right-wing *Heimatschutz* architects and everyday residential apartment builders. More all-embracing was the concept of *Entschandelung*, or 'removal of blots', stemming ultimately from Schultze-Naumburg's *Kulturarbeiten* and denoting removal of tawdry commercialism and 19th-century historicist architecture. These programmes were indirectly descended from the turn-of-century facade competitions, although their aesthetic emphases had shifted, away from spiky picturesqueness towards 'Biedermeier classicism'.[59]

The Third Reich's two most ambitious *Altstadt* restorations both concerned cities of high ideological significance: Danzig, the 'Free City' trumpeted by the regime as a frontier outpost of German community, confronting the supposedly rootless modern Polish city of Gdynia; and Nuremberg, designated *Stadt der Reichsparteitage* (National Party Congress City). Both programmes benefited from massive injections of government subsidy. The Danzig scheme, superintended from 1934 by Prof. Otto Klöppel, involved restoration of the main streets of the old town, or Rechtstadt, in a generalised 16th–18th-century style. Following the outbreak of war, this was to provide an architectural template for the *Eindeutschung der eroberten Ostgebiete* (Germanisation of the conquered eastern regions). Prior to 1914, the frontier outposts had been elite castles: Marienburg and Höhkönigsburg. Now, in the era of total national community, that role was played by an entire heritage-city. Ironically, the wholesale destruction of these streets by Soviet bombardment in 1945, and the transfer of Danzig to Polish sovereignty, would be followed by a 'restoration' very similar in visual character, other than in its 'Polish' Renaissance detailing and Socialist Realist painted decoration.[60]

Nuremberg was even more emblematic than Danzig, symbolising the German heartland rather than the outer frontier, being the former medieval imperial capital as well as the site of the annual NSDAP Congresses and of the 1935 proclamation of the infamous racial laws. This 'treasure-chest' of German history was developed as a National Socialist Rome, whose restored medieval *Altstadt* was visually juxtaposed with a vast new classical complex of party parade grounds and halls on the outskirts (the '*Reichparteitags Gelände*', mainly designed by Albert Speer). In 1937, Oberbürgermeister Willy Liebel hailed this project as the 'biggest building site in the world'. This juxtaposition was celebrated in Leni Riefenstahl's 1935 film of the previous year's congress, *Triumph des Willens* (Triumph of the Will). Here the *Altstadt* becomes a supporting actor, from its initial appearance beneath Hitler's plane, skimming down from the clouds, to its role as backdrop to the culminating parade through the 'Adolf-Hitler-Platz' (Hauptmarkt). The *Altstadt* was targeted for a comprehensive campaign of *Entschandelung*, to reshape the city's medieval image from the 19th century's spiky eclecticism to a more 'simple' character. This campaign extended down to the most subtle detail of shop signs but also included the complete demolition of the city's main synagogue, an ornate, neo-Romanesque building in the Hans-Sachs-Platz, designed by Adolf Wolff in 1874 and removed following sustained attacks in the anti-semitic newspaper *Der Stürmer*, which protested in early 1938 that

> the synagogue rears up ostentatiously, soullessly and impudently above the historic roofscape of Nuremberg, treasure-chest of the German Reich and city of the Party Congress. Lodged in the midst of the most German city ever known is this alien, oriental object, this lump of shame (*Schande*) set in stone.

Months later, in August 1938, it had completely disappeared. A few key buildings in Nuremberg were singled out for more thoroughgoing restoration treatment, above all the Kaiserburg, which

Figure 6.12 *Entschandelung* of buildings and towns

(a) St Georg, Cologne, following its 'purification' by Clemens Holzmeister in 1928–30;
(b) The showpiece late-1930s' facade *Entschandelung* scheme at Semlowerstrasse,
Stralsund – before/after elevations annotated with 'defects' and improvements;
(c) Propaganda poster about Danzig, c.1939–40

217

underwent a comprehensive post-1933 campaign of *Entschandelung*, overseen by Bavarian state heritage architect Rudolf Esterer under Hitler's personal patronage. The castle's colourfully eclectic, castellated 19th-century restoration as a Bavarian royal residence (by von Heideloff in 1833 and August von Voit in 1850–60) was branded 'weak academicism'. It was replaced in 1933–4 by a 'pure and unfalsified' scheme, including an '*Ehrenwohnung für den Führer und hohe Gäste*' and reception apartments for Hitler-Jugend leader Baldur von Schirach, all in a massive, rough style of rubble and jutting timber balconies that anticipated the new SS 'Ordensburgen' of the Third Reich. The high-roofed 1494 castle granary and imperial stables block was converted to a youth hostel in the same style, in time for the 1936 Party Congress.[61]

To celebrate the completion of the regeneration of Nuremberg, Oberbürgermeister Liebel commissioned in 1935 a huge wooden scale-model of the *Altstadt*, painstakingly constructed by a team of four elderly craftsmen in the conservation section of the city building department, led by sculptor Alexander Hehl. The remarkable model was officially completed only in November 1939, after the commencement of war. Unlike the generic detailing of the famous 1:250 model of Constantine Rome commissioned from Italo Gismondi for Mussolini's 1937/8 Augustan Exposition of Roman Civilisation (*Mostra Augustea della Romanità*), the Nuremberg model showed a precise representation of each individual building in its actual state – including the empty site of the city synagogue. A commemorative article of February 1940 in a local newspaper boasted that it provided an 'exact record of the city's 1939 condition, which would bequeath posterity a unique living testimony of its time'.[62] The authors could hardly have foreseen how quickly and comprehensively that prediction of uniqueness would be realised, as the old city itself perished in a hail of bombs, leaving the model (today housed in the Fembohaus City Museum) as the sole surviving three-dimensional representation of the Third Reich's most lovingly crafted *Altstadt*.

Nuremberg also featured on the front cover of Werner Lindner and Erich Böckler's 1939 book, *Die Stadt: ihre Pflege und Gestaltung*, an official commemoration of the success of six years of 'fanatical zeal for urban order' in expunging the degraded urban legacy of bourgeois liberalism. The regime's favoured conservation policies and solutions were set out in typically rhetorical language – *Entschandelung, Wiedergutmachen, Gesundungsmaßnahmen*, etc. – and illustrated with hundreds of Schultze-Naumburg-style 'good versus bad' photographs, captioned appropriately in *Fraktur* ('good') and Roman ('bad') typefaces. Subject-matter was very eclectic, ranging from 'sound' building materials to Geddes-style urban greenery or conversions of redundant medieval buildings into Hitler-Jugend hostels. The overriding message was that the intrinsic integrity and authenticity of the typical small town should serve as an example for conservation in large cities, to help in purging the 'race' of discordant and diseased elements: 'The small town is romantic, but the large city is romantic too . . . all towns and cities should, and must, play an intimate part in the life of the *Volk*.'[63]

The dominance of political propaganda aims in this conservation strategy also, as in Italy, allowed the capitalist world's tension between heritage authenticity and tourist exploitation to be

facing page

Figure 6.13 '*Nürnberg, die deutsche Stadt*'

(a) Playing cards from 1930s' set depicting historic towns; (b) Front cover of 1930s' children's book ('Gertrud Grows up into the Third Reich'); (c) Late 1930s' poster publicising the NSDAP National Party Rally

(a)

(b)

(c)

Figure 6.14 *Entschandelung* of Nuremberg's synagogue

(a) Pre-World War I postcard view; (b) View of the synagogue under demolition in August 1938

Figure 6.15 Modelling the 'Reich treasure chest'

(a) 1939 wooden model of the Nuremberg *Altstadt* (by Alexander Hehl and others), viewed from 'south-east', with the vacant synagogue site at the bottom; (b) 1950 plaster model of the ruined city, viewed from the same perspective

easily reconciled: here they were seen as one and the same thing. 'Picture-postcard' Rothenburg ob der Tauber, where overnight visitors reached a new peak of 134,000 in 1938 through the boom in KdF (Strength through Joy) domestic tourism, was always in the forefront of enthusiasm for the NSDAP, recording Germany's highest support (87 per cent) for Hitler in the 1932 presidential election. After 1933, the town's leaders put its heritage directly in the service of the government's extremist policies, including the mounting of anti-semitic plaques on the town's gates in 1937, and the expulsion of Rothenburg's remaining 17 Jews in October 1938; the *Fränkischer Anzeiger* local newspaper enthused that 'Our Rothenburg is now Jew-free: the centuries-long defensive struggle of our ancestors has found its fulfilment.' But, despite the rhetoric, here as elsewhere the main thrust of the regime's physical conservation policy was cosmetic and gesture-orientated, emphasising quick fixes rather than systematic building-preservation work. And, within a handful of years, this world of propaganda-driven heritage, and its loudly trumpeted values, would be largely reduced to rubble and disgrace in a war unleashed by its political patrons.[64]

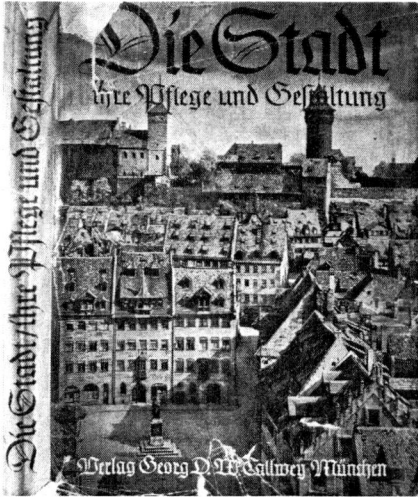

(a)

Figure 6.16 *Die Stadt: ihre Pflege und Gestaltung*

(a) The front cover of Werner Lindner and Erich Böckler's 1939 book; (b) Page from *Die Stadt* showcasing the conversion of a 16th-century granary in Dinkelsbühl into a Hitler-Youth hostel

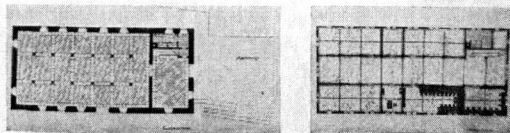

(b)

A 'curate's egg': diverse interwar initiatives across Europe

Iᶠ the Third Reich was unable, owing to its war preoccupations, to implement its grander urban visions, the same applied even more forcibly in Franco's Spain. Here, the earlier 1930s, despite the economic difficulties, had witnessed an extensive Republican government conservation campaign, superintended by the Zonal Conservation Architects (instituted in 1929) and a National Board established under the 1933 Law on Artistic Treasure. This programme had shifted gradually in emphasis from conservative repair to more assertive restorations. For example, Leopoldo Torres Balbas, renowned for his writing and research, oversaw the conservation and restoration of the Alhambra of Granada from 1923 to 1936, and Alejandro Ferrant in 1930–2 supervised the dismantling, renewal and restoration of the 8th-century S. Pedro de la Nave church, threatened by a hydro-electric scheme.[65] During the Civil War, from 1936, all that descended into chaos, first in the outburst of iconoclastic violence against religious and feudal monuments by socialist and anarchist revolutionaries, and then in the aerial bombardments by Franco's forces.

After the fascist victory, official propaganda portrayed the iconoclasts and the Republic as the same thing, vilifying or ignoring its conservation efforts, while pragmatically appropriating its legislative base: the 1933 Law on Artistic Treasure and the zonal organisation of state conservation stayed in force for over 50 years. However, the conservative repair standards of earlier years were now abandoned. Restoration of war-damaged monuments, to standards that fluctuated wildly between caution and exaggeratedly Viollet-like creativity, was overseen by a newly instituted General

Directorate for Devastated Regions. Its most symbolically important project was the largely 16th-century Alcazar of Toledo, a massive quadrangular towered palace crowning the acropolis of one of Spain's oldest cities and twice rebuilt after fires in 1710 and 1887. In mid-1936 the building, by then a military academy held by Nationalists, was subjected by Republican militias to a famous siege, culminating in devastating bombardments and mine explosions that reduced it largely to rubble: the siege was ended on 27 September by the arrival of Franco's army. After the war, the Alcazar became a martyrdom and victory site: initial debates tended towards keeping it as a ruin, but it was eventually fully reconstructed in 1948–62, with only the damaged office of commandant Colonel Moscardo preserved in its 1936 state. Another new national body, the Commission for the Artistic Heritage, began ambitious enhancements of symbolically prominent complexes, including the Escorial Monastery, the Cathedral of Santiago de Compostela and the palace of Charles V in the Alhambra – a campaign that continued until the late '50s. However, some Republican initiatives unobtrusively continued: for example, an innovative 1930 programme to convert disused castles and monasteries into state-owned hotels (*paradores*), serving both tourism and conservation.[66]

While, in the fascist countries, the existing heritage was aggressively accentuated, an almost opposite pattern prevailed in states newly emancipated by the Versailles settlement. In post-1918 Poland, for example, strenuous attempts were made to expunge all traces of foreign rule. War-damaged towns such as Kalisz (burnt in 1914 by German troops) were rebuilt, and the almost brand new Alexander Nevsky Orthodox Cathedral and bell-tower campanile in Warsaw (by Leon Benois, 1894–1912) were demolished in 1924–6 as a symbol of Russian domination. The nationalist-inspired campaign to reconstruct the Wawel palace in Krakow in the Italian Renaissance style associated with Poland's golden age, although opposed by Max Dvořák during Austro-Hungarian rule, was now pushed to completion under chief conservator Adolf Szysko-Bohusz, with new interiors created in former barracks spaces.[67]

Away from the epicentre of wartime destruction and postwar political extremism, more innovative policies could be developed, despite the economic troubles of the times. Some countries

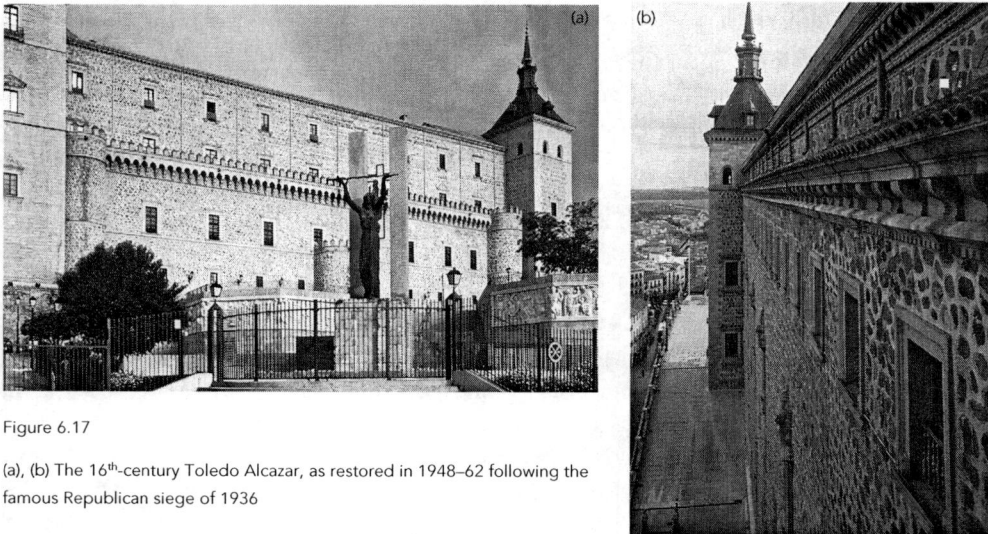

Figure 6.17

(a), (b) The 16th-century Toledo Alcazar, as restored in 1948–62 following the famous Republican siege of 1936

witnessed a combination of pioneering government inventorisation, with voluntary efforts to preserve urban ensembles. In the Netherlands, 1918 saw commencement of a pioneering urban inventory of Amsterdam, instituted by a new government-backed National Committee for Preservation of Historic Buildings. In 1928, this published a preliminary list of 4,200 properties, and civic restoration initiatives began for smaller towns, such as Amersfoort, rehabilitated from 1937 under chief building inspector J Kapteijn.[68] But at the same time a range of voluntary preservation initiatives flourished, including a new trust aimed at buying and restoring old buildings, the Hendrick de Keyser Society (founded in 1918) and a *Heimatschutz*-like grouping, the Delft School, dominated by architect M J Granpré Molière. In Denmark, a 1918 Preservation of Historic Buildings Act enabled an inventory programme of 1,258 buildings, drawn up by Ministry of Education inspectors; this inspired a similar Norwegian law two years later. In Stockholm, conversely, the initiative was seized by the established voluntary trust, the *Samfundet S:t Erik*, which expanded its activities from its pre-1914 inventorisation of the Gamla Stan, to repair of individual tenements (from 1923) and, finally, full-scale city-block rehabilitation, in programmes to restore the street-blocks Kvarteret Cepheus (from 1934 onwards, by Gösta Selling and Albin Stark), and Kvarteret Cygnus (in 1938–9). These schemes resembled the *Entkernung* and *Gesundung* projects in Cologne or Frankfurt, but lacked their propaganda overtones. Across the Atlantic, a French-flavoured system of national inventorisation of monuments and ensembles was pioneered in Brazil by the nationalistic Getulio Vargas regime. A 1937 federal law established a new Institute for National Artistic and Historic Heritage (IPHAN), charged with inventorisation, designation, and restoration. The focus of IPHAN's wide-ranging efforts, unlike Mexico's archaeological orientation, was colonial-era architecture.[69]

From Williamsburg to Charleston: urban restoration and preservation in the USA

In the English-speaking world, the war's passions seemed less immediate, and the established traditions of private voluntary initiative could continue and develop in innovative ways. Their links to commercialism, and embrace of more modern heritage, significantly anticipated the Conservation Movement's shift towards a capitalist orientation in the later 20th century – unlike the totalitarian states' more aggressively politicised, yet more traditionalist, vision of heritage tourism. These commercialised tendencies were naturally most accentuated in the USA, where the interwar years saw an explosion in personal mobility: in 1921, few roads were paved and Americans drove only 55 million miles, whereas, two decades later, following vast road-building campaigns to combat the Depression, car use had increased eightfold, and visits to national parks tenfold, to four million a year. The role of government in this campaign was pervasive, yet focused always on the facilitation of capitalist activity. The National Park Service, established in 1916 within the federal Department of the Interior, had a mission as wide as any continental *Heimatschutz*: 'to conserve the scenery and the natural and historic objects and the wildlife therein and to provide for the enjoyment of the same'. The Depression also fuelled a range of specific federal initiatives, including the Historic American Buildings Survey (HABS, founded in 1933), harnessing the scholarly rigour of the Fiske Kimball tradition of architectural history to the systematic recording of the national heritage, and the 1935 Historic Sites Act, instigated by the Roosevelt Administration to help coordinate state and federal preservation work. The established movement to preserve great men's houses gained material

Figure 6.18 Plaque commemorating the Kvarteret Cepheus block-restoration in Stockholm (from 1934, by Gösta Selling and Albin Stark)

reinforcement from this new scholarly approach, and from the growing fashionability of private houses in the classical Colonial style – as seen in the restoration of Robert E Lee's Virginia birthplace, the early-18th-century Stratford, from 1929 onwards by a specially created Robert E Lee Memorial Foundation Inc., in a scholarly project directed by Kimball.[70]

But the most startling advances in interwar American preservation – advances in the heritage appreciation of entire neighbourhoods that made the European living-*Altstadt* discourse seem repetitious and stale by comparison – were the work not of government departments but of voluntary and philanthropic agencies. Behind them lay partly the open-air museum movement, which began in 1920s' America to shift its focus towards the creation of complete museum-towns, focused naturally not on the Middle Ages but on the 18th and 19th centuries. Also important was the continuing tradition of old-town Expo displays, as in the 18th-century-style 'high street' included in the 1926 Philadelphia Sesquicentennial Exposition. This process was catapulted forward by the intervention of two magnates, J D Rockefeller Jr – patron of France's great reconstruction schemes – and Henry Ford. They financed two showpiece ensembles of a hybrid museum/Expo character – Rockefeller at Colonial Williamsburg, Virginia (opened in 1932), and Ford at Greenfield Village, Michigan (1933) – that radically challenged the now hackneyed European polarisation between the 'living' and 'museum' town, and inspired numerous imitators (such as George Francis Dow's 'Salem Pioneer Village' in Salem, Massachusetts, opened in 1928). Theirs was a vision of a modern Conservation Movement that would mingle old and new as freely as did *schöpferisch* fascist conservation, openly welcoming tourist commercialism as an equal partner rather than integrating it with totalitarian propaganda. Whereas illustrations of 'living' *Altstadt* set pieces, like those in *Die Stadt: ihre Pflege und Gestaltung*, showed environments purged both of commercialism and of any human life, the promotional literature of the American museum-recreations typically did the opposite, if necessary using costumed actors just as in Expo displays. Ford famously protested in 1916 that 'History is more or less bunk – we don't want tradition, but to live in the present, and the only history that is worth a tinker's damn is the history that we make today.' The effect of this doctrine on architectural heritage, in its Riegl-style emphasis on the primacy of reception over intrinsic authenticity, would in some ways prove energising and liberating.[71]

Williamsburg and Greenfield were significantly different in character and in importance. Williamsburg was by far the more influential and for long served as an international shop-window of American preservation. It was not an open-air museum in the narrow meaning of the term, but an existing town, an old colonial centre that was radically 'restored' to an idealised perfection in the Viollet sense, in the interests both of nationalist celebration and, of course, of tourism. What was expunged here, however, was not 'impure' phases of church architecture, but anything later than the mid-19th century, as well as poor housing occupied by black people. In a way, this was not unlike *Entschandelung* in European old towns, but its ambitions were far more radical.

Williamsburg's chief promoter, a local Anglican churchman, the Rev. Dr. W Goodwin, successfully persuaded Rockefeller to support the project and begin anonymously buying all the town's historic properties. Godwin insisted the town should be restored to 'its moment of principal significance' as a 'cradle of the Republic' during its time as Virginia's state capital in the mid- and late 18th century, but that this ideal should (in best Viollet fashion) be underpinned by exhaustive historical research. The aim was 'to restore a complete area and to free it entirely from alien and inharmonious surroundings'. The town was renamed 'Colonial Williamsburg', rather like an Expo display, and a master-plan was prepared by the Boston architectural firm of William G Perry,

Thomas M Shaw and Andrew H Hepburn. They were engaged as consultants to the project, which was implemented in 1926–33 by two linked organisations, Colonial Williamsburg Inc. (a non-profit educational body) and Williamsburg Restoration Inc., a subsidiary concerned with 'business operations'.[72]

Over the following four decades, the Colonial Williamsburg strategy was relentlessly pursued: the historic zone itself totalled a relatively modest 173 acres (eventually, 301 acres) but was surrounded by 3,000 acres of *cordon sanitaire*. In its first decade, as endowments accumulated, no less than 477 buildings were demolished, 67 restored, and 91 facsimile buildings added: visitors soared from 31,000 in 1934 to 90,000 in 1936. In a scholarly review of 1959, Frederick D Nicholls claimed that 'the most extensive restoration of the 20th century, comparable in magnitude and influence to those of Viollet-le-Duc in the last century, is Colonial Williamsburg'. And in 1958,

Figure 6.19 The marketing of heritage: 'Colonial Williamsburg'

(a) Rev. W Goodwin and J D Rockefeller Jr seen inspecting the site in 1926; (b) Map of the interpretative facilities of Colonial Williamsburg, c.1960; (c) HABS survey photograph of the reconstructed Governor's House; (d) HABS photo of the timber-frame structure of the Taliaferro-Cole Shop under reconstruction in 1940; (e) Mid-1960s' street view of Williamsburg

historian Daniel J Boorstin (author of the famous 1958/73 trilogy, *The Americans*) hailed it as 'an American kind of sacred document', celebrating mobility and freedom from the stuffy elitism of Europe.[73] Certainly, in true Viollet style, the eventual result, despite all the scholarship, was an essentially modern creation, whose brick plantation-houses were treated far more elaborately than ordinary colonial architecture had been. And as with Viollet, its combination of restoration radicalism with scholarly research provoked controversy – a problem that war reconstruction projects such as Ypres sidestepped by pleading enemy *force majeure*. A flashpoint at Williamsburg was the 1928–34 project to rebuild in facsimile the demolished Capitol, backed by Rockefeller as a tribute to the 'great patriots' of the revolution, but which involved the inaccurate recreation of a building that had only occupied the site in the period 1701–47. The Capitol project was dogged by clashes over authenticity between the Williamsburg promoters and the Association for Preservation of Virginian Antiquities, who had originally acquired the site in 1897. The association advocated a more asymmetrical 'vernacular' arrangement, whose authenticity was indicated by archaeological excavations, whereas Perry, Shaw and Hepburn insisted on a formal, Beaux-Arts arrangement and implausibly sumptuous fittings.[74]

Greenfield was more like a normal open-air museum/Expo display, its role being mainly educational. Ford wanted it to be an 'animated text book' recalling 'the everyday life of ordinary folk' and celebrating 'the American Dream'. It comprised an assemblage of relocated buildings intended to evoke a New England settlement, but presented in a more generalised way as an

Figure 6.20 'El Poble Español'

View of the 'main square' of the heritage zone built for the 1929 Barcelona Expo, by architects Francesco Folguera and Ramon Reventós

'American Village'. It included buildings associated with Americans whom Ford had admired, such as Edison and the Wright brothers, as well as his own birthplace house, with more unexpected elements, including buildings from London and the Cotswolds. This inspired successive later initiatives, such as Sturbridge Village, Massachusetts, opened in 1946, or the 'Plimoth Plantation' Pilgrim Fathers theme park (1947), and anticipated aspects of Disneyland (see Chapter 8). In contemporary Europe, such ensembles were still confined to Expos, but one key example was retained as a permanent tourist attraction: the 40,000-square-foot 'El Poble Español' display built on Montjuic hill for the 1929 Barcelona Expo, by architects Francesco Folguera and Ramon Reventós, with regional displays derived from a research tour of 1,600 old towns across Spain.[75]

The movement of 'historic neighborhood' preservation was now moving at a furious pace, and these interwar efforts were soon left behind. In 1931, only five years after the start of Colonial Williamsburg, similar principles were applied to an existing old city, Charleston, South Carolina, through collective civic action and philanthropic initiative. Charleston's 18th- and 19th-century vernacular classical architecture had twice been devastated, in the 1861–5 Civil War and in an 1886 earthquake. During the 1920s, its economy and general confidence were seemingly locked into a long-term cycle of decline. In response, local civic leaders, encouraged by pioneering heritage activist Susan Pringle Frost, launched a new cultural-economic promotion strategy, rooted not in the unpleasant recent realities of war and racism but in generalised antebellum imagery of 'Historic Charleston' and the 'tradition' of the 'Old South'. This was bound up with other local initiatives such as the Azalea Festival of flower processions (begun in 1933) and national media phenomena such as the 1936 book and 1939 film, *Gone with the Wind*.[76]

Vitally, the Charleston initiative was not just a matter of promotion and marketing, but rested on a second pillar, of public planning and development control. In 1922, a landmark ruling by the US Supreme Court in the case of *Euclid v Ambler* guaranteed planning a role within local government. From 1929, Charleston's city council enacted a succession of measures, including creation of a Planning and Zoning Commission, restriction of discordant gasoline stations and inventorising of the historic area. Finally, in October 1931, the council passed America's first governmental planning and zoning ordinance for an entire historic area – the 23-block, 800-acre 'Old and Historic District', containing 400 buildings. Now, all development was controlled by a Board of Architectural Review, headed by architect Albert Simmons, who aimed to incrementally revive a 'sanitised yet highly marketable' image of a pre-1860 golden age. A comprehensive survey of Charleston's historic buildings, financed by the Carnegie Foundation, was implemented in 1941–4. The Charleston achievement inspired similar initiatives elsewhere. In New Orleans, tentative efforts had been underway since 1921, when the Louisiana constitution was amended to allow the state government to sponsor restoration of the Vieux-Carré district: 1936 saw establishment of a Vieux-Carré Commission, empowered to make tax-exemption recommendations to the city council, and architects Thomas Tallmadge and Earl Reed spearheaded a detailed survey by unemployed architects, covering 350 buildings. The city's white elite, especially the New Orleans Association of Commerce, were determined to build a new tourist image for the city, foregrounding the architecture of the French quarter. After World War II, the same policy spread to other cities, notably Alexandra, Virginia (in 1946), Winston-Salem, North Carolina (1948), Annapolis, Maryland (1952) and Beacon Hill, Boston (1955); by the 1970s more than 200 American cities had enacted Charleston-style municipal preservation ordinances. Voluntary pressure groups, however, remained vital: the 1935 demolition of Savannah's Montgomery Street Squares for highway construction provoked a postwar

Figure 6.21 Reviving Charleston and the South

(a) The zone of Charleston destroyed by the 1861 'Great Fire'; (b) Robert Mills Manor, 1939–40: the city's first public-housing project, designed 'in keeping' with the colonial heritage; (c) HABS aerial view of Charleston's South Battery, placed under preservation ordinance in 1931; (d) Stratford Hall Plantation, Virginia, built in the 1730s, was acquired and preserved from 1929 by the Robert E Lee Memorial Association: the house had been the family home of the Confederate commander-in-chief (1970 photo)

wave of preservationist sentiment, supported by private campaigners and city councillors, and no further squares were lost.[77]

Georgians and mass observers: private interwar efforts in Britain

CONSERVATION in interwar Britain was a hybrid of these European and American extremes, with the state playing a restricted, but gradually increasing role. The 1931 Ancient Monuments Act excluded occupied buildings, but a more expansive role was welcomed by the town and country planners: a 1932 act encouraged municipalities to preserve historic buildings. Regulation of new developments in sensitive urban and rural settings was assigned to advisory Royal Fine Art Commissions established in 1924 in England and 1927 in Scotland. The English body (RFAC) was empowered to take proactive initiatives as early as 1933, the Scottish commission (RFACS) only from 1953. As late as 1960, by which date only 200 cases had been considered by RFACS in the 33 years since its establishment, it complained that many architects and planners were still unaware 'of its very existence'.[78]

It was in Scotland that British interwar conservation came closest to the continental pattern. Here, the growing economic centralisation in southern England provoked the growth of a mildly separatist nationalism. The contrasts with *Heimatschutz* were stronger than the similarities, however. Most conservation-minded architects, such as the young Robert Hurd, also favoured modern architecture. There was also a contrast with the collectivist politicisation of heritage by continental nationalists: with the continued high prestige of the monarchy and aristocracy, the landed classes maintained a remarkable hegemony within Scottish interwar conservation. The National Trust for Scotland (NTS) had only two chairmen, both aristocrats, in its first 36 years: Sir Iain Colquhoun (1931–46) and Lord Wemyss (chair 1946–67 and president 1967–91); Sir John Stirling Maxwell of Pollok was a key office-holder and author of the 1937 book, *Shrines and Homes of Scotland*.[79]

Dominant behind the scenes in the NTS, as in most interwar heritage matters, was Scotland's wealthiest aristocrat, the 4th Marquess of Bute – a fervent conservationist. In a well-publicised polemical speech of 1936, Bute attacked the menace of municipal socialism, a philistine 'bull in a china shop', whose 'mischievous destruction' of heritage would leave 'our towns and villages denuded of their archives of stone, their individuality destroyed', so that 'all will present the same unintelligent appearance of new Government settlements'.[80] To him, tourism and commercialism seemed a lesser threat, even an opportunity: 'What is the use of advertising Scotland for tourists

CULROSS, FIFE.
The-Study: *Property of the National Trust, and recently restored by H.M. Office of Works.*

A Plea for Scotland's Architectural Heritage

On the subject of the preservation of our domestic architecture I would like at once to put forward as a general proposition that all old houses have an architectural value, and there must be few which have not equally a domestic value. For example, I noticed in a village lately a but-and-ben cottage with the date cut on one stone, 1746. It·may not seem very old, but it is a living example of what type of house was going up at the time of the '45, far better built stone houses than are going up to-day. Especially are rows of houses or streets of some particular period of value. I don't mean only aesthetically but intrinsically.

The reason that has caused the wholesale destruction that is taking place of Scotland's unique domestic buildings, and the working with such desperate anxiety to destroy what has been called the " old dignity " of our country, is not a pleasant one to contemplate and far less to discuss.

But it is difficult to see any other basic reason

Figure 6.22 Bute's 'Plea'

(a) The front pages of Bute's 1936 pamphlet; (b) Culross, Fife, restored from 1932 by the National Trust for Scotland

when everything intrinsically Scottish is to be destroyed?' The solution, Bute argued, was for the government Office of Works to 'schedule' all houses over 150 years old – an argument he followed up by financing a pioneering inventory of old burgh houses by the young architect Ian G Lindsay. Although the NTS's early acquisitions, such as Glencoe (1935–6), focused on landscape preservation, it soon became preoccupied with old town preservation, especially in the decayed Fife burgh of Culross, where it started in 1932 a rescue programme that would last 40 years. In other historic towns, local initiatives were prominent, especially in St Andrews, where a decade-long agitation against clearance of old houses culminated in the 1937–8 foundation of the St Andrews Preservation Trust by Annabel Kidston and Ronald G Cant.[81]

Edinburgh remained a special case, with Hurd, Bute, Lindsay and others collaborating in successive Old Town preservation initiatives, along with Geddes's architect-planner son-in-law Frank Mears and City Architect Ebenezer McRae. Their conservative-surgery efforts, like the Giovannoni school in Italy, or German cities like Cologne and Frankfurt, combined fairly radical renewal of background housing with restoration of key monuments. However, the public popularity of the classical New Town had never really disappeared, although early efforts to conserve it, unlike Charleston, were piecemeal. In 1930, under a 1925 planning act, Edinburgh Corporation enacted a special protection scheme for the late-18th-century Charlotte Square, whose north side, designed by Robert Adam, was subsequently acquired and restored by Lord Bute.[82]

In England, the contrast with the Continent was even more extreme. It was one of the few countries from which overt heritage nationalism was largely absent – perhaps partly due to the relative success of economic recovery in southern England. Although there were a few mildly fascist groups, such as 'The Ruralists', the incessant debates about commercial despoliation of the rural environment were dominated by liberal architect-planners, some of an establishment character, such as Patrick Abercrombie, and others more flamboyantly individualist, notably the indefatigable Clough Williams-Ellis, author of the polemical *England and the Octopus* (1928). In 1925, Ellis founded Portmeirion, a picturesque ensemble of slightly miniaturised buildings generally recalling an Italian hilltop town, transplanted into the lush landscape of rural north Wales and painstakingly developed over the following half-century. With its mixture of new and relocated buildings (including a new campanile bell-tower, in 1928, and the 18th-century bath house from Arno's Court, Bristol, relocated in 1958), Portmeirion echoed Expo displays and museum-towns such as Colonial Williamsburg.[83] Inspired by Abercrombie's 1926 book, *The Preservation of Rural England*, a new Council for the Preservation of Rural England was formed in that year, advocating modern planning doctrines such as curbing of 'ribbon development' or advocacy of green belts around cities. During the post-Depression years, the English countryside began to experience a growing prosperity based on motor tourism and walking: the Youth Hostels Association was founded in 1930. Correspondingly, there was an explosion of popular literature on heritage and environment, ranging from J B Priestley's influential *English Journey* of 1934 to popular compendia such as Arthur Mee's *The King's England* series (from 1936) or the Shell Guides, founded in 1933 by journalist John Betjeman.[84]

The rise of tourism, and of a recreational property market in country cottages, distanced the cultural status of the English village from its politicised continental counterparts, even in places of similar appearance, such as the Suffolk village of Lavenham, many of whose timber-framed houses were bought and restored by well-off urbanites from 1937 onwards. The first English parish church to charge for admission, from the 1920s, was, predictably, Holy Trinity, Stratford-upon-Avon. More

Figure 6.23 Portmeirion and Elm Hill, Norwich

(a) Mid-1960s' view of Portmeirion (developed from 1925 by Clough Williams-Ellis); (b) 1972 view of Elm Hill, Norwich, purchased for preservation by Norwich Corporation in 1927

problematic was the growing plight of the country house: many landowners had been sharply impoverished by successive agricultural recessions. From 1937, the National Trust began refocusing towards this building type: its Country Houses Scheme encouraged owners to stay in residence. In Ireland, in sharp contrast, the many Ascendancy houses gutted during the civil war were left to moulder forgotten, just as Georgian Dublin languished in decay.[85]

Between the wars, there was a growing contrast between the thrusting modernity of the largest English municipal councils, whose modern slum redevelopments of flats (e.g. in central Liverpool by city architect Lancelot Keay, or at Quarry Hill, Leeds, by R A H Livett) were matched in few other parts of Europe, and the conservation lobby, which remained dominated by voluntary organisations. Among these, SPAB was still pre-eminent, although rivalled from 1924 by the new Ancient Monuments Society, a Manchester-based group founded by John Swarbrick to press for tougher government enforcement of Ancient Monuments legislation. In general, the early 20th century saw an expansion of the voluntary civic sector. As in Scotland, local authorities often became involved in conservation through these groups as intermediaries. One of the first historic-street preservation schemes, when Norwich Corporation bought the old houses in Elm Hill in 1927, was brokered by SPAB. SPAB support also underpinned some early building-preservation trusts: Bath (in 1934), Blackheath and Cambridgeshire (both in 1938).[86] England's first building preservation trust was established in 1928 in Oxford, underlining the persistent hierarchy in English historic towns, with London, Oxford and Cambridge ranking above smaller towns such as Warwick and Durham. Such towns were dominated by 18th- and 19th-century buildings, constructed with 'real' facing materials (brick, stone) rather than plaster, and thus Britain saw few German-style attacks on applied decoration.[87]

Because of this more modern urban heritage, coupled with the English tradition of polemical architectural debate, English conservationists were as adventurous as their American cousins in expanding the scope of the Conservation Movement: a few avant-gardists were now even enthusing about the mid- and late-19th-century 'Victorian' age. The catalyst, as usual, was demolition and the focus was naturally London, the subject of development pressures from which Oxford and Cambridge were immune. Those pressures began with an abortive 1919 proposal to demolish 19 Wren churches; 1934 saw unsuccessful resistance to the demolition of Waterloo Bridge and

How we Celebrate

THE CORONATION

By Robert Byron

Figure 6.24 Robert Byron's 1937 protest article in the *Architectural Review*, 'How We Celebrate the Coronation'

I

THE Cæsars have fallen in the east and west; echoes rule the wave-lapped thresholds on the Bosporus; tourists eat lunch in the Forbidden City. But the King of England still goes to Westminster to be crowned, mightier than the Cæsars or the Sultan and the Son of Heaven. Dear London! You are the mother of the occasion. You saw it inaugurated by the Norman bastard, and have seen it grow till all the nations of the earth now throng your streets to cheer.

"*But who, London dear, has given you such a pretty new dress ?*"

"*The Government and the local authorities.*"

"*For what purpose? They don't generally care what you look like.*"

"*To please our visitors and honour the King.*"

"*Is it possible, perhaps, that these guardians of yours want to hide your poor old body ?*"

"*Having no vanity, I had not thought of that.*"

"*But your body is hidden, is it not, under the flags and flowers and streamers and the stands and crowds ?*"

"*Of course it is. They would be ashamed of it naked.*"

"*Why is that ?*"

The answer is not as simple as it seems, gentle visitor from overseas. We all know London is ugly—muddled, smoky, and lacking in those grand vistas and symmetrical spaces which are the proper adornment of a great capital. It is not this that shames us. We like London's ugliness, just as one likes the ugliness of an old woman because it tells of a long life and strong character. Confused as London is, her streets and buildings commemorate our history and institutions more vividly than any other town or monument. And since our history and institutions have affected a large proportion of the world's population, we

successful resistance to the rebuilding of Nash's Carlton House Terrace by the classical architect, Sir Reginald Blomfield. The threats provoked a succession of protest publications, such as *Disappearing London* (1927) and Robert Byron's *How We Celebrate the Coronation* (1937).[88]

The new campaign to save Georgian London exploited all the extreme language of the Pugin tradition: Byron thundered that the demolition of Adam's Adelphi showed 'a barbarity hardly equalled since the devastations of Jenghis Khan', and that 'the Venetians when they fired on the Parthenon, the Germans on Rheims Cathedral, had an excuse of an emergency – even if posterity has not accepted it'. He indicted an excessive commercialisation of the built environment in England, 'The only nation in Europe that destroys its birthright for the sake of a dividend.'[89] In this fight, the SPAB was seen as outmoded and ineffective by a new generation of young dilettante-agitators, including Douglas Goldring and the 3rd Baron Derwent, as well as Robert Byron. In April 1937, they founded a splinter group of SPAB, the Georgian Group, which defended London's Georgian squares from redevelopment by campaigns of publicity-conscious militancy. Although this offended SPAB traditionalists and provoked libel threats from building developers, it also reinvigorated England's Pugin-Morris tradition of knockabout civic-amenity polemics.[90] In these increasingly aggressive pressure groups and causes, we see an early instance of the leapfrogging process that would characterise the Conservation Movement in the late 20th century.

The English city in the late 1930s was also the incubator for another equally far-reaching development within urban conservation – the emergence of a new concept of popular participation. During those years, dominated by totalitarian discipline in historic cities from Rome to Leningrad, England witnessed several experimental attempts to engage with 'real' popular life and planning challenges in the mainly Victorian city fabric. These began with Mass Observation, a vast, ramshackle recording and collecting initiative of 1937–8, in which nearly 3,000 volunteer observers gathered *ad hoc* anthropological information and took photographs: 1936 had seen publication of photographer Bill Brandt's influential book, *The English at Home*, and Mass Observation was also strongly pervaded by artistic and poetic concepts of everyday life.[91] Responding to the same climate of elite attempts to engage with 'ordinary life', progressive planners and students proposed replanning concepts involving public participation and conservation of existing urban landscapes, along with scientific Modernist planning techniques of survey and analysis. The Architectural Association-trained planner Max Lock began in the late 1930s to devise new, participatory urban planning procedures, culminating in 1943 in 'Civic Diagnosis', a Geddes-influenced regional survey of Hull, while in 1937–8 a group of Associate of Arts undergraduates, supervised by E A A Rowse, proposed a new city ('Tomorrow-Town') of 50,000 inhabitants in the Oxfordshire countryside, preserving existing historic villages and natural landscape features.[92]

As early as 1925, a heritage manual by writer William Harvey had presented itself as *A Text Book on the New Science of Conservation*, rejecting 'the usual opportunist patching of old buildings' for a methodical, proactive, planning-based approach, based on comprehensive survey and respect for the intrinsic character of the fabric. Could the same approach now be extended to the fabric of entire cities, social and cultural as well as physical? In Chapters 8 and 9, we will trace how the ideal of a systematic, socially committed conservation was pursued in the Western countries after 1945. But first, as we will see next, the Conservation Movement had to undergo six years of trial by fire, which would purge from it a century's accumulation of militant nationalism and reorientate it towards a peaceful, low-key approach, collaborating constructively with the reconstruction strategies of Modern architecture and planning.[93]

Total war and cultural bombing

1939–45

When you look round a city like Nuremberg, with its streets and districts reduced to charred ruins, you inevitably ask yourself – how long will it take, with the modern building methods at our disposal, to get things in order again? I can only say this: it'll hardly even take a year – and what's one year in the long history of Nuremberg? Nothing at all – it doesn't matter a bit!

Josef Goebbels, 1944

Travelling through the destroyed areas, you are struck by architectonic impressions of unprecedented power: all superfluous trimmings have been purged away.

Richard Zorn, 1943[1]

Urban bombing: history and morality

THE night of 13 February 1945 began uneventfully and cloudlessly in the Saxon capital of Dresden, one of Europe's foremost centres of 18th-century Baroque architecture. Hitherto, owing to distance, the city had been spared the destruction visited on its west German counterparts. That, however, was about to change. Already a major communications hub and centre of munitions manufacture (as well as of anti-semitic persecution, with only 4 per cent of its prewar Jewish population now surviving), the city had been designated, on New Year's Day 1945, as a 'redoubt' for last-ditch defence of the Third Reich, and troops and refugees from Soviet-occupied Silesia were pouring into it. Correspondingly, in a last-minute demonstration of the brutal fortunes of war, the 'Florence of the Elbe' at last reached the top of the target list of the British and American bomber forces. At 22.03, the 'marker' planes arrived over the city, spotlighting with eerie red and white flares the skyline of bridges, cupolas and towers that had only minutes left before the destruction began. At 22.13, the first wave of 245 Lancaster heavy bombers began to rain down high explosive and incendiaries, leaving the city a sea of flames. From 01.23 to 01.55 a second wave of 528 bombers sealed the destruction, followed next day by two smaller waves of American bombers. Not only were about 18,000–25,000 people killed (according to a city council investigation of 2006), but the entire city centre, totalling five square miles, was left completely in ruins; the skyline was symbolically decapitated by the collapse of the dome of the 18th-century Frauenkirche, newly repaired in the 1930s. The late-18th-century Kreuzkirche, whose striking Jugendstil interior scheme had been

Figure 7.1 Watercolour painting by Otto Griebel of the destruction of Dresden on 13–14 February 1945

installed following its earlier fire in 1897, was gutted yet again, with only the reredos and crucifix by Anton Dietrich (1900) surviving.[2]

The bombing of Dresden, in the suddenness of its projection of mass destruction into a renowned cultural centre, symbolised everything that had previously been feared about aerial warfare. In 1933, Paul Clemen had tempered his praise of the new regime with apprehension that vast fleets of enemy aircraft could reduce German cities to rubble overnight. Now, that vision had become reality. For the Third Reich's propagandists, the bombing of Dresden was not only a fearful event but a criminal atrocity: Josef Goebbels protested that Dresden had no war industries and that the raid was an act of deliberate cultural desecration and mass murder of up to 200,000 refugees: the claim of hundreds of thousands of casualties, and Goebbels's expression, 'Anglo-American air gangsters', was repeated by communist East Germany's rulers, and by western writers such as David Irving.[3]

One can, of course, debate endlessly the ethical rights and wrongs of the strategic bombing campaign: was it a 'war crime'? From a heritage viewpoint, what is incontrovertible is that it formed part of a long process of escalating urban destruction.[4] We saw earlier how the bombardment of relatively small areas by cannon fire escalated in the artillery shelling of World War I, while still remaining organically linked to the ebb and flow of battle on the ground – a pattern still generally maintained in the Spanish Civil War. World War II decisively broke from that precedent in two areas of relevance to the heritage. First, its large fleets of bomber aircraft exposed places far from the front line to danger, but in a curiously asymmetrical way, owing to the concentration of the German air force on light tactical bombers and the British and Americans on long-distance heavy bombers. Second, the politicisation of heritage by National Socialism legitimised destruction of cultural

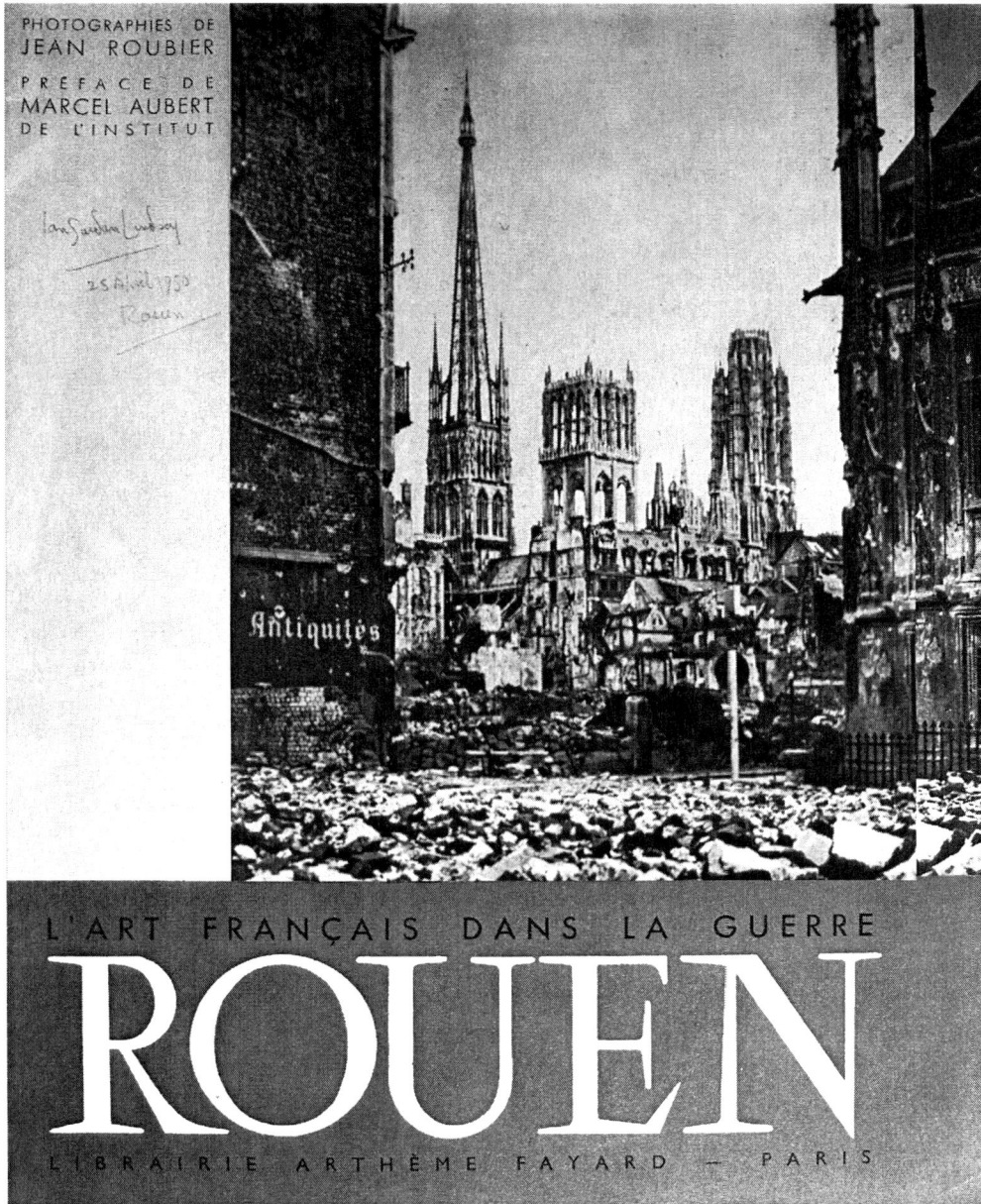

PHOTOGRAPHIES DE
JEAN ROUBIER

PRÉFACE DE
MARCEL AUBERT
DE L'INSTITUT

Antiquités

L'ART FRANÇAIS DANS LA GUERRE

ROUEN

LIBRAIRIE ARTHÈME FAYARD — PARIS

Figure 7.2 1946 magazine cover featuring wartime damage to central Rouen (copy belonging to Ian Lindsay)

property as a direct aim rather than a careless by-product of warfare – an attitude reflected, for example, in art historians' increasingly florid rhetoric about the border-zone of Silesia, described by Dagobert Frey in 1938 as an artistic 'combat zone against the east', and by Hans Tintelnot in 1943–4 as a 'gangway of attack for the forces of German form'.[5] But this new cultural warfare was selective, and conditioned by a hierarchy of ethnic and architectural esteem and an anxiety to avoid repetition of the embarrassment of Reims. An east–west differentiation was perceptible from the beginning, in the violent September 1939 attack on Warsaw, a largely 19th-century industrial Polish city, and the massive destruction later inflicted both on Belgrade in April 1941 and on many Russian cities, culminating in the onslaught on Stalingrad in 1942. These contrasted with restraint at first shown towards historic towns in Western Europe, other than in isolated cases, such as the massive May 1940 attack on Rotterdam, launched partly by accident, or the June 1940 tactical bombing of Amiens, Beauvais and Rouen.[6]

England versus Germany: early exchanges

THE confrontation that would end in the destruction of Dresden, however, was between Germany, the home of *Altstadt*, and Britain, the home of Anti-Scrape. It began almost gingerly, with both sides showing elaborate restraint, and focusing solely on military and industrial targets. Gradually, the scale of operations in Western Europe had broadened into full-scale area bombing, with the Rotterdam attack of May 1940 followed by sporadic RAF attacks on German cities; these in turn provoked ever larger-scale bombing of London from 22 June onwards and a full-scale campaign of strategic bombing of British cities from September 1940 to May 1941, including attacks on London for 76 consecutive nights and the destruction of much of central Coventry on 14–15 November, a raid which inspired Propaganda Minister Goebbels to coin a new verb, '*coventrieren*'.[7] By mid-1941, however, it was becoming clear that the destruction of urban England was reaching its limits, owing not to lack of willingness on the part of Hitler, but to the small size of the German aircraft and the need to divert them to support the land campaign in the east. The sharp separation in England of industrial and historic cities also minimised damage to the latter, given the focus of German attacks on industrial or strategic targets. In heritage terms, the chief victims of this strategically indecisive campaign would ultimately be Germany's own historic cities, devastated in the inevitable riposte.

It was, appropriately, in England that practical responses to area bombing were first consistently developed, including the policy of emergency clearance of ruined facades capable of being rebuilt, often at the instigation of city engineers or the local military, anxious to tidy bombed areas to maintain 'morale'. Ten days after the Coventry attacks, for instance, a raid on the port city of Bristol damaged the early-17th-century 'Dutch House' – a spectacular, five-storey timber-framed structure on a central intersection, recently reinforced with a steel frame. Easily repairable, it was instead pulled down by the army two days later.[8] The rush to clear away damaged buildings also sometimes stemmed from an architectural and planning zeal for modern reconstruction – a motive that, as we will see shortly, transcended the differences between the combatant nations.

It was England, too, that saw the first debates about how far bombed buildings should be 'recreated', a debate that echoed the Ypres controversies 20 years before, but also reflected the special strength of the Anti-Scrape doctrine there, and the wartime artistic movement of the New Romanticism, with its love of ruins and decay – in contrast to the highly negative 1860s' response to towns ruined in the American Civil War.[9] The New-Romantic artist John Piper painted a celebrated picture of the Coventry cathedral ruins after the 1940 raid. A 1942 book, The *Bombed Buildings of Britain*, contrasted city engineers' pursuit of 'tidy clearance' with the accidental beauty of ruins: an 'aesthetic of destruction [that] bears no relationship to any architectural merit the building may have possessed in life'. Modernist architect J M Richards argued that

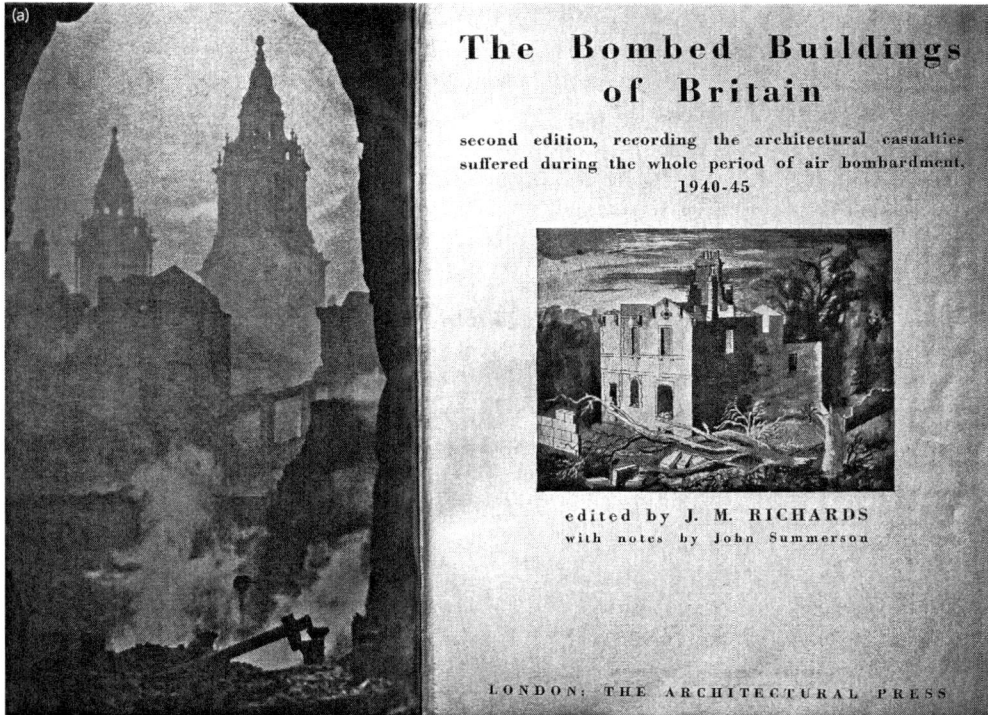

Figure 7.3 Heritage destruction in the 'Blitz'

(a), (b) The title page and frontispiece of *Bombed Buildings of Britain* (including John Piper view of damage in Bath), and the page describing the destruction of the 'Dutch House' in Bristol; (c) The front cover of *Bombed Churches as War Memorials*, a 1944 *Architectural Review* publication advocating preservation of some bombed-out English churches as commemorative ruins one day we may even need reminding of the intensely romantic character that the manner of their destruction brought to bomb ruins, and of their poignancy, which is of a kind not possessed by monuments that have undergone a slower, more merciful process of dismemberment and decay.

More straightforwardly, art historian and National Gallery director Kenneth Clark argued that 'bomb damage is in itself Picturesque'. We will see shortly that bomb ruins in Germany prompted some rather different architectural responses, emphasising the Sublime rather than the Picturesque.[10] As with Coucy in World War I, there was a movement (here initiated in 1944 by the *Architectural Review*) to preserve some bombed-out English churches as commemorative ruins. Correspondingly, any move towards facsimile reconstruction in England was met with the full force of the Anti-Scrape movement: the SPAB weighed in with a strongly worded pamphlet.[11]

Ironically, in view of the strength of feeling against facsimile building, the precautionary surveying of buildings in advance of potential bombing was especially systematic in Britain. This campaign indirectly stemmed from the late-19th-century 'recording movement' but was, more immediately, a brainchild of Lord Reith, appointed reconstruction minister in 1940. Unable to secure emergency legislation to protect historic buildings, he invited Lord Greene to set up a new National Buildings Record, in 1941, dedicated to emergency photography and drawing of noteworthy buildings: a similar record was set up in parallel in Scotland. Soon, too, the Anti-Scrape orthodoxy was challenged in places, with heated debates about reconstruction of some key monuments, including the Temple Church in London, gutted in May 1941 and eventually built in virtual facsimile from 1947, to the designs of architect Hubert Worthington. Here, the army was only narrowly restrained from destroying the church's medieval tower, mistakenly believing that its centuries-old slant was the result of the bombing.[12]

In most of Western Europe, the early wartime years were dominated not by the aftermath of bombing but by precautions against possible attack. Following Italy's entry into the war in 1940, furious efforts began to protect the country's artistic treasures: in Florence, for example, a brick dome was built over Michelangelo's David.[13] Far more immediately at risk, however, following the Third Reich's provocative onslaught on London, were the *Altstädte* of Germany's principal cities: the only exceptions were the university towns of Heidelberg, Tübingen and Göttingen, which, according to persistent rumour, were protected by a 'gentleman's agreement' in exchange for German avoidance of Oxford and Cambridge. One of the clearest potential targets, for ideological as well as strategic reasons, was Nuremberg. Here, safeguarding work began as early as July 1940, before a single bomb had fallen, in a campaign masterminded by city conservation chief Julius Lincke, working with local churches and the Germanisches Nationalmuseum to store the city's artistic treasures safely in its underground network of *Felsengänge* (tunnels). Works successfully secured ranged from Veit Stoss's renowned 'Englische Gruss' Annunciation carving from the Lorenzkirche to the newly completed city model. Fixed treasures were more problematic and largely ineffective structures of brick or sandbags were built round them.[14]

The German invasion of the Soviet Union in June 1941 triggered a frenzy of protective measures in Leningrad and the nearby palace complexes. From the city itself, over 45 railway carriage-loads of treasures were evacuated and key pieces of public sculpture were buried in parks: the statue of Lenin in front of the Finland station was sandbagged, and the golden spires of the Admiralty and the Peter and Paul fortress were put in giant mufflers. At Pavlovsk, over 12,000 works of art were removed or hidden in the ground or in cellars, and at Peterhof 7,000 out of 40,000 works were extricated before German forces overran the complex in September: many pieces were stored in the basements of the Hermitage and St Isaac's Cathedral. In some of the Russian palaces, the panelled construction of interiors allowed entire rooms to be removed, although the renowned Amber Room in the Catherine Palace at Tsarskoe Selo could not be dismantled in time by the Russians, and a vain attempt was made to conceal it by covering it with wallpaper instead. Insulated from the tensions of cultural looting, the same approach proved far more successful in the isolated German case of the 1751–3 rococo Cuvilliés-Theater in the Munich Residenz. It was taken down in 1943–4 and moved in pieces to safe storage in the basement of the Befreiungshalle at Kelheim – which allowed it to escape the destruction of the Residenz, and to be eventually re-erected in 1956–8, on a slightly different site within the complex, in a protracted restoration programme masterminded by Esterer.[15]

Few other German cities attempted anything on this scale – until it was too late. For by late 1941, researches were already underway within the Royal Air Force as to how the new generation of heavy bombers about to be introduced could be most effectively used: in July 1940, Churchill had written to Lord Beaverbrook, Minister of Aircraft Production, that an 'absolutely devastating, exterminating attack by very heavy bombers from this country upon the Nazi homeland' was the only way to victory. Exiled German scientist Prof. Frederick Lindemann (later Lord Cherwell) studied the heavy air raids on Hull from May 1941 onwards and reported that the inhabitants had been demoralised above all by losing their homes.[16]

'Reaping the whirlwind': *Altstadt* destruction from 1942 to 1945

IN February 1942, the newly appointed head of Bomber Command, Air Chief Marshal Arthur Harris, persuaded the Cabinet to sanction a massive experimental raid upon the 'soft target' of a historic city: he remarked that 'I wanted my crews to be well blooded, as they say in fox hunting, to have a taste of success for a change.' Selected for initial attack were two easily identifiable old Hanseatic towns on the Baltic, Lübeck and Rostock – with the aim of building up to a massed attack on Cologne in May. In 1941, evoking Old Testament imagery, Harris warned that

> the Nazis entered this war under the rather childish delusion that they were going to bomb everyone else, and nobody was going to bomb them. At Rotterdam, London, Warsaw, and half a hundred other places, they put their rather naive theory into operation. They sowed the wind, and now they are going to reap the whirlwind.'

And in a harshly ironic twist, it was the *Altstädte* cherished by *Heimat* nationalists that would experience the first force of that hurricane of nemesis.[17]

The Lübeck attack, in the early hours of Palm Sunday (29 March) 1942, was carried out by a continuous wave of 234 bombers, dropping 25,000 fire bombs and 800 high explosive bombs and mines along the western side of the *Altstadt*, which was highly visible from the air owing to its island shape: a British propaganda poster later misleadingly portrayed aircraft raining bombs down on the docks, with the old town silhouette in the far distance. With only nine flak batteries available, the raid was effectively unopposed. Firefighters were impeded by severely cold weather, which froze many hoses, and so the flames spread largely unimpeded from one timber-framed building to the next. Eventually the entire area around the Marktplatz, the Cathedral and the Petrikirche, amounting to some 10 per cent of the *Altstadt*, was reduced to gutted ruins.[18]

In this instance, the restricted death-toll of 320 (around half the Coventry total) and the fact that in this experimental operation only a relatively small part of the *Altstadt* had been destroyed, were almost incidental to the raid's effect in Germany. Whereas Harris had targeted the timber Lübeck houses merely from a utilitarian desire to maximise destruction, the hyper-politicisation of the *Heimatschutz* philosophy in Germany led to immediate public outrage and calls for revenge attacks specifically targeting the English architectural heritage. On 14 April, Hitler issued an order for retaliation bombing of historic cities. These attacks were carried out by Luftflotte 3 between April and June 1942. On 23 April, the cathedral city of Exeter was attacked, followed by similar attacks on Bath, Norwich and York (where the Guildhall was destroyed). The German radio claimed that 'Exeter was a jewel: we have destroyed it', and foreign ministry spokesman G. Braun von Stumm pledged on 24 April that 'we shall go out and bomb every building in England marked with three stars in the Baedeker Guide'. This statement earned Stumm a sharp rebuke and disavowal from Goebbels (doubtless mindful of the Reims debacle in 1914–18), but the nickname 'Baedeker Raids' stuck in the popular imagination.[19]

Figure 7.4 'Knocking out teeth' in 1942

(a) Lübeck Cathedral ablaze on Palm Sunday, 1942, prior to the collapse of its spires; (b) Collapse of the south tower of the Lübeck Marienkirche

facing page

(c) Damage map of the Lübeck *Altstadt*, showing the relatively localised destruction (in light shading); (d) British propaganda poster showing the attack as directed against port facilities; (e) St Martin le Grand, York, burnt out in the April 1942 'Baedeker Raid' and reconstructed in 1961–8 by architect George Pace as a partially preserved ruin

(c)

(d)

BACK THEM UP!

(e)

In any case, Stumm's prediction was highly over-optimistic. The fact that the German air force only had light bombers, most committed anyway to the eastern front, meant that these retaliations could only be on a pinprick scale (only 25 planes, for example, in the case of Exeter) compared to what was inexorably coming in the opposite direction. For example, as German bombers headed for Bath on the nights of 25 and 26 April, a massive formation of RAF bombers crossed their path en route for Rostock, where three nights of successive attacks gutted not only the city's Heinkel aircraft factory but also much of the *Altstadt*. At a meeting three days later, Hitler raged against the destruction of Rostock, and Goebbels recorded that 'he shares my opinion absolutely that cultural centres, health resorts and civilian centres must be attacked now. There is no other way of bringing the English to their senses. They belong to a class of human beings with whom you can only talk after you have first knocked out their teeth.'[20]

But most of the historic teeth being knocked out in this competition continued to be German ones. Churchill warned in a May 1942 broadcast that 'we have a long list of German cities. . . . All these it will be our stern duty to deal with, as we have already dealt with Lübeck and Rostock.' On 30 May came the first RAF 'Thousand Bomber Raid', 'Operation Millennium', by 1,047 aircraft against the Cologne *Altstadt*, where no less than 600 acres were devastated. In retribution, the German air force mounted token attacks against Canterbury on the nights of 31 May and 2 and 6 June; in neither case was the cathedral seriously damaged, although some 300 old houses were destroyed around St George's Street and Burgate. Following the Canterbury attack, there was a small but unprecedented raid on Cambridge, in which 100 houses were destroyed. Aware of the heightened cultural significance of the targets, both sides unleashed a Reims-style barrage of rhetoric against 'barbarism'. Subtly emphasising the special English esteem of the 18th-century classical heritage, the *Picture Post* (4 July 1942) illustrated bomb ruins in Bath as 'what the Nazis meant by a Baedeker Raid . . . like looking into an elegant drawing room after a grimy, spiteful urchin out of the gutter has tried to do his worst to it'. From the German side came accusations that the Lübeck attack was a calculated act of cultural vandalism by the 'Jewish leadership in England and the USA'. And Senator Dr Böhmcker, Staatskommissar für Lübeck, insisted that 'Despite everything, the British have lost the battle of Lübeck. Their terror attack came to nothing against the immovable, incredible resistance and iron victory-will of our courageous people.' Other nationalist commentators argued that the loss of built fabric was almost incidental to the survival of the 'national will'.[21]

Both Lübeck and the English Baedeker targets subsequently experienced the familiar pattern of differential clearance, with most street facades still standing after the raids, but subsequently demolished on grounds of public tidiness and safety, leaving churches and major public buildings stranded in somewhat Haussmann-like open spaces. In Lübeck, the gabled facades of the side streets around the cathedral square were cleared in mid-1942, while in Exeter, the classical facades of Bedford Circus and Southernhay largely survived the attack, but were later demolished. In Canterbury, the ruins of some seven hundred houses and other medieval structures also later vanished.[22]

Paradoxically, Lübeck itself survived the war as one of the least damaged west German cities, owing to a remarkable intervention by Hamburg-born allied air force liaison officer Eric Warburg. In June 1943, learning that Lübeck was targeted for another massive air raid, Warburg unofficially alerted Max Huber, president of the International Committee for the Red Cross in Switzerland, and persuaded him to reroute all British prisoner of war letter traffic through the city. This prosaic arrangement, upheld by Huber's successor from 1944, Jacob Burckhardt, successfully insulated the

Figure 7.5 The onslaught on urban Germany

(a) Postwar map showing the degree of destruction in individual cities; (b) July 1943 cartoon, 'Who talked about Huns?' from the *Völkischer Beobachter*

city from the vast and escalating destruction elsewhere in the Reich, and Lübeck eventually emerged in 1945 in a condition not dissimilar to Exeter.[23]

On the Allied side, it seemed that cultural destruction was a by-product of a bombing campaign mainly intended to demoralise civilians by destroying their homes and, more generally, to exact vengeance for the protracted German bombing campaign of 1940–1. The aim was, simply, to inflict maximum overall destruction – an approach exemplified in the terrifying onslaught in July and August 1943 ('Operation Gomorrha') that incinerated much of inner Hamburg. Unlike England, most historic German cities were ringed by dense 19th-century tenement housing zones and thus highly vulnerable to this approach. In the case of Nuremberg, the danger was exacerbated by the fact that it was not only a large industrial city and centre of munitions production, but was also widely vilified in Britain and America as the spiritual home of Nazism; it thus became a special target. It was subjected to 59 separate attacks, most from 1943 onwards, culminating in the complete devastation of the *Altstadt* and the inner suburbs on 2 January 1945; over 38 per cent of the city's building fabric was destroyed (compared, for example, to 17 per cent in Munich). However, reflecting the frequent disjunction between human and architectural casualties, as well as the city's network of underground tunnels, the civilian death-toll of all Nuremberg raids combined (just over 5,000) was only a small fraction of that inflicted by single attacks on Hamburg, Dresden or many Japanese cities. Emergency repairs could only help protect isolated prestige buildings, as with the temporary roof erected over the Lorenzkirche after an earlier raid of 10–11 August 1943.[24] The repeated bombing of Berlin also had overtones of cultural decapitation: a September 1943 *Newsweek* article showed a map of its historic monuments, captioned 'Heart of the Reich: the buildings and monuments of Central Berlin formed a target for the RAF'. Perhaps appropriately, considering its

Figure 7.6 The destruction of Nuremberg

(a) Painting of the 2 January 1945 attack by M M Prechtl;
(b) View looking south from the Pellerhaus across the
Egidienplatz towards St Lorenz after 1945; (c) Plaque on the
site of the destroyed Moritzkapelle: 'Its reconstruction is a task
for future generations'

strong Nazi sympathies, even Rothenburg ob der Tauber became a late, accidental victim of the Allied bombing campaign. In 1938, its foremost local historian and preservationist, Martin Weigel, had boasted of its martial heritage as a 'peasant and warrior town'. That offered no protection when, on 31 March 1945, an American bombing mission against an oil storage depot at Ebrach was diverted by fog and instead unloaded its bombs on the 'secondary target' – Rothenburg – reducing to ruins the eastern section of the *Altstadt*, or about 28 per cent of the town's built fabric. The town was saved from further destruction in April when US Assistant Secretary of War, John F McCloy, asked army commanders to avoid any artillery bombardment of it.[25]

Even amidst this vast destruction, there were those who, as previously in London, found unexpected beauty in ruination. In Germany, the appreciation was subtly slanted towards a Schinkelesque Sublime rather than the English Picturesque. For example, in fire-bombed Hamburg, an architect (and NSDAP member), Richard Zorn, part of the emergency clearance team of city planner Konstanty Gutschow, prepared in December 1943 an extraordinary photographic survey of ruined facades, whose captions were permeated with an exalted rhetoric of architectural essence, as well as the usual condemnations of *Gründerzeit* 'vulgarity'. Zorn rhapsodised that, 'when travelling through the destroyed areas, you are struck by architectonic impressions of unprecedented power: all superfluous trimmings have been purged away': the gutted facades of early-19th-century buildings were 'abstemious . . . the proportions are simply perfect. . . . The building commands its devastated setting more proudly than ever.'[26]

Abb. 36. Ist dies eine Ruine ? Nein – niemand wird zweifeln, das Mauerwerk ist standfest geblieben. Stolzer denn je beherrscht das Gebäude die zerstörte Umgebung.

Abb. 5. Dieser Portikus war unbeachtet, solange links und rechts anschließende Ladengeschäfte schreiend übertönten.

Figure 7.7 The evocative power of ruins

(a) Photograph of Admiralitätstrasse from Richard Zorn's 1943 photographic survey of Hamburg: the caption reads, 'Is this a ruin? No – nobody can doubt that the masonry is still intact. The building commands its devastated setting more proudly than ever'; (b) View of Gröninger Strasse, captioned: 'No-one noticed this portico while garish shopfronts shouted out on either side of it'; (c) 1947 cartoon of the ruins of the Rynek Starego Miasta in Warsaw

(c)

Wartime destruction and conservation across Europe

THE mounting devastation of the German heritage was raised in a British parliamentary debate in February 1944, by the Bishop of Chichester. Reviving the older monument justification of universal European values, something last seen in the Athens charter of 1931, but since swamped in nationalistic hatred, the bishop pleaded for restraint:

> In the fifth year of the war it must be apparent to any but the most complacent and reckless how far the destruction of European culture has already gone. We ought to think once, twice and three times before destroying the rest.[27]

Less measured claims of outraged internationalism were increasingly made by German commentators, shifting agilely from the previous championing of the sacred *Heimat*. For example, following the 1944 destruction of the Goethehaus in Frankfurt, the official NSDAP newspaper, the *Völkischer Beobachter*, declaimed that 'the ruins of Germany's cities show just what the British and Americans think of Europe. This is a devastation that dwarfs any natural catastrophe.' At last, the uncomfortable spectre of Reims was banished, as Nazi propagandists warmed to their new cause of defending world culture against the jealously philistine Anglo-Americans: Dehio's 1905 *Kaiserrede* had already grumbled that

> of all the races, the Anglo-Saxons have created the smallest amount of art; now they do the . . . spiritually richer nations the honour of plundering their art works, and since America has joined in, they have become a serious danger to the artistic patrimony of historic Europe.

In July 1943, the *Völkischer Beobachter* published a cartoon, captioned, 'Who talked about Huns?' showing a giant, demoniacal RAF pilot stamping on Cologne Cathedral; and in a radio talk on the 'Life of Monuments' the same month Pinder contended that, even if the substance of Germany's peerless historic treasures was lost, their 'inner image' would survive, accusingly, in the consciousness of the whole world. Writing with more unrestrained gusto following Munich's worst raid in April 1944, the Reich press bureau advised editors that

> in the light of the blatant and ruthless terror-attack by the murderous English arsonists against the world-renowned art treasures of Munich, the wanton barbarism of the British culture-destroyers must be exposed and branded in the most searing terms, expressing the outrage of the entire German people and shaking awake the cultural conscience of all humanity against these infamous criminals.[28]

Much of the European heritage losses of World War II, however, were of a more incremental character, stemming from the pressures of occupation or from attacks of a more traditional tactical sort, supporting the campaigns of land armies. On the Eastern Front, these campaigns and occupations sometimes amounted to attempted cultural decapitation. In Poland, 1939 witnessed the appointment of Austrian art historian and SS-Sturmbannführer, Kajetan Mühlmann, to oversee a programme of *Sicherstellung* (safeguarding) of works of art – a campaign whose character and scale recalled Napoleon's depradations of around 1800. At first, under the 1940/2 'Pabst Plan', the intention was to protect the historic core of Warsaw as a 'German'-style *Altstadt* but, in 1944, the unsuccessful Warsaw Uprising was followed by a deliberate campaign to destroy the Polish cultural patrimony. This included the dynamiting of the Royal Palace in Warsaw, the burning of libraries and archives, the blowing up of statues and the ploughing up of parks. Hitler ordered that, as Polish culture was an 'oxymoron', central Warsaw must be 'pacified, that is, razed to the ground'. In reaction, Polish art historians such as Stanislaw Lorentz and Jan Zachwatowicz secretly recorded and salvaged fragments for future reconstruction: of the 957 prewar classified historical monuments in Warsaw, 782 were totally destroyed, 141 partly destroyed and only 34 left undamaged.[29]

A more complex case was that of Leningrad and its nearby royal palaces, whose international-status classicism survived the war relatively intact. During the 900-day siege between September

Figure 7.8
Some wartime
bomb-sites still survive
even today: this 2009
view shows the ruins of
the Viceroy's Palace in
Cagliari, Sardinia,
destroyed in a 1943
air raid

1941 and January 1944, following a secret instruction that 'the Führer is determined to eliminate the city of Petersburg from the face of the earth', the urban area of Leningrad was subjected to incessant artillery and aerial bombardment, with over 150,000 shells and 107,000 bombs. However, despite the huge citizen casualties from starvation (reputedly several hundred thousand), according to Soviet sources most artillery damage was confined to the suburbs. Only a few historic buildings in the centre, such as the 18th-century Gostinny Dvor, were completely destroyed. The outlying palaces fared much worse, as during their occupation there was widespread devastation and looting. At Tsarskoe Selo, for example, the Amber Room was quickly discovered and dismantled for 'repatriation' to the Reich, the Catherine Palace was burnt out and the landscaped

Figure 7.9 Wartime destruction in the USSR

(a) 1941 view of refugees fleeing the bombardment of Leningrad; (b) The ruins of Petrodvorets in 1944, following its recapture by Soviet forces

parks were ravaged. Ironically, the Amber Room was originally designed by Prussian sculptor Andreas Schlüter in 1701–9 for installation in Charlottenburg Palace and was only moved to Russia in 1716; in a further irony, the panels seemingly perished in a Soviet bombardment of Königsberg in early 1945.[30]

In Western Europe, by contrast, token attempts at heritage protection in occupied countries were made by both sides. Echoing Clemen's role in the previous war, May 1940 saw the appointment of an eminent Rhineland conservationist and art historian, Franz Graf Wolff Metternich, by OKH (Army Command) as a civilian art protection official in occupied Paris; similar measures were taken in historic Italian cities during the retreat of 1944, with placards attached to Siena's monuments proclaiming, '*Kunstdenkmal unter deutschem Schutz*' (Monument under German protection).[31] In the Netherlands, too, the worst excesses were avoided. A ban on demolitions or alterations of listed buildings, imposed on 21 May 1940 by the Dutch armed forces commander for the duration of the war, was upheld by the occupying Germans and later perpetuated in the 1950 historic buildings act. Amsterdam architects were exempted from forced labour, to make precautionary drawings of the city's heritage. Sometimes, unofficial bribery secured protection: the exiled owners of Cuypers's restored Kasteel de Haar saved it from damage or confiscation by arranging substantial payments to local German commanders.[32]

Following accidental wartime damage, the occupied and occupiers in Western Europe often cooperated to repair damage to historic property. In the historic port of Bergen, the apparently accidental detonation of a ship carrying high explosives in April 1944 devastated the adjacent Bergenhus Castle as well as much of the 18th-century timber-built Bryggen harbour zone. Before the accident, municipal planners had proposed part-redevelopment of Bryggen, but that was now forgotten as an unofficial coalition of German staff officers and Norwegian conservationists (including cultural protection officer Gerhard Körner as well as Prof. J Bøe, director of Bergen Museum) worked feverishly to consolidate the ruins and prevent impulsive naval officers from blowing up the medieval Rosencrantz Tower to access communications equipment thought to be located beneath. In appealing to the military, they exploited the fact that the area had originally been built by Hanseatic traders, and was still popularly known as 'German Bryggen' (a designation dropped after the war). Ironically, the remainder of Bryggen was subsequently devastated by serious fires in 1955 and 1958 before finally (in 1979) being designated a World Heritage Site.[33]

Under pressure from the allied advances in 1944, the level of 'tactical' destruction in Western Europe sharply worsened, with historic towns across northern France and northern Italy, such as St Malo, Caen, Bologna or Verona severely battered by shellfire and tactical bombing. To minimise these attacks' effects, from 1944 a network of Allied cultural damage monitors was organised in Italy, Germany and elsewhere, under Sir Leonard Woolley. American art protection officers also played a key role, notably in the successful efforts by art historian Lt John D Skilton (Monuments, Fine Arts and Archives Officer) to save the daring vaulted ceiling and Tiepolo fresco of Balthasar Neumann's staircase hall at the Würzburg Residenz, which had miraculously survived a devastating RAF air attack on 16 March 1945: Skilton personally organised the sourcing and installation of timberwork to protect the ceiling from further inundation by rain.[34]

Other destruction stemmed from defensive measures by German forces. In Florence, following failure to agree an open-city status in August 1944, the army systematically blew up a swathe of houses north and south of the historic Ponte Vecchio, as well as bridges on each side, fearing allied assault. And in Marseille, following persistent partisan activity in 1943 in the dense Vieux-Port area, the army demolished much old housing in the area, although carefully leaving standing the outer facades. This area was already dogged by prewar controversy between the Ministry of Public Works, which had advocated sweeping demolitions in 1935, and the Historic Monuments Commission, which had refused the proposal. The postwar French authorities eventually demolished the facades left standing by the Germans.[35]

Paris narrowly escaped a similar fate, when the city's wartime commander, General von Choltitz, surrendered in defiance of an order from Hitler to blow up key public buildings. That near-debacle followed a period, under the Vichy government, when significant advances had been made in urban conservation – advances tainted, however, by indirect links to genocide. After the 1940 defeat, the initial drift of government policy was somewhat anti-urban: with Marshal Pétain's appeal for a 'return to the land', French regionalism broadened into a full-blown anti-socialist, anti-Modernist, *Heimat*-style reconstruction philosophy; 1942 saw the setting up of a *chantier d'études rurales*. Le Corbusier, already shifting from a white-walled to rubble vernacular modernism, made strenuous efforts to contribute to the Vichy programme and promoted *le folklore* and projects for a *ferme radieuse*. But these rural reveries were paralleled by increasing efforts at coordinated protection of urban ensembles. Here, the most important and enduring contribution of the Vichy government, the law of 28 February 1943, embodied a very simple idea – to piggyback a new system of area conservation onto the existing system of individual *monuments classés et inscrits*, by creating a 500-metre protected radius around each monument, within which both demolitions and alterations would be controlled. Instantly, every designated monument became a kilometre-diameter conserved area! This radius concept, strengthening permissive powers contained in the 1919 *loi Cornudet* and in a monument law of 1930, was so effective that it was simply made permanent in 1951 by the postwar government. And it also anticipated the more comprehensive philosophy of area conservation, through *secteurs sauvegardés*, pioneered by France in the 1962 *loi Malraux*. The 1943 law also boosted state grants to lesser monuments.[36]

However, there was a darker side to this wartime initiative, which emerged in the first rehabilitation project promoted by the Historic Monuments Commission in Paris – the *Ilot 16* scheme in the Marais, where (following rejection of an engineer-led 1942 plan for radical

redevelopment) a meticulous programme of '*curetage*', reminiscent of conservative surgery or *diradamento*, was proposed in 1944. The distinguished architect Michel Roux-Spitz, as *architecte en chef des bâtiments civils*, prepared a careful, low-cost, building-by-building renewal schedule, stressing the 'silent' task of replacing back-court accretions with communal gardens, rather than exuberant, Viollet-like restorations: the aim was to 'revive the essential texture' of the area. But the Marais was also Paris's principal Jewish quarter, and these proposals for purging physical imperfections from the area, with their potential overtones of social and racial 'cleansing' (as with some *Entschandelung* schemes in Germany), coincided with a growing wave of deportations from the area, with thousands of adults and children shipped off to concentration camps.[37]

A blessing in disguise? Reconstruction debates in the mid-1940s

THE war's end left the *Heimat* idea of nationalist heritage, not only in Germany and France but elsewhere, comprehensively discredited, both by its own implicit complicity in genocidal warfare and by the catastrophic destruction it had brought down on its own cherished treasures: in the closing days of the war, Hitler had wildly threatened a policy of heritage scorched-earth within the *Reich*, demanding that, in the face of the invaders, 'even works of art that had survived the bombing attacks would have to be entirely swept away – historical monuments, palaces, castles, theatres, opera houses, all would have to be destroyed'. With egregious exceptions, such as the dynamiting of much of 'Fortress Breslau' in 1945, that proved to be empty rhetoric.[38] But the concept of the Heritage Militant, marching as a soldier in the national struggle, certainly now seemed both repugnant and absurd. Increasingly, people focused on the opportunities of peaceful, progressive postwar reconstruction, and a complex relationship between modern architecture and conservation took shape, beginning in England with its early experiences of area bombing, but soon developing along consensual lines that strikingly linked the adversaries.

Modernising commentators in England, just as later in Germany, secretly welcomed the destruction, as an unexpected boost to comprehensive replanning: Coventry's City Architect and Planning Officer, Donald Gibson, argued in 1940 that the raid on the city had been 'a blessing in disguise'. And a 1941 leader in *The Builder* claimed that the wartime 'stiffening of national character' was enhanced by destruction of old buildings, which would otherwise have choked the emergence of a vibrant modern architecture: 'We shall do better to look upon the war ruins not so much as the graveyard of the past but as the cradle of the future.' The appropriate response was not to rebuild in the old style, but to make records of what was lost. The same sentiment was echoed by German propaganda chief Goebbels, who welcomed the bombing of Germany's 'old and used-up past', a past that would in any case largely have had to be liquidated in the process of National Socialist renewal.[39]

Of course, it was really the destruction of the vast areas of late-19th-century built fabric that was really being welcomed in statements like these. Earlier buildings and areas, whether medieval or 18th century, were as cherished as ever – despite Goebbels's flippant dismissal of the devastation of

facing page

Figure 7.10 The Würzburg Residenz

(a) External view; (b) Balthasar Neumann's vaulted ceiling and Tiepolo fresco in the staircase hall, saved from water destruction in mid-1945 by US art protection officer John D Skilton

Nuremberg – but bombing of industrial zones could be seen as a kind of accelerated *Entschandelung*. Even in Britain, commentators contrasted the two in forcible terms. W H Godfrey, secretary of the National Buildings Record, hailed the medieval and Georgian town as 'a composite work of art, a precious national possession', whose destruction risked sapping 'the moral values from which the nation must draw its future strength', whereas Victorian architecture, 'a dementia that turned skill into a blind alley and made invention a mockery', was actively expendable. A wartime text by Scottish architects Robert Hurd and Alan Reiach, *Building Scotland, Past and Future*, pleaded in Schultze-Naumburg-like terms for a wartime crusade to purge Victorian 'chaos and ugliness' and 'trashy vulgarity' from historic towns, through a modernity inspired by the 'homely virtues' of vernacular architecture: 'There is a new Scotland to be built.'[40]

But what form should that reconstruction take? Here, Coventry's Gibson and Germany's Goebbels would have differed in their public pronouncements, favouring respectively a restrained modernity and a Sublime classicism. But behind the scenes, different coalitions were forming, with English solutions preceding similar German ones by about 1–2 years, in some cases directly influencing them. Regarding the city as a whole, there was a wide consensus on the need for planned urban decentralisation and land-use zoning: Abercrombie's *County of London Plan* was strikingly echoed in Hamburg in the *Siedlungsgruppe als Ortszelle* concept devised by Gutschow's planners in the mid-1940s. But in the old towns, conservation-orientated planners favoured a different policy, of elaborated conservative surgery, including functional modernisation and selective insertion of traffic infrastructure, but avoiding aggressive, *tabula-rasa* replanning. In England, debates were well advanced by 1943, when an Ecclesiological Society exhibition and publication, *The Continuity of the English Town*, harnessed various national planning and conservation figures, including F J Osborn, Esher, and A E Richardson, to the challenge of how national traditions could be reconciled with reconstruction and modernisation. In a striking display of the speed of ideological shifts during wartime, the Georgian Group had now changed from radicals to reactionaries, with Richardson denouncing the Modernist 'Babylonian creed [of] gigantic scale'. With the continuing strength of Anti-Scrape in England, large-scale Ypres-style facsimile rebuilding was ruled out on principle, even by historicist architects such as H S Goodhart-Rendel, who argued in the 1940 SPAB *Annual Report* that, in rebuilding one must

> avoid every form of fake, every form of sham antiquity, since, like a human being, a building that is born old is nothing but a horrid deformity . . . to make a sham antique building can be not only inartistic but unpatriotic.

Already, the rebuilding of Coventry Cathedral was emerging as a litmus test of the balance between modernity and heritage. SPAB representatives argued the destruction should be remedied not by a

facing page

Figure 7.11 Lübeck: from destruction to rebuilding

(a) 1943 plan by architect Carl Gruber for rebuilding in accordance with the 'Lübeck spirit'; (b) Inspection party in the ruins of the Fischstrasse, Lübeck, in mid-1942: Stadtbaudirektor Hans Pieper is second from left and Prof. Georg Rüth of the Technische Hochschule, Dresden, is in the middle; (c) The rebuilt Marktplatz in Lübeck: 1975 view looking north towards the Marienkirche and (on the extreme right) the Rathaus

(a)

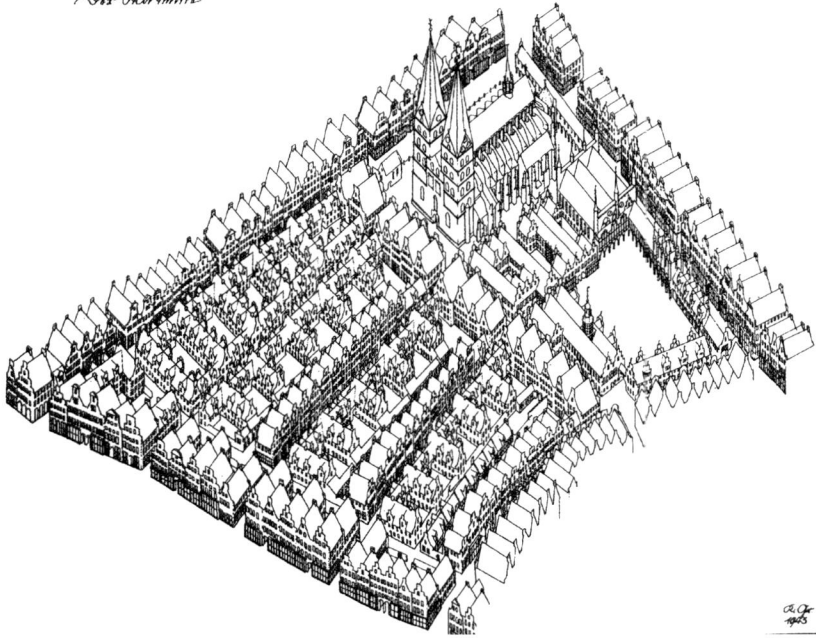

(b)

(c)

'purely modern building attached to the tower and spire', but through a 'harmonious' modern rebuilding within the existing shell – a concept that was initially endorsed by the Cathedral authorities. Participants agreed on the vital importance of maintaining the primacy of voluntary societies in safeguarding the English architectural heritage. And as we will see in Chapter 8, the mid- and late 1940s duly saw a rash of planning reports for small historic cities, envisaging a broadly conservative-surgery treatment.[41]

The continuing power of the New Romanticism in England helped reconcile these debates with continuing expansion of the heritage: J M Richards's influential 1945 book, *The Castles on the Ground*, for example, pleaded for the inclusion of the 19th-century suburb within the tradition of the Picturesque, something that would still be unthinkable in Germany, where the *Heimat* ended with the *Biedermeier* period.[42] German conservationists were rather more ready than English Anti-Scrapers to sacrifice old fabric in pursuit of an instrumentalised *schöpferische* and *lebendige* heritage ideology. There, wartime *Altstadt* reconstruction debates were torn between pressures for conservation and for classical reordering, with Lübeck the first testing ground for the two positions. As early as May 1942, Stadtbaudirektor Hans Pieper was busy not only demolishing surviving street facades in the central zone, but also drawing up plans for a stately reconstruction of the city under the 1937 law, including substantial demolition within the unbombed zone of the *Altstadt*, arguing opportunistically that wider streets and large building blocks would make the city more resistant to air attack. His plans met with furious opposition, for example later in 1942 from Hamburg architects Fritz Höger and Konstanty Gutschow, who argued for a sensitive reconstruction 'without slavish copying' or 'pseudo-Gothic' style, but which would still 'perceptibly' evoke the 'Gothic spirit'. Höger drew a sharp distinction between beneficial destruction of 'bad and ugly' eyesore buildings and destruction of old towns, which should be put right in a sympathetic way. Responding to their call, in 1943, architect Carl Gruber designed a counter-proposal for rebuilding in a 'Lübeck spirit', incorporating courtyards and picturesque agglomerations of gabled facades. After the war, in 1947, referring especially to Lübeck, Heinrich Tessenow would protest against 'sacrifice of the cultural riches of our old towns to the gods of the so-called modern world'.[43]

Until the end of World War II, the Conservation Movement had preserved much common ground in Europe, transcending even the confrontation lines of the war. But all that began to change in 1945. With the emergence of the sharp ideological polarisation of the Cold War, urban conservation fragmented into three separate strands, comprising not just the capitalist West and communist East, but also a new, increasingly powerful strand of explicit internationalism. In the next three chapters, these three separate strands will be traced separately, before they, and the story of the Conservation Movement, finally converge again in Chapter 12. During that half-century, World War II was pushed out of the foreground, becoming a refracting backdrop to the varied reconstruction efforts. All the post-1945 strands of the Conservation Movement insisted on an exclusively peaceful approach and (with relatively few exceptions) avoided exaggerated politicisation. Only after the end of the Cold War in 1989–91 would World War II once again become the focus of explicit conservationist commemoration, often under the umbrella of 'intangible heritage' – a postmodern transformation of the old concept of all-embracing *Heimat*.

Heyday of the Movement

Parallel narratives of postwar preservation

– 8 –

Parallel lives

New and old in the West, 1945–68

We shall do better to look upon the war ruins not so much as the graveyard of the past,
but as the cradle of the future.
The Builder, 19 September 1941[1]

THE sweeping geopolitical changes after 1945 transformed the wider context of the Conservation Movement. Previously, the heritage world had been divided by violent national rivalries and by ideological differences such as the balance between private and state initiative, although key values were held in common. Now, there was a new balance of forces. The disappearance of aggressive nationalism allowed the Conservation Movement to become a more peaceful, measured place, developing its ideas and policies rather out of the public limelight. But there were new divergences. The Cold War boosted opportunities for building and planning, but also opened up a far sharper set of ideological differences, in which the role of old buildings was often seriously called into question.

Between 1945 and 1989, there were several parallel narratives of conservation: a Western-bloc narrative, a socialist-bloc narrative and finally a growing internationalist narrative that mediated between the others. Each of these blocs had a marked tendency towards internal homogenisation. Each also had its own distinctive internal ideological differences and internal chronology, often different from the others – for example in the sharp Soviet lurch from socialist realism to orthodox Modernism in the late 1950s and equally sharp Western rejection of orthodox Modernism by protesters, including militant conservationists, a decade later. Yet key values were still held in common, above all the Conservation Movement's fundamental narrative of progress; and by the 1970s and '80s there was a more open convergence, as the internationalist bloc gradually gained the ascendancy.

After 1989/91, these blocs disappeared altogether and a more confused combination of capitalism and globalised internationalism took its place – embracing parts of the world previously excluded from the scope of the heritage discourse. Conservation once again became more homogeneous, but the sense of a dynamic 'Movement' had begun to diminish, leaving a sense of disorientation typical of postmodern culture. For this reason, the next four chapters trace the 1945–89 story in a sub-divided way. Chapters 8 and 9 cover the West from 1945 to 1968 and 1968 to 1989; Chapter 10 the socialist bloc from 1945 to 1989; and Chapter 11 the ever-growing international infrastructure of conservation. Finally, in Chapter 12, we review the more unified, yet more uncertain years since 1989.

From hard to soft heritage nationalism

ESPECIALLY on the continent of Europe, the postwar decades saw marked convergence in the practice and principles of conservation. The aggressive outward projection of the nation ceased almost everywhere, as the state refocused its built-environment interventions towards welfare-state modernisation. The modern social-democratic state deployed a more subtle, humane propaganda than its militant predecessor, and the Conservation Movement assumed a consensual, moderate form. Although heritage nationalists were banished from the central circles of political power, an apolitical concern for historic built landscapes remained prominent within the concerns of the social state. In the view of even J M Keynes, preservation of historic buildings was 'properly' the affair of 'public-spirited individuals' rather than the state, especially given the 'perverted' excesses of state power under totalitarianism.[2] Where previously the English detachment of heritage from overt nationalism was exceptional, now this became the norm throughout Western Europe, especially in strongholds of 'Christian Democracy' – a statist alliance between (mainly Catholic) religion and moderate economic liberalism that underpinned much welfare-state prosperity.

This shift from bellicose to peaceful conservation was especially prominent in West Germany, where immediate postwar debates were dominated by concerns of guilt and expiation. Karl Jaspers's 1946 book, *Die Schuldfrage*, bemoaned the nation's 'criminal, political, moral and metaphysical guilt', while a new genre of 'ruin films' explored the social and moral collapse spawned by National Socialism. Tainted by the prewar politicisation of *Heimatschutz*, the conservation establishment was sidelined: *Deutsche Kunst und Denkmalpflege* was suspended until 1952 and provincial conservators could only react case by case to local threats. The confident certainties of nationalist heritage set out in Dehio's 1905 *Kaiserrede* were now in ruins; his birthplace (Tallinn) was now in the USSR and his academic seat (Strasbourg) in France. With the rehabilitation of former Nazis in West Germany, some key figures recovered their influence, such as Rudolf Esterer, and refugees expelled from the eastern regions kept up raucous complaints. But even in *Heimat* strongholds, prewar extremists such as Rothenburg's Nazi mayor, Fritz Schmidt, could only resume their old policies (*Entschandelung*, etc.) in a depoliticised form.[3]

This collapse of the strong nationalist discourse left survivors disorientated. Paul Clemen, for instance, was bewildered by the vastness of urban destruction: in his native Cologne, only 2 per cent of built fabric had survived the attacks intact. This, he argued in 1946, was a European tragedy: 'for the first time ever, [Europe] finds herself confronted with a vast territory turned entirely into rubble'. He condemned the prevalent 'hurry to demolish or tear down': ruined areas should be cordoned off to

> await their hour of restoration, not as a dead relic, but as one which still speaks clearly to us, reminds us of old memories, and constantly warns us to rebuild it, and in doing so, to rebuild the *Heimat* as well.

But the strengthening Modernist ascendancy now pointed in a very different direction. Fritz Schumacher's famous Hamburg reconstruction speech of summer 1945 saw loss as outweighed by

facing page

Figure 8.1 *Kleine Bettlektüre für Heimattreue Schlesier*: front cover of 1975 bedside nostalgia book for Germans expelled from Silesia three decades previously. The illustration shows Breslau/Wrocław Town Hall being uprooted and carried off 'home'

Kleine Bettlektüre für heimattreue Schlesier

Leckerbissen für alle,
die sich ein Herz
für Schlesien
bewahrt haben

Scherz

hope, and the destruction of characterful old towns as counterbalanced by the promise of modern reconstruction. Conservation had, temporarily, slipped to a low priority: but the growth of Christian Democrat politics in West Germany would soon give defenders of 'tradition' a new, post-totalitarian political anchor.[4]

In countries not ravaged by destruction, there was less need for radical change. In Switzerland, whose central government had distanced itself from German *Heimat* extremism through a new arts and heritage agency, Pro Helvetia (founded in 1939), a restrained ideology of *Altstadt-Sanierung* continued unimpeded. Summer 1945 saw a special municipal exhibition in Basel, *Altstadt Heute und Morgen*, with striking before-and-after perspectives of *Entkernung* projects, their unified, high-roofed groupings resembling prewar work in Danzig or Frankfurt. In Scotland, too, the prewar nationalist enthusiasm for old-town conservation continued until the early 1960s; in 1961, restorer Moultrie R Kelsall lambasted the 'de-Scotticising' of architecture in a fierce rhetoric worthy of Schultze-Naumburg:

> It is surely obvious that a child brought up in a burgh or village where there is ample witness to the past in stone and lime is likely to have a much deeper sense of social continuity – of being part of a mature, yet still developing community – than a child brought up in a typical housing scheme, raw, rootless and lacking any individuality.[5]

Brandi and Lemaire: the modern 'art' of conservation

BY far the most remarkable escape from the sinking ship of militant nationalist conservation was that of Italy. Its timely change of sides from Axis to Allies in 1943 had had damaging short-term heritage consequences, in the widespread artillery and bombing damage to historic cities. In the longer term, though, it allowed Fascism's controversial legacy to be sloughed off and Italy once more to become the archetypal culture-nation and international heritage mecca – just in time to anticipate the Conservation Movement's turn towards internationalism in the 1960s and '70s. In the process, ironically, Fascism's administrative conservation machinery was not only perpetuated by Italy's continuously Christian Democrat-led governments from 1944 to 1981, but also became the foundation of a new international apparatus of conservation.[6]

Arising from the ruins of Mussolini's failed overseas empire, Italy's new, peaceful heritage 'empire' was partly based on the overseas archaeological work of organisations like the Italian Middle and Far Eastern Institute (ISMEO). In architecture, Italy's bid for leadership of the nascent international Conservation Movement depended chiefly on an attempt, led by Italian writers and architects, to re-establish Art itself as the foremost heritage value. At a theoretical level, anti-fascist philosopher-historian Benedetto Croce promoted the centrality of classicism and context in the modern philosophy of aesthetics. Others, notably Guilio Carlo Argan and Cesare Brandi, laid the institutional foundation for postwar developments, building on the Fascists' 1939 Law 1089 on

facing page
Figure 8.2

(a), (b) 1945 exhibition perspectives of the ongoing *Sanierung* of the Basel *Altstadt*, showing before/after views of clearance of congested courtyards

cultural heritage and establishment of a Central Institute of Restoration (first director: Brandi). This institute pursued a more scientific philosophy of restoration, strictly re-establishing objects' original aesthetic qualities: in its attitude to authenticity, a more rigorous echo of Viollet.

During his 22 years as director of the Central Institute, Brandi established himself as the foremost postwar theoretician of conservation at an international as well as Italian level, and laid the groundwork for Italian pre-eminence in many international conservation organisations and charters. In itself, Brandi's thinking was mainly consolidatory in character, reasserting the turn-of-century emphasis on art and the creative artist, and correspondingly rejecting Marxist-style base/superstructure thinking.[7] Published in instalments in 1950–6 and in its entirety in 1963, Brandi's main conservation theory (*Teoria del Restauro*) ran as follows. Artists first form out of external stimuli an ideal art-concept. They then give it specific historical reality, after which others can then form their own perception or reception. Contrary to Ruskin, art is specifically *not* an imitation of nature, but something higher and purer. Art thus comes before history, and the historical meaning, and the authenticity, of monuments emerges from the sequence of creation and reception. Architecture without art becomes mere 'passive constructiveness'. Thus, careful restoration is justifiable, but must be based solely on recognition of the building as a work of art or 'aesthetic creation', both originally and in subsequent creative interventions.[8]

Despite the narrow base of his theory by comparison with, for instance, that of Riegl, and its elitist irrelevance to historic buildings of little aesthetic significance (e.g. industrial archaeology sites), Brandi's writings had an extraordinary impact in the world of international conservation, which was just poised to enter a period of exponential growth. Their emphasis on 'civilised' artistic values, rooted in the universal legacy of Rome, seemed ideal to a generation revolted by the excesses of nationalist or social-materialist extremism. Other Italian writers helped flesh out this renewal of Boito-style 'critical restoration' (*restauro critico*) – all with slightly different emphases. Historian Renato Bonelli, for instance, argued that conservationists should either leave monuments totally alone or aim to restore their pure artistic essence – an operation that could in itself create a new work of art, as restoration was 'a critical process, and then a creative act, the one . . . an intrinsic premise of the other'. All this helped make post-nationalist, postwar Italy not just a significant conservation influence in other Western countries but a focus of special concern in the case of heritage dangers or disasters. That was resoundingly demonstrated, as we will see shortly, in November 1966, when freak flooding in central Florence, chiefly damaging works of art and books rather than buildings, sparked off an unprecedented international rescue effort, while a flood the same month in Venice provoked a more enduring campaign to 'save' the city from sinking into the Lagoon.[9]

In the new, post-nationalist world, the Ruskin-influenced concept of the strong, living heritage, driving forward the national destiny, was now marginalised. The general Enlightenment idea of historical progress in architecture remained strong, but had now reorientated itself unambiguously towards the future. The focus of state-sponsored efforts shifted to the planning and building of vast *new* environments for all, shaped not by accumulated traditional cultures but by a vision of bright new rebirth: in France, the first three postwar decades were known as the *trentes glorieuses*. During those decades, the chief ideological issue within the Conservation Movement was: how were monuments, and their protection, to relate to this modernity? Their target was not stationary: the postwar Modern Movement teemed with conflicting perspectives and ideas, like any phase of architecture. There was a shift from the rectilinear simplicity of blocks in flowing space, to more complex urban formulae. Attitudes in the West also reacted against developments east of the

Iron Curtain: the overt historical references of Soviet Socialist Realism (and of fascist grand classicism) provoked a Western reaction towards Modernism, which was now widely attributed democratic connotations, and commanded much broader support than before the war. Now, the clearest differentiation between old and new seemed to be called for – a polarisation that (ironically) almost resembled a Marxist dialectic. Anything in between the two seemed very awkward: in English, the word 'pastiche', originally of French and Italian derivation and denoting a derivative literary or musical work compiled from various sources, was now used (always negatively) to describe new architecture that supposedly accommodated historic surroundings too literally.[10]

Within the postwar Conservation Movement, an entire generation of younger architects took for granted this old–new dialectical tension. For example, the Belgian architect, Raymond Lemaire, who would become one of the foremost international leaders of conservation in the 1960s and '70s and co-authored the 1964 Charter of Venice, was a militant believer in the sharp differentiation of conservation and modern architecture. He authored in the same years a conservative restoration of the Grand Beguinage at Leuven (1963–4) and an advanced Modernist plan for the new town of Louvain-la-Neuve. And in individual restorations, he relished the juxtaposition of old fabric and frankly modern insertions, as at St Lambertus Chapel at Heverlee (late-1950s), where a rough stone tower and arched shell was stabilised and glazed across its openings. Equally expressive of the principle of old–new separation was Gottfried Böhm's 1959–68 rebuilding of Godesberg Castle near Bonn, ruined since 1583, as a restaurant, interpenetrating the castle rubble with concrete, steel and glass platforms.[11]

Internationally, this schizophrenic outlook was shaped by the 'other' Athens charter, the CIAM Modernist manifesto, with its eloquent respect for historic buildings' place within the planned modern city. In Britain, this world-view was reinforced by a century's rhetoric against 'false' historical-styled architecture, stretching back to Pugin through Anti-Scrape. For conservationists as well as for modern architects, the historian Nikolaus Pevsner's linking of Morris and the Modern Movement seemed only logical. Ralph Tubbs's 1942 Modernist propaganda book, *Living in Cities*, argued that, while 'sham classic' buildings must be rejected, modern architecture did not 'ignore tradition'.[12]

During the *trente glorieuses*, this new orthodoxy spread irresistibly, even in places where authoritarian regimes remained in power. One weathervane of the shift in the new/old relationship was the concept of the vernacular. Reacting against Soviet Socialist Realism, the *Heimat* concept of integrated old and new vernacular was now rejected in favour of a respectful separation, with old vernacular inspiring new architecture spiritually rather than literally. Ironically, Italian Fascist architecture had pioneered much of this thinking: Giuseppe Pagano (editor of *Casabella*) argued in 1935 that vernacular was 'a great repository' of 'expressions of honesty, clarity, logic' that could inspire modern architecture without direct copying. The Spain and Portugal of the 1930s–'50s saw a vigorous move away from the Salazar and Franco pursuit of traditional folk-architecture styles (e.g. in the designs of Raul Lino) towards new concepts of Modernist vernacular, as in the work of Fernando Tavora in Portugal, or Alejandro de la Sota and J L Fernandez de Amo in Spain. And in Brazil, the architecture of early colonial times was cited in the late 1930s and '40s as a potential stimulus to the growth of an indigenous Brazilian variant of International Modernism.[13]

Many younger postwar English architects combined early Modernist tendencies with a later turn towards heritage. Conservation specialist Donald Insall, for example, began his training in the mid-1940s at E A A Rowse's Modernist School of Planning and Research for National Development

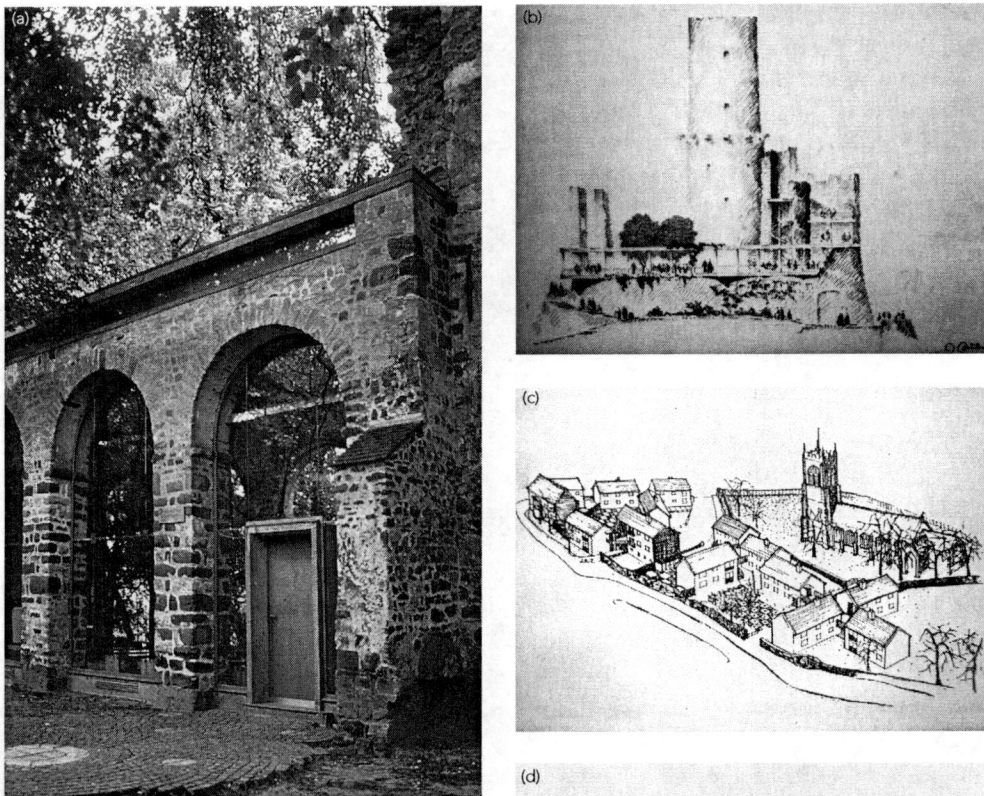

Figure 8.3 Juxtaposing the Old and the New

(a) St Lambertus Chapel, Heverlee, restored in the late 1950s by Raymond Lemaire; (b) Godesberg (Burg), rebuilt 1959–68 by Gottfried Böhm; (c) Alderson Place, Norwich – a public-housing infill redevelopment of 1958, by the staff of City Architect David Percival; (d) Unidad Habitacional Nonoalco-Tlatelolco, Mexico City, an extensive social housing complex built in 1960–5 by architect Mario Pani around the Tlatelolco archaeological area (the Plaza de las Tres Culturas), sharply juxtaposing modern public-housing blocks with colonial and Aztec remains. The inscription commemorates both the indigenous and Spanish heritage of Mexico. The complex subsequently had a troubled history, including a 1968 massacre of student protestors and a 1985 earthquake

(SPRND) at the Architectural Association, but combined this with Morrisian values as a student SPAB scholar; subsequently he remained faithful to that mixture of Modernism and Morris.[14] Bernard Feilden, later renowned as an international conservation leader, spent the first 14 years of his career (1949–63) as a contextually sensitive Modern architect in Norwich before receiving his first major conservation commission, to repair Norwich Cathedral (from 1963). He responded to this challenge, as we will see later, with a typically Modernist strategy of high-technology structural intervention. Feilden's 1963–4 conversion of the medieval King's Manor in York to house the Institute of Advanced Architectural Studies (which taught both Modernist and conservation courses) was a textbook example of the same approach, expressed with English Arts and Crafts sensitivity, juxtaposing a medieval bottom layer with Modern upper storeys and concrete stairs. The same pattern applied to conservation-orientated planners: Frank Tindall, later the regional planner of East Lothian and restorer of the burgh of Haddington, began his career working with Berthold Lubetkin on Peterlee New Town.[15]

In some cases, during the 1940s–60s' reconstruction years, this polarisation became an outright opposition, with one side openly criticising the other. The establishment Modern architect Sir Hugh Casson argued in 1962 that Britain 'runs the risk of becoming a nation of museum keepers . . . this is the only country in the world where people say, "What a pity they are going to build." Everywhere else, they say, "How wonderful. They are going to build."' Other architects pursued an active reconciliation of old and new – even if the two were kept distinguishable from each other. In Edinburgh, the Geddes tradition of planning encouraged people to look both forwards and backwards without any strong sense of polarisation: in a Geddes-like lecture of 1964, Modernist architect/planner Robert Matthew argued that Le Corbusier's 'greatest mistake was to isolate himself from the immediate past', which had inspired 'planners to yearn for a clean sheet, a new start, unimpeded by inconvenient relics from the past'.[16] In England, old–new reconciliation was encouraged by the widely shared love of the Picturesque. This received a powerful reinforcement from 1940s' New Romanticism: writers such as Nikolaus Pevsner tried to anchor Modernist set pieces like the London County Council (LCC) housing programme or the Festival of Britain in the complexity of the English past. A typically Picturesque integration of 'human-scale' housing and historic context, the Alderson Place development in Norwich (1958), designed by City Architect David Percival, mingled low-rise flats with a medieval parish church.[17]

Eventually, in the 1980s and '90s, a fresh strand of old–new reconciliation emerged, when Modernism itself became heritage, but for now that connection was only latent, hinted at in the growing veneration (from the '50s) for rediscovered 'pioneers' such as C R Mackintosh. Le Corbusier also counted as a pioneer, as shown in the international protest campaign of 1957–60 against demolition of his Villa Savoye, finally blocked by French culture minister Malraux.[18]

For obvious reasons, it was in Christian Democrat-dominated Italy that the reconciliation between modern architecture and the historic environment was pursued most comprehensively: Brandi lambasted CIAM open-planning orthodoxy as an 'architectural cancer' in the city. It was assumed within postwar Italian architecture that new urban projects should respond explicitly to the historical setting – a sentiment celebrated in the very title given to one leading Italian architectural journal, *Casabella-Continuità*, when Modernist Ernesto Rogers assumed its editorship in 1954: at an urbanists' convention in 1958, Rogers asserted the indivisibility of the Conservation and Modern Movements. However, there was much creative diversity in expressions of that indivisibility. Sometimes the outcome was exact rebuilding, as with Montecassino Abbey, destroyed in 1944 and, following impassioned debate, re-erected in 1945–64 by Giuseppe Breccia Fratadocchi, But the past could equally be evoked through modern buildings styled 'traditionally' – as in the vernacular-styled INA-CASA housing schemes of the early 1950s, or the neo-Liberty castle-style tower, the Torre Velasca, designed in Milan by Rogers's firm, BBPR – even though, to many foreign critics, this approach seemed like 'pastiche'.[19]

In other cases, especially in conversion of historic buildings into museums, the emphasis was on a refined, occasionally quirkish juxtaposition of old fabric and steel/glass 'interventions' (a term actually invented in Italy). This approach was exemplified in BBPR's conversion of Milan's Castello Sforzesco into a museum in 1956, and Carlo Scarpa's Castelvecchio conversion in Verona (1956–64) and Venice museum schemes (e.g. the Querini Stampalia Foundation, 1959–63 and the Accademia galleries, 1945–59). A notable series of municipal museums in Genova on the same lines, by Franco Albini, included the Palazzo Bianco (1949–51), a pioneering white-space museum with blue slate pavements, embedded in a Baroque palace; the Palazzo Rosso Museum (1951–62), with its range of loggias and courts; and the Museum of the Treasure of San Lorenzo (1952–6), set in a dark, sombre crypt beneath the Cathedral Archbishop's Palace and focused on four enclosed 'tholos' spaces clad in grey limestone.[20] Other architects approached the issue from an explicitly conservationist perspective, continuing the Giovannoni principle of *diradamento*. In 1955–8, Giovanni Astengo prepared an influential *piano regolatore generale* for Assisi, which avoided new construction or roads altogether and stressed protection of vistas of the town's skyline. And in 1958–79, Pietro Gazzola, who had restored Verona's Castelvecchio bridge (Ponte Scaligero) in 1945–51 in his capacity as city conservation chief, oversaw a remarkable and protracted rehabilitation of the ruined Citadella in Cagliari, Sardinia, into an archaeological museum. The result was a fantastic agglomeration of modern glass and concrete walls, castle fragments and archaeological elements unearthed during restoration, embedded in the acropolis-like setting of the outer citadel walls. As we will see later, Gazzola, with other eminent Italian architects such as Roberto Pane, Piero Sanpaolesi and G De Angelis d'Ossat, would play a major role in the growth of the international Conservation Movement in the 1960s and '70s, grounding it in Italian norms and experience.[21]

Early reconstruction efforts: the Goethehaus debate

Aʟʟ that, however, lay in the future. In the mid- and late 1940s, just as in 1918, the first and foremost task was that of reconstruction. The damage, rather than limited to a front-line zone, was dispersed in cities and towns right across Europe. However, the rebuilding challenges were not greatly different from those of Ypres and Reims, and the solutions fell within the same range of permutations, from faithful reconstruction to complete modern replacement. Often, these were accompanied by exhibitions and propaganda campaigns, orientated towards collective mobilisation rather than overt citizen participation. In the built projects, as with the post-1918 rebuildings, there was a contrast between the lavish recreations of key individual monuments and the freer treatment of urban ensembles: only in the case of Warsaw (see Chapter 10) was the reconstruction of an entire *Altstadt* attempted, and even then with considerable imaginative licence.

In Western Europe, increasingly dominated by Modernist ideas, even restoration of major monuments sometimes met with controversy, especially in West Germany. In Frankfurt, the years 1945–7 saw impassioned debates, not only about whether to rebuild the Paulskirche as an expiatory monument (eventually completed in 1948) but also about the more thorny issue of what to do with the ruins of Goethe's birthplace house in the Hirschgraben, destroyed in early 1944 down to its ground floor. For the postwar Conservation Movement in West Germany, this was an exemplary debate, like that about Heidelberg in 1905, although it could not compete quantitatively with the mountain of verbiage generated by the latter.

268

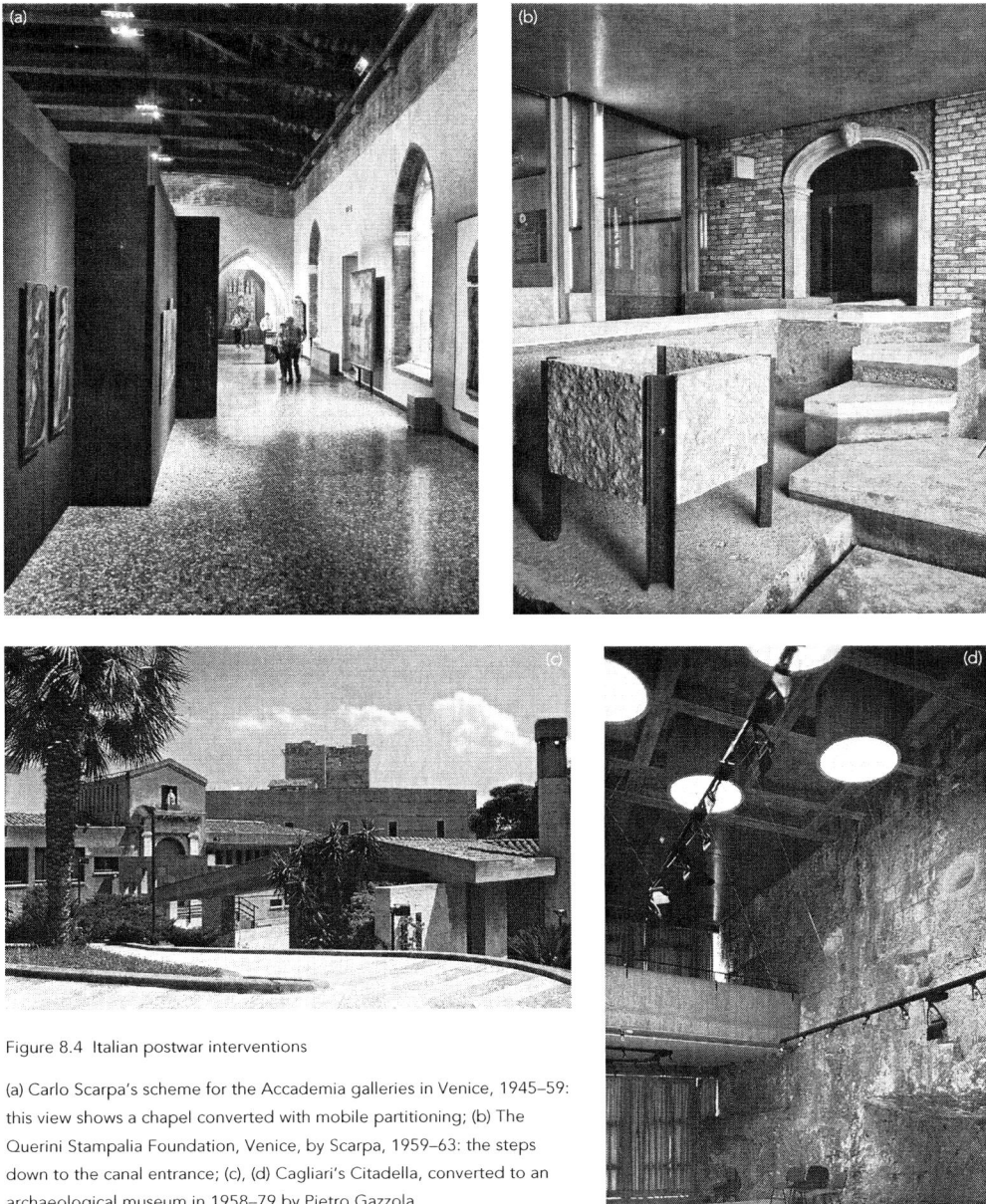

Figure 8.4 Italian postwar interventions

(a) Carlo Scarpa's scheme for the Accademia galleries in Venice, 1945–59: this view shows a chapel converted with mobile partitioning; (b) The Querini Stampalia Foundation, Venice, by Scarpa, 1959–63: the steps down to the canal entrance; (c), (d) Cagliari's Citadella, converted to an archaeological museum in 1958–79 by Pietro Gazzola

Following the Goethehaus bombing, the National Socialist press had demanded that the ruins stay as an eternal accusation against the 'terror attacks' – which would also conveniently have muzzled the activities of the Freies Deutsches Hochstift (FDH), distrusted by Gauleiter Jakob Sprenger as a hotbed of liberalism under its director, Ernst Beutler. However, Beutler had prudently organised a bevy of local schoolgirls to measure every conceivable detail of the Goethehaus in 1942 and so, as early as April 1944, the regime reluctantly authorised a reconstruction, in the form of a

269

facsimile, built on the foundation walls. After 1945, Beutler and the FDH revived the project, reframing it as a monument of 'understanding between peoples' and 'common humanity'. A vigorous public debate flared about the rebuilding scheme. Modern architects lined up in opposition: in early 1947, the local branch of the Werkbund argued for a modern reconstruction, and Catholic theologian Walter Dirks advocated 'having the courage to say farewell', marking the site by leaving it empty. This he justified partly in conventional anti-restoration terms: a facsimile would 'kill truth . . . one would go through the house feeling [only that] . . . Goethe did *not* live here!' But more important still, Dirks contended, was the link between war guilt and the nationalist politicisation of heritage. In a more refined echo of 'Bomber' Harris's hubris-nemesis argument, he claimed that

> the house wasn't destroyed by some chance event, like a lightning flash, [but] by a historical event that was intrinsically bound up with its nature. This destruction had a bitter logic about it . . . it wasn't a mistake or a quirk of history, but a just outcome, one that has to be acknowledged and marked. . . . The German nation of poets and philosophers, under the influence of idealism and classicism, under the influence of Goethe himself, had allowed commerce and power to get out of hand, [causing] a division of the German spirit which . . . led directly to the catastrophe that has overtaken us.

And thus 'the great devastation stands, consequentially, at the end of a way which leads from the Goethehaus'.[22]

As with the opposition to rebuilding the Venice Campanile, these were poignant but unworldly arguments. Eventually, Beutler successfully secured his facsimile restoration, appealing over the architects' heads to other elite circles: poets, academics, politicians. The City Council voted in April 1947 for rebuilding and, after a massive fundraising campaign, André Gide laid the foundation stone in 1947. The FDH argued cogently that this would be 'certainly [not the same

Figure 8.5 Rebirth of the Goethehaus

(a) Ernst Beutler and his son in the ruins in August 1945; (b) Brochure for the rebuilt museum (third edition, 1969)

THE GOETHE HOUSE
IN FRANKFURT AM MAIN

FREIES DEUTSCHES HOCHSTIFT
FRANKFURTER GOETHEMUSEUM

house] in which Goethe was born', but a 're-marking of the site of a monument', containing the original furniture, and stressed that the upper walls had largely been replaced when the house was opened to the public in 1885, and a steel frame had been inserted in 1931.[23]

In the case of another renowned urban monument – the spectacular timber-framed, gabled 1529 Knochenhauer Amtshaus (Butchers' Guildhall) in Hildesheim, the outcome was the opposite from the Goethehaus. Widely seen as the town's architectural signature since the beginning of the *Heimatschutz* era and as Germany's 'most monumental timber building' (Dehio), it was burnt to the ground in March 1945 along with almost all the Old Town Square. There followed a protracted and bitter debate. Local citizen demands for reconstruction were stymied by an alliance of architects and planners seeking enlargement of the square and anti-restorationist regional conservation officers. They argued that, while it was legitimate to build a facsimile of a cultural monument such as the Goethehaus, that did not apply to the Knochenhauer Amtshaus, as an architectural monument. And whereas the restoration of gutted Romanesque churches in the town was seen as acceptable because they had partly survived, the Knochenhauer Amtshaus had been completely destroyed. Eventually, the square was widened and a six-storey, towered, Modernist hotel designed by Dieter Oesterlen was built in 1962–4, but (as we will see later), this ensemble was then in turn redeveloped in 1988 with a traditionally built copy of the Knochenhauer Amtshaus. The postwar resurgence of Dehio-style Anti-Scrape in West Germany impeded other key reconstruction projects, such as the Golden Hall in the 1615/20 Augsburg Rathaus, gutted in 1944. It was initially reinstated in a very plain style, on the insistence of the Bavarian government conservation architects; only in 1980–96 was the hall fully restored in all its splendour.[24]

A different variant of reconstruction 'honesty' emphasised the ideal of legibility, restoring damaged elements in a rough or simplified manner designed to stand out from surviving original sections. Hans Döllgast's influential 1946/52 restoration of the Alte Pinakothek in Munich infilled a 45-m destroyed section of a classical facade with similarly proportioned bays in primitive brickwork and slender columns inside supporting the new roof. More frequently, destroyed interiors were reinstated in a simplified manner generally recalling the style of the original. One of the earliest examples was the plan for rebuilding of the fire-damaged House of Commons in London, already approved in December 1943 and published in 1944. After an overtly Modernist reconstruction was vetoed, Sir Giles Gilbert Scott was engaged to design a replacement on the same intimate scale, but in a simpler Gothic style with decoration 'altogether fresher, lighter and more alive' than Pugin's ornate 19th-century original designs.[25]

Predictably, the closest professional and public attention was given to the restoration of bombed churches. Here, pressure for immediate clearance was low, and they were often restored as fully as scarce resources allowed, incorporating fittings that had been kept in safekeeping. As we saw in Chapter 7, this approach was initially rivalled by the concept of leaving ruins unrestored as reminders of war, sometimes enfolded or adjoined by a new church building. A seminal example of this approach was the new Coventry Cathedral, designed in 1951 and built in 1954–62 to the designs of Basil Spence. Set at right angles to the roofless shell of the old cathedral, it cleverly balanced modernity and tradition in its sandstone cladding and modern Gothic column-and-vaulting structure.[26] This formula had almost limitless variations. In Rudolf Schwartz's St Alban and Gürzenich rebuilding in Cologne (1952–5), a ruined church was enfolded on three sides by a rebuilt public hall. Similarly, in architect George Pace's 1961–8 rebuilding of the 'Baedeker'-destroyed St Martin-the-Grand church in central York, part of the ruins was roofed in as a shrine of

Figure 8.6 'Legible' restoration in Munich

(a) The south facade of the Alte Pinakothek, showing Hans Döllgast's influential 1946/52 restoration; (b) The stair hall located immediately behind that facade; (c) The Siegestor, restored from 1956 in accordance with architect Josef Wiedemann's controversial 'didactic' proposal

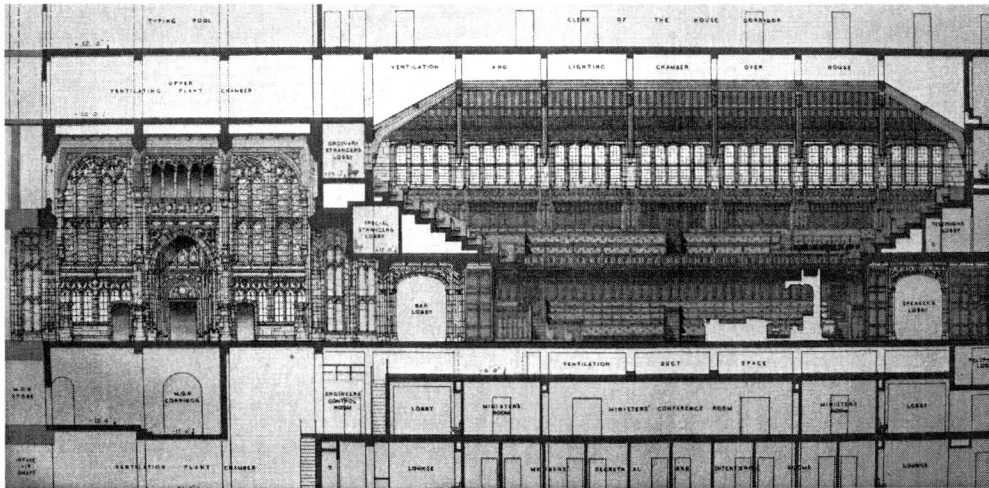

Figure 8.7 The 1944 House of Commons rebuilding report: long-section of Sir Giles Gilbert Scott's new design

remembrance. In Egon Eiermann's executed 1958 scheme for the Kaiser-Wilhelm-Gedächtniskirche in Berlin, the stump of an 1890s' church spire was sandwiched between new glazed, polygonal blocks (an earlier proposal for total demolition having been blocked by massive public protest). The most eccentric offshoot of the theme of juxtaposed ruin and new church was built slightly earlier, in 1949–54: architect Raffaello Fagnoni's reconstruction of San Domenico in Cagliari, Sardinia, reduced to roofless ruins in a 1943 bombing raid. Fagnoni, with engineer Enrico Bianchini, conceived the unusual solution of building a new church on top of the damaged older one: the new upper space had a lightweight roof supported by half-portal concrete frames converging at a single point on either side, and opened at the east end into a parabolic dome.[27]

Rebuilding the old towns: street-pictures and simplification

IN the reconstruction of the wider fabric of historic centres, the solutions adopted across Europe as a whole varied widely, from the facsimile reconstruction of Warsaw to the explicitly Modernist rebuilding of central Rotterdam.

Probably the widest diversity in rebuilding solutions was found in France, with recipes varying from semi-facsimiles to total renewal. In 1944–5, across the combat zone of the northern and western coast, from Dunkerque to Brest, the same story was heard over again: damaged but salvageable buildings were torn down, despite the protests of local preservation enthusiasts. But this destruction was followed almost immediately by a flowering of diverse reconstruction philosophies. The *tabula-rasa* approach, most famously exemplified in Perret's plan for Le Havre, was also applied to Brest, Lorient and St Nazaire. In Marseille, following the 1943–6 demolitions in the Vieux-Port, the area was almost entirely rebuilt, with a classico-modern facade to the harbour. Of its handful of surviving historic buildings, some were actually repositioned, Moscow-style, to new locations in the renewal plan – as with the Hôtel de Cabre, shifted a short distance on rollers in 1954. Greco-Roman remains,

273

Figure 8.9 The Hôtel de Cabre, Marseille, in its 'new' location following its 1954 removal: 2008 view

facing page

Figure 8.8 Church ruin juxtapositions

(a) St Alban's Church, Cologne, 1952–5, by Rudolf Schwartz (with sculpture by Käthe Kollwitz) – a ruin enfolded by the rebuilt Gürzenich hall; (b) The Kaiser-Wilhelm-Gedächtniskirche, Berlin (1958, by Egon Eiermann); (c) San Domenico, Cagliari, rebuilt in 1949–54 by Raffaello Fagnoni and Enrico Bianchini: external view of the 'new' church; (d) San Domenico, interior view of the 'decapitated' lower church

275

unearthed in the demolitions, inspired a new stress on Marseille's archaeological heritage, showcased in 1963 in a new Musée des Docks Romains. A very different approach was attempted in the walled city of Saint-Malo, which became Western Europe's closest approach to Warsaw. Here, although almost 90 per cent of the 'intramuros' zone was destroyed, 750,000 tons of rubble were marked and removed in 1947–8 and, in 1948–50, the entire area was painstakingly rebuilt, on a pattern that replicated the prewar street layout fairly closely.[28]

More typical in France was a simplified recreation of prewar street patterns, with buildings attempting to echo local vernacular in their style – often introducing strange inconsistencies on the way. In the rebuilding of Dunkerque, for example, directed by chief architect Jean Niermans and planner Theo Leveau, a low-density, open layout was combined with a pitched-roofed brick Modernism that differed radically from the reality of prewar Dunkerque, dominated by stuccoed facades. Echoing Viollet's concept of ideal authenticity, Niermans wrote that

> Brick being the customary material in the north, I pictured the whole city of Dunkerque rebuilt of brick, exposed brick, without any stucco, which denatures buildings, soils easily and at the end of the day makes the buildings look grim.

In the Netherlands, the modern rebuilding of central Rotterdam contrasted with the highly integrated campaign of planner Pieter Verhagen for the restoration of historic Middelburg, whose reconstruction in a consistent 17th-/18th-century golden-age style commenced as early as 1941.[29] In Amsterdam, the apparently intact fabric concealed a more insidious dilapidation, with several thousand empty dwellings (mostly belonging to Jews sent to concentration camps) stripped of timber and left as ruins during the 'hunger winter' of 1944/5. The early '50s saw mounting threats from modernising highway and canal-infilling schemes. An influential artist and antiquarian, G Brinkgreve, played an energising role from 1954, organising public protest meetings and stirring up support among key municipal leaders and private philanthropists, such as the Amstel Brewery chief, Six van Hillegom. These efforts culminated in 1956 in the formation of Stadsherstel (City Restoration Company). Coordinated by Amsterdam's municipal conservation bureau, Stadsherstel's first restoration scheme, for a typical mid-17th-century canal house at Brouwersgracht 86, was implemented in 1957–8. By 1966, 26 schemes were complete and a further 58 in prospect. These selective initiatives were backed by wider policy initiatives, including 1957 municipal building ordinances enforcing contextual conformity in new Amsterdam buildings; under a gentleman's agreement after the war, the entire historic centre was treated as if 'listed'.[30]

Across Western Europe, the same pattern of reconstruction predominated, for example in Malta, where a plan by British colonial technocrats modernised the layout of the heavily bombed old city by introducing a 'girdle' road, while restoring damaged buildings in a generally traditional style. In West Germany, where all states except Bavaria had passed reconstruction laws, solutions ranged from R Hillebrecht's modern, automobile-orientated rebuilding of Hannover, where surviving timber-framed buildings were gathered into a 'tradition island' recalling the prewar Hamburg plans, to projects in other towns for rebuilding set-piece streets in a generalised traditional style. In Münster, for instance, the partly intact facades of the main square (Prinzipalmarkt) were

facing page

Figure 8.10 The Kaasjager reconstruction plan of 1955 for central Amsterdam, envisaging the infilling of some historic canals

daarmee alle vreemdelingenverkeer | plaatsen en wat ruim baan te ver-

Zo zóú het worden

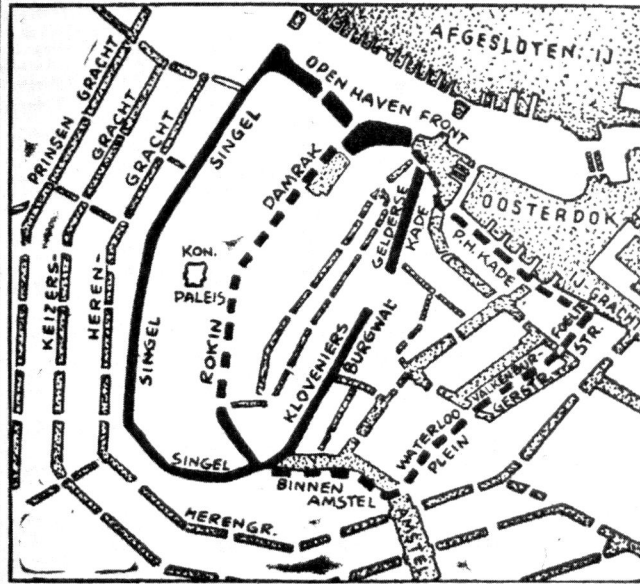

bioscoop

overbodig

10

eft de telefoon de laatste dagen niet
bioscopen en theaters wilden weten
it de fooien waren afgeschaft. Deze
taan en het publiek begon al minder

bestuurders van de „Algemene Be-
isverstand in het spel, dat gemakke-

rantieloon door de wekelijkse ver-
diensten overschreden. In de klei-
nere plaatsen echter is het bioscoop-
personeel aangewezen op het garan-
tieloon en daar profiteert het dus
direct van de verhoging.
    Verder is het maximum van het
basisloon afgeschaft. Het basisloon
is het vaste bedrag, dat elke week
door de werkgever moet worden be-

*Onze tekenaar heeft de beide
verkeersringen in kaart ge-
bracht, die in de plannen voor
Amsterdams toekomst voorko-
men.*
    *De gesloten lijn geeft aan de
ring, die hoofdcommissaris
Kaasjager zich denkt als Open
Havenfront, Singel, Kloveniers-
burgwal en Geldersekade zou-
den zijn gedempt. Daarenboven
wil hij de oude Raamgracht
dempen ten behoeve van par-
keerterrein.*
    *De stippellijn is de Binnen-
ring van Publieke Werken, die
ontworpen is in de jaren dertig
en die voor het grootste gedeel-
te is voltooid. Via de Valken-
burgerstraat—Foeliestraat zal
hij aansluiten op de tunnel
volgens Plan V.*
    *Dempingen heeft deze ring
niet gevraagd. Hij wordt ge-
vormd door Prins Hendrikkade,
Damrak, Rokin, Binnen Am-
stel, Blauwbrug, Waterlooplein
zuidzijde, een doorbraak, Val-
kenburgerstraat en Foeliestraat.*

Bedrijfsbond" nog de 6% loonsver-

demolished in 1949, and replacements in a generic gabled style, designed by the City Planning Office, were built instead, in 1950–8. In artillery-damaged Freudenstadt, the famous market-place was rebuilt in 1949–56 in a traditional-looking way, but with continuous horizontal cornices as opposed to the authentic prewar gabled facades. In Munich, a similar general approach applied, aided by appointment of key prewar figures to influential positions. Rudolf Esterer, restorer of the Nuremberg Kaiserburg in the '30s, became Bavarian government architect in charge of all state-owned palaces, and not only once more restored the Burg, in much the same style (from 1947), but also oversaw a comprehensive multi-phase restoration of the Munich Residenz – including the somewhat 1930s-style classicism of the Herkules-Saal (1949–53) and the re-erection of the dismantled Cuvilliés-Theater in 1949–51 on its slightly different site. But even in Bavaria, the infrastructure of support was significantly democratised: in the 1951–2 campaign to rebuild the 1789 Chinese Pagoda in Munich's Englischer Garten, a citizens' association raised a large sum of money through public appeals, augmented by donations from the American-controlled Radio Free Europe.[31]

The problems of facsimile building and area reconstruction substantially overlapped in the emotive case of Nuremberg. This was because, although 90 per cent of the *Altstadt* was in ruins in 1945, most of the stone-built basement and ground floor walls remained standing, especially in the north-west section. Thus, the facsimile issue was diluted, yet also generalised throughout the city. The influential postwar local and national politician, Dr Oscar Schneider, recalled in 2009 that 'Nuremberg as a living city had ceased to exist in 1945, but its memory remained, and that memory was powerful and universally shared.' Of course, that shared memory now expressly excluded the nationalism that had brought calamity to the city: its first postwar cultural event, on 29 July 1945, was a performance of Haydn's oratorio, the 'Creation', in the charred ruins of St Sebaldus's Church. There was a consensus from the start among the citizens to try to retain as much original fabric as possible. However, following two successive competitions in 1947–8 (won eventually by the contextual proposal of Heinz Schmeissner and Wilhelm Schlegtendal) and a 1950 city council decision, a compromise approach prevailed. There were lavish restorations of public buildings, fortifications and churches, but a freer approach with private houses, recreating the street-picture and steep roof-profile rather than the details of all individual buildings: some buildings, even here, were demolished that could have been rescued. For example, following a 1952–3 competition, the early-17th-century Pellerhaus, an ornately gabled Renaissance palace on the Egidienplatz, was rebuilt not as before the war, but in an idiosyncratically arched, semi-Modernist style above its surviving ground floor; the reconstruction by local architects Fritz and Walter Mayer, completed by 1957, preserved fragments of the ruined mansion around its back court. Away from the *Altstadt*, a freer, more modern approach prevailed: 100,000 dwellings were built in the first postwar decade. The position was similar in the bombed eastern sector of Rothenburg, where plans by Munich architect Fritz Florin (appointed in 1947) aimed at creating a 'new unity' via paraphrases of the previous fabric. In response, the prewar boom in tourism resumed at an even more spectacular rate, eclipsing the 1930s' maximum by 1960 and eventually reaching 500,000 overnight stays by 2000.[32]

All in all, Nuremberg is the only worthy rival to Warsaw (see Chapter 10), and perhaps Saint-Malo, in attempting a high level of historical authenticity in a post-1945 project of urban area-rebuilding. Its impressive achievement can be gauged by comparison with the approach adopted in the destroyed western sector of the Lübeck *Altstadt*, where Pieper's planning office had been busy for several years clearing surviving facades. Here, although the city council in 1949 endorsed general preservation of the old street layout, the small-scale site parcels and lanes were largely hollowed out,

Figure 8.11 The Prinzipalmarkt, Münster, showing 'standard' gabled reconstruction blocks of 1950–8, designed by the City Planning Office

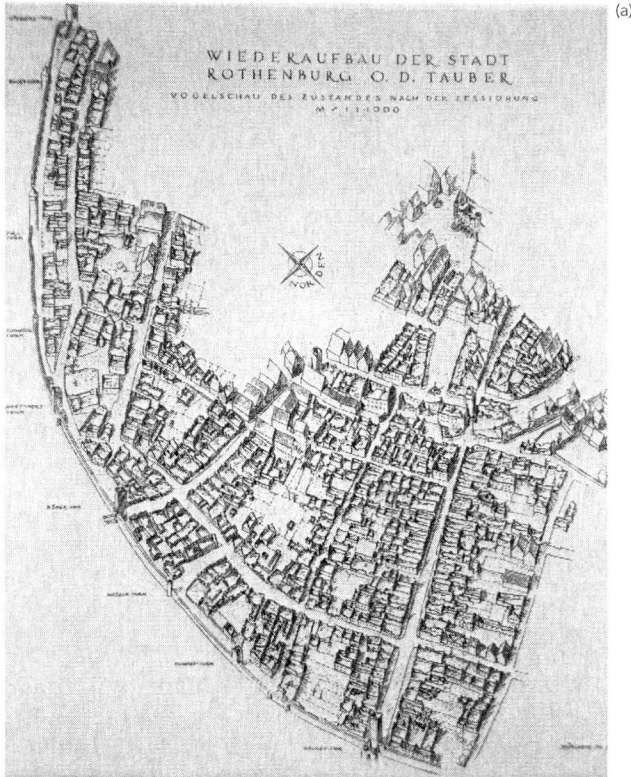

Figure 8.13 Rebuilding Rothenburg

(a) July 1948 perspective by Fritz Florin of the devastated eastern zone of the *Altstadt*;

(b) Present-day view of the rebuilt Galgengasse, looking towards the Weisser Turm

facing page

Figure 8.12 Creative reanimation of the Nuremberg *Altstadt*

(a) 'Die Altstadt von Nürnberg: Wiederaufbau von Heilig Geist', etching by Friedrich Neubauer, 1961; (b) Brigitta Heyduck, 'Blick zur Fleischbrücke', colour etching, 1962; (c) 1970s tiled mural of the restored *Altstadt* in the underground concourse at Nürnberg Hauptbahnhof; (d) 2010 view of the Sebalder Pfarrhof (left foreground) and the rebuilt Albrecht-Dürer-Platz (background): the Pfarrhof survived the bombings together with its renowned oriel window (Chörlein) – in reality a copy of an original relocated in 1902 to the Germanisches Nationalmuseum

281

the lovingly restored churches were isolated in large new spaces, new roads were punched through and new street facades were freely designed, while keeping generally 'in scale'. The present-day appearance of the Lübeck *Altstadt* is deceptive, however, as a fresh wave of site redevelopments by large banks and stores swamped the city in the mid-1960s.[33]

In Britain, the position in 1945 was slightly different from these continental cases, as (Baedeker Raids notwithstanding) the severest bombing had been concentrated in industrial cities such as Hull or Coventry, and the two set pieces, Oxford and Cambridge, had emerged unscathed. Many great English industrial cities had no pre-18th-century fabric of any importance, and the influential 1940s' Modernist plans for their reconstruction were preoccupied with demography, land-use, transport and slum-clearance, and not heritage. An exception was James Paton Watson and Patrick Abercrombie's plan for Plymouth, striking in its combination of stately Beaux-Arts modernity with a 'tradition-island' of relocated buildings. Their postwar plan elaborated the ringing principle of reconstruction advocated by Reith in 1941, arguing that the prewar centre, with its maze of narrow streets, was 'ripe for rebuilding' into a 'great, modern city'. This was complemented by the restoration of a small historic district by Barbican Quay, which Abercrombie proposed should be intensified, Williamsburg-style, into a tourist magnet. Plymouth's 17th-century Charles Church (burnt down in 1941) was also one of relatively few executed examples of the policy of 'Bombed Churches as War Memorials'.[34]

In smaller cities and towns, the late 1940s witnessed a frenzy of proposals for improvement, chiefly in places undamaged by bombing, such as Durham, Warwick or Salisbury. Mostly by eminent planning consultants such as Thomas Sharp, these envisaged a careful, conservative-surgery approach, with overtones of *Entschandelung*. Sharp's wartime Durham report (1944), which pioneered the term 'conservation area', urged removal of 'unharmonious' buildings. The only severely bombed town among these was Exeter, where the Baedeker Raids stimulated a 1946 reconstruction plan by Sharp, *Exeter Phoenix*. Sharp argued that, although the destroyed classical streets had been 'near perfection', any 'Carcassonne'-like recreation would only create a 'dead museum': what was needed was 'not restoration, but renewal', combining retained historic elements with newly opened vistas of the cathedral towers. The eventual built results curiously resembled the outcome in Lübeck. In Canterbury, Charles Holden's 1945 plan proposed a 'Civic Way' linking the cathedral and a new civic centre, but this was largely abandoned in 1948. In most historic English towns, however, the main enemy, even in the 1940s, was the car. In Oxford, Sharp's 1948 plan proposed to safeguard the 'kinetic townscape' of the High Street from traffic by a new by-pass road south and west of the centre. Few such schemes were actually executed: for example, while Minoprio and Spencely's 1946 plan for Worcester envisaged sweeping redevelopment of the historic centre, in the event almost nothing was done.[35]

The heritage challenge of the reconstruction of London fell outside the well-established continental old-town discourses. Its old core, the City of London, was discernible only on plan, and London featured few grand, Haussmann-style planning schemes. The haphazard dynamism of the city's previous development had left a legacy of multiple overwriting and juxtapositions, and damage was confined to limited areas. There was thus no question of any general restoration to some previous state: even among the homogeneous facades of the Georgian terraces and squares, the haphazardly rebuilt bomb sites added a new layer of picturesque variety. Within this diversity, wildly contrasting solutions were closely juxtaposed with each other. At the bomb-scarred 17th- and 18th-century classical ensembles of the Inns of Court, the influential users insisted on faithful restoration of the prewar state – a project completed in 1947–58. Of the city's bombed Wren churches, some were lavishly restored, such as St James's, Piccadilly (built in 1676, gutted in 1940 and rebuilt in 1947–54 by classical architect Sir Albert Richardson), while others were treated more freely: a 1946 report suggested relocating some to provincial towns. In the case of St Mary Aldermanbury, a desire to celebrate the Western alliance led to a bizarre solution: the gutted walls were shipped in 1964 to Fulton, Massachusetts, where the church was re-erected in Westminster College, scene of Churchill's famous 'Iron Curtain' speech ('Sinews of Peace') of March 1946: in 1990 the ensemble was augmented with a relocated stretch of the Berlin Wall.[36]

In the area of most concentrated destruction, the northern part of the City of London, early postwar debates headed in a different direction from those in cities such as Nuremberg. In London, the conflict was between advocates of rapid, practical rebuilding and advocates of long-term replanning: should the cluttered area round St Paul's be filled up piecemeal, or pierced with Haussmann-like vistas? During the war, a 1942 report by the Corporation of London's City Engineer, which proposed to hem the cathedral closely with buildings as before, to leave 'the waves of the world's traffic beating upon its steps', had been strongly opposed in 1944 by a committee headed by architect Sir Giles Gilbert Scott: they argued in Haussmann-like terms that the destruction would allow grand new open spaces to give the cluttered city 'civic dignity'. Eventually, after a 1951 council report, illustrated by Townscape-picturesque artist Gordon Cullen, the area round St Paul's was laid out in the spacious manner familiar from Lübeck and elsewhere, whereas the humdrum, industrial devastation zone further to the north was left until the late '50s. Then, from 1959 to 1970, the latter was developed with an ambitious high-density Modernist project of housing and cultural institutions, the Barbican, by architects Chamberlin Powell & Bon, echoing the traditional city in its network of courtyards but dramatically crowned with three huge tower blocks.[37]

The role of the expert in the *trente glorieuses*

THE European postwar reconstruction drive was overwhelmingly concentrated in the late 1940s and early '50s: from the later 1950s, different and more subtle priorities began to gradually push

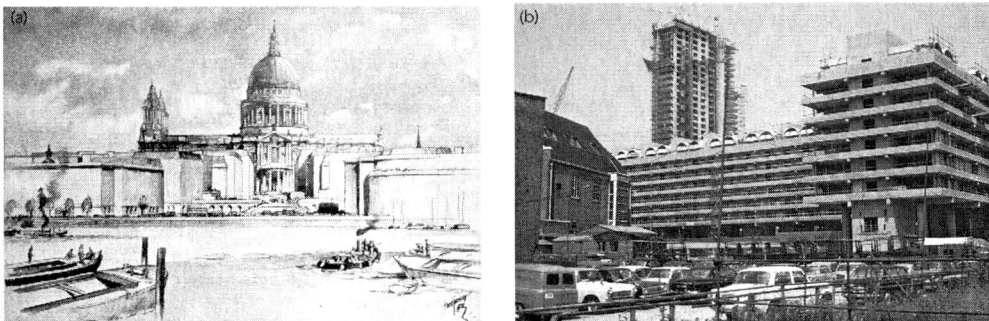

Figure 8.14 Reconstruction in the City of London

(a) Perspective of the grand new southern vista of St Paul's, proposed in the 1944 report of the City's Improvements and Town Planning Committee; (b) June 1968 view of the City Corporation's Barbican redevelopment under construction – the tallest public-housing blocks (45 storeys) in Europe

283

forward instead. But one thing remained constant within the Conservation Movement in the West during the postwar decades: the ever-growing involvement in heritage by the state and the professional expert. Partly, this was a response to the state's urgent and forceful interventions in Modern Movement planning and building – activities whose extremism provoked growing counter-reactions within the machinery of government. But it also reflected a general postwar upsurge in the consensual ethos of professionalism, a movement fusing scientific rationalism and humanistic idealism in the service of the welfare state. Although the technological advances of the war had reinforced the prestige of the scientist and specialist, the *trente glorieuses* linked all this expertise to very different community ideals from the aggressive nationalism of the '30s, focusing instead on peaceful, social-democratic (or Christian Democratic) progress.[38]

In the longest established area of professional involvement in heritage, the restoration and maintenance of cathedrals and other elite buildings, private architects still reigned supreme. An exception was that of monuments in state care, where state-employed archaeologists and architects were in charge.[39] But these private architects were generally now Modernist-trained and preferred a more scientific approach, as presaged in the prewar Athens charter. It was now assumed that the most authoritative technology should be deployed to ensure scientifically optimal solutions. In England, although Scott's 19th-century restorations and repairs had depended behind the scenes on engineers, the new wave of cathedral restorations masterminded by Bernard Feilden from the mid-1960s (architect/surveyor to Norwich Cathedral from 1963, York Minster from 1965 and St Paul's Cathedral from 1969), emphasised the spectacular, highly publicised use of bold structural techniques. At York, following a £2 million appeal to 'save' the cathedral, Feilden's interventions included steel reinforcement and (with Arup's advice) a scheme to secure the 16,000-ton central tower by a reinforced concrete collar and rafts beneath; and at Norwich, his project protected the central spire from collapse. Donald Insall, with his SPRND Modernist training, applied the new professionalism to the different task of area conservation schemes, producing numerous methodical consultancy reports, including his influential Chester volume in the 1960s' 'Four Towns' report series.[40]

Classify or list? Government inventorisation programmes from the 1940s

THE real growth area for state and professional heritage intervention in Western countries was not restorations of individual monuments, but overall regulation and inventorisation of built heritage – an area requiring a lighter official touch. After 1945, country after country introduced ambitious heritage legislative and regulatory bureaucracies, generalising the system pioneered in France a century before. There was a sharp differentiation between conventional protection of individual monuments, common to virtually all countries, and the growing concern to safeguard historic areas.

The principles of protection of individual monuments had changed little from their 19th-century origins, but each country had its own specific recipe. Organisationally, the two chief distinctions were between centralised and decentralised administration, and between 'extensive' and 'intensive' protection systems. The latter involved designation of a small, closely regulated elite of buildings with intensive state support, while extensive systems designated large numbers of buildings for protection through development control – in effect, through the planning system. The two extremes in this approach were represented by France and Britain, with most other Western countries distributed in between. Yet, despite this huge organisational difference, both systems seemed equally capable of producing highly ordered, well-maintained historic environments.

In France, the early-20th-century rate of around 25 buildings classified annually had accelerated, by the 1960s, to roughly 80–100. Designation required recommendation by the *Commission supérieure des monuments historiques* and the relevant Minister – from 1958, the Minister of Cultural Affairs. By 1970 the total exceeded 11,000 *classés* and 18,000 *inscrits* – all with their own 500-m protected radius zones, or *abords*. Designation criteria slowly widened: by 1950, the works of architects born over a century beforehand were eligible. Automatic state subsidy was central to this system, which required annual visits to all classified monuments by the government architects inspectorate (*architectes des bâtiments de France*). Monuments were eligible for 50 per cent grants, and 50–100 per cent tax exemptions. The system also allowed designation of natural landscapes, by local *commissions des sites*, under 1930 legislation. Local civic plans in historic cities, such as the 1947 plan for Toulouse by Charles-Henri Nicod, also emphasised conservation alongside urban renewal and grand axes. An ancillary role was played by voluntary organisations, including *Demeure Historique*, an exclusive association of 500 owners of *monuments classés*, founded in 1924, and *Vieilles Maisons Françaises*, a larger public association established in 1958, which grew to over 15,000 members. In Toulouse, for example, vehement public protests led by local academics succeeded in blocking the only major municipal slum-clearance proposal, for the Saint-Georges area, in 1959: professor Pierre Babonneau attacked the 'architectural mastodons' of 'HLM' (low-rental public housing) tower blocks and raised the bogeyman of Americanisation: 'if we do not take care, Toulouse will have the inhuman aspect of a New York or Chicago'.[41]

France's 'intensive' designation system was followed in numerous other places, including Scandinavia: in Denmark, a 1918 law (updated in 1966 by a Building Preservation Act) authorised a highly selective listing programme by the Ministry of Education, amounting by 1966 to only 2,300 properties (30 per cent in Copenhagen) and by 1980 to 4,800 buildings (of which 37 per cent were churches). A similar system applied in Norway, while in Sweden there was even greater selectivity. France's integrated financial support system was also imitated, for example in Italy, where the 100,000 privately owned monuments designated by the Ministry of Education Fine Arts Department by 1978, under the provisions of Mussolini's comprehensive 1939 law, were all entitled to substantial grant support.[42]

These systems all required a centralised administrative regime, inspired ultimately by the French arrangement. A centralised heritage system prevailed even in some federally organised countries, notably Austria, where federal legislation of 1920 set the main principles. Elsewhere heritage administration was devolved to the regional level. In Switzerland, each canton had its own historic monuments commission, and in West Germany, each state had a separate *Landesamt für Denkmalpflege*, mostly operating under resuscitated pre-1933 legislation (e.g. a 1908 law in Bavaria). By 1970, only one state (Schleswig-Holstein) had passed new postwar heritage legislation, while as late as the 1950s another (the Rhineland) still had no heritage law in force at all. West Germany's highly dispersed, variegated system supported around 80,000 listed buildings by the mid-1960s. Only in the 1970s did all the *Länder* adopt modern conservation laws.[43]

The system in Britain was radically different from all these continental countries. It relied not on direct statist control over elite monuments, but on large-scale regulation through the planning system. The per-capita total of statutorily designated buildings in Britain was far higher than elsewhere: eight times higher than the French level in England, and 15 times higher in Scotland (even including both *monuments classés* and *inscrits* in the reckoning). This striking divergence was a paradoxical by-product of the British heritage tradition of voluntarism, which had prevented any strong CMH-style government organisation from emerging in the 19th century and deferred any decisive state intervention until the post-1945 era of planning-led reconstruction. At that point, a regulation-based state control system suddenly emerged within the space of two decades, linked to a vigorous system of voluntary national and local pressure groups. Listing in Britain began no earlier than 1936, with the National Trust for Scotland's historic burgh lists. In England, matters got under-way more slowly: initial pressure in 1938 from Angus Acworth of the Georgian Group was followed in 1940–1 by an emergency Ministry of Works listing programme. Further development of the 40-year-old system of inventories by the Royal Commissions on Ancient Monuments, established in Scotland, England and Wales in 1908, was ruled out: the glacier-like pace of their painstaking county surveys was incompatible with the new, planning-based British system, stressing flexibility, speed and sensitivity to threat. As the late-1960s' English heritage Minister, Lord Kennet, drily pointed out, the inventories would take another 500 years to complete at their present rate and so

> if we went on losing listed buildings at the rate of 400 a year, they would all be gone in 250 years, and the last parts of the country would still have another 250 years to wait before the Commission got round to recording where they had been.[44]

In both England and Scotland, the new 'extensive' system made lightning progress. A government designation programme was launched by the reformist Town and Country Planning Acts of 1944 and 1947 (1945 and '48 in Scotland). The 1944 Act, which established the principle of listing, was followed in 1945 by establishment of a corps of government investigators under S J Garton and a government advisory committee on listing, chaired by Sir Eric Maclagan, director of the Victoria and Albert Museum, and including pro-Modernist figures such as William Holford as well as traditionalists.[45] A new Historic Buildings Division within the Ministry of Town and Country Planning, headed by Anthony Wagner, issued in 1946 a seminal set of *Instructions to Investigators*. These established the guiding principle of a core statutory list, comprising a small elite of Grade 1 and a long 'tail' of Grade 2, as well as a supporting list of lesser buildings. Garton's successor as Chief Investigator from 1961, Antony Dale, typified the new British heritage regulation system, in his eclectic combination of official and voluntary activity. A historian and solicitor from Brighton, Dale worked for 30 years as a government official, while chairing the local Regency Society and authoring numerous books about his native town – a place whose heritage, typically for England, stemmed largely from the 19th century. One of the chief preservation controversies in 1940s' England concerned the Nash Regency terraces around London's Regent's Park, left dilapidated rather than destroyed by the war: rehabilitation was agreed by the Cabinet in 1946–8.[46]

The British designation system was piggybacked onto the draconian planning powers of the 1947/8 Town and Country Planning Acts. As Kennet argued in the 1960s, this was

> the first introduction into the law of any democratic country the principle that society might forbid a man to do what he would on his own land, without compensation. That principle is the foundation of . . . any practical system of land-use planning. It is also the foundation of our system of preservation law, and of any practical system of preservation law.

By 1969, a quarter-century from the start of government designation, there were no less than 111,000 statutorily listed buildings in England (5 per cent Grade 1, 95 per cent Grade 2) and 131,000 on the supplementary, non-statutory list; by 1974 the statutory total had mushroomed to 170,000.[47]

In Scotland, listing made even faster progress. The *Notes for Guidance of Investigators*, produced by Ian Lindsay in 1948, emphasised that large numbers of listed buildings were the aim: the act 'does not say they should be old, nor that they should be beautiful. The field is thus open to a very wide survey indeed'. Unlike the English 1946 instructions, Lindsay embraced the previously unfashionable mid-19th century: 'We may not like revival baronial but future generations may. Even if they don't, it plays its part in the history of architecture.' Whereas Lindsay's 1936–8 NTS survey had only listed 1,158 buildings, by 1956, eight years after the 1948 act had put him in charge of a seven-person Department of Health for Scotland (DHS) listing branch, 7,000 buildings had been designated; by 1969, there were 18,161 statutorily listed buildings (Categories A and B) and 11,874 non-statutory (Category C) – 12 times more, per capita, than similarly sized Denmark! Furthermore, many were group-listings of vernacular ensembles, pushing the actual total still higher. Only a few Continental countries remotely approached these levels of coverage, notably the Netherlands, whose Ministry of Education, advised by the State Council for Care of Monuments, oversaw an extensive designation programme of buildings over 50 years old, encompassing around 45,000 properties by 1974 (17 per cent located in Amsterdam).[48]

The planning roots of British postwar conservation tied it closely to modern development, whether the latter was seen as a threat or opportunity. At a 1949 conference at Oriel College, Oxford

---

LISTING OF HISTORIC BUILDINGS

7. Under Section 52 of the Town and Country Planning (Scotland) Act 1972 the Secretary of State for Scotland is required to compile lists of buildings of special architectural or historic interest. The administration of both local and national conservation policies is based on these lists, which are constantly under review.

How the Buildings are chosen

8. Most buildings of the early 19th century and earlier whose interesting character remains substantially unimpaired are included. Later buildings must be of definite character and quality to qualify.
Special regard is paid to:
  1. planned streets, villages or burghs;
  2. works of well-known architects;
  3. buildings associated with famous people or events;
  4. good examples of buildings connected with social and industrial history and the development of communications.

The Statutory List

9. After a survey by the Department's Historic Buildings Investigators* a list is drawn up and issued of all the buildings of special architectural or historic interest within the area of each of the planning authorities. This list is divided, for ease of identification and administration, into, for example, districts, parishes and burghs. All buildings included on this statutory list are subject to the legal provisions described in this booklet.
10. The buildings on the list are assigned to one of three categories according to their relative importance. The categories are as follows:

Category A
  buildings of national or more than local importance, either architectural or historic or fine little-altered examples of some particular period or style.

Category B
  buildings of primarily local importance or major examples of some period or style which may have been somewhat altered.

Category C
  good buildings which may be considerably altered, other buildings which are fair examples of their period, or in some cases buildings of no great individual merit which group well with others in categories A or B.

* Historic Buildings Investigators are professional staff who have made a special study of architectural history with particular reference to Scotland. They advise on buildings to be included in the lists and on any subsequent casework relating to them — alteration, restoration, demolition etc.

(Note: the majority of buildings in category C, which were not previously included on the statutory list are likely to be afforded statutory protection within the near future.

The Descriptive List

11. Details of a building's category are not at present given on the statutory list itself: this information, along with descriptions of the buildings included, is given in a separate descriptive list. Work has already started, however, on amalgamating the statutory and descriptive lists, and single all-purpose lists will be issued for each planning authority area during the next year or two.

Where to see the Lists

12. You can inspect the lists at the offices of:

  The Historic Buildings Branch, Scottish Development Department, presently at (Room H605) Argyle House, 3 Lady Lawson Street, Edinburgh EH3 9SF.

  The Royal Commission on the Ancient and Historical Monuments of Scotland, 52/54 Melville Street, Edinburgh EH3 7HF.

  The planning authorities (the district councils of the Central, Fife, Grampian, Lothian, Strathclyde, and Tayside regions; the Borders, Dumfries and Galloway and Highland Regional Councils; and the Island Areas Councils).

How many buildings have been listed?

13. The approximate number of listed buildings in Scotland at 30 June, 1975 was:

  Category A: 2,100
  Category B: 19,600
  Category C: 11,200
  Total: 32,900

PROTECTION

14. The fact that a building is listed as being of special architectural or historic interest does not mean that it will be preserved intact in all circumstances, but it *does* mean that demolition must not be allowed unless the case for it has been fully examined and that alterations must, as far as possible, preserve the character of the building. This is done under a procedure known as 'listed building consent', which applies to all statutorily listed buildings regardless of whether they are category A, B, or C.

Figure 8.15 1976 Scottish Development Department leaflet (originally published 1972) explaining the new national designation and heritage-management system in Scotland

287

for Ministry listing investigators, their administrative head emphasised the modernity of their task: they were compiling 'a Domesday Book of our architectural heritage', not 'a collection of curiosities', and must 'integrate it into the future of England . . . and keep it alive and in use to the maximum extent we can'. Indeed, 'in some cases, we shall have to let [monuments] go, because it is better to have a new harmony than a jumble of old and new'.[49]

Preservation and the Great Society: public expansionism in postwar America

IN Western countries outside Europe, these years also saw a sharp rise in state interventions in the heritage, mostly concerned not with direct regulation but with facilitating and educating. The Conservation Movement often found itself within a federal system, sometimes as a national and sometimes as a regional responsibility.

In the USA, this was the period that banished the idea that historic preservation was no concern of government. The first steps towards nationally coordinated preservation were led by a philanthropic organisation, the National Trust, founded with congressional backing in 1949. The Trust, established partly at the instigation of the National Park Service, was modelled on the English National Trust (itself partly inspired by US precedents), although it was not concerned with landscapes, which were seen as the province of the NPS. The Trust's aim, like its English and Scottish counterparts, was a double one. It actively preserved individual buildings, while educating and lobbying via a mass membership of thousands, even hundreds of thousands, its first acquisition being Woodlawn Plantation, Mount Vernon, in 1951. In some ways the US Trust, although a voluntary organisation, was closer to the government than the British Trusts – although one should also bear in mind the way that the NTS and DHS had worked in partnership over Lindsay's historic-burgh survey in 1936–8.[50]

The work of designating individual monuments was expanded by a fresh federal measure in 1960, creating a National Historic Landmark Program: 600 buildings were registered by 1966, qualifying for tax benefits. Designation was significantly developed at state or city level, too – as in Virginia, where the Virginia Historic Landmarks Commission earmarked numerous landmarks (again, qualifying them for tax concessions), or Savannah, Georgia, where the Historic Savannah Foundation in 1964 established a revolving fund for house preservations. In New York City, building on a legacy of preservation battles since the 1890s, and where a 1965 Landmarks Law founded a Landmarks Preservation Commission, by 1967 some 750 buildings in the city were under consideration for designation, involving restrictions on demolition or external alteration, as well as the usual financial incentives. Some of these potential landmarks comprised entire districts, such as sections of Brooklyn Heights or Greenwich Village.[51]

But it was ultimately the federal government that made the greatest strides in conservation – especially in the landmark year of 1966, when the 'Preservation Congress' enacted no less than seven federal preservation laws. The years since 1945 had seen a vast amount of urban redevelopment: publications like *Lost New York* bemoaned the consequent losses. In response, the Johnson Administration unleashed a barrage of initiatives, as part of its 'Great Society' strategy enunciated in 1964. Preservation enhancements began modestly, with 1964 proposals for beautification of the White House (led by Mrs Johnson) and a President's Task Force on Sites of Natural Beauty. The following year, President Johnson established a Special Committee on Historic Preservation, whose report, *With Heritage So Rich* (published privately), prompted the 1966 laws. The foremost of these, the National Historic Preservation Act, established a coordinating national body (the National Advisory Council on Historic Preservation); initiated a National Register of Historic Places; enhanced the powers of the National Trust; aided municipal and state historic building surveys; and obliged federal departments to acknowledge heritage concerns in their development programmes. The latter was undeniably a piece of restrictive legislation, albeit modest by European standards. A further 1966 law, the Demonstration Cities Act, obliged the federal housing department (HUD) to include preservation in urban renewal schemes. In a February 1966 'Message on Preserving our National Heritage', Johnson stressed that these initiatives brought over US $32 million in funds to the historic environment.[52]

Although this system was initiated by the federal government, it was rapidly (in 1967) devolved to the states, with the National Trust the only national body fully involved: it and state governments were eligible for 50 per cent matching grants. And in 1976, 1981 and 1986, the financial aid system was upgraded with new powers of tax credits for rehabilitation. By the end of the 1960s, with the tangle of interventionist measures at state and federal level, it was no longer true that US preservation was an exclusively private affair. However, these measures continued to focus on preservation of individual sites and buildings, especially through manipulation of the property system by tax credits. From a European perspective, preservation in the USA still remained more restricted in scope and more bound up with commercialism. In 1963, it was estimated that over two-thirds of all US preservation activity was privately financed. In a 1965 article in the *Journal of the Society of Architectural Historians*, Alan Gowans argued that

> In what has come to be the Big Business of preservation and restoration of historic architecture in the US . . . it is nonsense to try to save every bit of our heritage; some old buildings, even very worthy ones, simply must go.

The job of preservation was 'survival of the fittest', as defined by architectural historians; the criterion of a good restoration was not archaeological accuracy, but how well it conveyed historical 'meaning'. On the whole, though, Gowans argued, most buildings before 1830 should be 'sacrosanct'.[53] To some extent, the USA and Britain were at opposite ends of the spectrum of state intervention, with Britain emphasising development restrictions and the USA financial assistance, with continental Europe somewhere in the middle.

In Commonwealth countries, full-blown state intervention was normally avoided by establishing a National Trust on the English/US model – for example in New Zealand (in 1954) or in the newly independent anglophone Caribbean states (e.g. Barbados, in 1961). In Canada, owing to the stronger tradition of government intervention and the dearth of large private institutions to finance heritage works, the government took a more activist line. A restricted federal heritage programme was inaugurated in 1919 (via a Historic Monuments Committee), but legislation was mainly devolved to the provincial level: Ontario passed the first antiquities law in 1953 and Quebec took the lead in designations, while a 1964 Historic Monument Act established a Historic Monuments Service and Commission (followed in 1972 by a Cultural Property Act). In Canada, national political motives for heritage were polarised between anglophone loyalists, who cherished historic towns as imperial relics, and the francophone Quebec position, which was the reverse. A comprehensive federal list and institute of historic buildings (the world's first computerised

(a)

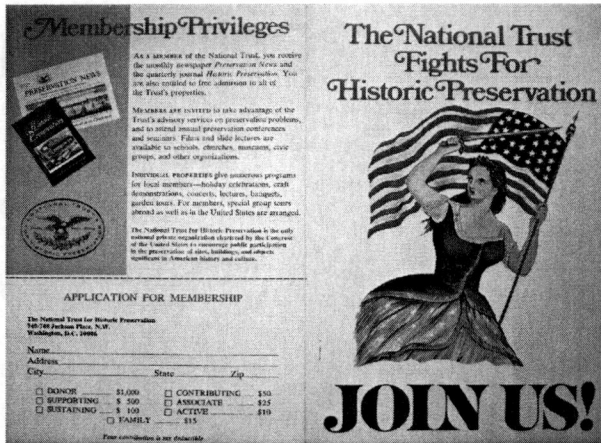

Figure 8.16 Heritage organisations of North America

(a), (b) Publicity leaflet of the US National Trust, c.1970; (c) Front cover of Heritage Canada's annual report for 1973–4

(b)

(c)

inventory) was finally initiated in 1970, and a national charitable foundation, Heritage Canada, was established in 1973, aided by a C\$12 million federal government grant. Initially modelled on the English NT (like some provincial trusts, such as that of Nova Scotia, established in 1959), within a year Heritage Canada boasted 22,400 members and had charted out a distinctive strategy focused on area conservation initiatives in collaboration with local municipalities.[54]

Shades of Disneyland: heritage, interpretation and tourism

THE rise of a distinctive American system of collaborative private-public support for historic preservation allowed the USA to take a decisive lead in the integration of heritage and capitalism. The interwar experiments in conservation planning, at Charleston and New Orleans, had not been forgotten, and in 1954 the Supreme Court authorised legislatures to impose planning schemes.[55] In Western Europe, the general movement to commercialise planning from the 1960s onwards stemmed both from postwar economic recovery and from ideological Americanisation. 'Out' went socialist command planning; 'in' came modernising economic planning, intended to facilitate capitalist economic growth. Within this broad movement, Europe and America differed in the relative value attributed to the state and to the free market, a distinction particularly relevant to the Conservation Movement. Eventually, by the 1990s, the balance would swing everywhere towards the free-market end of the spectrum, but for now the position was finely balanced. For conservation, the focal area of contention was that of tourism and its mixed effects: beneficial, in bringing financial help and public enthusiasm; and harmful, through compromising monuments' authenticity.

These conflicting issues of heritage tourism were first explored in the USA, where the interwar years had witnessed the emergence of a hybrid touristic-educational discourse of 'interpretation', initially focused on the National Park movement and with a marked nationalistic bias. As in interwar Europe, the cult of nature had been initially bound up with right-wing propagandising, a key proselytiser being Charles M Goethe, a Californian white supremacist and eugenicist who had introduced the concept of 'nature guiding' from Switzerland and Norway to the National Park Service in 1919–21 at Yosemite National Park. The postwar years saw the modernisation and depoliticisation of NPS interpretation, through the Mission 66 programme, involving building of visitor centres in a strikingly Modernist style.[56]

At Colonial Williamsburg, the postwar decades witnessed a similar shift from nationalistic jingoism to a sophisticated commercial and educational ethos: aided by the continuing boom in car ownership, annual visitors soared from 166,000 to 496,000 from 1947–57, and a range of pageant-like activities was instituted: for example, the 1961 Colonial Williamsburg Centennial saw a 'militia muster' in the town several times weekly. Likewise, at Jamestown, Virginia, ten miles from Williamsburg, site of the first permanent English settlement in America (1607), the years from 1930 saw an NPS-aided movement for recreation of the settlement. This culminated in the 1957 Jamestown Festival, a multi-agency initiative with federal and state support, which provided extensive transportation and reconstruction infrastructure, along with more ephemeral trappings such as replica ships and costumed '17th-century' staff.[57]

By the mid-1960s, an array of museum-villages spanned the eastern USA, with annual visitor figures approaching 850,000 at Greenfield Village and 950,000 at New Salem State Park. In anglophone Canada, an equivalent project emerged at Louisbourg, site of an early French settlement razed by the British in 1760. Here the years from 1962 witnessed an ambitious, C$25million, 20-year re-creation project, financed by the federal government (later by Parks Canada) to help counterbalance the slump in Cape Breton's coal and steel industries: the project used original archive records in France to achieve a high standard of historical fidelity.[58]

From here to the creation of wholly fictitious 'historic' environments was a short step, through a movement led by film producer Walt Disney. In 1954–5, at a cost of US $17m, he constructed 'Disneyland', the world's first dedicated theme-park, at Anaheim, California. This innovative amalgam of funfair, Expo and open-air film-set, designed by animators (such as Herb Ryman) rather than architects, comprised an array of self-contained 'realms', each a self-contained, kinetically conceived performance zone crewed by 'cast members'. One of the realms at Anaheim – the so-called Main Street USA, flanking the main entrance route – had a markedly 'heritage' character. Entered through a 'Victorian railroad station', and leading axially to a 'Sleeping Beauty Castle', the street was lined with structures in the style of a Midwest town of c.1900 (supposedly inspired in specific detail by Disney's boyhood home of Marceline, Missouri), slightly miniaturised and ingeniously exploiting 'forced perspective'. Another realm contained a futuristic 'Tomorrow-Land', designed in a style

Figure 8.17 Museum townships of
North America

(a) Jamestown, Virginia: reconstructed
chiefly in the years from 1957; (b)
Louisbourg, Nova Scotia, recreated from
1962 onwards; (c) Disneyland, Anaheim,
California (1954–5): 2010 view of 'Main
Street USA'

resembling the 1939 New York or 1958 Brussels World's Fairs. In Disneyland, with its exuberantly eclectic display strategies, capitalism finally overwhelmed and devoured the old nationalistic Mussolini concept of tourism as mass propaganda. Significantly, Disney himself was fascinated by the subject of city planning and before his death in 1966 planned a futuristic theme park-cum-experimental town, Epcot, in Florida.[59]

In many European countries, the 1960s' economic boom made possible a diluted version of the American combination of heritage and educational tourism. From 1960, for instance, the National Trust for Scotland, under its modernising secretary (1949–83), Sir Jamie Stormonth-Darling, imitated the National Park's Mission 66 concept in a new strategy of 'history on the spot', to be communicated to visitors in a 'chain link' of interpretation centres from central Scotland up to the Highlands. At the Bannockburn battle site, traditionalist architects and sculptors were swept aside in 1960 and Modernist architect Robert Matthew was appointed to plan a more dynamic landscape and visitor centre (opened in 1967). Stormonth-Darling argued in 1966 that 'all the agencies connected with Scotland's economic strength will come together one day to project a single, more realistic image of Scotland'. In Stratford-upon-Avon, the infrastructure of the Shakespeare Birthplace Trust was radically upgraded with American help. A 'Shakespeare Centre' adjoining the Birthplace House, containing a new headquarters and interpretation centre, was built in 1962–4. Designed by architect Laurence Williams, it was largely financed by an American 1964 Shakespeare Committee, chaired by the Honorable Eugene R Black (long-standing president of the World Bank).[60]

Likewise, in postwar Sweden, controlled by the Social Democrats since the 1930s, a comprehensive range of interpretative activities was developed, using 'neutral' or scientific heritage presentation to highlight the enlightened progressiveness of the present day. In 1949, the National Heritage Board began a 20-year programme to restore the Vasa Castles – Uppsala, Örebro, Gripsholm – to their condition in Sweden's 17th-/18th-century golden age, stripping away their more recent stucco covering to reveal the layers of building construction beneath in a didactic, documentary way. More boldly, 1961 saw the state-financed salvage of the 1626/8 warship, *Vasa*, and the move of the wreck in 1962 to a new glass and concrete shelter building in Djurgården, near Skansen – the Wasavarvet, designed by Hans Åkerblad and built in 1959–61. This project mingled scientific daring and experimentation with mildly nationalistic educational interpretation.[61]

Conservation and economic expansion in the modernising sixties

ACROSS Western Europe, the modernising fever of the 1960s most significantly impacted on heritage in the campaign to actively integrate historic towns into frameworks of planned development – a planning parallel to the architectural collaboration between Modern Movement and Conservation Movement. This shift in thinking was especially marked in Britain, whose traditions of radical planning interventionism and private activism gave a double stimulus to conservation: positively, by facilitating its bureaucratic spread within government; and negatively, by encouraging the sweeping redevelopments that provoked the growth of activist pressure groups. In Britain, the 1950s' years of Conservative Party (right-wing) rule had been a negative time for both planning and the general heritage interest. A polarised debate of 1952 between architects and planners in the *Architects' Journal* concerning civic amenity revealed sharp differences: even a

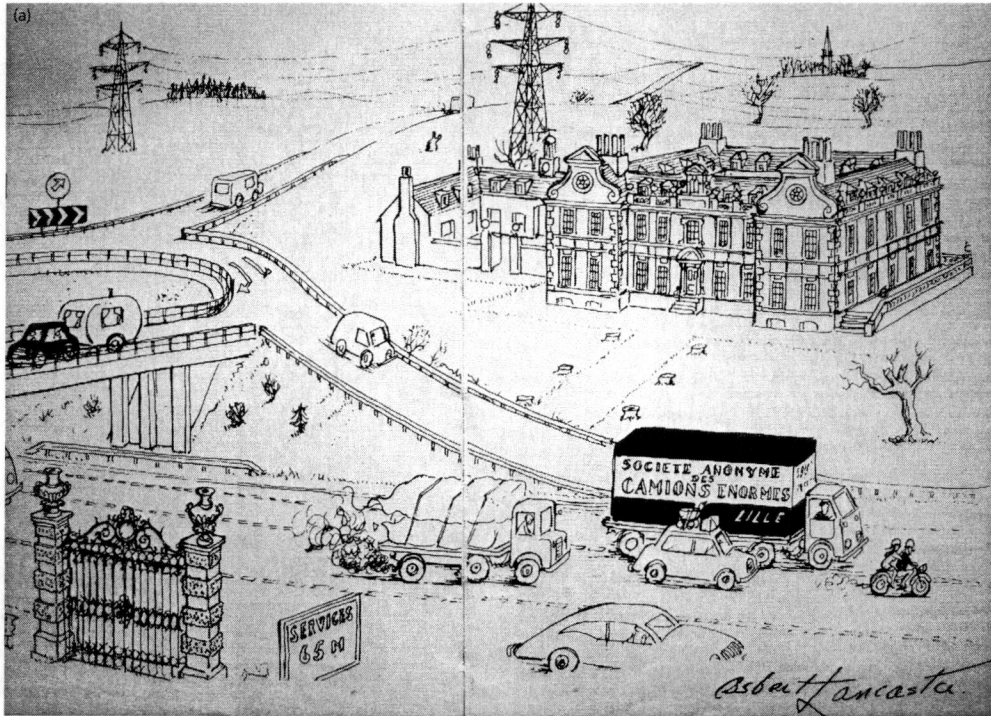

Figure 8.18 The disjunctions of modernity in postwar Britain

(a), (b), (c) Three 1960s' cartoons by Osbert Lancaster

'Gentlemen, let us get our priorities right! Historic buildings must not be allowed to stand in the way of expensive accommodation for the tourists who come to see these historic buildings!'

'All over the country the grime, muddle and decay of our Victorian heritage is being replaced and the quality of urban life uplifted' – Harold Wilson

renowned public authority architect such as C H Aslin of Hertford County Council argued that 'we don't want the planner – he is going to be "out" very shortly'. The 1954 Crichel Down scandal, about the abuse of compulsory land purchase, further discredited the planning cause.[62]

By the early 1960s, with growing confidence in economic growth, the fortunes of planning soared again. In Scotland, the 1961 Toothill Report on planned economic growth provoked a 'frenzy of work' on regional planning, including creation of a new Scottish Development Department (SDD, in 1962), a devolved government ministry responsible both for planning and for heritage. Further acceleration of listing resulted: SDD's 1964 annual report boasted that three-quarters of the country was covered by provisional lists, and by 1971 the new statutory list for the city of Edinburgh contained 2,246 designated buildings – the same total as for the whole of Denmark. The 1964 election of a Labour government boosted the social-democratic rhetoric of progressive conservation: the planner Lord Holford in 1965 claimed that old towns embodied a 'recorded social experience' that provided a 'measuring rod' for planning. Some modernisers went further: the new technology minister, Anthony Wedgwood Benn, argued in 1964 that 'the most distinguishing characteristic of a vigorous society is one in which the future is more real and important than the past'.[63]

In England, the mid- and late '60s, with the economy shifting uneasily from boom to crisis, saw a blizzard of government initiatives to boost the heritage cause, begun in 1965 by the new housing and environment minister, Richard Crossman, and masterminded from 1966 by his deputy minister, the architect Lord Wayland Kennet. During Kennet's years in office, 1966–70, the number of listed buildings grew further, from 90,000 to 120,000, as did Historic Buildings Council grant-support. And a viable structure of controls was created, exploiting the shock of recent losses, notably the unauthorised demolition of the Grade 2 listed, 17th-century timber-framed Silhill Hall, Solihull. Kennet recalled that

> the owner, a speculative housing developer, was going home past the manor house one night. He saw it flapping a bit in the wind, and since he had his bulldozer handy, he thought it safest to bring it along and knock the manor down.

The owner could only be fined £100 for a demolition that, in effect, raised the site value from £15,000 to £60,000. The huge expansion of listed buildings in Britain under the planning system only required the loosest coordination, unlike the tightly centralised French tradition. Kennet was astonished, when he asked his civil servants how many listed buildings were designated annually, to be told, 'We don't know, and this is because nobody tells us; and this is because we never asked anybody to tell us.'[64]

Whipping up a climate of urgency and expansionism, and supported by forceful administrators such as Vivian Lipman (Historic Buildings Council secretary), Kennet boosted the 'downtrodden' heritage branch in the Ministry of Housing and Local Government (MHLG). A new Preservation Policy Group chaired by Kennet and staffed with expert advisers such as Nikolaus Pevsner devised an effective planning enforcement mechanism, Listed Building Consent, implemented by updated planning acts in 1968–9. Between 1969 and 1970, listed building demolitions dropped by 25 per cent in a single year.[65] But by then, as we will see later, the period of mutual respect between modern reconstruction and conservation was over, and the planning system was gearing up to throw in its lot wholeheartedly with the latter.

Retreat from the rural: conservation of villages and country houses in postwar Britain

T HE chief area in which these fluctuating relationships between the Conservation Movement and planned modernity were played out was that of urban development. Within Britain, postwar conservation saw a decisive move away from the countryside as a focus of concern, towards the cities. And on the Continent, the discrediting of the ideal of *Heimat* had diminished the heritage prestige of the rural environment. The position in North America was very different: there, the nature-conservation and wilderness movement retained its wide support.[66]

All this left rural heritage concerns in Britain increasingly fragmented. On the one hand, there were general planning efforts at coordination of the rural landscape, backed up by the long tradition of polemic against unplanned development. Such criticisms had driven forward the 1940s' movement for designation of National Parks in England, but the tirades of Ian Nairn against 'subtopia' shifted the focus of attack to the town and suburb.[67] In the 1960s' growth era, the drive for coordinated planning of rural regions culminated in a succession of integrated county plans, notably in the county of East Lothian in Scotland, where Frank Tindall, chief planner in 1950–75, methodically pursued a comprehensive regeneration strategy. Its linchpin was the sensitive regeneration of smaller old towns, including the depressed county town of Haddington, whose revival exploited a government programme of population and industry overspill from Glasgow. A second significant issue in rural architectural conservation concerned villages. Here, younger Modernist architects became eagerly involved. In Loddon, Norfolk, the young firm of Tayler and Green contributed sensitive infill housing in an empiricist modern style, while Donald Insall, at the Essex village of Thaxted, devised a Geddesian development and conservation plan in 1966, to safeguard the village's 'living history' and 'heart' against intrusion from motor traffic.[68]

The third element of rural conservation debate was concerned with the once prosperous country houses. Wartime requisitioning and punitive taxation had tipped many estates into insolubility, and a growing number of houses were abandoned or demolished, often by the army using explosives. Organisations such as the Georgian Group and the National Trust began an impassioned campaign, fuelled by the growing popularity of 18th- and 19th-century buildings in England. Following a 1950 official report, by the Gowers Committee, the government established in 1953 a special subsidy system for threatened houses, administered by separate Historic Buildings Councils (HBCs) for England and Scotland, with public access as a condition of grant. Although the HBC system was soon extended to other building types, country houses remained especially dependent on government aid. However, the ending of petrol rationing in 1953 prompted a minor tourist boom, and the National Trust stepped up its involvement with country houses: its annual visitor numbers rose to 600,000 in 1953 and 1.4 million a decade later.[69]

An alternative approach was championed by several innovative owners, inspired by American capitalist heritage-marketing. In 1949, the 6th Marquess of Bath, facing a £750,000 inheritance tax bill on succeeding to his title and ownership of the renowned Elizabethan mansion, Longleat House, opened the house and grounds commercially to the public, recording his millionth visitor in 1957 and later (in 1966) expanding his operation to include a safari park, set up by circus impresario Jimmy Chipperfield. He was imitated by the 13th Duke of Bedford at the classical Woburn Abbey: inheriting his title in 1953, Bedford offset death duties by opening house and grounds in 1955, adding a Chipperfield-run safari-park from 1970. In his 1971 manual, *How to Run a Stately Home*, Bedford argued that, 'if you provide stately loos, good teas and plenty of parking space, 87.3% of your visitors will not notice if you have no house at all'. Many houses, like Stratford or Abbotsford previously, acted as magnets for tourist intensification: by 1973, Blenheim Palace recorded 290,000 visitors, 15 per cent of whom were Americans.[70]

Conservation, 'urbanity' and mobility in the motor-city

WITHIN contemporary architecture and planning, all this seemed beside the point. There, the emphasis had been moving away from the countryside in any form, towards an ever more intense focus on urban fabric – a trend that created considerable scope for stimulating interaction between the Modern and Conservation Movements. At CIAM's seventh meeting at Bergamo, in 1949 – held, significantly, in the great hall of the Palazzo Vecchio – the focus was no longer on the broad sweep of regional planning but on 'The Heart of the City'. Henceforth the problems of the inner city would provoke ceaseless debate and stimulate violently conflicting solutions, ranging from radical renewal to various permutations of conservation. Among planners, the same applied. The 1954 congress of the International Federation of Housing and Planning saw fresh praise of Geddes's place-sensitive planning.[71] For modern architects, this represented a narrowing of focus, but for urban conservationists it was the opposite: a broadening of outlook from the stifling narrowness of the old *Altstadt* ideology to a wider world-view, integrating the modern concern to tackle bad housing in a large-scale, integrated way, with a new, broad concept of urban civic amenity.

For Modernist architects and planners, the essence of the new urbanity was rejection of the vast, flowing open spaces, isolated blocks and greenery of the CIAM ideal for more integrated solutions, linking buildings and transportation methods in complex, multi-level, decked infrastructures. By the late '50s, CIAM's dogmas came under criticism from groups such as Team 10, who demanded the intensification of existing human associations rather than the *tabula-rasa* creation of community. One crucial variable in the shifting balance between modernisation and conservation was the extent to which motor mobility was given priority over existing built environment. The pursuit of massive traffic infrastructures was seen at its most extreme in G A Jellicoe's *Motopia* of 1961, with its proposal for roads laid across the roofs of buildings. More usually, strenuous efforts were made to accommodate older areas between surgical road incisions, rather like the German *Traditionsinsel* concept. In Britain, a leading role in this movement was played by government planner Colin Buchanan, whose influential report of 1963, *Traffic in Towns*, presented case studies of increasing scale, from the small town of Newbury to a chunk of inner London. These set out a violently dialectical framework, with restricted motor access to historic zones enabled by radical road interventions to divert the traffic; even the Newbury plan envisaged planting a massive urban motorway along the flank of the High Street. Buchanan argued grandiosely that

> There is a great deal at stake: it is not a question of retaining a few old buildings, but of conserving, in the face of the onslaught of motor traffic, a major part of the heritage of the English-speaking world, of which this country is the guardian.[72]

More generally, in 1960s' Britain, a variety of planners and urbanists reacted against the spread-out greenery of CIAM and the Garden Suburb, and against the 'Failure of the New Towns', by advocating patterns of urbanity that partly acknowledged the past. An influential RIBA symposium on 'The Living Town' in May 1959 debated the balance between modern 'comprehensive' redevelopment and conservation: Norwich City Architect David Percival insisted that Geddes-style piecemeal conservative surgery was often the only viable approach in historic centres.[73] Acceptance of the need to maintain that balance united both the earlier writers of the Picturesque/Townscape school, such as Gordon Cullen and Kenneth Browne, and younger critics such as the Smithsons, who advocated 'human association' rather than any particular physical pattern (in the process, acting as a bridge to the radicalised, anti-Modernist conservationism of the post-1968 era). Some late Modernist critics, by contrast, proposed radical 'megastructural' utopian solutions disconnected from the heritage, either in the abstract (such as H de C Hastings's *Civilia* of 1971 – a hypothetical 'constructed hill-town') or in relation to specific places, such as Theo Crosby's proposal for a decked redevelopment programme in his *Fulham Study* of 1963–4. The administrative impact of all this urbanity began to be seen in the Planning Bulletins of the MHLG in the early 1960s, which promoted a balance between modernity and heritage.[74]

In towns up and down Britain, the expansionist early 1960s saw numerous attempts to actually put these ideas into practice. The Tory housing and planning minister, Sir Keith Joseph, told the Town Planning Institute in 1964 that the pace of modernisation would have to be furiously accelerated to keep pace with an impending 'era of physical development that is likely, in the end, to be greater than anything we have seen in this country'. Some public authorities tried to reconcile modernity and heritage by laying out bold new redevelopments in a way that indirectly evoked past patterns, as in Chamberlin Powell and Bon's vast Barbican development in London (from 1959), with its multi-level network of squares and towers, or in Manchester Corporation's redevelopment of the 19th-century slum terraces of Hulme with both squares and Bath-style crescents.[75] But behind all the insistence on a balance of bold urban renewal and cautious conservation, there still persisted economic uncertainty.[76] One of the most representative figures of this balanced world-outlook was Newcastle-upon-Tyne's City Planning Officer, Wilfred Burns, chief strategist of the bold redevelopment and urban motorway programme demanded by city leader (1958–63) T Dan Smith, and embodied in the city's 1963 Development Plan Review. Burns's 1963 book, *New Towns from Old: The Techniques of Urban Renewal*, argued that a balance of redevelopment and conservation could answer the needs both of prosperous growth areas, and dilapidated inner-city 'twilight areas' of low demand. Some younger municipal planners were inclined to tip the balance much more towards conservation: Leicester's City Planning Officer, Konrad Smigielski, had within five years of his appointment in 1962 blocked a radical redevelopment of the market place and construction of an inner ring road, and saved the classical New Walk.[77]

Only by the end of the 1960s (as we will see later) would this balance shift decisively, as the machinery of Modernist renewal became appropriated by the conservationists and 'comprehensive planning started to make possible comprehensive conservation'.[78] One foundation for this shift was the spread of pedestrian precincts in city centres during the 1950s and '60s – an international phenomenon spanning the whole of Western Europe. In fact, pedestrian schemes were nothing new: by the beginning of the 19th century, for example, the Via Nuova in Verona, connecting the Piazza Erbe (site of the Roman forum) and the amphitheatre, had been banned to vehicular traffic and paved in large stone blocks, developing subsequently into a high-class shopping street (now named the Via Mazzini). The first postwar examples were located in *Altstadt* reconstruction zones in Germany, including a scheme in Essen from 1952, and in Cologne's Höhe Strasse from 1959; in

Figure 8.19 Pedestrianisation pioneers

(a) Via Nuova, Verona (introduced early 19th century);
(b) Höhe Strasse, Cologne (implemented from 1959);
(c) Strøget, Copenhagen (from 1962): 1973 view

America, the most influential pioneer was probably the Main Street of the Anaheim Disneyland (with first concept sketches in late 1953), followed by 'proper' pedestrian malls in Lincoln Road, Miami Beach (1961) and Nicollet Mall, Minneapolis (1963). More extensive European schemes followed in Copenhagen (the 1.1-km-long Strøget zone, from 1962), Bologna and Verona (1968) and Munich, with a first phase of 50,000 sq. m. constructed in 1966–8 and the full project opened in time for the 1972 Olympic Games.[79]

1950s–60s' housing: from mass clearance to mass rehab?

DURING the early postwar years, the interaction of modern architecture and planning with the Conservation Movement was stimulating and generally amicable. The first significant strains between the two started becoming obvious in one of the focal areas of Modernist efforts, mass housing, when the pressure for radical redevelopments provoked a growing resistance. In some countries, such as France, most postwar mass housing was separated off in vast *grands ensembles* at the edge of the city. But in others the late 1950s saw the start of huge demolition programmes of 'sub-standard' 19th-century housing – even, incredibly, in West Germany, where areas of West Berlin tenements that had survived the war were condemned as 'stone deserts' and redeveloped from 1957 onwards. Those that remained were subjected to blanket campaigns of *Entstückung* – the stripping of late-19th-century stucco ornament from their facades – leaving them more austerely modern in appearance. For example, 60 per cent of all surviving tenements in the Kreuzberg district had their facade decoration removed between 1945 and 1979.[80] The year 1957, in West Germany, saw a key milestone in the Modernist drive for space and segregation in the city: the *Hansaviertel Interbau* International Building Exhibition project in Berlin, an 'architectural zoo' of housing blocks by eminent architects built on an area of redeveloped apartments (formerly inhabited by deported Jewish families). Simultaneously, Johannes Göderitz, Roland Rainer and Hubert Hoffmann's famous Modernist planning manifesto, *Die Gegliederte und Aufgelockerte Stadt* (*The Structured and Opened-*

Figure 8.20
The survival of *Sanierung*

(a) Johanneskirchenviertel, Göttingen, rehabilitation project of 1977–90, official photograph of 'congested' street block before commencement of work;
(b) Layout plan of Johanneskirchenviertel project

(a)

(b)

up City) was published as part of the *Interbau*; Göderitz planned the opening up of closed tenement street blocks in Wedding. In towns untouched by bombing, architects were busily planning radical sanitary redevelopments: in Regensburg, for example, the Munich Pinakothek architect Hans Döllgast put forward a 1959 plan to hollow out most street blocks in the old town (rescinded in 1975 and replaced by a plan to preserve the entire *Altstadt* intact). Indeed, the ethos of *Altstadtsanierung* would continue in force for the rest of the century: in the small town of Wangen im Allgäu, for example, 1979 saw the beginning of a 20-year programme to clear the 'haphazard accumulations of 150 years' from courtyards, and a similar project in the Johanneskirchenviertel area of Göttingen, in 1977–90, included back-court demolitions and relocation of timber-framed buildings.[81]

However, it was Britain, in the mid- and late 1950s, which saw the beginning of arguably the most radical urban clearances of any Western country, and some of the earliest plans for large-scale housing rehabilitation – plans that remained largely unfulfilled so long as the chief local authorities were preoccupied with large-scale housing construction. Some of these early rehabilitation plans stemmed from an active architectural preference for conservation, and others from an attitude of practical expediency, recognising that not all sub-standard houses could be demolished at once, and trying to make short-term improvements to those that remained.

The foremost example of the first approach was the London County Council (LCC) Architect's Department's project for the Brandon Estate in Southwark (from 1955), imaginatively combining an open layout of high towers and other Modernist blocks with preserved early-19th-century classical terrace houses converted into municipal housing. The scheme, designed by architect Ted Hollamby, was praised by LCC Council Leader Ike Hayward for preserving 'all that is best of the traditional character of the area' and in 1956, the LCC's chief planning councillor, Bill Fiske, called for its lessons to be systematised into a 'comprehensive rehabilitation procedure' matching the Comprehensive Development Area procedure of redevelopment. The second, more pragmatic strategy of short-life housing reconditioning inspired an ambitious programme of 1954 onwards in Birmingham and in 'Operation Rescue', launched in 1953 in Leeds by Councillor Karl Cohen, aimed at 'saving the back-to-back' by converting large houses to small flats.[82] Uniquely, Leeds accelerated its improvement drive after 1958 and by 1968 had improved 4,714 houses in 83 separate areas, largely by municipal direct labour. What seemed vital in these early improvement projects was to avoid over-ambitiousness: for every success story like Leeds, there was a disastrous failure, as in Newcastle, where an ambitious scheme by the city planning department from 1960 for 'Long-Term Revitalisation' (rather than mere short-term reconditioning) of a pilot area, Rye Hill, ended in abject failure, with rampant dereliction spreading out from the area across much of western Newcastle.[83]

Governments in Britain had long suspected that large-scale rehabilitation could save huge subsidies, but despite the steadily increasing grants available to municipalities and private owners, the continuing emphasis on new public housing sustained a vicious circle of neglect and decline in the 19th-century housing stock. A succession of government ministers struggled to break the cycle of decline of these so-called twilight areas – attempts that began in 1963 with Keith Joseph and intensified under his Labour successor, Richard Crossman. Within popular culture, a turning point came in December 1960 with the start of the television soap opera, 'Coronation Street', portraying the previously hated twilight areas with sentimental affection.[84] The tipping point was reached under Kennet, who unleashed a barrage of initiatives, beginning in Modern Movement style with systematic research. The Ministry's pilot study of the twilight areas focused on the depressed Lancashire textile town of Rochdale, where its Urban Planning Directorate carried out an ambitious

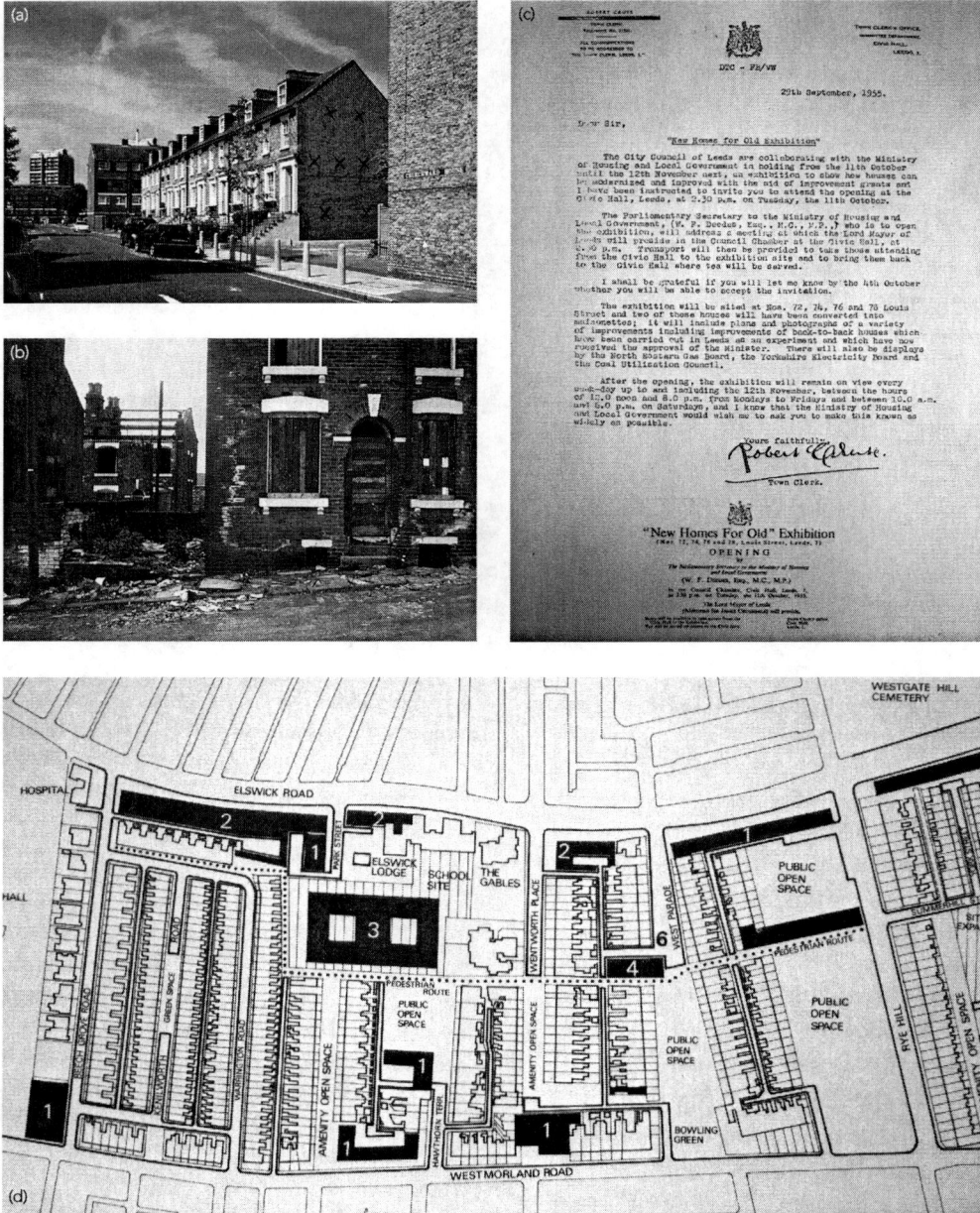

Figure 8.21 Rehabilitation in England

(a) Combined rehabilitation and modern multi-storey redevelopment at the London County Council's Brandon Estate (from 1955); (b) 1987 view of derelict back-to-back houses in Lincoln Road, Leeds; (c) 1955 invitation letter and ticket for opening of rehab demonstration scheme in Louis Street, Leeds, 1955; (d) 1960s' area rehab proposals for Rye Hill, Newcastle

interdisciplinary study of the Deeplish district in 1966.[85] It concluded that the twilight areas need not be doomed to a spiral of decay and eventual demolition, but could be revitalised into 'neighbourhoods of marked character and community feeling', through a new government strategy of 'environmental improvement'. This expansive strategy would supersede the old focus on internal facilities within dwellings, cooperatively involving the low-income owner-occupier residents through enhanced improvement grants. A follow-up pilot scheme covering four streets in the area (completed in 1967) put these principles into practice. In parallel with the Deeplish Study, an influential report chaired by Greater London Council housing chief Evelyn Denington signalled a reversal in the officially preferred approach to renewal, from demolition to rehabilitation.[86]

In 1968, with economic crises gathering, these trends in professional architecture and planning research decisively converged with the wider climate of political and public opinion in England, where there was a growing feeling that mass housing in tower blocks was both unaffordable and undesirable. We will trace in Chapter 9 the way in which a succession of MHLG initiatives presaged a policy and subsidy change from new to old housing. In Scotland, the years after 1968 would also see a dramatic policy shift towards rehab, but in a subtly different form, given the perception of inferior housing standards, and lower rate of improvements by private owners, than in England. Here a key official publication, the 1967 Cullingworth Report, would instead demand acceleration of both demolition and rehabilitation, overseen by local authorities through municipalisation of private-owned tenements.[87]

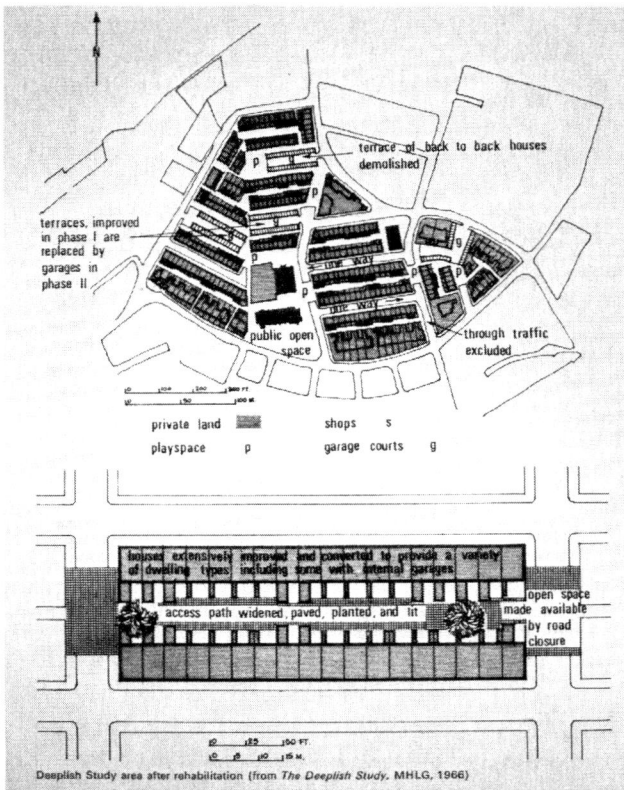

Figure 8.22 The Deeplish Study
Area plan from the 1966 report, showing balance of demolition and rehabilitation

The Malraux effect: civic amenity and conservation in the 1960s

ALTHOUGH the drive to radically renew 19ᵗʰ-century urban housing, by demolition or rehabilitation, was especially accentuated in Britain, the story of the 'rehab revolution' was similar in many places. Scattered initiatives of the 1960s were followed by large-scale action in the '70s, a trend driven both positively, by a passion for preserving dwellings and communities, and negatively, reacting against the growing unpopularity of mass housing construction. During the early 1960s, however, all that lay in the future. In those years, the most prominent element in the Conservation Movement's urban policy debates was something different: a new phase of the civic amenity movement, concerned not specifically with housing but with enhancing the overall visual and social environment of old towns. In the West, the Modern Movement's insistence on differentiating old and new, and its condemnation of 'bogus' and 'pastiche' styles, increasingly relegated Warsaw-style facsimile old towns to a taboo status that would last until at least the 1970s. With rare exceptions, such as *Vieux-Québec* (see below), old town recreations were now confined to museum-tourism and Disneyland-like theme parks. The insistence on the need for 'living' rather than museum-like environments continued, but was now interpreted to mean a vigorous interplay of modern architecture and old buildings – a formula, however, to be achieved not by popular participation but by the 'good taste' of elite civic leaders and designers, many of whom were now beginning to view the general upsurge in modern development from the late 1950s as a potential threat to urban amenity in its own right.

Although each country pursued its own interpretation of the general theme of civic-amenity enhancement of historic towns, an exemplary role was played by one particular initiative, stemming not from Italy but from the France of de Gaulle's presidency (1959–69): the '*loi Malraux*'. In his drive to revive French grandeur, de Gaulle exploited the time-honoured tradition of state coordination of the arts to enhance national prestige. In 1959, he created a new government post of Minister of Cultural Affairs, appointing as first post-holder the author André Malraux, whom he charged with communicating arts and heritage to the broad mass of the people, for the glory of France – rather like a lower-intensity variant of Mussolini's pursuit of *italianità* and *romanità*. Malraux was a slightly idiosyncratic choice for the job, as he had previously, in 1923–4, been imprisoned by French colonial officials in Cambodia over alleged theft of sculpture from a Khmer temple.[88] Now, in a law of 1962 (implemented by decree on 13 July 1963), which became known simply as the *loi Malraux*, he set out the world's first systematic national programme for rehabilitation of historic urban areas, to be designated under the formal title of *secteurs sauvegardés*.

The programme methodology sprang from the French tradition of in-depth, top-quality repair and conservation overseen by official architects. It also built on mounting area-conservation efforts in provincial cities such as Toulouse, where the entire historic core was declared a '*centre archéologique*' in a *Plan Directeur d'Urbanisme* of 1962. The Malraux programme was governed by a Paris-based *Commission nationale des secteurs sauvegardés* with representatives of government ministries and private interests. Each *secteur sauvegardé*, on designation by the Commission, was entrusted to a government-appointed local architect, and a detailed plan of conservation or restoration would be drawn up. Work would be initially focused on one or more pilot areas (*îlots operationnels*) of around 2 hectares. Given the poverty of most inhabitants of the areas designated, comprehensive government financial support was provided, through a combination of public loans from the *Crédit foncier*, and contributions from hybrid state/private local 'mixed-economy

304

organisations' (*sociétés d'économie mixte*, or SEM). This permitted up-front public financing of 80 per cent of expenses, topped up by the landlords after completion.[89]

In May 1964, the first *secteur sauvegardé* was formally designated: a 30-hectare site in Vieux-Lyon, a dilapidated old-town zone across the River Saône from the city centre. Public concern about the area had been provoked initially by an urban-motorway scheme promoted by the Mayor of Lyon (and blocked by Malraux through mass classification of individual buildings). A succession of conferences and fairs was organised by local protest groups, culminating in 1963 in a Lyon-based national conference on '*les quartiers anciens*' and the setting up of a *Société d'économie mixte pour la restauration du Vieux-Lyon* (SEMIRELY). Work began immediately in 1964 on the first two *îlots operationnels*, along the Rue St-Jean. The 47 constituent buildings in these areas comprised deep courtyard groups subdivided into over 560 dwellings: the rehabilitation scheme, superintended by local architecte-en-chef Prof. A Donzet, envisaged reduction of this total to only 269.[90]

The detailed, no-expense-spared approach pioneered in Lyon, focused on courtyard clearance and clarification of earlier architectural features, was comparable with the most meticulous of the prewar *Altstadt* schemes, such as Angelini's work in Bergamo or the programme in Nuremberg. But, almost immediately, two major problems emerged, both stemming from the architectural perfectionism and consequently high cost. First, progress was painfully slow; and second, despite determined aspirations to prevent gentrification or tourist colonisation, few of the original, mostly poor and elderly inhabitants could afford to move back into the pilot areas after completion.

One of the largest *secteurs sauvegardés*, a 126-hectare site in the Marais in Paris, was designated in April 1965. This already contained 176 *monuments classés* and 526 *monuments inscrits* (many added following publication of a ten-year conservation plan by the City Council in 1961). Public opinion in Paris was now overwhelmingly against redevelopment and demolition, and the Ministry of Cultural Affairs and the City Council had already acquired several key monuments for renovation, such as the Hôtel de Bethune-Sully (in 1962). Work began in 1967 in the Marais on several *îlots operationnels* at once, overseen by SOREMA, a new local SEM, and with plans by architects Maurice Minost, L Arretache, B Vitry and M Marot. Initial pilot areas included *Ilot* 1 Carnavalet (around Place Thorigny and Rue du Parc Royal), *Ilot* 9/33 (to the north-west) and *Ilot* 16 (Jardins St Paul) – the latter two both being *Entkernung*-style courtyard-clearance schemes. Echoing the role of 18th-century paintings in the reconstruction of Warsaw (see Chapter 10), 17th- and 18th-century hotels and tenement street-blocks were restored, where possible, to the state shown in the 1739 Turgot Plan, by removing workshops and other later excrescences from their gardens. The need to accommodate motor transport was ingeniously addressed by constructing small underground car-parks beneath some courtyards.[91]

The Marais was by far the most expensive of the first group of *secteurs sauvegardés* (costing over 10 million francs). Here the process of gentrification and rent inflation was seen at its most extreme, with old utility shops replaced wholesale by craft boutiques and antique shops. Over the whole country, contrary to initial hopes to extend the procedure to about 1,000 historic towns, only 60 *secteurs sauvegardés* had been successfully designated by 1977, whereupon (as we will see later) a simpler and easier system of designation was introduced.[92]

Cities across the Western world learned much from the Malraux experience, both for good and for bad. What it suggested was that, while lavish, architect-led rehabilitation projects were feasible in isolated set-piece cases, a choice would generally have to be made between quantity and quality. Other intensive projects similar to the Malraux programme included Stockholm's Gamla

Figure 8.23 Urban rehabilitation and redevelopment in France

(a) Saint-Georges, Toulouse: a patchwork of new and old, following the defeat of the city council's area-demolition plans in 1959; (b) The pioneering Vieux-Lyon *secteur sauvegardé* (from 1964) – 2008 view of an opened-up rear courtyard in the Place Neuve St Jean, in one of the two initial *îlots operationnels* off the rue St-Jean; (c), (d) The Marais, Paris: views of the hybrid new/old townscape created around the Place Thorigny in *îlot* 1

Stan, where the work by the Samfundet S:t Erik and others had pressed on into the postwar era; Copenhagen's meticulous municipal 1960s' scheme for the Christianshavn area, under the provisions of an enabling act passed in 1962; and Leuven's 15-acre Grand Beguinage, restored with intense care by Lemaire for Leuven University from 1964.[93]

Each country and city established its own balance between quantity and quality, dependent on local circumstances. In the Netherlands, for instance, the general protection of ensemble 'views', included in the 1961 *Monumentenwet*, was combined with the careful approach within Amsterdam that we traced earlier. In Italy, the vast extent of historic urban fabric imposed political limits on both intensity and extent of operations. Although there were, in theory, comparable powers to the *loi Malraux* by the late '60s, these were in practice little used: urban improvement remained scattered and *ad hoc*, with only isolated places benefiting from comprehensive, heritage-sensitive development plans: for example, that for Siena, prepared in 1953-8, or Giancarlo de Carlo's for Urbino (1954–8). In Rome, a general consensus had emerged by the '50s that the historic centre (i.e. the 14 *rioni* still built up in 1870) should be preserved in its entirety. A 1954 municipal resolution argued that only *risanamento conservativo* was now admissible. But in the absence of any general grant system, progress was painfully slow. A pilot rehabilitation scheme at Tor di Nona (from 1954) took decades in execution and was only completed in 1971.[94] Attracting most urgent attention was the situation in Venice, where successive special laws (1956, 1973, 1975) attempted to safeguard the city from decay and rising water levels. Local architects such as Saverio Muratori argued that new developments should reflect local traditions and, in the late 1950s, sought to devise 'typologies' of Venetian morphological patterns ('lagoon, courtyard, street') that could be echoed in new development. This was the first appearance of the Italian concept of 'type', which would strongly influence international architectural debates in the 1970s and '80s.[95]

In the USA, the new concept of 'landmark' designation was applied to quite extensive historic districts, notably in the New York City Landmarks Preservation Law of 1965. By 1966, one district, Brooklyn Heights, a mid-19th-century area of 50 street-blocks and 1,300 dwellings, had been fully registered as a landmark and other areas were under consideration, including 'SoHo', a dilapidated warehouse area colonised by Abstract Expressionist artists in the 1950s and finally designated as a landmark in 1973. The shift to urban preservation in New York was aided by the decline in power of the city's forceful planning chief Robert Moses, especially his resignation from the Mayor's Committee on Slum Clearance in 1960.[96] Just as before the war in Charleston, the key to progress in North American cities was a balance of public encouragement and plentiful private investment: like most other market-orientated interpretations of heritage, the concept of 'gentrification' was pioneered in the USA. The dilapidated historic Washington suburb of Georgetown, for example, was rediscovered as a desirable residential area in the 1940s, and local residents took action in 1951 to stop demolition of mid-18th-century houses for a parking lot by founding a stock-funded, for-profit company, Historic Georgetown Inc. An Old Georgetown Act, passed by Congress in 1950, helped deter demolitions, and HABS carried out survey work in the 1960s.[97]

The most influential postwar historic urban renewal area in the USA was the College Hill district of Providence, Rhode Island. Here, from 1957 onwards, various federal, state and civic agencies, led by the Providence City Planning Commission, had collaborated with local residents and private companies in implementing a multi-strand, market-led renewal programme. This combined effective architectural regeneration with social and racial divisiveness, as poor black families were displaced from northern Benefit Street by white gentrification. The 1956 foundation

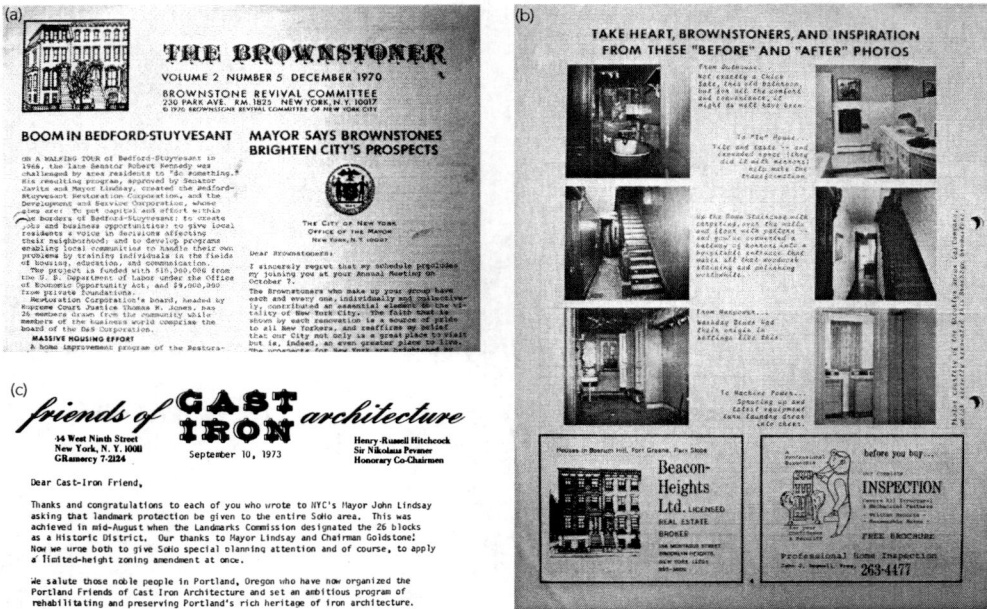

Figure 8.24 Urban revival in New York City

(a), (b), (c) The intimate three-way link in 1960s–70s' America between intense voluntary preservation efforts, housing 'gentrification' and gradually increasing official support for area preservation is underlined in these publication extracts (1970 and 1973) from two contrasting groups – the Brownstone Revival Committee and the Friends of Cast-Iron Architecture – both of which refer to area landmark initiatives in New York City by Mayor John Lindsay. The first of these groups, still thriving today as the Brownstone Revival Coalition, was founded in 1968 by Everett and Evelyn Ortner, veteran campaigners for preservation of the Park Slope area of Brooklyn

of the Providence Preservation Society was followed by designation of part of the area as a Historic District in 1957/60 (covering 500 out of the 1,700 buildings in College Hill, 20 of which were already restored) and a 'demonstration grant study' report of 1960 by Blair Associates, city planning consultants. Thereafter an ambitious regeneration programme was financed by a combination of private capital and federal Housing Act and urban-renewal funding from the Housing and Home Finance Agency (later from HUD). The area was incorporated from 1961 in a wider, 343-acre East Side Urban Renewal Project, under whose auspices 'disharmonious' buildings in the Benefit Street area were demolished and replaced by historic houses relocated from elsewhere by a range of for-profit preservation companies – converting the area into something akin to Colonial Williamsburg. More generally, during the 1960s and 1970s, the burgeoning power of HUD, under legislation such as the Demonstration Cities Act, encouraged a socially conscious, multi-disciplinary approach to urban environmental regeneration in the USA – a strategy constantly undermined by underlying racial tensions, as with Milwaukee's ambitious 'Relative Residential Status' strategy of 1974, which eventually exacerbated the decline of the city's black neighbourhoods.[98]

In strong contrast to these diverse urban-renewal projects in the USA, the old city of Quebec, in Canada, witnessed North America's only significant attempt at facsimile-(re-)creation of a dense, old, walled *Altstadt* in the classic European tradition – as a part of the postwar nationalist movement

Figure 8.25 Gentrification USA

(a) Georgetown, Washington DC, restored and gentrified from 1951 onwards (HABS survey photo); (b) Benefit Street, College Hill, Providence: the target for an ambitious gentrification and preservation scheme from 1957 onwards (HABS survey photo)

to reassert the 'traditional French' character, and political sovereignty, of Quebec. The early-17th-century core of Quebec City, centred on the intimate Place Royale, had seen extensive change following the British takeover in the 1950s, and haphazard modernisation in the 19th and 20th centuries: the area's traditional character had almost vanished by the 1950s. In 1953, Gérard Morriset, Director of the Art Inventory of Quebec, called for the restoration of the Hôtel Chevalier as a testimony of Quebec's 'golden age'. Its eventual reconstruction in 1958–60 presaged a vast programme of urban restoration, a 135-hectare '*arrondissement historique*' being declared in 1963/4. In 1967, the Ministry of Cultural Affairs undertook the reconstruction of 64 buildings in streets round the Place Royale, in the *style traditionnel canadien-français*. In its sharp juxtaposition of a newly created *Altstadt* with the spacious modernity of a 20th-century city, the ensemble curiously resembled postwar Warsaw, and indeed, in 1985, five years after the Polish capital, the 'Historic District of Old Quebec' was duly inscribed on the World Heritage List – although in this case, unlike Warsaw, the inscription text makes no reference to the reconstructed character of the 'property'.[99]

Breadth, not depth: civic amenity and urban conservation in 1960s' Britain

T HE antithetical positions of France and Britain in historic monument designation, between quality and quantity, between intensive and extensive, were extrapolated into the wider field of urban conservation. If the Malraux system of *secteurs sauvegardés* saw the postwar Conservation Movement at its most intense and meticulous, the wide-ranging 'conservation area' system in Britain was almost the opposite.

The origins of this system, however, were quite complex. In the immediate aftermath of war, the uneven effects of bombing further exaggerated the sacrosanctity of Oxford and Cambridge and the smaller historic towns. Within 'Oxbridge', the intense picturesqueness of the ensembles was carefully safeguarded, as a responsibility of the individual colleges. It was also owing to college initiatives that the 1950s and '60s witnessed an innovative succession of contextual modern interventions in the two cities, spearheaded by architects Powell & Moya: their projects at Brasenose

309

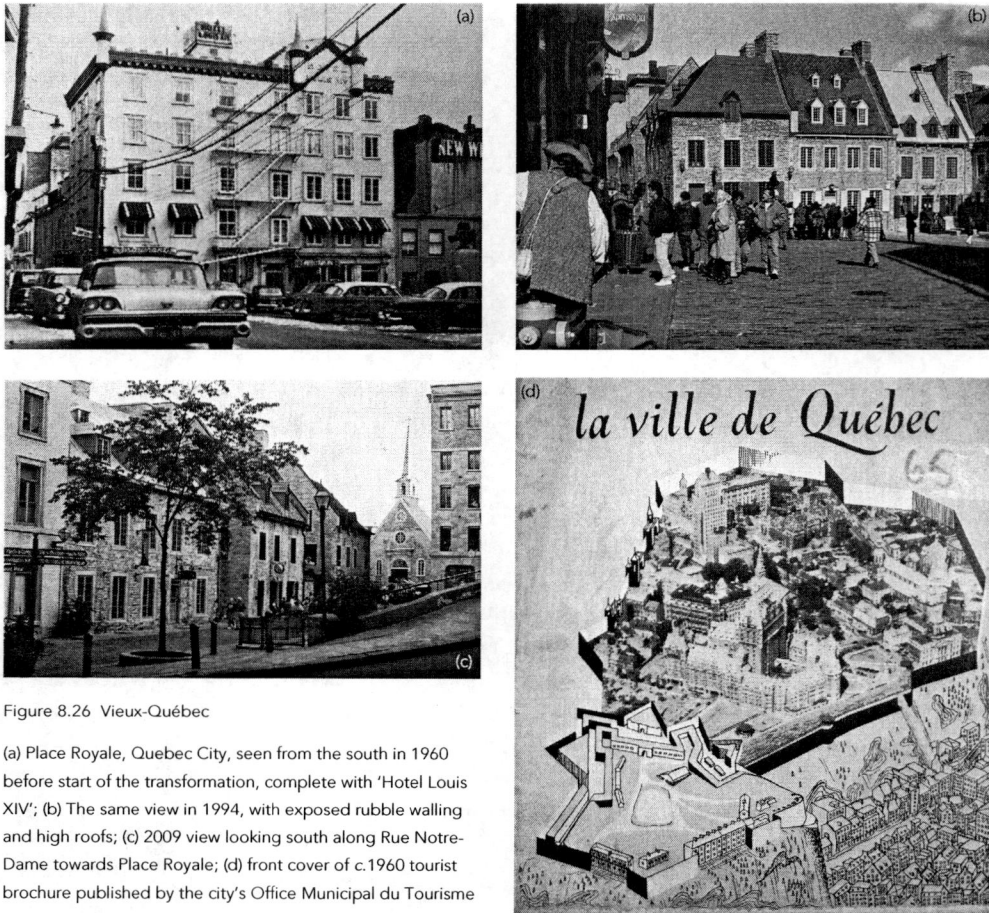

Figure 8.26 Vieux-Québec

(a) Place Royale, Quebec City, seen from the south in 1960 before start of the transformation, complete with 'Hotel Louis XIV'; (b) The same view in 1994, with exposed rubble walling and high roofs; (c) 2009 view looking south along Rue Notre-Dame towards Place Royale; (d) front cover of c.1960 tourist brochure published by the city's Office Municipal du Tourisme

College (1959–61) and Christ Church, Oxford (Blue Boar Quad, completed in 1964), and St John's College Cambridge (Cripps Building, completed in 1967) featured intricate plans, stone cladding and vertical articulation suggestive of oriel windows.[100] In that respect, Oxbridge was almost a microcosm of the wider structure of urban conservation in Britain, depending on individual local initiative and a collaboration of voluntarism and collective institutional pride, rather than detailed central direction by administrators.

Outside these hallowed precincts, most larger English towns and cities had suffered not the devastation and wholesale reconstruction widespread on the Continent, but instead a more piecemeal disfigurement by commercialism, industrial dilapidation or inappropriately thrusting modern buildings. To tackle this, extensive rather than intensive civic-amenity protection seemed to be required. The seminal postwar civic-amenity initiative at a national level in Britain was the brainchild of an enlightened and charismatic Conservative government minister, Duncan Sandys. In 1957, he marshalled sufficient private financial backing to allow the formation of a national 'Civic Trust', with a guaranteed annual income of £40,000 and tasked with fostering local civic engagement with planning across the country – despite the opposition of key Ministry civil servants.[101] Sandys's scheme harnessed local enthusiasm to national issues: the Civic Trust organised conferences and study groups on contemporary issues such as traffic management, town-centre redevelopment or obsolescence and initiated a programme of coordinated demonstration schemes – a modern *Entschandelung* of unsightly commercialism or dilapidation. The pioneer of this 'facelift' movement, Magdalene Street in Norwich (1959), involved some 80 owners and public bodies in visual coordination, landscaping and elimination of 'unnecessary clutter' from an 18th- and 19th-century streetscape. As the Civic Trust later recalled,

> The result was startling: when Magdalene Street was chosen for the experiment, people said, 'Why choose a run-down old street like that?' When the scheme was finished, they said, 'Of course, it's easy in a beautiful street like that. . . .'

Local authorities complemented these efforts with experimental traffic schemes, including a 1967 pedestrianisation of London Street, Norwich, prepared by City Planning Officer Alfred Wood on the model of the earlier German schemes.[102]

Gradually, the work of the Civic Trust built up a head of political steam: a Conservative Party pamphlet of 1960, *Let our Cities Live*, argued that municipalities up and down the country should declare environmental improvement-areas, and the mid-1960s saw an explosion of civic and individual activism aimed at saving local historic buildings from demolition. Kennet's 1972 book, *Preservation*, argued that the SPAB tradition of voluntary agitation had been effectively updated for the new Welfare State era. It related exemplary tales of local initiatives, 'when the brigadiers and poets, in uneasy alliance, descend upon the Town Hall and tell the council they are a lot of philistines'. In this volatile climate, even individual amateurs could play a prominent role, as in the startling 1965 campaign by 16-year old Portsmouth schoolboy Stephen Weeks to save an early-19th-century Gosport mansion already being stripped of its roof slates. His campaign included television appearances, confrontations with the Borough Engineer and a personal visit by listing chief Antony Dale to praise his efforts. In Scotland, there was a different configuration of organisations, stemming from the longer roots of the campaign of burgh conservation in the '30s and its continuities with

Figure 8.27 The Civic Trust in action

(a) Duncan Sandys pictured in the City of London in 1975, in his capacity as organising committee chairman of European Architectural Heritage Year (EAHY); (b) Magdalene Street, Norwich, scene of the Civic Trust's pioneering 'facelift' scheme in 1959: 1988 view

the Geddes era. That continuity was evinced in the 1930s' and 1940s' work of Edinburgh City Architect Ebenezer MacRae, and in the late 1930s' and 1950s' regeneration of the Old Town of Stirling by Geddes's son-in-law, Frank Mears, with a mixture of preserved 17th- and 18th-century houses and new infill 'in character'.[103]

In the 1950s, some Scottish advocates of burgh preservation still kept up the nationalist rhetoric of the '30s, fiercely rejecting the postwar tide of social housing. For example, George Scott-Moncrieff, an impassioned traditionalist, in 1956 lambasted mass housing as 'detestable', 'suicidal' and likely to spawn philistine 'children of concrete mixers or . . . of technological town planners'. Now, 18th-century planned villages (themselves originally 'modern') began to replace irregular old burghs as the focus of preservationist efforts. In 1957, the recently founded Historic Buildings Council for Scotland extended its grant-aid operations from country houses to the urban heritage, awarding £30,000 subsidy to the rehabilitation of the 18th-century classical planned town of Inveraray, originally developed by the Duke of Argyll to complement his new castle. Numerous dilapidated tenement buildings were acquired by the Ministry of Works at the instigation of the Minister, Hugh Molson, and were conveyed to the town council for rehabilitation: the entire Inveraray project, including repair of the Castle (opened to the public since 1953) was overseen by Ian Lindsay. By the mid-1960s, Scottish urban conservation had entered a much more self-consciously Modernist phase: 1966–7, for example, saw a 'multi-disciplinary research study in comprehensive planning' of the historic Scottish town of Perth, including computer analysis, masterminded by the SDD Urban Planning Directorate.[104]

From the Four Towns studies to conservation areas

IT was, however, in England, under Crossman and Kennet, that the Conservation Movement in Britain attempted the first large-scale application of civic-amenity concepts to historic towns, in a programme coordinated by the government, in close collaboration with private owners. Such a collaboration had been presaged in 1955, not in a medieval Old Town but in a pioneering 'Terraces Scheme' in 'Georgian Bath', providing for a 1:1:2 funding split between owners, central government and the city council.[105] Now, in the 1960s, Kennet argued that central government must set about systematic collaboration with the local authorities – in contrast to France, with its administratively dominated system of powerful prefects. Although the officially backed English system of civic-amenity conservation emphasised breadth as opposed to depth, the ground for it was laid by an intensive burst of case-study research, beginning at a January 1966 conference in Churchill College, Cambridge, chaired by Crossman, on preservation of old towns, and continuing with a series of studies of representative old towns, York, Bath, Chester, Chichester and King's Lynn – a series that became known as the Four Towns studies after King's Lynn dropped out.[106]

All four books were published in 1968, but care was taken to make them very different in character, ethos and methodology. Unsurprisingly, the most radically surgical proposals for existing built fabric were contained in the Bath report, by Colin Buchanan's firm. Their 1965 Planning and Transportation Study for the town had controversially suggested carving an inner ring-road through the centre, provoking a decade of bitter conflict between council and protesters. Their Four Towns study proposed the same approach, boldly slicing up the centre with road incisions around pro-tected environmental areas. Conversely, architect Donald Insall's Chester report was grounded in meticulous survey and analysis. It used a Geddesian and SPRND philosophy of regional and civic

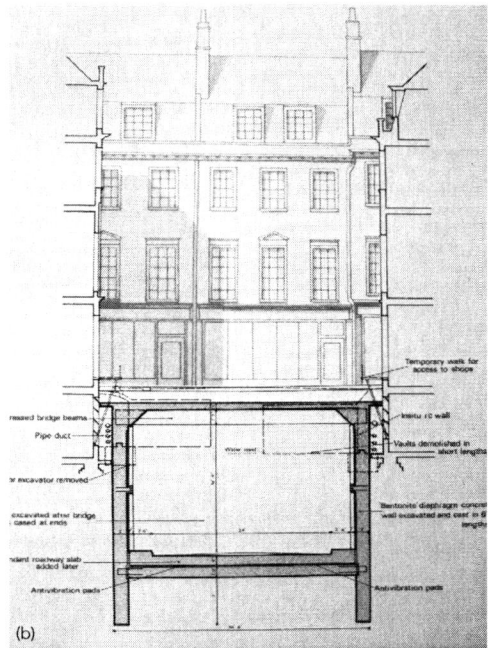

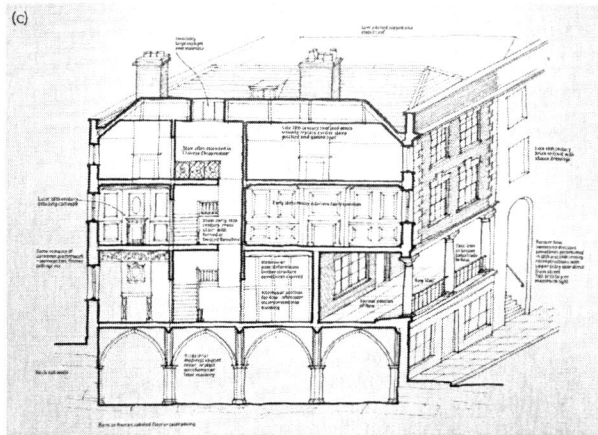

Figure 8.28 The Four Towns studies: 1968

(a), (b) Front cover of the Bath volume and cross section of Buchanan's proposed road tunnel; (c), (d) Chester volume, 'The Rows', existing survey section and proposed new 'intervention'; (e) 1973 view of the south end of Stonegate, a pedestrian-priority street in inner York

313

survey as a foundation for small-scale conservative surgery, focused on case-study areas. Insall explained that this was 'no windscreen survey but a matter of overalls, dirty knees, getting into the roof spaces and basements' of all buildings in the pilot areas. The York study, commissioned from architect-planner Lord Esher despite opposition from the city council, combined townscape analyses and research into the city's social patterns. The outcomes of the studies were also widely divergent, with York handicapped by tension between Esher and the council, Bath falling victim to the gulf between pro-redevelopment city council and protesters, but Chester leading to a highly effective, 20-year programme of conservative-surgery collaboration between Insall and the council.[107]

The individual impact of the Four Towns studies in England was less important than the way they helped shape a growing government predilection towards systematic reinforcement of the Conservation Movement at a national level. Radical change was already on its way, as demonstrated in Crossman's advocacy of a balance of preservation and change, for example in a major speech in Bath in May 1966. This tendency was reinforced by a stroke of good luck, when Duncan Sandys secured top place in the ballot to introduce a 'private member's bill' in parliament, and decided to use his opportunity to introduce a bill dedicated largely to urban conservation. Crossman, who had already committed himself to expansion of urban conservation, offered Sandys MHLG backing in drafting his 'Civic Amenities Bill' – overruling opposition from civil servants who wanted to limit its scope to individual listed buildings. In his diary, Crossman recalled that 'I was still determined to make sure the Bill dealt with townscapes . . . and not merely with individual listed buildings, though the Department had obstinately drafted my policy paper . . . excluding my concept'. What emerged was a straightforward extension of the listed building formula of extensive rather than intensive conservation, based squarely on the planning system of development control and emphasising 'good new architecture' as well as protecting the old. Esher, as president of the Royal Institute of British Architects, noted in his 1966 address that Sandys's bill would inject 'new impetus into the Conservation Movement' and argued that 'we have got to put a ring round our Conservation Areas and accept nothing within them below the standard of the Civic Trust Awards'.[108]

The new formula of the conservation area (CA, a term introduced at committee stage of the 1967 Civic Amenities Act, despite countryside preservationists' protests) was based on the straightforward principle of 'designate first and think later', with minimum justification required. Kennet argued, 'I hope that the local authorities will designate rather a lot of land . . . whole quarters of towns and, in some cases, whole towns within the walls . . . including Victorian quarters'. The result, quantitatively speaking, was spectacular. By 1970, over a thousand conservation areas had been designated; two years later there were 1,350 conservation areas in 130 municipalities, and by 1974 the figure had soared again to 2,750. At a ratio of some 40:1 compared with the *secteurs sauvegardés*, this presented an even more radical contrast with France than the listed buildings system. Within this national average, the fierce autonomy of the local-authority system led to exaggerated divergences between even next-door London boroughs, between (for example) Westminster, with 14 CAs, and the 'deplorable' (in Kennet's view) Hammersmith, with none. At first, however, CA designation conferred no additional security, with unlisted buildings remaining unprotected: its main impact was hortatory, a task helped by officially supported follow-on publications and events.[109]

'Heroic failures': the rise of Victorian heritage in 1960s' Britain

ONE of the most dramatic by-products of the British strategy of extensive rather than intensive conservation, and of collaboration between official agencies and voluntary societies, was the way in which, during the 1960s, it made possible a sudden boom in professional and popular interest in the architecture of the mid- and late 19th century.

The rise in popularity of the built environments of the modernising Victorian Age in Britain and elsewhere was, indirectly, at the expense of the status of the pre-modern monuments that had been the foundation of old-style nationalist heritage architecture: but the Conservation Movement lost none of its drive for Progress and 'movement' in the process. The shift in opinion only really began in the 1950s: as late as 1945, Kenneth Clark could still cite Ruskin in support of a claim that 'an age which produces a mass of ugly buildings and ugly objects of daily use must have something fundamentally evil in its composition'. Among establishment figures such as Sir Albert Richardson, who reputedly only read old 18th-century newspapers, the good–bad division still seemed to be around 1830: there was little argument in 1947 with the decision to rescue the Nash Regent's Park terraces, while Ministry listing chief Antony Dale combined reverence for Regency Brighton with distaste for High Victorian Gothic. And Scottish traditionalist ideologue Scott-Moncrieff conceded that Edinburgh's classical New Town constituted high architecture but thundered that, in the Victorian eclecticism of David Bryce, 'Scottish architecture plumbed the depths of meaningless ostentation . . . in which every source is arrogantly misunderstood'.[110]

But by the 1960s, these voices were distinctly in the minority, as a sea change in attitudes had taken place within the Conservation Movement – a change driven, typically for Britain, by a coalition of enthusiastic amateurs, including a flamboyant aristocratic element. They came together in February 1958 to found a new 'Victorian Society', inspired initially by Betjeman and Irish socialite/aristocrat Anne, Countess of Rosse. The group's first meeting was held in Rosse's sumptuous Victorian-decorated house at 18 Stafford Terrace, and its first headquarters were in premises loaned by SPAB in Great Ormond Street, with Ian Grant as Hon. Secretary. Like the Georgian Group before it, elite amateur beginnings preceded exponential membership growth, from 100 members in 1961 to 1,700 in 1968 and 3,000 ten years later.[111]

What distinguished the 'Vixoc' from the Georgian Group and SPAB – the 'secret weapon' that allowed the Conservation Movement in Britain to authoritatively tackle the vast Victorian heritage – was an importation from Central Europe: Nikolaus Pevsner, Britain's most authoritative champion of Victoriana, who became the society's chairman in 1963, eclipsing its amateur founders. A Leipzig-born and educated art historian who followed Gropius in his Modernism and Pinder in his nationalism and fervently advocated *Kunstgeographie* (an art-historical variant of *Heimat*, seeking out 'national character as it expresses itself in art'), Pevsner had found himself reluctantly forced into exile in England after 1933 owing to his Jewish ancestry. Nationalistic love of Germany gradually transformed into an equal passion for all things English, and his 'Prussian' industry and seriousness came to seem lovably eccentric. Exploiting this aura, Pevsner systematically developed the English tradition of overlapping amateurism and professionalism in built-environment matters. Astutely, he recognised that any meaningful *Kunstgeographie* of England would have to embrace not only long-standing stereotypes, such as the Picturesque, but also the broad built environment of 19th-century industrial society. Initially, after the war, he had turned his furious energies to the propagation of *Kunstgeographie* and Modernism in Britain, including lectures on the 'Englishness

Figure 8.29 The Victorian revival in England

(a), (b) The 1974–5 and 1977 annual reports of the Victorian Society (featuring 'threats' to London Liverpool Street Station and Sir G G Scott's All Souls, Haley Hill, Halifax); (c) The permanent headquarters of the Victorian Society, 1 Priory Gardens, Bedford Park; (d) Late 1960s' view of St Pancras Station

of English Art' and the largely single-handed researching and authorship of a Dehio-style topographical series, the *Buildings of England*. Then, following the death of his wife in 1963, Pevsner channelled his prodigious powers of research and organisation, and his fearless conviction of the rightness of his causes, into the work of the Victorian Society, becoming its 'benevolent despot' and lecturing tirelessly to local societies across the country. In 1967, he recalled that, in 1949, when he became Slade Professor of Fine Art, students laughed when he lectured on the need to preserve Victorian buildings: 'I had to stand there and say to them, "This is not funny."' Throughout the

1940s, '50s and early '60s, the prevalent insistence on strong contrast of old and new allowed him to embrace both Victorian architecture and architectural Modernism.[112]

The rise of the Vixoc to a position of influence followed the time-honoured conservation pattern of exploiting heroic failures. In a 1959 lecture to its first annual general meeting, Lord Esher complained, 'No-one listens to what we say, and "Oh, it's only Victorian" means it can be ruthlessly destroyed. But it is exciting, I think, to be just in time to save what will be admired tomorrow.' That year saw the beginnings of the Vixoc's most decisive cause célèbre: the campaign to save the massive, neo-Greek 'Euston Arch' (fronting the London terminus), scheduled for redevelopment in a programme of station modernisation projects by the British Railways Board (BRB). More emotive still was the proposal (first mooted in 1938) to demolish G G Scott's spectacular neo-Gothic St Pancras Station and merge it with next-door King's Cross. In 1961, a high-powered delegation pleaded vainly with Prime Minister Macmillan to spare the Euston Arch, which was eventually demolished in 1962. And in 1963, another key monument of the industrial age, the London Coal Exchange, was demolished for road widening.[113] These high-profile losses galvanised the new consensus in favour of Victorian architecture and in 1967 Kennet duly listed St Pancras Station at Grade 1, finally ending the BRB's redevelopment hopes – although John Summerson still felt compelled to denounce its 'mass of ecclesiastical detail' as 'nauseous in the unstately context of what is mainly a huge block of bedrooms'.[114]

The focus of outrage in this new phase of heritage in Britain was not elite buildings such as palaces or churches, or medieval old towns, but monuments of Victorian industrial and commercial power. Partly, this also reflected a new phase of democratisation of the built heritage, something whose consequences we will trace in Chapter 12. An essential building block of this movement was the growing appreciation of Victorian architecture, as a built environment and as a system of values. This permeated Asa Briggs's monumental 1963 book, *Victorian Cities*, as well as Pevsner's detailed coverage of Victorian architecture in the *Buildings of England*. The new discipline of Industrial Archaeology played a key role, presenting a broad, inclusive overview of the industrial cultural landscape. Books such as Kenneth Hudson's *Industrial Archaeology* (1963) inspired the project for an open-air museum dedicated to industrial rather than rural folk themes at Ironbridge Gorge, Shropshire (from 1968); a similar but more eclectic project, at Beamish, south of Newcastle, gathered momentum from around 1968–70, under the leadership of Frank Atkinson, director of the Bowes Museum.[115] At a popular level, the railway preservation movement harnessed growing popular enthusiasm to the preservation both of engines and of old lines as linear museums, the pioneers being the Talyllyn Railway (opened in 1951), the Bluebell Railway in Sussex (opened in 1960 following a bitter, four-year legal wrangle between residents/enthusiasts and British Railways – BR) and the Keighley and Worth Valley Railway (opened in 1966–8); the first preserved railway locomotive to run on BR, J.52 '1247', was in 1962. The movement for preservation of railway stations, inspired by the Euston and St Pancras campaigns, blossomed in parallel with the locomotive preservation movement: between them the two embodied precisely the 'moveable'/'non-moveable' heritage distinction first set out in post-1789 France.[116]

By the 1960s, similar patterns had begun to emerge in other Western countries, especially the USA. Chicago, for example, saw the emergence of a Chicago Heritage Committee of architect enthusiasts dedicated to saving turn-of-century works such as an Adler and Sullivan theatre, or early skyscrapers, forcibly opposing the old preoccupation with 'a few precious national shrines'. The two decades from 1954 saw a protracted but ultimately triumphant campaign to restore Adler and Sullivan's

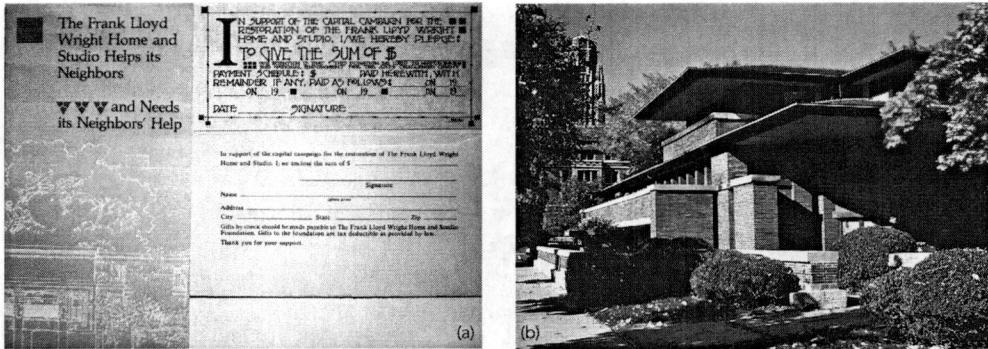

Figure 8.30 Frank Lloyd Wright

(a) Publicity material produced by the Frank Lloyd Wright Home and Studio Foundation c. 1980 for the campaign to restore Wright's former home at 951 Chicago Avenue, Oak Park, Illinois; (b) 1980 photo of Wright's Robie House, Oak Park

Auditorium Building for university use, steered by architects Crombie Taylor and Harry Weese, although Wright's Robie House was threatened with demolition as late as 1969. Early cast-iron architecture, too, was targeted for energetic preservation efforts by voluntary enthusiast groups.[117]

An example of the changing climate at a local level towards 19th-century heritage was the position in the Scottish city of Glasgow. Here, acceleration of slum clearance and comprehensive redevelopment in the early '60s went hand in hand with a municipally promoted urban conservation policy focusing on the city's spectacular 19th-century heritage, including the exotic classical eclecticism of Alexander 'Greek' Thomson. Beginning in 1961 with a demonstration Civic Trust scheme to facelift the classical early-19th-century Carlton Place, and prodded by a civic group, the New Glasgow Society, founded in 1965, the Corporation's Planning Committee initiated the conservation of the Park Circus area – with its sumptuous Italianate terraces, one of the most intact surviving mid-19th-century classical areas in Britain. The committee appointed architect A Lochhead and landscapist W Gillespie to assess the area: they produced a report (1967) stressing the need to protect the 'unity' of these townscapes, duly declared a Conservation Area in 1970.[118]

By 1969, with opinion turning against comprehensive development and mass housing in Glasgow, Lord Esher was imported by the Planning Committee to develop a full conservation strategy, to defend this, 'the finest example of a great Victorian city'. Reflecting the late-Modernist insistence on balancing preservation and change, he recommended in 1971 the designation of a range of Conservation Areas to augment Park Circus, but also praised, Buchanan-like, the municipality's surgical urban motorway plan, as 'in itself a major civic improvement, a powerful agent of urban renewal', that would allow rescue of further decayed areas such as the 'Merchant City' in the south-east city centre.[119]

But, by 1971, these ideas were already obsolete. When John Summerson agonised at a Vixoc annual general meeting in 1968 on the 'problem of failure' posed by the Victorians' 'horribly unsuccessful' architecture, almost no-one was listening any more.[120] As we will see in Chapter 9, progressive heritage opinion in Glasgow, and elsewhere in the West, was now racing beyond this measured balance between Conservation Movement and Modern Movement, towards a more militant confrontation between the two and a conservationist rejection of modern redevelopment in any form whatsoever.

318

Figure 8.31 'Greek' Thomson
Alexander Thomson's Caledonia Road U.P. Church, Glasgow (1856–7) seen immediately following the 1965 fire that gutted it after its acquisition by Glasgow Corporation

319

– 9 –

From counter-culture to control

Western triumphs of conservation, 1968–89

> *Now houses are 'units' and people are digits*
> *And Bath has been planned into quarters for midgets.*
> *Goodbye to old Bath. We who loved you are sorry.*
> *They've carted you off by developer's lorry.*
>
> John Betjeman, 1973[1]

Conservation and modernism: from coexistence to conflict

In Chapter 8, we saw how, in Western countries, following the self-immolation of the Heritage Militant between 1914 and 1945, the Conservation Movement became successfully contained within the consensual framework of the welfare state, whose governing ethos of peaceful progress allowed it to flourish as never before. The doctrine of sharp separation of old and new, agreed by both conservationists and modern architects, allowed the Movement to coexist with the dominant forces of radical reconstruction and to begin spreading its scope unobtrusively across wider areas of the built environment, including the vast environments of the 19th century that still formed the main target for modernising redevelopment.

That new status quo survived for the first two postwar decades. But in the late 1960s and early 1970s, in one of the sudden explosions characteristic of architectural debates, this stable, balanced system was suddenly convulsed by confrontation, and peace was replaced by conflict. When, after 5–10 years, things settled down again, and a stable system of heritage values began to re-emerge, the overall landscape of the Conservation Movement in Western countries had changed significantly – and in ways that further exaggerated its own internal dynamic of driving progress. Modernism in architecture had now fallen resoundingly from professional and popular favour, and the old collaborative and subordinate relationship of the Conservation Movement to the Modern Movement had been replaced by a renewed anti-modernity, both among conservationists and contemporary architects. As a result, conservation and new architecture actually moved much closer together: for a decade or two, conservation seemingly became *the* dominant force in built-environment policies and debates. Behind the scenes, the administrative machinery of modernist comprehensive planning survived and was simply appropriated by conservation: 'Planning was both the problem and the solution' (John Pendlebury).[2]

This was 'Progress' indeed, progress beyond the wildest dreams of conservation's 19th- and early-20th-century pioneers. In 1982, international conservation leader Sir Bernard Feilden could argue, in his standard textbook, that the discipline was 'still in a violent stage of growth and expansion': he felt like 'a geographer trying to map the caldera of an active volcano before seismic activity has ceased'.[3] This new conservation radicalism had nothing to do with the old warlike ethos. It was a creative chaos, from which a wide variety of new positions emerged, ranging from a political agenda of participation and democratisation, only indirectly concerned with the built fabric, to a new architectural movement of postmodernism, which accorded preserved old buildings a central and honoured role.

These permutations were largely confined to the Western world: as we will see later, the position in the socialist bloc was quite different, and the reunification of the two in 1989–91 would provoke a fundamental shift everywhere towards a more homogeneous market-driven heritage. The years 1968–89 in the West laid the groundwork for this wider change, especially through a shift in conservation's political affiliations from the leftism of the 1970s towards the neo-capitalism of the late 1980s and '90s.

Riots, radicalism and the 'disembedding' of consensus

THE most public face of the late 1960s' destabilisation of welfare-state consensus was unambiguously left-wing – an international storm of riots and strikes that erupted first in Amsterdam in the 'Provo' radical disturbances of 1966, and then in other Western countries in 1968, including the nationwide May *évènements* in France and radical-led riots in West Germany and the USA. By 1969, a general climate of radical, student and anarchist agitation prevailed, and the early 1970s added recessions, environmental discontents and energy crises to the mix. In Britain, waves of strikes made the country 'ungovernable', and even the august Cambridge University witnessed a student riot in February 1970 against the Greek dictatorship; in Northern Ireland, the Civil Rights movement of 1968–9 presaged full-scale ethno-political conflict during the 1970s.[4]

In an echo of the utopian anti-modernity of Ruskin and Morris a century earlier, a vast new landscape of environmental activism sprang up, united by hostility to the ethos of unbridled science-driven progress. It was fronted by voluntary amateur groups such as Friends of the Earth and individual critics such as E F Schumacher, whose *Small is Beautiful* (1973) argued passionately for 'permanence' and husbanding of resources, or Rachel Carson, author of the earlier *Silent Spring* (1962); many advocated a general change in public ethics away from materialistic utilitarianism. By the early 1970s, a broad genre of popular environmentalist agit-prop was established, including book titles like *Battle for the Environment* or *The Environmental Revolution*. Governments and traditional professional organisations followed in the wake of this movement, galvanised by voluntary pressure groups and international events such as the Stockholm Environment Conference of 1972 (see Chapter 11).[5]

In the built environment, the outcome of these agitations was a complex new world-outlook that rejected the massive, state-led redevelopment schemes underway everywhere, and questioned the authority of established professional groups vis-à-vis users or local communities. Everywhere, fiercely polemical publications and campaigns flared up, stemming partly from the Pugin polemical tradition as developed by critics such as Ruskin and Schultze-Naumburg, but vastly expanded in scope and number. In France, since the late 1940s, Henri Lefebvre had indicted the supposed corruption and devaluation of *la vie quotidienne* by capitalist consumerism. In the anglophone countries, a bridge to the new anti-modernist militancy was provided by the writings of Jane Jacobs in America and by the prolific publications and broadcasts of Ian Nairn in England, extending his 20-year-old 'Outrage' campaign to the everyday environments of the welfare state. In 1975 Nairn blasted public housing as a 'loveless landscape . . . built without love, they have inspired none in their occupation'. Prior to Jacobs, America had seen some early community campaigns against urban redevelopment and in favour of densely 'mixed' quarters, including agitation in 1956–9 in New York City against the prestigious Lincoln Center urban renewal scheme, but the tide began to really turn with Jacobs's outspoken campaign from 1961 against Robert Moses's West Village redevelopment project.[6]

For the Conservation Movement, this new and passionate ethos represented a major opportunity, but one which implied a change in its own practices, away from dominance by elite experts and towards greater social inclusivity. Each country developed its own variant on this theme. In some places, such as France or West Germany, a somewhat Marxist perspective was common, condemning postwar environments as the alienating offspring of capitalist developers and bankers – while also, paradoxically, heaping praise on the ornate eclecticism built by the capitalists of the 19th century. Wolf Jobst Siedler and Elis Niggemeyer's 1964 picture-book, *Die Gemordete Stadt*, turned the *Heimatschutz* and *Entstückung* framework on its head, praising the previously condemned *Gründerzeit* plaster ornamentation and contrasting photographs of lively children's play in tenement courtyards with bleak scenes from modern redevelopments that 'look as if an atom bomb has exploded there'. And Alexander Mitscherlich's 1965 book, *Die Unwirtlichkeit unserer Städte* ('The Inhospitability of our Cities'), argued that the functional separation of Modernist planning had created faceless anonymity.[7]

In the proselytising of conservation, angrily worded protest publications in the SPAB tradition played a central legitimising role. Some were generalised, others were profession-specific, against 'the planners', or 'the [modern] architects'. Anti-modernist critic Gavin Stamp wrote that 'the juggernaut, the Terror of comprehensive redevelopment inflicted on every city is being challenged'.[8] Outrage-texts ranged widely, synthesising architectural and environmental rhetoric and supplying guidance for protest groups: the 1970s across Europe saw a flood of rhetorical titles such as *How to Pull a Town Down*, or *Sind die Dörfer zum Sterben verurteilt?* (*Are our villages condemned to death?*). Colin Amery and Dan Cruickshank's *The Rape of Britain*, hailed by John Betjeman as 'a devastating book', raged against 'this vicious attack on the nation's cities', and argued that wartime destruction 'pales into insignificance alongside the licensed vandalism of the years 1950–1975'. Some environmentally slanted critiques focused on the relationship between heritage and energy resources, which now had to be safeguarded: *Architectural Review* editor Peter Davey simply stated, 'Old buildings are irreplaceable, like oil.'[9]

These generalised onslaughts were complemented by local outbursts, railing against the menace of redevelopment in individual towns: for example, *Marbach im Abbruch*, in 1972 (which argued that an *Altstadt* that had survived World War II had now been 'systematically annihilated'), *Goodbye London*, in 1973, or the self-explanatory *Die Zweite Zerstörung Münchens* (1978). In West Germany, these polemics coexisted uneasily with continuing celebrations of reconstruction modernity, such as the much-republished Swiridoff photo-books, *Porträts aus dem geistigen Deutschland*. Some causes célèbres achieved an international notoriety that helped legitimise local responses elsewhere – for instance, the widespread protests against New York's World Trade Center towers, denounced in the *Architectural Review* as 'the latest and most terrifying stage of modern

(a)

(b)

Figure 9.1 The downfall of Modernism

(a) Cartoon by Louis Hellman in the *Architects' Journal*, 19 February 1975; (b) Front cover of Wolf Jobst Siedler and E Niggemeyer's 1964 polemical picture-book, *Die Gemordete Stadt*

redevelopment'. Within England, the most influential single outrage-text was Adam Fergusson's *The Sack of Bath* (1973), culmination of a torrent of media campaigning about threats to that city, including an *Architectural Review* special issue of May 1973 ('Bath – a City in Extremis'). Complete with introductory poem by Betjeman, Fergusson's book hit a raw public nerve and went into a second edition within months. Its concern was not with the great set pieces like Royal Crescent. They were now safe, but that very fact was turned into an accusation. Echoing Giovannoni, Fergusson inveighed against losses of 'minor architecture' or 'artisan Bath' – demolitions of ordinary classical streets that left the masterpieces 'like mountains without foothills, like old masters without frames'. Partly, of course, these polemics reflected the increasing *success* of the Conservation Movement: Bath's chief planner, Roy Worskett, argued

Figure 9.2
The Sack of Bath:
front cover of Adam
Fergusson's 1973
polemical book

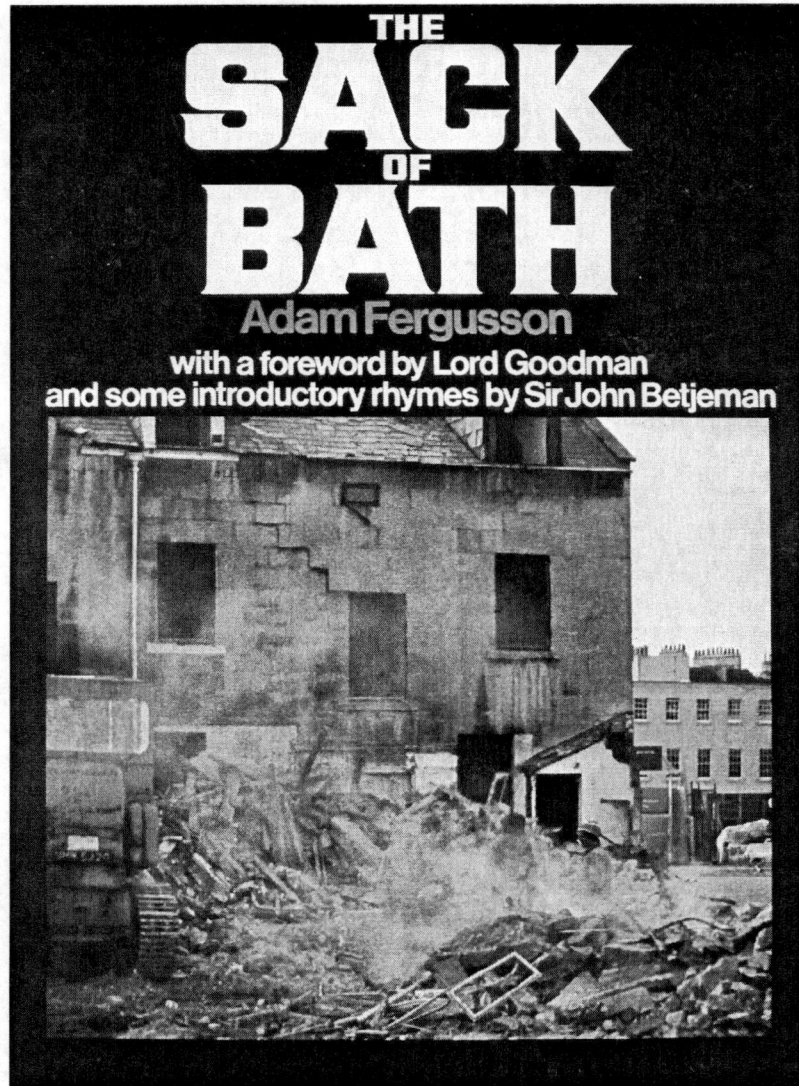

that 'once you got "over the hump", the arguments got hotter and hotter, about smaller and smaller things'.[10]

The emphasis on local campaigning and polemic was accentuated by the way that, within new architecture, the universal forms of Modernism were being rejected and replaced by a new proliferation of regional or vernacular expressions – all part of the vast visual and ideological eclecticism of the 1970s, and following the Schumacher principle of 'think locally and act globally'. In England, for example, there was a surge of interest in brick neo-vernacular styles, exemplified in housing publications by the 1973 Essex Design Guide and extended more uneasily to monumental public buildings in RMJM's Hillingdon Civic Centre, 1973–6.[11] Neo-vernacular was supported by a new phase of the Puginian polemical tradition, led by writers such as the young architectural journalist Nicholas Taylor. Following a special *Architectural Review* issue in November 1967 in which he roundly condemned Modernist housing, Taylor correspondingly eulogised the 19th-century English suburb in his 1973 book, *The Village in the City*. Downplaying the urban heritage that excited the modernists, he argued that

> within these quiet suburban villages there is a way of life which, in its freedom, diversity and individuality – the ability of ordinary families to do their own thing – is more sophisticated than either the Tuscan piazza or the Oxford quad.[12]

As the polemic grew more and more heated, many of the Grand Old Men of modern architecture and planning began to modify their views and adopt conservation values. Ex-RIBA president Sir Robert Matthew became a vociferous environmentalist and 'human-ecology' campaigner in the early 1970s and, in 1972, Colin Buchanan acknowledged that planning had erred in trying to 'dictate the style of life to people'.[13]

Conservation and participation: temporary bedfellows?

THE torrent of locally based protest rhetoric, although often appearing purely negative and destabilising in character, also contained the seeds of a restabilisation of the Conservation Movement. For a short time, in the late 1960s and '70s, radical heritage in Western cities had been a significantly disruptive force, especially through its links to the near-anarchic counter-culture of those years. But architecture, with its intrinsic dependence on the patronage of Authority, can never be truly avant-garde or radical, and the urge towards stability soon asserted itself – at first still in a generally socialist form, but soon assuming a politically right-wing rather than left-wing slant.

The initial 1960s–70s' move towards radical democratisation of the Conservation Movement was bound up with a revulsion against professionalism and experts in general. At an elevated intellectual level, the advances of structuralism and semiotics during those years had begun the process of deconstructing the postwar grand narratives of social progress. Now, in place of rational management, people talked of 'routinisation of bias'; in place of scientific progress there were 'technological shortcuts to social change'; in place of 'objective public interest in planning', there was 'struggle for resources and power'. Even as the fashion for corporate management reached its height, as in the 1969 Redcliffe-Maud Report on English local government, the legitimacy of professions such as architects and planners was increasingly challenged, whether through full-frontal attacks, such as the polemic of Ivan Illich, or indirectly, through Trojan-horse transformation of welfare state processes from within.[14]

The radical fringe of built-environment debate was occupied by a miscellany of groups. In the USA, for example, radical writers in the early 1970s condemned establishment-dominated 'advocacy planning' and instead argued for a 'guerrilla design process'. In England, 'A.R.s.e.' (Architects for a Really Socialist Environment) devoted themselves to organising demos, squats, self-help and anti-landlord agit-prop. By the early '70s, there were said to be several thousand squatters in Inner London and as many as 100 in even a smaller historic city such as Norwich. Even the established professional structures of architecture became for a time virtually mouthpieces of environmentalist and conservationist propaganda: the 1972 RIBA conference at Lancaster was dedicated to an environmentalist agenda of 'Designing for Survival', with the President, Alex Gordon, passionately proclaiming the necessity of 'long life, loose fit, low energy'.[15]

More immediately significant, and easily appropriated by the Conservation Movement, was the 'user-participation' movement, a doctrine of involvement of ordinary citizens in built-environment decision-making, whose roots stretched back to the prewar days of Mass Observation – or even, indeed, to the mass protests and direct action of the commons preservation movement of the 1860s. As early as 1951, it was argued that conservation was one of the few areas of the built environment that consistently engaged the interest of 'lay people', and the researches of Young and Wilmott cited the strength of the traditional community patterns of Bethnal Green as evidence against uprooting slum-dwellers in redevelopment schemes. The movement was encouraged by the spread of popular media representations of present-day life in ordinary 19th-century environments, such as 'Coronation Street'. Within elite architectural discourse, the Smithsons' writings on 'patterns of association' and Rudofsky's *Architecture without Architects* (1964) both challenged the hegemony of elite architectural design.[16]

In the emergence of the architectural participation movement, the two key issues were first, how literally the participation process should be reflected in architectural forms and second, whether it should be seen ultimately as an opponent or ally of the ruling system. In the first case, the early '70s saw attempts by architects ranging from Lucien Kroll, in Belgium, to Ralph Erskine, in the Byker redevelopment in Newcastle, to enlist users into the design process and to express the results literally in an anarchic building style. An *Architects' Journal* writer chided in 1979 that 'whereas participation should normally be about uncovering conflict, it has been used at Byker to prevent conflict'. In 1972, Italian architect Giancarlo de Carlo condemned these highly aestheticised formulae of participation: what was important was the process, which should be non-linear and dominated by 'disorder'. In the USA and Britain, anarchists/architects Robert Goodman and Colin Ward also proselytised generalised counter-culture views of participatory urban life.[17]

Concerning the issue of participation's relationship to 'the system', the picture was also mixed. The 1950s and 1960s in several countries had witnessed pioneering attempts to encourage user-participation within the Modernist structure of technocratic urban renewal – notably in Philadelphia, by city planner Ed Bacon, and in France, where the building of massive suburban housing complexes such as Sarcelles was accompanied from 1966 by the formation of articulate local residents' councils. But these efforts were soon left behind by more radical critics, who criticised them as insufficiently detached from establishment interests. The 1969 Skeffington Report on public participation in planning in Britain provoked similar accusations of establishment manipulation. Others argued in the opposite direction: Tony Aldous's *Battle for the Environment* (1972) claimed that, although some people assumed participation was a matter of 'left wing residents fighting helmeted policemen with water cannons', such conflicts were a sign of failure. And the establishment Modernist architect Sir Hugh Casson argued that participation

> can be babyish. . . . If you want a pot moved, you say to Fred, "Fred, let's have a cup of tea and think about that pot". Fred knows you want the pot moved, and you know he knows. It's a charade. Participation weakens the discipline of taking responsibility.[18]

Despite the uncertainties over these issues of principle, the effect on the ground was a veritable explosion of individual campaigns and efforts, varying between different countries in intensity and character. In some places, such as France or in southern Europe, there was a tradition of relatively low public activism in the built environment – although, as we will see below, the communist-controlled city of Bologna was able to devise a potent recipe in the early '70s that combined municipal planning initiative with effective local participation in urban area conservation.[19] In other places, such as West Germany, conservation protest movements were deeply bound up with wider class-political radicalism, echoing at one remove the violent anarchist and revolutionary groupings of those years, such as the Red Army Faction. The many German urban protest movements from the mid-1960s were sharply polarised between right and left approaches – between top-down, establishment efforts at conflict management

Figure 9.3

(a), (b), (c), (d) Illustrations from the 1969 Skeffington Report on public participation in planning; (c) and (d) show a hypothetical townscape improvement scheme

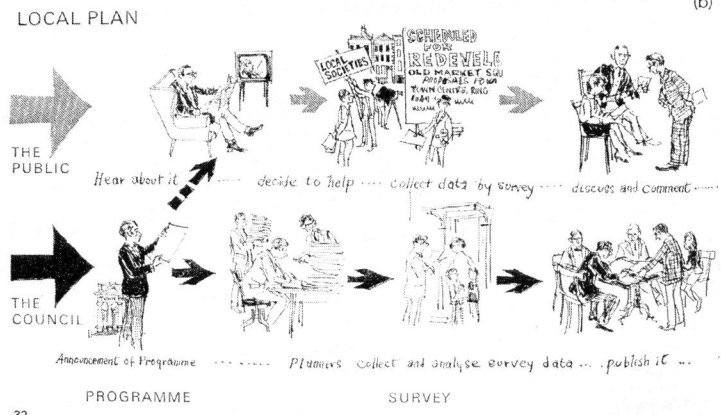

327

(*Bürgerbeteiligung*) and bottom-up, critical movements (*Bürgerinitiativen*). The latter sprang up in a range of major cities threatened by redevelopments. Sometimes, the threat stemmed from mass-housing redevelopment, just as in Britain: for example, in the Lehel area of Munich, where 1965 proposals for clearance of *Gründerzeit* tenements spurred a vigorous protest movement. More often, the provocation stemmed from road schemes or capitalistic developments, such as a proposed office scheme of 1969 in Eppsteinerstrasse, Frankfurt-Westend, which inspired Germany's first organised squat and in turn provoked establishment attacks on the city's 'Wild West'. In the federal capital, Bonn, 1972–3 saw no less than eight *Bürgerinitiativen* emerge, mostly opposing the redevelopment of 19th-century tenements in Poppelsdorfer Allee by insurance company Deutscher Herold: by 1974 there were huge protests, backed on party-political lines by the social-democrat SPD, and in 1976 author Roland Günter produced a *Handbuch für Bürgerinitiativen*, setting out ground rules for successful confrontation and campaigning.[20]

In the anglophone countries, with their vociferous built-environment pressure groups and journalists, revolutionary socialism was much less prominent in urban protest and participation movements. With outstanding exceptions such as the violent 1969 confrontations over the 'People's Park' in Berkeley, California, protesters were usually concerned less with political radicalism than with environmental improvements in their own right – causes in which the Conservation Movement inevitably played a key role. Equivalent English texts to Günter's, such as Charles McKean's *Fight Blight* (1977), were concerned chiefly with combating urban decay through participatory improvement initiatives. There were fewer German-style occupation protests against property developers – a notable exception being the 17-year controversy (1959–76) over the Tolmers Square redevelopment in Euston, central London, in which local working-class residents and journalists collaborated to frustrate successive redevelopment projects for a grimy pocket of terraced houses ringing a derelict cinema.[21]

Within heritage circles in Britain, the absence of a Franco-German-style split between lay public and government experts encouraged a more integrated civic climate of participatory debate and engagement. This broadly constructive relationship was already evident by the mid-1960s in the successful protest campaigns in many cities, such as the 1966–8 campaign in Edinburgh against municipal plans for a six-lane ring motorway, defeated by an alliance of civic organisations with establishment links (notably the Cockburn Association). An ambitious scheme by the Greater London Council for three 'Ringways' round London followed it into oblivion in 1974.[22] Around 1970, a typical outcome was a sharp reduction in the scale of reconstruction. For example, in Newcastle, a decade of controversy over proposed redevelopment of the Georgian Eldon Square ended in a significant scaling-down in 1970, from a towering Arne Jacobsen hotel to a medium-rise megastructural complex, approved in 1972 and built in 1973–6; critic Colin Amery grudgingly agreed that the architects, Chapman Taylor Partners, 'had to tame a monster in the city, and they have done so with consummate skill'. The same years saw the clear-cut defeat by protestors of a 1971 proposal for a massive hotel development in Clifton, Bristol: the victory of STAG (Save the Clifton Gorge) was celebrated in a 1980 book, *The Fight for Bristol*. And in 1974–5, a combination of local protesters and economic crisis forced the abandonment of a proposal by the Hammerson property company to demolish the historic market place of the town of Chesterfield for a megastructural shopping development. The coup-de-grace was delivered by high court writ; in 1975, the borough council instead commissioned Feilden & Mawson to undertake a conservation study, and the following year the town centre was designated a conservation area. In his 1982 conservation textbook, Feilden repudiated prima-donna design individualism, while still claiming a central coordinating role for the architect: 'In executing a conservation project, the architect has a role similar to that of the conductor of an orchestra.'[23]

Conservation campaigners faced a more complicated task in preserving an entire city centre or urban quarter. Here, too, in Britain, an incremental, collaborative relationship prevailed: the turn-around from redevelopment to conservation was either gradual or sudden, depending on local circumstances, and the administrative machinery of comprehensive renewal was taken over by conservationists by stealth or by a sudden coup. In the key battleground city of Bath, the city council remained in confrontation with heritage groups over the Buchanan plan until 1975, when, following the publication of Fergusson's *Sack of Bath*, the pro-development City Architect and Planning Officer, Howard Stutchbury, was replaced by former Ministry conservation adviser Roy Worskett. During Worskett's four years in the post, until 1979, a local policy revolution ensued and the city became a 1975 set piece of European Architectural Heritage Year (see Chapter 11); a comprehensive conservation study was prepared in 1976–8.[24] Very often, too, initial campaigns by radicals and protesters were followed, after a period of years, by tourism-led gentrification of the areas rescued.

The foremost exemplar of the constructive-engagement approach to participatory conservation in 1970s Britain was the highly publicised struggle over Covent Garden in central London, following the mid-1960s' decision to relocate its fruit market and the publication of a 1970 local-authority plan for comprehensive redevelopment of the run-down area with megastructures and urban motorways. After intensive lobbying by a coalition of local residents and academic historian-activists, and the holding of a public inquiry in 1971, a decisive point was reached in 1973 when the Conservative Environment Minister, lawyer Geoffrey Rippon, approved the redevelopment in a way that 'effectively sabotaged' it, by also listing 245 buildings in the area. The list was prepared secretly in the basement of the *Architectural Review* office by two architectural journalists, Colin Amery and Dan Cruickshank, who passed it to a civil servant in Rippon's department; it was adopted in its entirety by Rippon following an after-dinner-party tour of Covent Garden by Lady Dufferin (Lindy Guinness), a flamboyant patron of the arts. From that point, the battle for Covent Garden was won: a conservation area was declared, and a new, heritage-orientated plan was drawn up, inspired partly by the sketches of Kenneth Browne, 'townscape editor' of the *Architectural Review*; in 1980, the refurbished market buildings were opened as a cultural and tourist complex.[25] But the gentrified result was curiously similar to the Marais, even though the Covent Garden result had been achieved by very different means, through constructive campaigning by national and local activists, as with the Georgian Group and the Bute Lists in the 1930s.

Towards community-based housing improvement: the late 1960s and early '70s

INTERNATIONALLY speaking, the principal area of conservation where protest shaded into concerted stabilising action, especially at a community level, was housing rehabilitation, or 'rehab'. Just as new mass-housing blocks in place of shabby 19th-century slums had been an internationally shared recipe of postwar Modernism, so now the rejection of these blocks and revitalisation of 19th-century housing swept into universal favour within the Conservation Movement in Western countries. In economic terms, rehab suddenly seemed much cheaper than new housing: by around 1970, similar rehab programmes were springing up almost everywhere, not just in Europe or

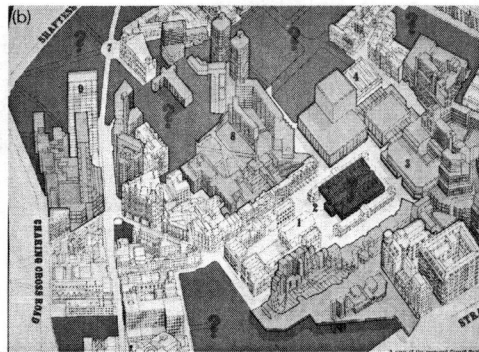

Figure 9.4 1970s' area demolition controversies

(a) Poppelsdorfer Allee, Bonn: this developer's publicity leaflet emphasises the project's involvement of local citizens; (b) Covent Garden, London: the megastructural redevelopment plan of the Greater London Council, reproduced in the July 1972 *Architectural Review*; (c) Perspective by Kenneth Browne of the conservationist counter-proposal for Covent Garden – in the event, substantially implemented; (d) Trefann Court, Queen Street East, Toronto, scene of the defeat of the city council's urban renewal strategy in 1966–72 by a local coalition (aided by Jane Jacobs) and implementation of a selective infill formula instead; (e) The Milton-Parc district of Montreal, scene of a similar decisive battle by residents and protest groups (notably Save Montreal, co-founded 1974 by Michael Fish and Heritage Montreal, founded 1975 by Phyllis Lambert); this view shows 'old' and 'new' in Rue Hutchison

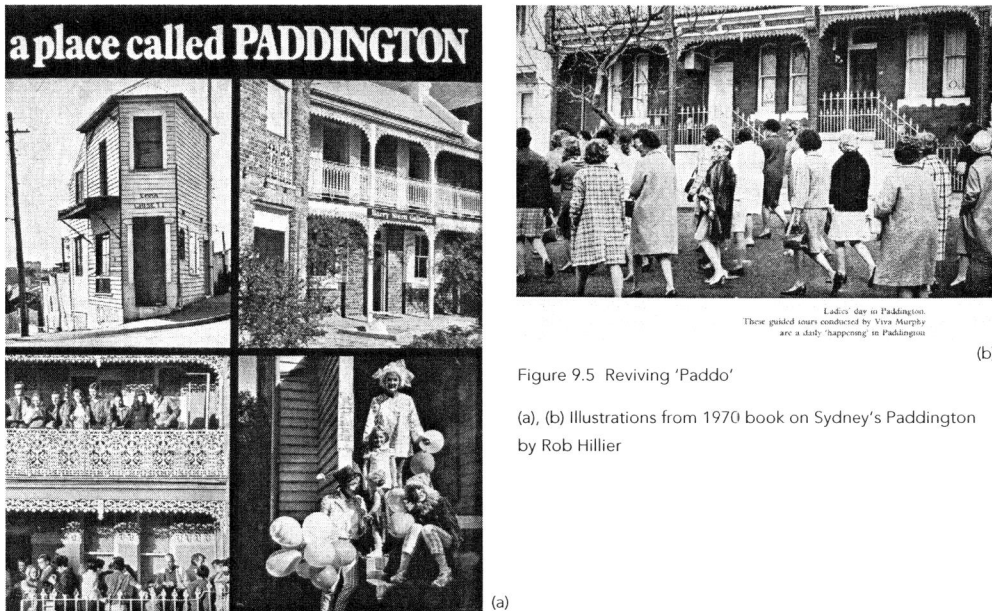

Figure 9.5 Reviving 'Paddo'

(a), (b) Illustrations from 1970 book on Sydney's Paddington by Rob Hillier

America. In Sydney, Australia, the late-19th-century terrace-house suburb of Paddington witnessed an early example of resident-led area rehab and bohemian gentrification in the shadow of redevelopment threats, starting as early as 1956–8 (when the area's first antique shop and art gallery opened) and culminating in the mid- and late 1960s: in 1967, one commentator wrote that 'a lot of people have their hopes pinned on Paddo, or rather the Paddo that is about to be when the present wage-earning tenants have largely moved out . . . and the new owners have completed renovations'.[26]

Within Europe, large-scale area rehab was often still seen as complementary to modernist housing construction. In Amsterdam, the gargantuan Bijlmermeer development of honeycomb-plan 11-storey deck-blocks formed half of an early 1970s' municipal planning policy that also embraced pedestrianisation in the historic core and rehabilitation of the decayed Jordaan area (laid out in the 17th century) on its western fringe. As early as 1955, the Jordaan had been proposed for *rehabilitatie* as opposed to *sanering*, and several pilot schemes were started in the late '60s (notably the Claes Claesz Hofje project of 1968–73, a courtyard-clearance scheme of Lilliputian intricacy, sponsored appropriately by the Madurodam 'miniature city' theme-park in Scheveningen). Large-scale rehab in the Jordaan finally began in the early '70s and in 1975 the area became a demonstration project for European Architectural Heritage Year (see Chapter 11). Other rehab schemes of the period, such as Oude Westen in Rotterdam, involved consultant architects collaborating with local communities. The years 1968–71, in the Netherlands, saw housing policy shift dramatically from a 3:2 bias in favour of demolition (375,000 for demolition, 250,000 for rehabilitation) to a rehab-dominated policy.[27]

France and West Germany saw similar policy shifts. France's sixth national housing plan, in 1970, redirected subsidies to rehab of old neighbourhoods, and a 1971 federal urban-renewal and town-planning act in West Germany authorised participatory 'social plans' for neighbourhood renewals and large-scale housing improvement aided by subsidised loans: by 1972, it was clear that *Sanierung* now meant not demolition but conservation. In Munich, preparations for the 1972

331

Figure 9.6 Amsterdam infills

(a), (b) Claes Claesz Hofje courtyard restoration scheme, Amsterdam (1968–73); the commemorative plaque highlights the sponsorship role played by the Madurodam theme-park; (c) Early 1970s' infill in the Jordaan area rehab project, Amsterdam

Olympic Games boosted rehabilitation of *Gründerzeit* tenements to about 300 or 400 schemes per annum in the early '70s, in a programme developed by the city's planning department (as also in West Berlin). Although most German rehab concerned *Gründerzeit* tenements, innovative cities such as Hamburg also began mass upgrading of 20th-century social housing. Another prominent early 1970s' effort to preserve 19th-century planned workers' communities, orchestrated by Roland Günter, was the successful '*Rettet Eisenheim*' tenant campaign of 1972–3 to save a pioneering 1844 mining settlement in Oberhausen, threatened with demolition since the 1950s: one of the first *Bürgerinitiativen* in the Ruhr.[28]

Within this great wave of housing rehabilitation across Europe, the programme in Britain was one of the most vigorous – unsurprisingly, given the unparalleled scale of urban redevelopment in the country, and the mounting public disquiet it had provoked. This discontent was harnessed to direct political action in the English local government elections of 1968, when many cities passed from Labour to Conservative control on the basis of pledges to curb urban renewal and multi-storey flat-building. Unlike the listing programme,

Rettet Eisenheim

Die Kommunikation innerhalb der
Siedlung geht über ein übliches
Nachbarschaftsverhältnis hinaus.
Es ist zur Selbstverständlichkeit
geworden, daß man sich hilft:
im Krankheitsfall
bei der Beaufsichtigung
der Kinder
im Haushalt
bei Durchführung not-
wendiger Reparaturen

housing rehab followed markedly different paths in England (and Wales) and Scotland, owing partly to the sharp differences between standard 19th-century housing types – terraced houses and tenements respectively. Northern Ireland, racked from 1969 by mounting political-sectarian violence, saw even more exaggerated swings against modernist redevelopment, with widespread resistance in the inner-city ghettoes to any invasive population displacements.[29]

In both England and Scotland, the early 1970s witnessed a sharpening focus on tackling sub-standard housing by government grants, combined with elaborate collaborative relationships between local authorities and community

Figure 9.7 Fighting the demolition juggernaut in West Germany

(a) Demolition of *Mietskaserne* tenements in Wedding, West Berlin, around 1970; (b) Eisenheim industrial village, Oberhausen (built 1844), safeguarded from demolition by Roland Günter's community mobilisation campaign from 1972 onwards: the left-hand house in this view, Werrastrasse 1, served as a *Haus der Initiativen* (community initiative centre) from 1974 to 1980; (c), (d) Front cover and community-participation page from 1973 agit-prop booklet, 'Rettet Eisenheim'

groups; but the two countries arrived at this outcome by different routes. In England, housing improvement policy grew organically out of internal debates within the Crossman and Kennet regime in MHLG, especially among the Ministry architects, headed by Whitfield Lewis, who oversaw government housing design research. A 1968 conference convened by Kennet at New Hall,

Cambridge, to investigate whether higher improvement grants could equalise new construction and improvement costs, coincided with publication of a House Condition Survey suggesting that 1.8 rather than 1.1 million houses in England were 'unfit'. With the number of slums apparently rising as the costly new-building programme declined, a boost in improvement seemed imperative. Accordingly, within months, a seminal government report, *Old Houses into New Homes*, proposed a 250 per cent increase in improvement grants and a targeting of both the external environment and internal facilities of sub-standard housing, through 'General Improvement Areas' (GIAs), declared by local authorities rather like conservation areas.[30]

This new English system was, at first glance, spectacularly successful, with over 260 GIAs designated within two years. Crucially, it operated quite separately from conventional slum-clearance. This subtly segregated it from the main thrust of housing improvement need and ensured a strong emphasis on design-led gentrification of the private 19th-century housing stock, creatively exploiting the complexity of public and private spaces in the terraced-house environment. The first GIA to be completed (in 1970), the 358-dwelling Newtown GIA in Exeter, established the theme of aspirational environmental enhancement, with planting and hard landscaping apparently lifted straight from Gordon Cullen's *Townscape*. A self-improving individualism characterised much GIA advocacy literature. For example, a leaflet by the North-Western Gas Board gave home-owners guidance: 'How do I start?' With the aid of gas-fired appliances, 'Yes, your old house can become a new home.' And in the London Borough of Lambeth, borough architect Ted Hollamby, veteran former project architect of the Brandon Estate, boldly proposed a 1,500-acre programme of 50 GIAs: one pioneering Lambeth GIA in Vassall Road included a show house converted into three flats by the council in collaboration with *Ideal Home* magazine and the London Electricity Board, with results trumpeted as 'amazingly chic'. Hollamby's meticulous work in Lambeth was England's closest equivalent to the detailed architectural planning of the Malraux *secteurs sauvegardés*.[31]

In their overall environmental strategies, many GIAs focused more on the problems of affluence than on slum poverty: the 1,500-dwelling Arlington GIA in Norwich, for instance, was designed by chief planning officer A A Wood to curb car traffic and parking clutter through road closures and demolitions to create off-street car parking, all at an estimated cost of only £1 million (as opposed to £3 million for conventional clearance and redevelopment). Plans such as these were achieved through a continuation of traditional local-authority dirigisme: one 1970 reviewer compared Hollamby's Lambeth programme to slum-clearance in its 'authoritarian . . . clean-sweep' approach, including decanting of all tenants during work. And in the event, the Arlington residents in Norwich rebelled successfully against Wood's plans, arguing that a plan calculated to 'produce a landscape free of cars' was unbalanced and unworkable.[32]

The most obvious way to protect the emergent English housing improvement programme from criticisms of elitism was to encourage community participation. Accordingly, in London's Barnsbury, where a residents' association had been fighting planned mass-demolition of classical terraces since 1964, the GLC in 1967 launched a programme of combined rehab and new infill, followed in 1968 by compilation of a 'Barnsbury Environmental Study'. Prepared by a team from various local authorities, this included an exhibition in Islington Town Hall combining Geddes-style conservative-surgery rehab proposals and cautionary sketches of urban motorways and megastructures. More radical were the efforts in Liverpool from 1969 by 'SNAP' (Shelter Neighbourhood Action Project), a project initiated by Liverpool Corporation and Shelter, headed by Des McConaghy. Here a small professional group, including architects, a housing manager and sociologist, helped stimulate the local

Figure 9.8 GIA: catalyst of environmental regeneration in England

(a), (b) Newtown GIA, Exeter: before/after layouts from 1969 master plan, showing landscape improvements and infill of scattered bomb sites (reproduced in Department of the Environment, *Area Improvement Note 5*, 1972); (c) East John Street, Newtown GIA, in 1970 following first phase, as pictured on front cover of North Western Gas Board 1970 booklet on home improvement; (d) Perspective of proposed bridge-link across street (from DOE, *Note 5*); (e) Corresponding view of East John Street in 1991, showing townscape enclosure effect created by bridge-link (built mid-70s)

community to make their own plans, arguing that 'as an essential preliminary to rehabilitation, people must reassert their communal identity'.[33]

Perhaps appropriately, the most influential of all GIA initiatives was located not in a big city but in the historic industrial town of Macclesfield, quaintly picturesque rather than harshly sublime in its townscape. This project, at Black Road, originated in 1971, when a small area of 34 terrace houses, very early in date (begun in 1812), was scheduled by the borough council for demolition. A young architect, Rod Hackney, stepped in and bought one house (222 Black Road) for £1,000, and persuaded the residents to set up a community association and lobby the municipality to designate a GIA. Having established a collaborative relationship with the council, Hackney secured a grant of £250 for each dwelling, converted his own house into a project office and masterminded the initial GIA (opened by the Mayor of Macclesfield in November 1974) and a follow-up phase, completed in 1979.[34] Black Road seemed like a microcosm of the English GIA movement, in its sylvan intricacy and its focus on environmental improvement. Indeed, the GIA programme overlapped significantly with the conservation area movement. As Worskett argued, 'Lots of local authorities designated conservation areas, at first, simply to stop housing redevelopment – which was a misuse of the system, I thought!' Many of the councils that most favoured GIA designation, such as Lambeth, were equally precocious in their conservation area programmes, and – hardly coincidentally – were also places where architects played a prominent role in housing and conservation policy.[35]

But where, in all this, was the concern for rectifying poor housing conditions – of rehab as a successor to mass slum-clearance? To be sure, some GIAs in England and Wales still continued older traditions of basic, short-life 'patching', especially the continuing improvement programme of back-to-back houses in Leeds, driven not by architects but by the Chief Public Health Inspector – or the late 1960s' rehabilitation drive in the Rhondda Valley, south Wales, intended to prolong the lives of terraced houses for only 30 years. By 1973, however, just 24 per cent of the 210,000 houses in GIAs had received grants for rectification of sub-standard amenities.[36]

In Scotland, a very different government strategy emphasised the basic approach almost exclusively. Here, policy developed from an authoritative report of 1967, *Scotland's Older Houses*

Figure 9.9

(a), (b) Macclesfield, Black Road No 1 GIA (1971–4)

(the Cullingworth Report), which advocated a vast acceleration both of slum-clearance (to 466,000 demolitions in 15 years – 25 per cent higher than the target for the Netherlands) and of temporary patching. All the focus was on interior improvements, with tenements' austere, masonry street facades and rear courtyards left little touched. The 1969 Housing (Scotland) Act introduced a new mechanism, the Housing Treatment Area (HTA), which, unlike GIAs in England, combined both patching and clearance and perpetuated the dirigiste municipal ethos of new public housing. Saddled with this inflexible organisational framework, the HTA programme was plagued by delays, especially after a devastating storm in January 1968 inflicted widespread damage on tenements across Clydeside, with many roofs damaged by falling chimney-stacks. By 1974, only 140 HTAs, covering some 8,000 houses, had been declared, a tiny amount compared with the 210,000 dwellings in GIAs – the reverse of the higher Scottish level of new council house-building. Glasgow's two prototype schemes, at Oatlands and Old Swan, had proved painfully slow and crudely inflexible in their use of compulsory purchase powers.[37]

The remedy for this Scottish rehab quandary came, as at Black Road, from community participation and enabling architects, but in a very different form, orientated towards the monumental environment of tenement Glasgow and benefiting from close research links with the architectural department of Strathclyde University. There, lecturer Jim Johnson orchestrated a remarkable series of student initiatives from 1969. One graduate student, Peter Robinson, rented a one-room flat in Springburn for a month and built a mock-up of a potential £680 modernisation scheme, including installation of a separate WC and running water in a 'box-bed' space. Other students carried out a survey of tenements commissioned by Scottish government research architect David Whitham. In 1970, one of Johnson's final-year students, Raymond Young, went a stage further, by launching a community housing scheme in Taransay Street, Govan, facilitated by a government research grant. Young exploited his initial initiative, the 'Tenement Improvement Project', to persuade Glasgow City Council to designate Taransay Street a HTA, and in 1971–2 the project was reorganised as a university research centre, named 'ASSIST', financed by grants from Whitham's department and philanthropic trusts. A show house was completed in 1972 to persuade local residents to choose improvement rather than compulsory purchase and demolition, and an ASSIST community shop in Govan Road was opened by Glasgow's Lord Provost in November 1972: the new Central Govan Housing Association was the first community-based housing association in Britain. Unlike Hackney in Macclesfield, Young initially had to circumvent considerable hostility within the municipality. The result, however, was an effective formula of area rehab without gentrification.[38]

By around 1972–3, it was becoming clear throughout Britain that the current improvement policy was too divided and polarised to be effective. Although the numbers of private improvement grants in England were soaring, from 116,000 in 1970 to 260,000 in 1973, government surveys suggested that the GIA programme had avoided the worst houses and encouraged gentrification. Ultimately, what seemed to be needed throughout Britain was a policy that combined slum-improvement and environmental enhancement. In response, a new system, the Housing Action Area (HAA), armed with much higher subsidies, was introduced in 1974. It was implemented largely by Govan-style community housing associations, overseen by a Britain-wide body, the Housing Corporation, headed by Arnold Goodman. In response, the proportion of total rehab carried out by housing associations soared (in Scotland, from 7 per cent in 1971 to 81 per cent in 1979). Simultaneously, to deal with decayed twilight areas, a new policy of 'gradual renewal' was announced

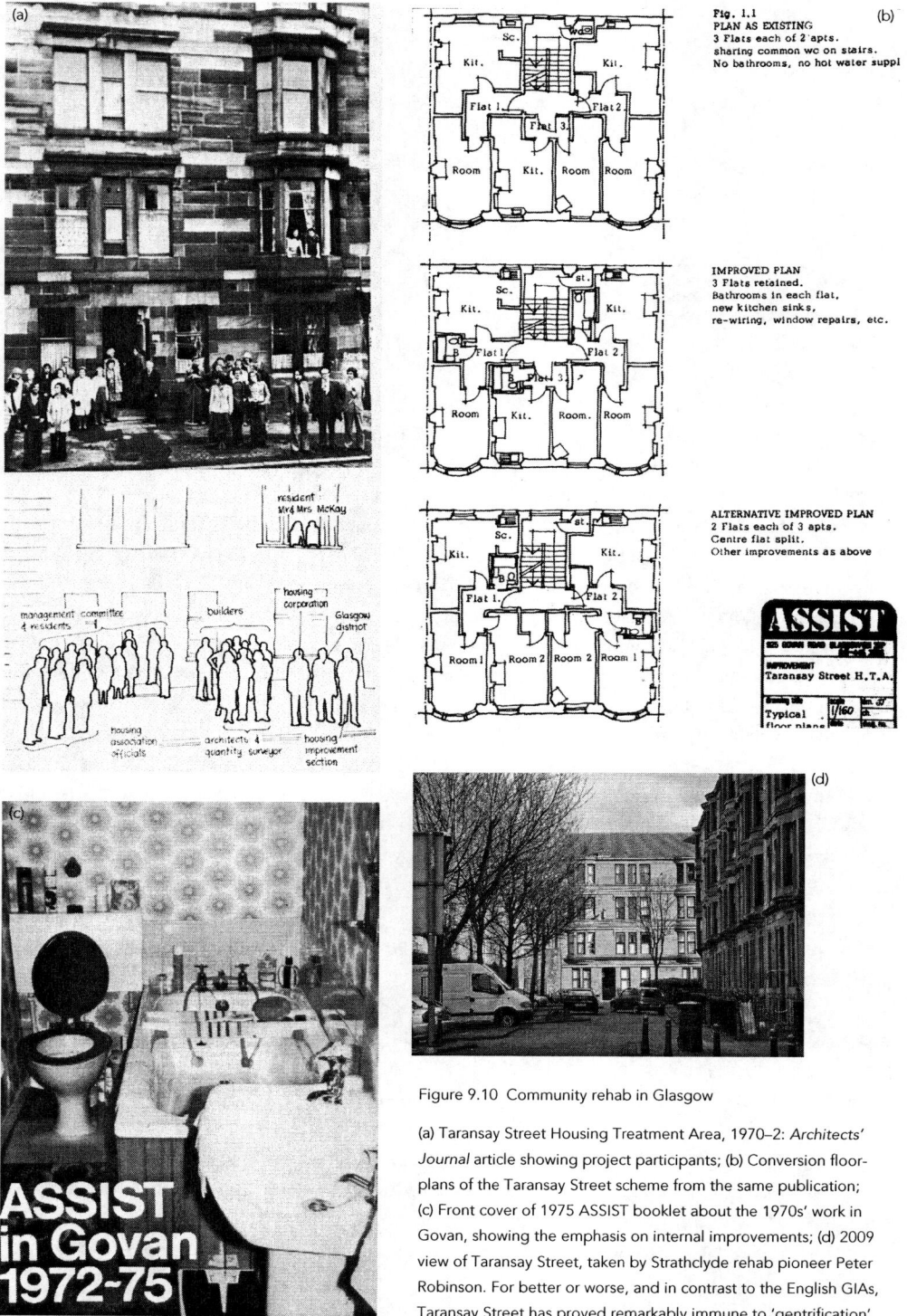

Figure 9.10 Community rehab in Glasgow

(a) Taransay Street Housing Treatment Area, 1970–2: *Architects'*
Journal article showing project participants; (b) Conversion floor-
plans of the Taransay Street scheme from the same publication;
(c) Front cover of 1975 ASSIST booklet about the 1970s' work in
Govan, showing the emphasis on internal improvements; (d) 2009
view of Taransay Street, taken by Strathclyde rehab pioneer Peter
Robinson. For better or worse, and in contrast to the English GIAs,
Taransay Street has proved remarkably immune to 'gentrification'

by the government in 1975 – a Geddes-like concept of incremental rather than comprehensive improvement, whose first showpiece scheme was the inner Oxford area of Jericho.[39] This programme was massively augmented in 1976 in England by an inner-city regeneration initiative devised by environment minister Peter Shore, and in Glasgow by a huge urban-renewal project, GEAR (Glasgow Eastern Area Renewal). Here, and in the research programmes of SLASH (Scottish Local Authorities' Special Housing Group), the focus of concern began shifting away from 19th-century housing altogether and towards decayed 20th-century public-housing schemes. Overall, however, the burden of proof and argumentation had now decisively shifted from the demolishers to the preservers. As late as 1970, journalists had criticised Leeds's ambitious rehab programme as 'indiscriminate', but by 1981 heritage campaigner Ken Powell could argue that 'the marvellous character of the surviving townscape of Leeds demands to be recognised and conserved for its inherent quality'.[40]

The years of conservation triumph: late 1970s' urban renewal in Europe and America

IN its concern for renewal of twilight areas, housing improvement policy in Britain overlapped significantly with the wider international narrative of urban conservation in the mid-/late-1970s, which was a time of consolidation and expansion of scope after the upheavals of the previous few years. Within France, dissatisfaction with the slowness and social/architectural elitism of the Malraux policy had provoked successive initiatives of radical expansion, combining housing and environmental improvement in the GIA manner, although still at a much smaller national scale. First, in October 1974, came a 'Hundred Town programme' of diluted *secteurs sauvegardés*, followed by the 1975/6 report of the Simon Nora Commission. This inspired the 1976 establishment of a *Fonds de l'amenagement urbain* (FAU – a national coordination and information service for urban improvement) and a programme of *opérations programmées d'amélioration de l'habitat* (OPAH) from 1977, emphasising community participation and avoiding population displacement. The OPAH programme, in effect, replaced the *îlots operationnels* of the Malraux law, the last of which was designated in the Petite Venise district of Colmar in 1975. By 1980, some 500 OPAH projects had been begun.[41]

Across Western Europe, the same picture of overlap, consolidation and expansion applied throughout the urban Conservation Movement. One of the most lively centres of late 1970s and 1980s' conservation and renewal was West Berlin. There, surviving 19th-century tenement districts, notably 'Kreuzberg SO36', became hotbeds of cosmopolitan counter-culture, complete with state-aided community bureaux and self-help rehab groups: a Western European equivalent to the inner-city tenement sociability praised by Jane Jacobs in New York. West Berlin acted as a bridge from 1970s' rehab to the next evolutionary phase of the urban Conservation Movement, the postmodern era of 'intervention'. The main vehicle for this was the *Internationale Bauausstellung Berlin* (IBA), an umbrella programme for regenerations of the city's still ragged tissue, prepared over a long period from 1979 and culminating in 1984–7. Uniquely, this programme was divided into parallel 'new' and 'old' strands, the *IBA-Neu* directed by Prof. Josef Paul Kleihues, and *IBA-Alt* by renowned conservation architect Prof. Hardt-Walther Hämer. Hämer already had to his credit a string of pioneering community-participation tenement rehabilitations, including a 1968–74 project at

Putbusser Strasse 29–31 in Berlin-Wedding and a more comprehensive rehab scheme of 1974–80 for an entire street-block (Block 118) of over 400 flats in Berlin-Charlottenburg: here, old-style radical *Entkernung* of the central courtyard was avoided in favour of a more complex pattern, with extensive tenant participation. Interestingly, as we will see in Chapter 10, these projects were influenced in some respects by some pioneering rehab projects just across the Wall, in East Berlin. From 1979, Hämer applied his philosophy of '*behutsame Stadterneuerung*' (cautious renewal), with its supporting 'Twelve Principles', across the whole of West Berlin through the *IBA-Alt* programme. This was an updated version of the time-honoured conservative-surgery approach, applied now to the 19th-century environment and linked, as at Charlottenburg Block 118, to systematic user-participation.[42]

The same pattern of rapid citywide spread of area-conservation initiatives in the 1970s applied across Western Europe. In Rotterdam, a newly elected socialist town council in 1974 designated 11

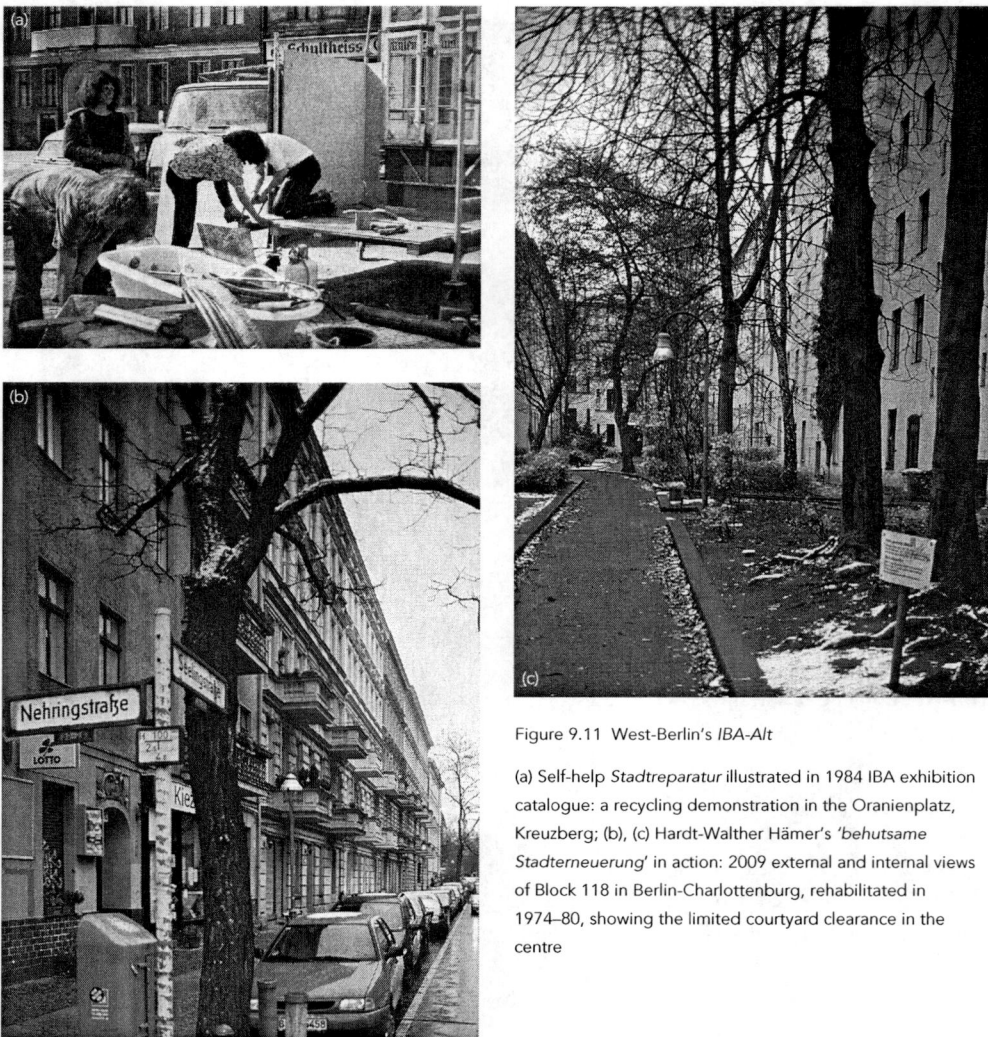

Figure 9.11 West-Berlin's *IBA-Alt*

(a) Self-help *Stadtreparatur* illustrated in 1984 IBA exhibition catalogue: a recycling demonstration in the Oranienplatz, Kreuzberg; (b), (c) Hardt-Walther Hämer's '*behutsame Stadterneuerung*' in action: 2009 external and internal views of Block 118 in Berlin-Charlottenburg, rehabilitated in 1974–80, showing the limited courtyard clearance in the centre

housing improvement areas, with the aim of boosting overall urban renewal. And in Austria, an extensive apparatus of area protection developed alongside the growing stock of some 40,000 individual protected monuments, which included by default all historic buildings in public ownership.[43] The Federal Monument Office developed the unique mechanism of an 'Atlas of Historical Zones of Protection', a non-statutory inventory that coded areas by degree of significance. In 1972 the City of Vienna amended its building law to permit multiple zones of protection: within a decade over 60 zones, covering the entire inner city, had been designated. In the inner western suburb of Ottakring, the first area-rehabilitation projects commenced. There, the city's first community area trust was established in 1974 in a converted bus and a local pub and began a conservative surgery-style renewal of several adjacent street-blocks, including new municipal housing in time-honoured Red Vienna fashion. Vienna's ambitious programme was partly financed by the proceeds of a so-called 'cultural schilling law' – a 10 per cent tax on radio and TV fees.[44] Only the most laissez-faire countries stood aside from this expansion of area conservation: in Belgium, for instance, the old town of Bruges almost entirely lacked statutory protection until 1970, and only a strictly limited range of restoration projects were carried out from 1956 by the private Marcus Gerards Foundation; even after the founding of a city department of Historic Monuments and the preparation of a structure plan in 1971–2, only 149 listed buildings and sites were designated over the following 20 years.[45]

In North America, private initiative of course remained very important, and the 1970s saw a proliferation of societies, projects and publications on exemplary schemes. Underpinning the Conservation Movement's voluntary-dominated US activities, however, was a vast programme of government-subsidised and regulated initiatives, spearheaded by the HUD and a range of other public bodies, especially in the later 1970s under the 'New Localism' policy endorsed by President Carter.

Figure 9.12 Ottakring, Vienna: the city's first area rehabilitation zone, commenced in 1974, financed by the 'cultural schilling law'

Following a 1970 conference in Pittsburgh, hosted by the National Trust and the National Advisory Council on Historic Preservation (with findings published in 1971), various initiatives were developed, including financial mechanisms such as revolving funds and participatory schemes to involve low-income or ethnic minority residents. An even greater challenge, just as in Glasgow, was to avoid displacement of residents during area upgrading. In the late 1970s, for example, HUD spent US $100 million over two years on 100 preservation-dominated urban renewal initiatives under the Urban Development Action Grant programme, mostly as subsidies to leverage private development projects, and HUD grants also fuelled bold attempts at large-scale rehab, including Savannah's 'Victorian District', a 45-block area of some 1,200 sites.[46]

One of the most ambitious expressions of the New Localism was the Neighborhood Housing Services (NHS) programme, launched by the Urban Reinvestment Task Force, a programme which was bound up with a new generation of powerful community housing leaders. Foremost among these was Mrs Gale Cincotta, founder of the West Side Neighborhood Housing Coalition in Chicago, and subsequently (in 1972) of 'National People's Action', a nationwide alliance of 39 local groups dedicated to training and empowerment of community rehab initiatives. Yet, despite all these diverse efforts, many projects were frustrated by community dissension and distrust, as with the pioneering mid-1970s' proposal to rehab an area of late 19th-century houses in a black area of east Detroit. In his 1972 book, *After the Planners*, critic Robert Goodman argued that all these efforts were pervaded by a 'conservative bias' and manipulated the opposition into submission through offering an illusion of false choices.[47]

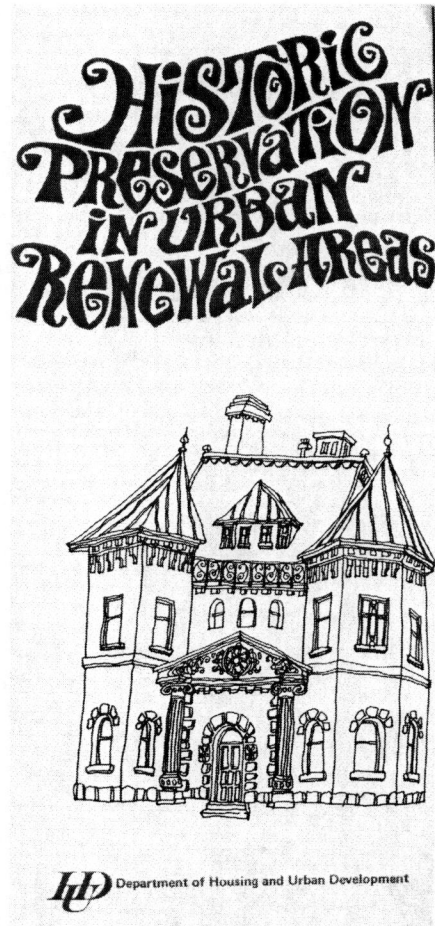

Figure 9.13 HUD leaflet of March 1970: 'Historic Preservation in Urban Renewal Areas'

Civic amenity to managerialism: the heyday of urban conservation in 1970s' Britain

As the number and scope of protected buildings and zones increased elsewhere, in 1970s' Britain the scale of conservation expanded still further: by 1976, the number of listed buildings had doubled in eight years (to 230,000), while the number of conservation areas in England alone, having passed the 1,000 mark in mid-1970, had now reached no less than 3,600. These conservation areas were all now bolstered with powers of planning control over alteration and demolition that, Worskett remarked, would have seemed 'almost revolutionary' in 1967. Was this a case of local civic initiative carried too far? One Yorkshire local-government official argued that

> [A conservation area] costs the council nothing to designate. A hundred planning applications go through, and a dozen conservation areas. The meeting starts at 2 pm, it's finished at 4.30 pm. It costs nothing, it implies nothing, it means nothing. Bash it off to the Secretary of State and the local civic society will stop screaming at our heads!

It was telling that this official talked of civic societies with such apprehension, as this period was the culmination of their power and influence. In England, the Civic Trust (CT), with its strong roots in the 1950s' Modernist ethos of contrasted old and new, celebrated its national umbrella role by holding a vast 1971 conference on the lessons of the Civic Amenities Act, 'Conservation in Action', in the Royal Festival Hall in London. The event was introduced, naturally, by Sandys, who also hailed conservation as a possible way of recovering the national 'sense of purpose . . . sense of mission' and 'distinctive identity', lost in Britain's 'difficult period of readjustment since the war'. The conference featured arrays of case studies of local civic surveys, all balancing preservation with new architecture; the renowned Modern landscapist Peter Shepheard argued that 'the first principle of conservation is to keep the good parts of cities and rebuild the bad parts'.[48]

The mid- and late 1970s saw a blizzard of Civic Trust initiatives in urban conservation, not all by any means focused on elite subjects. For example, 1978 saw the start of an ambitious revitalisation scheme for a depressed quarrying town in Yorkshire, Wirksworth, building on earlier urban renewal and GIA experience.[49] The greatest achievement of this civic-amenity heritage system, however, was somewhat hybrid in character and earlier in date: the campaign to rescue the decayed 18th-/19th-century classical New Town of Edinburgh. This was largely the work of the CT's Scottish offshoot, the Scottish Civic Trust (SCT), founded in 1967 with support from the Bank of Scotland and other industrial and charitable sources, and chaired by TV producer Maurice Lindsay, with architect John Gerrard as executive deputy (ensuring a visible casework presence and energising local societies from Orkney to Stranraer).[50]

In Edinburgh, the shift in focus from the medieval to the 'Georgian' had begun as early as the 1930s. In the late 1950s, like London in the 1930s, Edinburgh had seen the emergence of a new protest society, the Scottish Georgian Society, provoked by the part-demolition of an early-18th-century classical square, George Square. When the SCT was formed in 1967 (the CT having previously, in 1962, masterminded a Magdalene Street-style facelift of High Street, Haddington), one of its key trustees was the redoubtable modernist architect, Sir Robert Matthew. In a striking example of the dialectic collaboration assumed by modernist architects to apply between new and old, Matthew took up the rescue of the New Town as a personal crusade. At the very first meeting of the SCT, he insisted it should be the new organisation's top priority. The New Town, at 318 hectares, was far larger than most Malraux *secteurs sauvegardés*: even the Marais was a mere 126 hectares. In the same year, Matthew had also persuaded the Royal Fine Art Commission Scotland (RFACS) to press for the rehabilitation of the New Town, 'a national asset of world significance', as the 'greatest task of urban conservation . . . in Great Britain'. A younger protégé of Matthew's, the architect John L Paterson, had designed a popular bicentennial exhibition on the New Town during the 1967 Edinburgh Festival, 'Two Hundred Summers in a City', as a mildly avant-garde piece of 'environmental theatre'.[51]

The centrepiece of the SCT's decisive New Town campaign was a major conference at the Edinburgh Assembly Rooms in June 1970. The chief international perspective was provided by Francois Sorlin, the French CMH's chief inspector. In a paper on the Marais, he acknowledged the limitations of any comparison (given the three-times-larger area, eight-times-lower density and greater affluence of the New Town) but argued that Edinburgh could still learn from France the need for a strategic approach and avoidance of tourism-led commercialisation. The success of the conference enabled the 1971 establishment of a joint local/central government support organisation, the Edinburgh New Town Conservation Committee (ENTCC), headed by Irish architect Desmond Hodges, previously the head of the Ulster Architectural Heritage Society. The ENTCC disbursed over £2.5 million over the following decade, focusing not on elite showpiece streets but on the 'tattered fringe' that had degenerated into slum conditions, as epitomised in its first major project, the rehab and complete refacing of 23–24 Fettes Row (coordinated by architect/resident John Knight). Colin McWilliam, Edinburgh's most prominent preservation campaigner, argued in 1985 that 'easily the most important architectural development in Edinburgh since the war has been at 13A Dundas Street' (address of the ENTCC's modest office in a New Town basement).[52] Closer to the CT's Wirksworth initiative was a parallel campaign to advance rehabilitation efforts at the pioneering 18th-century planned industrial community of New Lanark, a campaign inspired by Provost Harry Smith of Lanark and coordinated by the SCT with New Lanark Conservation Trust.[53]

In the wake of triumphs such as this, however, there were signs, at an international level, that the urban conservation system was gaining in efficiency but losing in vitality. The French local residents' councils, for example, evolved rapidly through a brief radicalised phase around 1968 to a

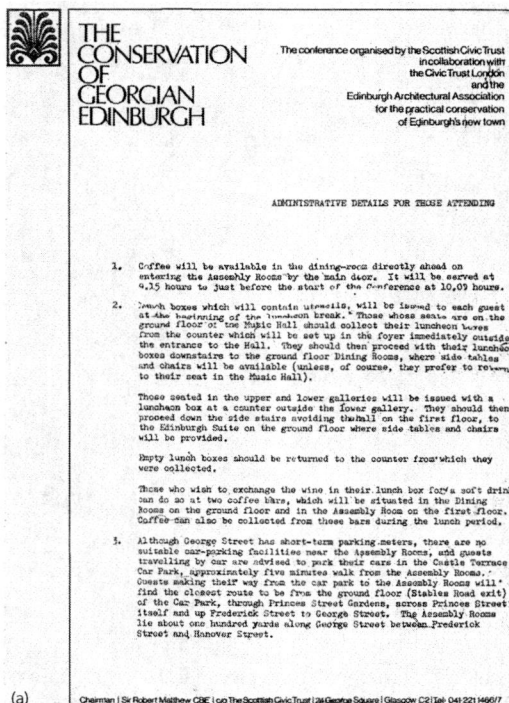

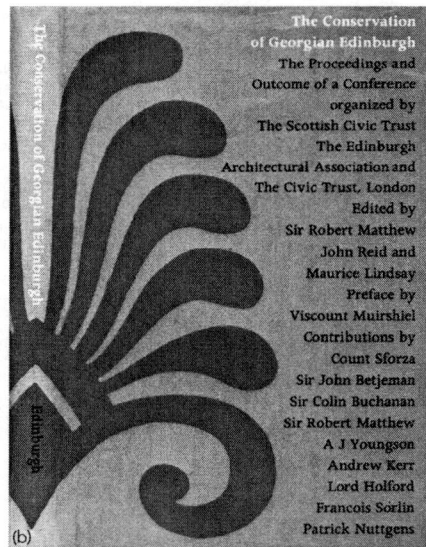

Figure 9.14

(a), (b) Literature from the Edinburgh Assembly Rooms New Town Conference of June 1970

highly institutionalised position in the 1970s, in which user experience was channelled into planning policy and urban sociological research sponsored and controlled by the central state. In Britain, too, urban conservation in the 1970s became dominated not by counter-culture radicalism but by ideas of professionalism and multi-disciplinary corporate efficiency: an initiative such as the Liverpool Inner Area Study, with its 1977 final report, *Change and Decay*, called not for radicalism but for a 'total approach to the inner city', with regional government coordinating agencies dealing with economic, social and housing factors. Radical activists became torn between satisfaction and unease at the increasingly establishment status of their causes, arguing in 1975 that 'the massive financing of radical projects makes it difficult to decide just what is Alternative'.[54] In some cases, there were informal and highly productive collaborations within the government between planners and professional historians. Within the Scottish Office, a 1971–9 Working Group on Urban Conservation effectively prioritised historic towns for development-control purposes through a 'star' system devised by Chief Historic Buildings Inspector (from 1976) David Walker, an expert on Victorian architecture. This internal government collaboration allowed Walker to keep a close eye on developments not covered by listed building controls; he pleaded successfully with chief planner Derek Lyddon in 1972, for instance, to block a proposed comprehensive redevelopment in the historic town of Kelso, which threatened to 'ruin one of our finest towns' with a 'fussy, broken-up scheme' in the largely unlisted centre.[55]

Bologna: 1970s' stronghold of socialist conservation

INTERNATIONALLY, the main trend was also one of moderation and institutionalisation, especially following that grand triumph of the Conservation Movement, European Architectural Heritage Year 1975 (see Chapter 11). Radicalism in mainstream urban conservation was increasingly marginalised. But the world of leftist heritage had one final contribution to make to the history of the Conservation Movement before it began to fade from view: the unique programme of socially integrated urban conservation undertaken from the late 1960s by the Italian city of Bologna.

Ruled by the Communist Party from 1945 onwards, the municipality of Bologna began during the 1960s to conceive a radical new plan that would avoid the enervating effect of unplanned peripheral growth, by upgrading the city's large but decayed core of 17th-, 18th- and 19th-century housing, and – equally importantly – converting the many redundant religious complexes for secular community use. Mastermind of the programme was the architect-planner Pier Luigi Cervellati, city council member and commissioner responsible for building and preservation since 1965, a designer-planner-politician-humanist who was just as much at home publishing historical studies of the evolution of Bologna's city fabric as he was drawing up phasing programmes for redevelopment and rehabilitation.[56]

The original rehabilitation programme, prepared in 1969, was based on a thorough survey of the city's buildings and open spaces, and exhaustive documentary and historical research (including a photographic survey by Paolo Monti). It proposed starting on rehab in five out of the 13 central zones (totalling 32,000 properties out of 80,000 in the central zone) and restricting any intrusions by concentrated modern commercial complexes. Originally, in 1972, it was intended to implement this plan by expropriating all land in the centre. This, however, provoked such fury among small property owners that, in mid-1973, an alternative formula was agreed, under which property would instead be acquired by agreement with owners. The housing element in the programme would be financed from the *Piano per l'Edilizia Economica e Popolare* (PEEP), a national scheme instituted by legislation in 1962 but normally focused on modern suburban developments. And in an experiment in public participation, the city centre was subdivided into 18 'quarters', each run by a community council.

The Bologna scheme's key architectural features survived the radical organisational switch from expropriation to collaboration. The built fabric to be rehabilitated was classified into four typological categories, identified by the architects and planners from the 1969 survey. These ranged from social and housing groups up to large monumental volumes (redundant churches, etc.) or 'historic containers' (*contenitori storici*), all unified externally, in their relation to the street, by the Bologna tradition of pavement arcading. The use of historically grounded 'types' to convey meaning in architecture was partly a reaction against Modernist Functionalism and was developed strongly in Italy in the mid-1960s by architects such as Aldo Rossi – as well as by anglophone authors such as Alan Colquhoun.[57]

In Bologna, the housing rehab areas, developed from around 1975 (beginning in areas such as Solferino and San Leonardo), mainly comprised dilapidated rows of houses above ground-floor shops and arcades, sometimes interspersed by collapsed or damaged buildings. Methods of rehabilitation were partly determined by the 'types'. Most sub-standard housing was converted initially to flats (with services flanking the access stairs), while preserving terraces' externally 'typological' appearance (*restauro tipologico*). There was a sharp contrast between the treatment of everyday areas and elite buildings. On cleared sites, new, standardised terrace-type houses were built, externally indistinguishable from the typologically restored houses. These echoed the 'typical' facades rebuilt in some war-damaged cities, such as Münster or St Malo or Gdansk, and anticipated the post-1989 explosion in facsimile-building in Germany and other countries. Slightly confusingly, the city's literature referred to these rather radical schemes of intermingled new–old building – so different from the sharply differentiated old and new demanded by Modernist conservation orthodoxy – as '*restauro conservativo*'. The reconstructed large monuments were treated differently. Redundant palaces, convents and churches were creatively converted into hives of community activity: the prototype, a palace on the via Pietralata, was reconstructed as a local community centre for one of the city's neighbourhoods, including a library, lecture hall, crèche and medical centre, set around a relandscaped courtyard. In the San Leonardo area, plans prepared in 1973 and executed in 1975–85 envisaged restoration and infill of new housing and conversion of a large monastery complex for community use.[58]

Bologna was the climax of the post-1945 social strand of the Conservation Movement in the West, but it was also its swansong. In aspects such as its extensive replica-building and the prominence of 'type', it pointed to the next phase – that of capitalist commodification of urban heritage to help in branding and image. We saw above how the libertarian radicalism of '1968' had paved the way for a phase of systematised expansion of conservation in the '70s. Now, in the mid-'70s, a new phase of libertarian radicalism got underway, helping adapt this corporate structure to new values of capitalist competition in the built environment. In country after country, the decline of the welfare state and its grand narratives of progress allowed new values of market modernity to pervade the world of architecture, with heritage increasingly taking its place as an element of tourism and city branding. We will trace in Chapter 12 the full-blooded development of these ideas in the 1990s onwards, but even in the '80s they were already powerfully at work.

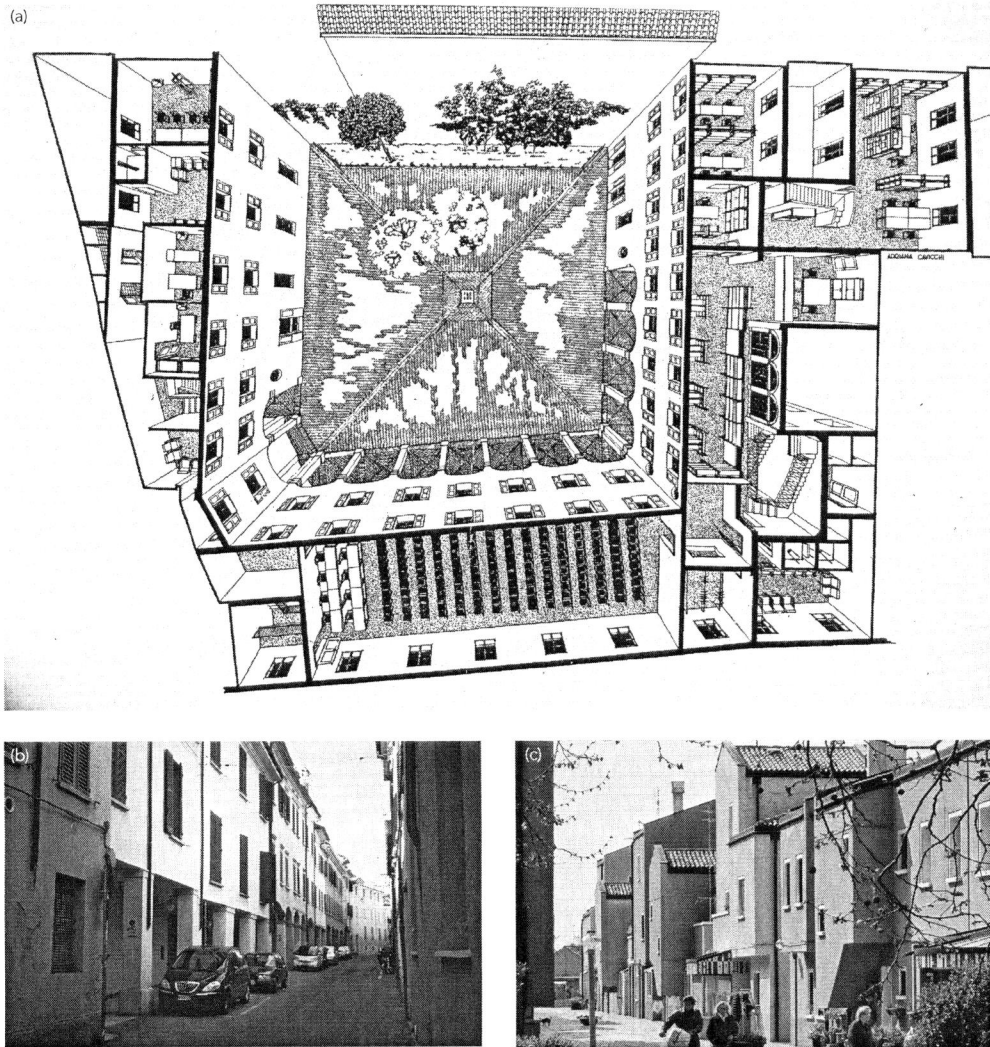

Figure 9.15 Old and new in 1970s' Italy

(a) Bologna, Malpighi community centre: prototype for the conversion of 'large containers' for multi-purpose use under Cervellati's 1969/73 programme; (b) Typical improved and infilled Bologna street from the mid-1970s: via Denisella, Area 7 Fondazzi; (c) New social housing complex in a historic context: Piano per l'Edilizia Economica e Popolare (PEEP) project by Giancarlo de Carlo on the island of Mazzorbo, Venice, 1980

Conservation and postmodernist urbanism in the 1980s

In the cultural field, the 1980s saw the onset of 'postmodernism' or 'postmodernity' – a new, frankly relativistic outlook that rejected the idealistic certainties of modernism for a more ironical world-view. The architectural expressions of this movement were highly diverse. Some designers and writers stressed homely or vernacular styles while retaining the modernist emphasis on integrity and wholeness. Others rejected the latter for a new love of decoration, ornament and design driven by show and image rather than by tradition or principles. Some modern architects saw this as a retreat: Lionel Esher's 1981 overview of postwar British architecture, *A Broken Wave*, ended with a chapter called 'At Low Tide' and speculated whether this was 'a defeat or a victory' for architecture.[59] In any event, the place of conservation seemed to have been unambiguously enhanced.

The jettisoning of the clear postwar differentiation of New and Old had left a more open, amorphous landscape in which heritage was both more pervasive yet also somehow less secure. The 1980s were also a time of rapid ideological shifts, most obviously from left to right affiliations. Within the Conservation Movement, a range of neo-conservative discourses flared into brief prominence, before once again subsiding into corporatist stability. In different countries, this conservative reaction took different forms. In Italy, it was architecturally led: the 1960s' ideas of urban typology had rejuvenated the long-standing preoccupation with aesthetic integration of modern development into historic contexts, and the early 1980s added to this a new and flamboyantly postmodernist Italian doctrine of 'urban intervention', exemplified in Paolo Portoghesi's *Strada Novissima* group exhibit staged in the Corderie dell' Arsenale at the 1980 Venice Biennale. At an international level, the many postmodernist architects in Europe and North America, such as Charles Moore (designer of the 1977/8 Piazza d'Italia in New Orleans) or Heinrich Klotz (who famously argued in 1985 for '*nicht Funktion, sonder Fiktion*'), further developed this new, overtly scenographic urbanism.[60] From now on, advanced architects ceased to be interested in integrating cities with their rural or regional context, or in planning in general, and started to use the term 'The City' as a synonym for architecture.

The Italian vision of intense, contextual urbanism was reinforced by a cautious German revival of the concept of *Heimat* in the 1980s.[61] Postmodern designers such as Rob and Leon Krier dusted down the townscape ideas of Camillo Sitte and updated them for a largely anglophone international audience. The renowned intervention projects by Karljosef Schattner in the small Bavarian cathedral city of Eichstätt, from the 1960s to the 1990s, echoed and transcended the Italian precedent of Scarpa and the Neo-Liberty designers, by actively engaging with postmodern urbanism. And the traditional *Altstadt* defenders, quiescent during the years of postwar reconstruction, began to raise their heads above the parapet and articulate the widely felt feeling that the reconstruction drive had unnecessarily sacrificed many salvageable remains and that something could still be done to put that right, above all by facsimile-rebuilding of key monuments. In Nuremberg, the historian Erich Mulzer, leader of the city's influential *Altstadtfreunde* society, fomented public discontent at the generic character of some of the rebuilt streets and at the failure to rebuild principal monuments such as the Pellerhaus. The same features proudly trumpeted in the 1930s as 'uniquely German' were now reinterpreted as both German and European: Mulzer argued that

> prewar Nuremberg was one of the most beautiful cities in the world: everyone was proud of its unique towered skyline and townscape, its shady streets and intimate squares, dotted with fountains, courtyards, bay windows – and its thousands of old houses. Together, these constituted a unique German expression of European culture and art, just as Florence and Venice are, still, its unique Italian expression.[62]

With the rejection of Modernism, the general taboo against facsimile-building as a pastiche violation of the sacrosanct contrast of old and new faded away.[63] What was being demanded in Germany was not, of course, a recreation of the everyday street environments of bombed cities, with

Figure 9.16 Schattner's Eichstätt: postmodern *diradamento* in action

(a) Journalism Institute, Ostenstrasse 20, 1985–7

(b) Theological Library, Ulmer Hof, 1977–80: courtyard reading room

their 19ᵗʰ-century tenements. Nor was there, just yet, any realistic aspiration to a Warsaw-style recreation of entire historic quarters. What was now at issue was the systematic reconstruction of lost set pieces and focal street facades – an agenda which, in effect, now took over the *Bürgerinitiative* movement, diverting it from left-wing radicalism to the cause of *Heimat*. The result was a new kind of conservative local activism, dedicated to the pursuit of facsimile-reconstruction projects, integrated with rather than opposed to the commercial establishment and often supported by the Christian Democrat conservative parties.

The first ranging-shot of this new campaign occurred as early as 1968–70, when, after much controversy, the war-destroyed *Steipe* in Trier, a castellated, four-storey building of 1430/1483 in the main square, was rebuilt in facsimile on its original site following a citizens' campaign and finance from a regional life-assurance society.[64] In Aschaffenburg, 1984 saw a successful citizens' campaign to ensure that the site of the 16ᵗʰ-century, timber-framed *Löwenapotheke* was filled not by a modern block but by a near-facsimile (by architect Christian Lauffs). A more tricky problem was posed when the site of a war-destroyed monument had already been filled by a modern replacement. One solution, especially where copious sculptural fragments or internal fittings had been salvaged, was to build a replica on a different site, incorporating those fragments. This approach had been ruled out on principle at the Goethehaus in the 1940s, but in Hannover, where Hillebrecht's radical reconstruction strategy had been modified in 1959, following a visit to Warsaw, to accommodate a *Traditionsinsel* of reassembled timber buildings, 1981–3 saw the next logical step: a facsimile facade incorporating fragments of the city's foremost Baroque mansion, the 1649–52 Leibnizhaus (whose restoration on its original site had been refused after the war by the *Landeskonservator* as a 'falsification'), was built not far from its old situation. Evoking Riegl's concept of transience, one prominent local conservationist argued in defence of the project that 'the last memories of the Leibnizhaus on its old location will be lost in a generation. Then the reconstructed Leibnizhaus will become a monument in its own right.'[65]

In a further permutation of the replica formula, early postwar buildings constructed on the site of a bombed monument could themselves be demolished to allow construction of a facsimile. The early 1980s saw the beginnings of a pioneering programme of facade reconstruction around the Römerberg square in Frankfurt, beginning in 1980–4 with the east side, following extensive debate; similar work began in the Mainz Marktplatz from 1978.[66] The decisive step, however, was the recreation of the Marktplatz in Hildesheim, including the renowned, gabled Knochenhauer Amtshaus, burned to ashes in 1945. In the 1960s, the whole complex had been replaced by an open, modern layout, including a towered hotel (the Hotel Rose, opened in 1963), in a scheme supported by the local conservation officers: they argued that, whereas reconstruction of Frankfurt's Goethehaus, as a cultural monument, had been legitimate, that of the Knochenhauer Amtshaus, as an architectural monument, was inadmissible. From 1978, though, a Christian Democrat-endorsed local activist campaign began to undermine the legitimacy of this solution, and a 1980 competition advocated the restoration of both the old, narrow plan of the market place streets, and of the Knochenhauer Amtshaus itself. In 1983 the regional culture ministry declared the old street layout a monument (*Baudenkmal*). Following furious fund-raising, the site was designated an improvement area (*Sanierungsgebiet*), the hotel was demolished in 1985, and the replica Knochenhauer Amtshaus (designed by local architect Heinz Geyer) was duly inaugurated in 1989. It should be borne in mind that the pre-1945 Knochenhauer Amtshaus was itself largely a reconstruction of 1886, following a devastating fire, at a time when tourist promotion of the town and the building itself had already started.[67]

Heritage's 'New Right' in America and Britain

IN the anglophone countries, the Conservation Movement was buffeted in the 1980s by a more overtly political rightward shift. The USA, during the Reagan years, witnessed much rhetoric against the interventionism of the Johnson era. In 1983, one *JSAH* (*Journal of the Society of Architectural Historians*) reviewer criticised James M Fitch's 1982 textbook, *Historic Preservation*, for over-emphasising 'the bureaucratic role of preservationists and "the responsibility of the state" to tell the people what to do'. Yet, in practice, behind the scenes, the late 1970s and early 1980s saw continued expansion of state-sponsored preservation, including the radical upgrading of the system of tax credits for rehab (in 1981 and 1986) along with the National Trust's Main Street Program (from the late '70s) for the revitalising of historic shopping streets. The size of the NT surged from 70 staff and 40,000 members to 300 and 140,000, respectively, during the years 1973–8 alone.[68] In Britain, where the 1980s saw a similar rolling-back of government controls in areas such as housing or nationalised industry, the heritage position was more complex, with a new network of more vociferous ideologues helping to prod the Conservation Movement to the right and to shift its preoccupations from the social to the economic sphere. The ingrained English tradition of furious architectural polemic and sharp swings in opinion made sure that this new right-wing radicalism would promote itself in a fiercely combative way.

The first harbinger of this change in climate in Britain came in the early '70s, with the eruption of an angry new protest movement focused not on 'The City' but on the alleged continuing neglect of country houses. Doubtless reflecting the continued social prestige and cohesion of the landed classes in Britain, country houses in Britain had commanded an increasingly wide cultural support in the postwar austerity years. As we saw above, the more prominent houses had reorientated themselves from a private to a public role, whether through takeover by the NT or by Longleat-style commercialisation. This development was framed presentationally by a new style of interior

350

Figure 9.17 Mansions and castles

(a) The 1974 exhibition, 'The Destruction of the Country House', at the Victoria and Albert Museum, London; (b), (c) Ballone Castle, Easter Ross – ruinous since 1830, but restored as a private house in 1989–99 by architect/owner Lachie Stewart of Anta Architects: before (1980) and after views

decoration, owing as much to American as to British tastes. In the 1950s and '60s, US-born decorator Nancy Lancaster, and her partner John Fowler (Colefax and Fowler) developed a chintzy, informal aesthetic of 'romantic disrepair', underpinned by US-style standards of comfort and conveniences. Linked to the rise of aspirational, glossy home-improvement magazines, this style helped foster a coordinated country-house brand in the public mind. Visitor numbers to country houses soared, from 2 or 3 million a year in the late '50s to 28 million in 1971. Yet the 1960s and '70s brought new financial pressures on owners, including the 1965 introduction of capital gains tax and the effects of a rise in land values. The economic crisis of 1972–4 provoked a fresh outburst of activist alarm about the supposed threat to British country houses, especially after the 1972 granting of permission for demolition of the eminent Greek Revival house, The Grange – averted when the ruin was taken into public Ancient Monuments guardianship.[69]

A much publicised exhibition at the Victoria and Albert Museum in 1974, *The Destruction of the Country House*, brought to the fore a new generation of leaders within the Conservation Movement in Britain. Its organisers, Roy Strong (V & A Director), journalist Marcus Binney and John Harris, proselytised a right-wing interdisciplinary approach, straddling architectural history, museums and connoisseurship. In true Puginian fashion, the exhibition painted an apocalyptic picture of threat, arguing that over 1,000 country houses had been demolished over the previous century and that 'unless drastic action is taken now the rest will follow'. The climax of the exhibition was a 'Hall of Destruction' inspired by Guilio Romano's painted room in the Palazzo del Te, Mantua. A linked, British Tourist Authority-sponsored report by historian/decorator John Cornforth (*Country Houses in Britain: Can they Survive?*) also painted a doom-laden picture, although Cornforth's statistics did not completely support his stance, suggesting that, of 2,000 notable surviving country houses, only a quarter faced even minor threats. Following a 1977–9 scandal about disposal of the contents of a 19th-century time-capsule house, Mentmore, a new charitable foundation, the National Heritage Memorial Fund, intervened with further support.[70]

The 1970s/'80s' controversies about threatened country houses in England were mirrored in a campaign for restoration of ruined castles in Scotland. Here, the threat was usually long in the past: owing to Scotland's uniquely forceful campaign of Improvement and its dynastic wars of the 17th/18th centuries, many castellated houses had been left in ruins across the countryside, in a manner hardly paralleled in Western Europe. It was estimated in the 1960s that only 1,250 out of an all-time maximum of 3,000 castles had survived roofed and intact. Here, the prestige of the Scottish lairds had outlasted the challenge of socialism and they easily dominated heritage organisations such as the National Trust for Scotland. With the 1979 advent of the Thatcher Conservative government, the cause of the 'Baronial castle' in Scotland made a dramatic come-back. The flamboyant laird, castle-owner and Tory government minister Sir Nicholas Fairbairn recalled with a shudder the welfare-state 1960s–70s, when

> those who strove to save our heritage, grand or humble, were branded as outcasts and pariahs . . . determined to live in the grisly past, not appreciating the liberal worth of contemporary concrete brutalism. We were like the 600 at Thermopylae against the infidel hordes.[71]

In the late 1970s and 1980s, sidestepping the doctrinaire Anti-Scrape ethos of the Scottish Development Department's Ancient Monuments branch, the department's Historic Buildings inspectorate, under chief historic buildings inspector David Walker, sanctioned a veritable explosion of castle-rebuilding, fuelled by mounting economic prosperity as well as Historic Buildings Council (HBC) grants. This work was mostly undertaken not by lairds but by middle-class enthusiasts, such as Graham Carson, restorer of Rusco Tower, who decided in 1970 that 'I had to restore a tower house, now', and picked Rusco from a short list of 11 ruined castles suggested by the popular castle historian, Nigel Tranter. Restorations ranged from conservative repair of castles disused for only a few decades, to Pierrefonds-like recreations of buildings that had been ruins for centuries – as with the 1991 rebuilding of Forter Castle, Perthshire, a 1560 tower, roofless ever since it was burned in 1640 by the Duke of Argyll. The programme gradually petered out in the later 1980s owing to a decline in numbers of easily restorable towers and an escalation of costs in some prominent schemes such as Fawside, near Edinburgh – a castle abandoned in the 18th century and restored in 1976–82 with a £104,000 government grant.[72]

In a sense, the facts about the country house 'crisis' were irrelevant. What was more important was that it signalled a new, right-wing activist upsurge within the Conservation Movement in Britain. This development was symbolised by the rise to prominence of 'SAVE Britain's Heritage', a society founded in 1975 by country-house activist Binney and various helpers. SAVE positioned itself from the start within the SPAB and Georgian Group tradition of publicity-seeking protest, although it had a simpler agenda than SPAB's anti-restorationism, straightforwardly attacking Modernist redevelopments and demolitions. Fronted by the charismatic Binney (hailed by followers as 'the man with the most aargh in Britain') and organised by secretary Sophie Andreae, SAVE rapidly expanded its polemical campaigns across the entire historic environment, collaborating with a new generation of right-wing urban activists, whose focus of operations was the crumbling early-18th-century inner-London neighbourhood of Spitalfields. In 1977, with some 35 per cent of the 240 listed buildings in the area already demolished, a group of activists led by Dan Cruickshank, Mark Girouard and others staged a sit-in in demolition-threatened 5–7 Elder Street – a radical gesture sanctified in a visit later that year by Sir John Betjeman. Following that, the area was rapidly transformed by a wave of restoration and gentrification, masterminded by a new Spitalfields Historic Buildings Trust.[73]

SAVE and Spitalfields were the two foci around which coalesced a new generation of aspirational right-wing activists, steeped in the values of gentrification and commercial, eclectic connoisseurship. The two decades before 1978 saw a rise in turnover at auctioneers Christies from £1.6 million to £57 million, and at Sotheby's from £3 million to £80 million. The flamboyant diversity of this network, and its victory in what its leaders called 'the 'Great Conservation War' of 1970–80, was celebrated in books such as Alexandra Artley and J M Robinson's *New Georgian Handbook* (1985). It was linked to a postmodern classicism in new architecture, whose idols, like Quinlan Terry or Robert Adam, designed with an irony-laden intricacy that only faintly reflected the monumentality of 19th-century classicism: Edinburgh's Colin McWilliam scornfully argued in 1985 that their approach 'has no place in a city like Edinburgh, where classicism is not yet a joke'.[74]

The bogeyman that drew all these writers and designers together was the Modern Movement, denounced wittily by historian David Watkin in his critique of the work of Nikolaus Pevsner and more thunderously by journalist/philosopher Roger Scruton. He argued that Modernism was 'one of the greatest catastrophes the world has ever known in peacetime', a sinister conspiracy that 'seized control of our cities and shook them free of human significance'. The forward march of conservation now foregrounded not only the Arts and Crafts and classical architects of the early 20th century, above all Edwin Lutyens, but also early interwar modernism. This mixture was celebrated in the

353

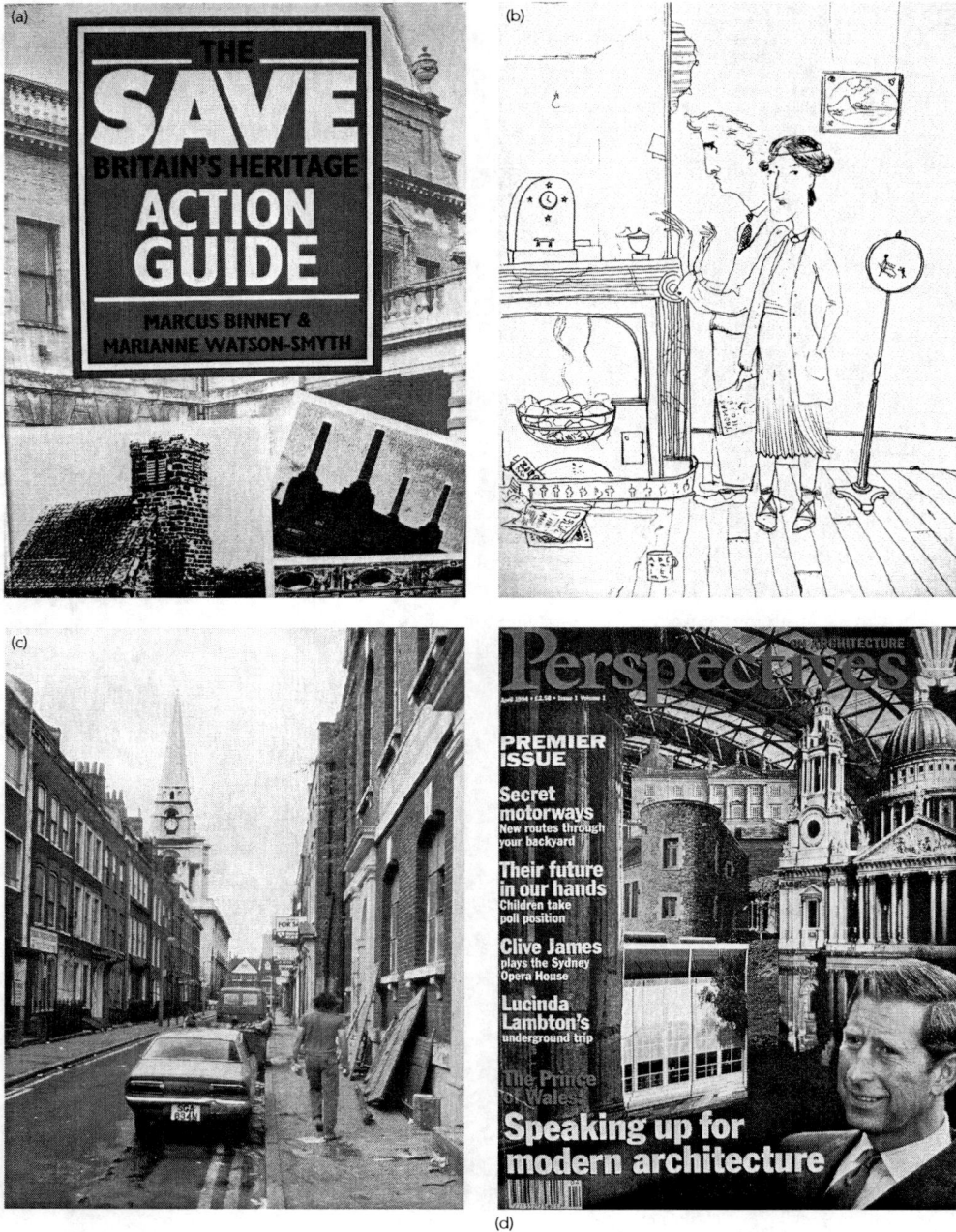

Figure 9.18 'New Right' conservation in England

(a) Cover of SAVE *Action Guide*, 1991; (b) Drawing of 'The Soanies', from the 1985 *New Georgian Handbook*; (c) 1982 view of Fournier Street in London's Spitalfields, epicentre of elite gentrification; (d) The front cover of the first issue (1994) of Prince Charles's short-lived journal, *Perspectives In Architecture*

founding of the Thirties Society in 1979 by Gavin Stamp, Alan Powers and others – a light-hearted echo of the Georgian Group in the 'real' 1930s. Its most famous campaign (from 1983, before it mutated into the more serious Twentieth Century Society) was for preservation of red Post Office telephone kiosks – a category of monument that blurred the boundary established in Revolutionary France between moveable and fixed heritage.[75]

The greatest achievement of the New Right coalition was to infiltrate and take over the community participation movement. Ironically, the instincts of many New Right leaders were firmly against participation: Leon Krier argued in 1985 that the only admissible expression of participation was 'a vote of NO CONFIDENCE in Modernist Architecture and Planning'.[76] But the takeover of the participation movement stemmed not from their efforts but from a highly publicised external intervention: a campaign launched by Prince Charles against modernist architecture. In successive speeches stretching from 1984 until the *annus horribilis* of royal scandals in 1992 (and continuing for several years in the work of the Prince of Wales's Institute of Architecture and the short-lived journal, *Perspectives on Architecture*) the prince exploited his royal position to proselytise a 'Vision of Britain' combining violently anti-modernist architectural ideas with rhetoric of community participation and 'sustainability'.[77] Improbably, Prince Charles presented himself as a popular crusader against underhand 'interest groups' and a 'sycophantic architectural press' and expressed astonishment that he was 'accused of abusing my power and, can you believe it, of acting undemocratically'. In 1987–8, he intervened in the City of London to block a plan to build a posthumous tower design by Mies van der Rohe, in Mansion House Square, arguing against modernist redevelopments that 'you have . . . to give this much to the Luftwaffe – when it knocked down our buildings, it didn't replace them with anything more offensive than rubble. *We* did that', and that the Mies plan (eventually rejected at a 1985 public inquiry and replaced by a postmodernist design by James Stirling) would produce 'a stunted imitation of Manhattan'.[78]

As an exemplar of an architecture that could help the outraged populace put right the evils of Modernism, the prince singled out the community participation, or 'community architecture' movement. He focused not on the controversial schemes in riot-torn (from 1982) Liverpool, but on the sylvan setting of Rod Hackney's Black Road. In 1985, the prince made a pilgrimage to Hackney's scheme, and Hackney, by now a bustling organiser-entrepreneur controlling a string of companies, was hailed in the *Observer* newspaper as 'The Hackney Phenomenon. . . . Architect to Prince and People . . . bulldozing a new order into the way Britain builds'. The fortunes of this new, right-wing 'community architecture' rose in parallel with the British government's radical shift from social to private housing provision. Although Hackney was elected RIBA President in 1987, by then the movement was on the wane: architectural journalists dismissed it as 'perfect camouflage for the political chameleon' and 'no longer a force for real political change'.[79]

The elitist hostility of the New Right movement towards libertarian free-market capitalism ensured that it would have only a short-lived effect on both the Conservation Movement and on contemporary architecture. Far more enduring, indeed, was the impact of the contemporary movement towards democratisation and commercialisation of high-culture, through such phenomena as the cult of youth and pop celebrity in general – a movement reflected across Europe, for example in French Cultural Minister Jack Lang's extension of 'culture' to include modern, popular themes.

From social heritage to heritage industry

INTERNATIONALLY, all the conservative 1970s/'80s' movements traced above were less important in their own right, than in the way they fuelled the commodification of the urban heritage. The starting point and driving force of this movement was provided by tourism. Where the old-style tourists of the 19th or 20th centuries visited elite highlights such as cathedrals, castles or reconstructed Old Towns, now entire cities could be branded for tourist promotion purposes (a trend anticipated in Kevin Lynch's 1960 book *Image of the City*). This was attempted sometimes through conventional 'great buildings' (as in Paris with President Mitterrand's *grands projets*), but more often through the fostering of an elusive 'spirit of place'. Even centres of long-standing tourism recorded sustained postwar growth in visitor numbers. In Rothenburg ob der Tauber, for example, overnight stays soared from 54,000 in 1950 to 144,000 a decade later and almost 500,000 by the end of the century – a rise aided by the town's seminal role in the invention and proliferation of a new international genre of tourist shop, the 'Christmas market' or 'Christmas village' (a kind of miniaturised bazaar-cum-theme-park, pioneered by Käthe Wohlfahrt GmbH from 1977 in a succession of shops in the town centre). In August 1978, an English Tourist Board survey showed that over 1 million people visited the top 26 attractions in England, including over 17,000 in York Minster in a single day; by 1992, the National Trust was the top earner of all charities in Britain (£55 million). The attitude of traditional postwar modernist commentators towards heritage tourism had been dominated by suspicion or con-demnation of its potentially falsifying effects, as shown in Hugh Casson's attack in 1971 on the

Figure 9.19 The heritage industry

(a) Käthe Wohlfahrt GmbH 'Christmas village' in Rothenburg ob der Tauber (2011 view); (b), (c) The Zollverein industrial heritage and regeneration complex, near Essen – centrepiece of the IBA-Emscher Park project of 1989–99

Figure 9.20 Replica of L. Mies van der Rohe's 1929–30 Barcelona Pavilion, built in 1983–6 by the city council and the new 'Mies van der Rohe Foundation', architects Cristian Cirici, Fernando Ramos and Ignasi de Sola-Morales. Paradoxically, the structure is only a stone's throw from the 'Poble Español' heritage theme park, a 'genuine' survival from the same exhibition

'degrading disease' of 'cuteness' under which 'old buildings [are] corseted and pomaded into a travesty of themselves'. Increasingly, though, these critiques started to seem both churlish and irrelevant.[80]

The process of branding of heritage cities now extended much further than before, embracing previously neglected places. The once stodgy city of Barcelona, for instance, was rebranded and restyled as a city of fashionable, youthful contemporary art, with a diverse heritage combining refurbished turn-of-century '*Modernisme*' with reconstruction of a copy of Mies's 1929–30 Barcelona Pavilion – a project carried out in 1983–6 by the city council and a newly constituted Mies van der Rohe Foundation, with architects Cristian Cirici, Fernando Ramos and Ignasi de Sola-Morales. The 'elevated' reuse of derelict industrial and commercial buildings was a key feature of this branding process: a showpiece was the conversion of the Ruhr's vast Zollverein mining and steelworks complex into a centre of regional cultural regeneration under the IBA-Emscher Park project of 1989–99, covering 800 square kilometres and 17 towns and cities. In New Lanark, too, the 1980s saw the end of threats to the complex.[81] During the 1980s and '90s, aided by government economic regeneration funds and cultural festivals, a post-industrial city such as Glasgow, still seen in 1977 by government planners as a 'desperate case of decline', where conservation 'could only act as a sticking plaster', attracted a bewildering range of cultural-economic promotion activities. These included conversion of decrepit city-centre warehouses into a branded 'Merchant City' of elite shops and cafes and opening of a tenement-flat museum by the NTS in 1983. In some other cities, such as Dublin or Liverpool, the dereliction and decline was more pervasive and a turnabout correspondingly longer in coming. In North America, too, numerous cities launched similar regeneration strategies: for example, Halifax, Nova Scotia, where a dilapidated 19th-century harbour zone was revitalised through a municipally sponsored rehab/intervention scheme of 1974–81 by architects Duffus Romans Kundzins Rounsefell.[82]

In Britain, the decisive beginning of the commercialisation of the Conservation Movement was signalled in the 1980s' emergence of new government heritage organisations dedicated to marketing as well as bureaucratic administration. In England, the heritage elements of the 1970s' technocratic super-ministry, the Department of the Environment, were converted into semi-commercial organisations by the 1983 National Heritage Act. Initiated by Conservative government minister Michael Heseltine, who had an acute sense of the potential interrelationship between heritage and tourism, this law created a new free-standing agency as from 1984, a Commission headed by the tourism-conscious Lord Montagu of Beaulieu and rapidly dubbed 'English Heritage'. It disbursed large-scale government subsidies, awarding £86 million in grants in 1988, including £1 million for the preservation of 'time capsule' Calke Abbey, and succeeded in establishing a membership base of 250,000 from nothing in seven years. The 1984 creation of an equivalent agency in Scotland, under director Tom Band, was accompanied by assurances that this would not lead to a 'Disneyland'.[83] As the reorientation of the heritage towards tourism and capitalism gathered pace in the 1980s, similar critiques became more and more common, claiming irreparable damage to authenticity. Robert Hewison's 1987 book, *The Heritage Industry*, argued that Britain's heritage had been corrupted by a 'miasma' of commercialised nostalgia.[84]

We will trace in Chapter 12 how this commercialisation process radically expanded in the 1990s and 2000s, until it began to conflict with key values of conservation. During the 1980s, however, the confidence of the newly triumphant Conservation Movement still seemed limitless, almost to the point of hubris: in 1980, Lord Montagu argued that conservation's victory was 'one of those fundamental shifts or watersheds in man's attitude towards the world'. A 1989 celebration of English Heritage's activity argued that the only dangers to the Movement now came from its own success: 'conservation as a whole is now a self perpetuating process with immensely hopeful benefits for society'; while David Lowenthal argued in a compendium volume that conservation was 'concerned with the whole fabric of the country', and oriented far beyond mere architecture to 'husbanding scarce energy and material resources'. At a 1986 conference on new building in historic towns, held in Bath by Bath Preservation Trust and the Georgian Group and appropriately called 'Conservation on the Attack', there was no trace of the old 1950s–60s' love of contrasting new and old. Instead, Royal Fine Art Commission (RFAC) chairman Norman St John-Stevas lambasted modern architecture as 'brutal, barbaric, philistine, arrogant', and demanded an uncompromisingly contextual approach.[85]

It was something of a paradox, then, that the late 1970s and early 1980s also began to see a growing interest in postwar modernism *as* heritage, as seen, for example, in the landmark designation of New York's Lever House in 1983. There were also the first attempts in Britain to list postwar architecture – although the GLC Historic Buildings department's 1979 attempt to designate the Royal Festival Hall, the National Theatre and Eero Saarinen's US Embassy was refused by city councillors and shelved. Conservation of modern architecture would become a key feature of international heritage debates in the 1990s and 2000s, especially after the 1988 foundation of an influential international pressure group, DOCOMOMO (Documentation and Conservation of the Modern Movement). Of course, the very idea of a 'Modernist heritage' only became possible following the discarding of the Modernists' own demand for contrasted old and new. Yet, in many ways, it also contradicted everything the Conservation Movement stood for – not so much because most conservationists still saw Modernism as the enemy, but because of the way 'Modernism-as-heritage' seemed to block off the process of Progress that had been central to conservation's character as a movement. We will unravel the complex ramifications of these developments in Chapter 12.[86]

– 10 –

Heritage complexities in the socialist bloc

1945–89

*The function of preservation is transformed into a medium for planning a
tradition-conscious renewal of the urban environment in a socialist society.*

Emanuel Hruška, 1975[1]

Conservation – class enemy or socialist showpiece?

IN the postwar socialist bloc, the Conservation Movement was as diverse as in the West. Here, however, the differences were not over the extent of state intervention in heritage, as between Western Europe and America. State oversight of conservation, in the interests of 'the people', was naturally taken for granted. Instead, the most contentious issue was how socialist principles of intervention and universalism could be reconciled with the traditional conservation values of national heritage and pride.[2]

Many socialist countries were in greater need of nation-building than Western countries, having experienced massive shifts of population and borders, and in Poland, the outcome was a political exploitation of heritage unparalleled anywhere else, before or since. And although the general trajectory of building was similar to the West, with postwar reconstruction followed by modernising urban renewal, the dominance of the authoritarian state under socialism ensured that there could be more dramatic fluctuations between demolition and preservation. In new architecture, too, there were big shifts of policy, not least the rejection of Socialist Realism for Modernism from the late 1950s. And there were wide disparities in approach and timing, with liberalisation in one country sometimes coinciding with aggressively anti-heritage policies in another. But these were not normally the subject of heated public protest or controversy: genuine private pressure groups or religious organisations were of little importance, and the economic reinforcement of urban conservation by 'gentrification' was lacking.

On the whole, despite the frequent East–West political crises, the trend in heritage over the socialist bloc from the 1960s was towards ever greater engagement and convergence with the outside world and international organisations. The Soviet Union often played a more exemplary role within its sphere of influence than did the USA in the West. Far less prominent was the other communist super-state, the People's Republic of China, where the 1950s saw a sharp division between continuing admiration for nature and traditional Chinese gardening, and aggressive, Soviet-style industrial developments; the Cultural Revolution of the 1960s inaugurated a period of introspection and anti-historical chaos.

Immediately following World War II, the Conservation Movement faced three main problems within the new Soviet European sphere of influence. Most immediate was the problem of conservation's complicity with fascism. Second, there was the issue of whether the built heritage was incorrigibly pervaded by hostile value-systems – religion, feudalism and the bourgeoisie. This applied both to its leading building-types (churches, palaces, public institutions) and to the way those structures had been woven into the structure of bourgeois history. And third, there was distrust of old buildings as incompatible with socialist love of the future – a sentiment that sometimes degenerated into outright nihilism, just as in 1790s' France. Initially, the advocates of an explicitly communist heritage briefly held sway. At a 1946 conservation symposium in Weimar, socialist ideologue Gerhard Strauss bitterly accused the *Denkmalpflege* movement of complicity in the crimes of the Third Reich and argued that the selection of monuments in socialist eastern Germany should be radically realigned around the values of class conflict. And at an attempted pan-German convention in Munich that year, opinion polarised over the issue of land reform, although unofficial contacts between eastern and western German conservationists continued right through to 1989.[3]

However, although certain buildings symbolic of 'class enemies' were indeed destroyed or left to moulder, there was no general political revolution in conservation in East Germany (the German Democratic Republic, or GDR) or any other socialist country. Former GDR conservationist Heinrich Magirius argued in a 2001 overview that just as much East German built heritage was saved by inertia and poverty as was lost by socialist reconstruction – especially in comparison to the radical rebuilding efforts of the *Wirtschaftswunder* in West Germany. And the scope of conservation focused on the same elite building types as before – a state of affairs legitimised by two ingenious and comple-mentary arguments.[4]

The first way in which post-1945 communism attempted to accommodate and appropriate the religious-feudal-bourgeois built heritage, again echoing revolutionary France, was by emphasising the artistic rather than historic-associative values of monuments: the universal power of Art would echo and reinforce the universal applicability of socialism. This emphasis emerged, as we will see shortly, in the postwar reconstruction efforts of the USSR. The second assimilatory argument went rather further, transforming reactionary heritage into an active asset through an ingenious dialectic logic which claimed that the dynamic of socialism allowed contradictory elements to play their part in the march of progress. As chief GDR conservation official Ludwig Deiters argued in 1975 at a key international seminar on Eastern European conservation held (intriguingly) in New York, the entire architectural history of a country was a 'chain of achievements' that collectively validated the progressive power of socialism: a monument of modern history such as Torgau bridge, where US and Soviet troops met in 1945, could stand in comradely solidarity, within the grand narrative of History, alongside superseded feudal and bourgeois relics. At that same seminar, the president of (the international conservation organisation) ICOMOS Czechoslovakia, Emanuel Hruška, expounded a comprehensive historical-materialist philosophy of heritage, fusing Riegl and Marx:

> Architectural structures, including historic towns, live, age and die like all organisms that must systematically renew their cells during their existence. We wish to find the most enduring measure of the regenerative process of the cells in the ensuing struggle between traditional and innovation, between preservation and destruction. The function of preservation is transformed into a medium for planning a tradition-conscious renewal of the urban environment in a socialist society.[5]

Figure 10.1 Heritage of revolution

(a) Cover of 1962 tourist guide pamphlet to the 'revolutionary sites' of the Lithuanian capital Vilnius. The painting, by Soviet historical artist Augustinas Savickas, depicts a turn-of-century demonstration in Gates of Dawn Street (later Gorki Street); (b) 2012 view of Gates of Dawn Street, now a key tourist thoroughfare in the Vilnius World Heritage Site

Alongside monuments of obsolete epochs, sites of socialist or egalitarian struggle could take their rightful place: of the 2,165 state-protected monuments in Leningrad in 1988, over 270 were listed mainly for their associations with the life of Lenin, including various museum-houses and flats. In Czechoslovakia, Prague's late-14th-century Hussite Bethlehem Chapel, destroyed in 1786, was rebuilt in 1950–3 as part of the new government's attempt to appropriate the Hussites as proto-radicals.[6]

Conservation and socialist reconstruction

IN the initial phase of postwar reconstruction, the role of conservation was problematised in eastern Europe by the omnipresent commemoration of the recent conflict, with the most devastated centres converted into vast war memorial landscapes. As with the interwar fascist countries, propaganda considerations meant that authenticity might become sharply divorced from the material substance of a building.[7]

Within the USSR, in both the Socialist Realist and Modernist phases of architecture, there was an overlap between heritage and the building of new propaganda-driven commemorative complexes. With the development of a secular socialist-cum-nationalist religion of the Great Fatherland Patriotic War, memorials proliferated across the country, shifting in style from Socialist Realism to forcefully expressionistic abstraction. But the implications of this for conservation, strictly defined, were at first glance limited. In the case of Volgograd (Stalingrad), for example, the building of a replanned city from scratch was combined with new, rhetorical monuments on war-sanctified sites, above all the vast Mamayev Kurgan complex, its grand staircases topped by the 52-metre-high statue, 'The Motherland Calls' (designed, like many other Soviet memorials, by the prolific sculptor Yevgeny Vuchetich). There was little explicit 'conservation' here: inaugurating Mamayev Kurgan in 1967, CPSU (Communist Party) General Secretary Leonid I Brezhnev declared that 'stones have longer lives

361

Figure 10.2 1979 tourist guidebook view of Lenin Square, Volgograd (formerly Stalingrad), with the remains of the 'Pavlov House' in the background

than people. But it is people and people alone who are capable of giving immortality to their achievements. The deeds of the heroes have made the stones of Mamayev Kurgan immortal.' Yet heritage was unobtrusively omnipresent: in some Stalingrad memorials, as in the early Christian era at Aachen Cathedral, small relics were preserved or embedded on a *pars pro toto* principle, for example in Lenin Square, where a new colonnaded ensemble, complete with Lenin statue by Vuchetich (1960), somewhat bizarrely framed the blasted stump of the Pavlov House, a four-storey tenement block held for two months by Soviet troops in 1942 against German attack. Still smaller were relics such as a shrapnel-pitted telegraph pole preserved outside the State Museum of Defence. A similar modern reconstruction formula was implemented in Skopje, Yugoslavia, following the 1963 earthquake: a few ruin fragments were retained, such as a railway station facade with frozen clock-face.[8]

With monuments of elite Art, however, Soviet propaganda concerns led towards faithful reconstruction: a national Central Scientific Restorative Workshop (TSNRM) was established in 1948. This movement culminated in the lavish recreation of war-devastated monuments in the Leningrad region after 1944. Leningrad was, of course, a city bound up with the October Revolution in Soviet lore and, as we saw earlier, had suffered less war damage than many other Russian cities.[9] Following the precedent of an 1837 restoration of the Hermitage following a fire, Stalin was persuaded by monuments director Anatoli Kuchumov that all key monuments should be restored to their prewar condition – a stupendous programme far removed from anything in the West, either in economic terms or in concepts of authenticity. Urgent repairs began in 1942, and in 1945 a Leningrad local monument department with around 1,000 craftsmen was established, perpetuating skills such as gilding and scagliola working. Mines in the Urals allowed destroyed precious metals to be freely replaced, and no expense was spared: for example, around 200 pounds of gold was used in the regilding of St Isaac's Cathedral and 'authentic' gas lighting was installed in the historic centre. The breakneck pace of repair of damaged buildings within Leningrad itself (with one-third of the ruined buildings in the centre repaired by December 1945 and by 1950 almost all destruction rectified) contrasted with the painstaking work at the devastated country palaces, where much facsimile ornamentation was necessary and initial restorations concentrated on a few showpiece elements. At Peterhof (patriotically renamed Petrodvorets), the efforts of craftsmen and citizens' work brigades from March 1944 ensured the main fountains were back in operation by August 1946, although restoration of the spectacular Neptune fountain (recovered from 'safekeeping' in Germany) took a further decade; Andreas Schlüter's Monplaisir palace of 1714, used during the occupation as a barracks, was reopened in 1961, whereas Adam Menelaws's Cottage was only fully restored in 1979. At Pavlovsk, photographs of whose devastation had been submitted as evidence to the Nuremberg war-crimes trial, over 40,000 fragments of decoration were incorporated in vigorous restorations, with most work finished by 1975, whereas at Pushkin (Tsarskoe Selo), following the gutting of the Catherine Palace, the initial work was mostly facade restoration. Only around a third of the ensemble had been restored by the 1970s, and a facsimile of the looted Amber Room, commenced in 1979, was only completed in 2003 with the aid of German company Ruhrgas. Although originally built by class enemies, such monuments, surviving or reconstructed, were accorded the highest respect by the socialist authorities.[10]

Poland: resurrecting a martyr nation

FURTHER west, the reconciliation of socialist principles and nationalist passions was more problematic, especially around the former confrontation line between Poland and Germany. It was only in Poland, with its passionately trumpeted wartime martyr status and its government policy of using cultural nationalism to defuse opposition to Soviet domination, that patriotic rhetoric was allowed decisively to get the upper hand in reconstruction – not least because the built environment had been specifically targeted in the German attempts at cultural decapitation after the 1944 Warsaw uprising. But this assertive stance had to tackle daunting inconsistencies stemming from Stalin's forcible westwards shift of Poland's territory, incorporating large numbers of 'enemy' towns.

The new Polish Communist administration quickly seized the initiative in heritage, establishing in 1945 a new Directorate-General of Museums and Monument Protection within the Ministry of Culture, in parallel with the nationalisation of all land. Before the war, state-sponsored conservation in Poland had been limited in scope, and dominated by Anti-Scrape suspicion of restoration. Now, the extent of the devastation, and its extreme politicisation, forced the abandonment of this bourgeois legacy, although a prewar inventory carried out by the Ministry of Education proved very useful, as did student measured surveys of the Warsaw Old Town.[11] Immediately, a grand programme of recreation of old-town centres was begun, above all in Warsaw. This programme revived the martyrdom/restoration discourse of 1920s' Belgium and France, embellishing it with an even more exuberantly pugnacious rhetoric. In 1963, for example, Stanisław

Figure 10.3 People's palaces

(a) Visit of 'ordinary Soviet citizens' to the restored Petrodvorets in 1953; (b) 1970s Intourist postcard view of the restored grand staircase of the Catherine Palace, Pushkin (Tsarskoe Selo)

Lorentz, head of the Directorate-General, explained at an international conference (held, appropriately, at Colonial Williamsburg) that

> the enemy had intended to raze Warsaw, and nearly did it. Therefore it was our duty to resuscitate it. The reconstruction of old Warsaw was the last victorious act in the fight with the enemy ... the finishing touch of our unbending struggle against enemy violence, and was so heroic in its very struggles for freedom and independence that it would be impossible to obliterate its historic aspect. We did not want a new city on the ruins of ancient Warsaw. We wanted the Warsaw of our day and of the future to continue the ancient tradition.

Lorentz's deputy, Jan Zachwatowicz, hailed the restoration as a reconciliation of nation, socialism and even religious duty: old Warsaw was a 'national reliquary' and 'the nation and its monuments are one'. His colleague Wojciech Kalinowski later recalled that the aim of the postwar restoration programme had been

> to educate and to develop social attitudes and patriotic feelings among the Polish people. . . . In 1945, we had two choices: to become a cultural mongrel among other culturally developed countries, or to be a proud nation with traditions and with visible traces of our thousand year old past. We chose the latter and we fulfil our patriotic obligations with zest and with obsession.

Attributing an almost Christ-like character to the Polish heritage, Kalinowski contended that the repeated invasions and devastations over the centuries had created a sacred memorial landscape of perpetual martyrdom and resurrection: 'Each time the country came back to life again and painstakingly rebuilt the ruined villages and towns, houses and churches.'[12]

The Polish restoration programme faced two major methodological challenges: first, how to reconcile its radicalism with the international Conservation Movement's entrenched concepts of authenticity; and second, how to reconcile the programme's explicit nationalism with the fact that much post-1945 Polish heritage had previously been 'enemy heritage'.

The first difficulty was skilfully sidestepped in the programme's centrepiece, the reconstruction of the Warsaw Stare Miasto (Old Town), led by the historic monuments department of Warsaw's Capital Restoration Office. As executed, this was a mixture of Viollet-style restoration with elements both of Socialist Realism and straightforward *Entschandelung*. Indeed, as Zachwatowicz openly acknowledged, the reconstruction of Old Warsaw was probably the all-time climax of Viollet's ideology: the provocation of Hitler's claim that Polish culture was an 'oxymoron', made it imperative to define a Viollet-like ideal of Polishness that could inspire the restoration. Before 1939, Warsaw had been a city of low standing in heritage terms, with a chaotic urban fabric of 19th-century tenements and factories that swamped its old town and lacked grand architectural ensembles. Now, the ruined city was reshaped on spacious and monumental lines, with a reconstructed Old Town taking its place in the planned order of socialism – a rediscovery-cum-modernisation celebrated internationally in a deluge of publications. In true Viollet manner, the historic zone was restored not to its 1939 condition, but to an idealised state purged of supposedly obtrusive late-19th-century buildings or 'bourgeois' plaster facade ornament, and enhanced by brightly coloured Socialist-Realist painted decoration. The strategy was to 'restore' the mid-/late-18th-century Baroque townscape of the heyday of Polish independence, as evoked in a series of topographical paintings of Warsaw by *vedutista* Bernardo Bellotto, nephew of Canaletto. These were thought by the Communist leadership to exemplify progressive 'Enlightenment'. Half a century later, another set of paintings by the same artist would shape the restoration of the German 'martyr city' of Dresden.[13]

In effect, Old Warsaw was remodelled around a two-dimensional image: but this was obscured, as in Viollet's case, by the restorers' meticulous craftsmanship and art-historical scholarship. The ruined fragments were carefully guarded, with boards displaying the national eagle, warning that they were 'monuments of national culture' that should not be demolished. The restoration had two main phases. The first, and by far the larger, in the late 1940s–early 1950s, included the Old Town Square (Rynek Starego Miasta) and the streets around, together with the classical streets running south, Nowy Świąt and Krakowskie Przedmiescie; this was superimposed above a modern east–west road in a tunnel, with an elaborate connection between the two by escalator. Mostly, the facades were rebuilt in facsimile, where possible as depicted by Bellotto, but containing modern flats, for occupation by members of the intelligentsia; the facades of Nowy Świąt were regularised and purged of the tall, ornate, 19th-century buildings that had dominated it before 1939. Also rebuilt were the previously overbuilt medieval walls, initially uncovered by Zachwatowicz in excavations of 1937–9 but now enhanced by construction of a spiky, largely hypothetical 'barbican': the juxtaposition of this and the baroque classicism of the streets created an entirely new eclectic mixture. The second phase of Old Town restoration concerned the chief architectural loss of 1944, the Royal Palace – a project deferred owing to its complexity, but always seen as an inescapable eventual duty. During the war, Lorentz's staff had rescued and hidden some 10,000 fragments of the demolished palace, at great risk to themselves. By 1969, Adolf Ciborowski's overview book, *Warsaw: a City Destroyed and Rebuilt*, predicted an imminent start on the project, which 'will strike the last chord in the symphony of the restoration of Old Warsaw'. In 1971, the new Communist leader, Edward Gierek, anxious to defuse popular unrest, finally authorised the reconstruction, which was triumphantly completed in 1984.[14]

Figure 10.4 The restoration of Warsaw

(a) 1987 view of Warsaw's restored Rynek Starego Miasta (Old Town Square); (b) 2009 view of the commemorative plaque at the square's south-west corner; (c) The 'Canaletto Room' in the rebuilt Royal Castle, featuring the Bellotto paintings used to guide the reconstruction of the Old Town. Photograph by Andrzej Ring and Bartosz Tropilo.

Immunised from doctrinal criticism through the city's martyr status, Warsaw's socialist restorers struck out in a direction that, at the time, seemed almost perversely divergent from the Conservation Movement's prevalent insistence on 'honest' differentiation of new and old, and which would only gain general international popularity 50 years later, in an age of global capitalist triumph. In its typically Socialist-Realist eclecticism and its bold juxtaposition with modern infrastructure, rebuilt Warsaw developed the principles of Viollet, and of 1920s' Franco-Belgian reconstruction, in a way that even anticipated the postmodern themed or branded historic city. In 1965, Zachwatowicz acknowledged that this had disconnected authenticity from the ruins' material substance and instead tied it to the ideological 'function of the object in the collective consciousness of society'. Faced with the inevitable criticisms at a 1979 conference that the result was a Disneyland, Kalinowski countered with the time-honoured rhetoric of the 'living' old town: 'the buildings are lived in and used'.[15]

A different situation applied in the previously German-controlled cities and monuments of western and northern Poland, places newly emptied of their inhabitants in 1945–6 and violently shifted from one national heritage to another. Regarding the conservation challenges of these cities' *Altstadt* areas, Polish conservationists had two responses: first, to sidestep the issue, by arguing that their significance was international (as when Lorentz hailed the Gdańsk Old Town as 'a valuable example of European town planning'); and second, to focus on 'restoration' to the earlier years of Polish domination – an architectural counterpart to the expunging of German signs and inscriptions (*usuwanie śladów niemczyzny*). Viscerally, their feelings were negative: Kalinowski claimed that Gdańsk and Poznań 'did not have the rich material and details of Warsaw', and Zachwatowicz argued of Malbork (Marienburg), in terms reminiscent of 1790s' France, that

> the former seat of the Teutonic Knights, whose deeds are inscribed in blood on the pages of Polish history, could not inspire the love of the Polish people. But, as an example of architecture, it should be preserved, despite the disastrous German restoration of the 19th century.

How could the architecture of these places be appropriated by Poland positively rather than destructively? To begin with, the architectural styles of their 'Polish' centuries, the Baroque and Renaissance, could be emphasised in any restorations – although the Gothic was also, perversely, revered as a symbol of Poland's Piast dynasty. As a bonus, all this would underline the innate cosmo-politanism of Polish architectural culture: in Kalinowski's words, 'Polish kings and magnates did not spare the money to bring talents from all parts of Europe to Poland.' From this could spring an equally inclusive conservation policy, aiming 'to give protection to all historic monuments, regardless if they were created by foreign occupants and regardless of the social class to which they belonged'.[16]

That, of course, was a rosy hindsight view. In reality, the 1940s and '50s were dominated by raucous and convoluted debates between nationalistic iconoclasm and preservation. In 1948, for instance, one provincial conservator (Zbigniew Rewski of Olsztyn) argued for the 'de-Prussification of architecture in the western territories' by 'removing towers and spires on particularly Prussian buildings', while another (Jerzy Güttler of Wrocław) pleaded that the Hohenzollern palace on Wrocław's pl. Wolności should not be targeted for demolition as a residence of Friedrich II, 'the Prussian king so hateful to our memory', but should be preserved as a set piece of 'cosmopolitan' classical architecture. Yet, on the ground, continuity ran alongside rupture. Thus, in Gdańsk, where Soviet bombing and shelling in 1945 had obliterated the careful 1930s' work of *Entschandelung* (other than two solitary houses, Brotbänkengasse/Chlebnicka 1 and Grosse Gerbergasse/Garbary 1 – the

367

latter still today displaying a telltale pre-1945 date plaque), the new Polish authorities, with unofficial help from outgoing German officials, embarked on a reconstruction whose facades were remarkably similar to the restorations of the 1930s, other than in their 'polonising' detail. The most important historic area, the 'Main Town', was tackled with furious energy by a department of 20 architects under Lech Kalubowski, who set out to recreate its 16th–18th-century appearance, stressing Italian Renaissance detail and supposedly Polish-style wall paintings, as (in Zachwatowicz's words) a 'visible symbol of the reunification of [the city] with Poland'. But as the historic centre of Gdańsk was far bigger than the Warsaw Old Town, a more thinly spread approach was adopted, beginning with chief public buildings and continuing to the everyday street-architecture, where the facsimile historical facades concealed rather modern layouts of new public-housing flats around wide inner courts: 'new houses from the 16th century'. Anything post-dating the Prussian takeover of 1792 was removed on principle and damaged building fabric on the margins of the Old Town was shipped off, Napoleon-style, for use in rebuilding Warsaw. In other ex-German cities, a similar pattern prevailed. In Wrocław, the destruction caused by the German designation of Breslau as a 'fortress city' (including the construction of an airfield in the city centre in March 1945) was followed by further devastation through looting and arson, and systematic Polish demolition of ruined buildings in the early 1950s, to allow millions of bricks to be sent to Warsaw. Yet the *Altstadt* itself was reconstructed with lavish care, ironically realising a scheme initiated in the 1930s by Breslau building commissioner Rudolf Stein.[17]

Figure 10.5 From Danzig to Gdańsk

(a) Panoramic view of the restored Old Town of Gdańsk in 1983, looking south-east from the town hall steeple: note the spaciously perimeter-planned layouts of the rebuilt street blocks and the surrounding expanses of cleared land around the Old Town; (b) 1983 detail of Old Town shop-front in Ul. Długa; (c) Ul. Biskopia, in an industrial inner-suburb of Gdańsk: a rare pre-1945 German shop sign (here denoting a dairy) that survived the *usuwanie śladów niemczyzny* ('removal of traces of Germandom')

The incoming Soviet authorities in the Russian sector of East Prussia and the city of Königsberg (Kaliningrad) dealt rather differently with the heritage left behind by the expelled Germans. There, the largely 16th–18th-century castle, reduced to ruins by 1944 RAF bombing and 1945 Soviet shelling (in which the dismantled Amber Room from Tsarskoe Selo probably perished), was eventually demolished in 1968 on the orders of Leonid Brezhnev, and replaced by a model socialist civic centre, including a Modernist, tower-like House of Soviets in the filled-in moat.[18]

Uncertainty and distrust: postwar East Germany

THE nationalistic aggressiveness and international assertiveness of postwar Polish restoration was mirrored in reverse in the policies of East Germany.[19] There, although the aspiration towards an instrumentalised socialist *Denkmalpflege* soon faded, conservationists were left with the overwhelming problem not only of repairing the wartime destruction (though that was generally less than in Western Germany) but also of their relegation from a core to a peripheral position within government priorities: the late 1940s saw the suppression of all *Heimat* organisations. The turn-of-century structure of provincial conservation offices and laws had survived the Third Reich but was distrusted by the central GDR government, which would eventually superimpose on it, in 1961, a central Institut für Denkmalpflege based in Berlin. In the meantime, the provincial conservators struggled on as best they could. An account of a November 1947 field trip by Hans Nadler, later (from 1949) Landeskonservator of Saxony, showed the heroic improvisations and privations needed in order to maintain any oversight over the countless dilapidated historic buildings in postwar eastern Germany, including all-night bicycle rides in the freezing cold, harassment at police checkpoints and dingy, *ad hoc* accommodation. Reconstruction could be concealed within other socialist programmes, as with Nadler's post-1956 rehabilitations in the undamaged but decayed *Altstadt* of Görlitz, which he represented as a housing programme. Paradoxically, although the early postwar Ulbricht leadership distrusted the classical heritage of Berlin for its supposed implication in Prussian militarism and derided Schinkel as a *Fürstenknecht* (lackey of the nobility), the Schinkel tradition was favoured by the Russians, as an ideal basis for a German variant of Socialist Realism.[20]

The immediate postwar situation in Dresden exemplified the wider problems the Conservation Movement faced in East Germany. Here, the official rhetoric of martyrdom was a milder version of Warsaw's, but without any prospect of immediate relief and restoration. Instead, the focus of the city council, from 1948, was on the socialist reconstruction of the city with grandiose Socialist-Realist axes. In the immediate aftermath of the war, some conservationists threw up their hands in despair: Dresden art historian Eberhard Hempel argued that the remains should be left to moulder for good as 'piety-imbued ruins – many parts of our inner cities will parallel the *Forum Romanum*'. For Nadler, the main preoccupation was with the safeguarding of Dresden's ruins, above all that of the Schloss – which, like all its cousins in the GDR, was potentially under threat of ideologically motivated demolition. His ingenious, and in the end successful, expedients included the establishment of temporary government offices or activities in the ruined buildings to fend off any sudden demolitions. Immediate restoration was confined to a few set pieces, above all the 18th-century rococo Zwinger, where architect Dr Hubert Ermisch, director of a comprehensive interwar restoration scheme, was now brought back to recover the site from shattered ruin – a task given enthusiastic official backing and eventually concluded in 1963. Rebuilding of some Dresden churches

also began, notably at the Kreuzkirche where the burnt-out ruin, denuded of almost all its *Jugendstil* decoration, was restored by architect Fritz Steudner with a harshly Sublime rough-textured internal finish, immediately after the war. Likewise, the Catholic Hofkirche of 1737–58 received an emergency roof in 1950. For the moment, the more emotive ruin of the Frauenkirche remained unrestored; indeed, some of its rubble was cleared away. And there was not yet any question, as in Warsaw, of attempted restoration of the Bellotto skyline-image.[21]

Linking Germany and Poland was the challenge of commemorating the genocide that had been planned in the one and committed in the other, an issue that rumbled throughout the postwar period. The normal socialist-bloc policy was to focus commemoration on political prisoners rather than on Jews or other persecuted ethnic groups. In Poland, the Communist parliament pledged in 1947 that the remains of Auschwitz would be 'forever preserved as a memorial to the martyrdom of the Polish nation and other peoples'. What was in fact mainly preserved was the section that housed Polish political prisoners, while the Birkenau zone dedicated to mass extermination of Jews fell into ruins. By the late 1950s, survivor organisations argued that exclusive reliance on the built remains was not explicit enough and pressed for a formal commemorative monument. In 1957–9, an international competition jury chaired by Henry Moore examined over 400 entries and eventually arrived at a joint committee design, collaging various tomb and sarcophagus motifs, which was completed in 1967.[22] In Warsaw, the treatment of the area of the former Ghetto contrasted strongly with that of the Old Town. Formerly dominated by drab 19th-century Berlin-like tenements, the area was largely destroyed in 1944, and so the new government decided to redevelop it as a planned socialist suburb, memorialising the Ghetto uprising instead through a variety of inscribed stones and monuments. This began in 1948 with a massive, somewhat traditional figurative memorial by Nathan Rapoport, originally set in a sea of rubble, but later surrounded by Modernist tower blocks.[23]

In the GDR, the centrepiece of the socialist memory landscape was the former concentration camp of Buchenwald, established in 1937 outside the historical city of Weimar, with its strong associations with Goethe. The camp witnessed in 1944 the murder of German Communist party chief Ernst Thälmann and then served as a Soviet internment centre (1945–50). When it became East Germany's Valhalla of anti-fascist sacrifice in the 1950s, following the dynamiting of a Bismarck monument in 1949, it already had a formidable history of multiple overlays of association. The GDR's memorial scheme at Buchenwald, completed in 1958, attempted to end this confusion with a highly monumental scheme, comprising a triumphal staircase ascending to a bell tower and heroic sculptural group of concentration camp prisoners.[24] This kinetic arrangement, leading upwards from darkness to light and redemption, ironically resembled Israel's equivalent, the Yad Vashem Martyrs and Heroes Memorial, begun in 1953. Schinkel's Neue Wache acted as a Berlin counterpart to Buchenwald, and Tessenow's 1931 scheme, with its roof oculus, was cosmetically upgraded. In 1956 the GDR Politburo decided the former war memorial should become a 'memory site for the victims of fascism and both wars', and in 1968–9 this role was enhanced through a new changing of guard ceremony and a plethora of ideological symbols, including a crystal-encased eternal flame, a GDR

facing page

Figure 10.6 GDR urban reconstructions

(a) The Zwinger in Dresden, as restored under Hubert Ermisch (completed 1963); (b) The Kreuzkirche in Dresden, showing Fritz Steudner's idiosyncratic early postwar restoration with 'primitive' internal finish; (c), (d) 1990 views of the still largely ruined Schloss in Dresden, showing GDR reconstruction scheme under way; (e) 1990 view of the Lange Strasse, Rostock, showing stately 1952–5 reconstruction in 'Hanseatic' variant of Socialist Realism

Figure 10.7 The two faces of Buchenwald

(a) Before 'die Wende': 1986 tourist postcard set; (b) The 'Neukonzeption': 1999 dedication ceremony of Nazi camp remnants unearthed by archaeological excavation

coat of arms set in the back wall and the remains of both a 'resistance fighter' and an 'unknown German soldier'.[25] The architectural alterations were designed by Lothar Kwasnitza and Gerhard Schaubing, with Heinz Mehlan as exterior consultant architect.

In socialist-bloc countries that had suffered less war damage than Poland and the GDR, a variety of attitudes to conservation prevailed, mostly involving stronger continuity with prewar ideas. In Hungary, where the Baroque period was especially prized, turn-of-century Austro-Hungarian concepts of urban conservation and opposition to restoration continued; and national monuments (MOB) chairman Tibor Gerevich steered through integrated legislation in 1949. However, Budapest itself had suffered extensive damage at the end of the war. The classical royal Buda Castle complex (ex-seat of regent Miklós Horthy) was ravaged by a lengthy Soviet siege, along with the surrounding *Altstadt* – which mainly comprised Baroque houses and churches. Here, postwar rebuilding was slower and more cautious than in Warsaw. The castle was eventually restored in a simplified form, the medieval walls archaeologically excavated, and the old town reconstructed from the '60s. The

facing page

Figure 10.8 The Neue Wache – weathervane of German heritage identity

(a) External view in 1982; (b), (c) Exterior views and (d) interior taken immediately pre-reunification in 1990, with ceremonial guard still in place but access now permitted to the entire interior; (e), (f) Exterior and interior in 2010, following the building's 1993 rebranding as a 'National Memorial to Victims of War and Tyranny'; the 1968–9 GDR scheme was expunged and a reproduction Kollwitz sculpture was installed

only structure immediately rebuilt was the emblematic 1848 Chain Bridge, blown up in 1945 but fully restored by 1949.[26]

In Czechoslovakia, the interwar years had witnessed an overall presumption towards repair rather than restoration, and the evolution of a progressive urban heritage policy based on the precepts of Riegl and Dvořak, including an ambitious 1935 rolling area-conservation programme of 35 'reservation-towns'. The 1938 Munich Agreement having ironically averted the threat of violent attack, the postwar socialist government was able to largely continue its predecessors' policies. It consolidated 30 of the reservation towns in a 1958 conservation law and set up in 1954 a well-resourced and non-politicised State Institute for the Reconstruction of Historic Towns and Monuments (SÚRPMO). This institute expanded from a specialist studio (1949–54) in the organisation 'Stavoprojekt', led by architect Vilem Lorenc, tackling not only conservation projects in Prague and other reservation towns, but also educational and consultancy work, and – more unexpectedly – new building projects in Kuwait and Iraq in the early 1980s: its staff expanded eventually to include over 750 design and construction specialists.[27]

In these countries, the socialist bloc was at its closest to 'normal' Western Conservation Movement practice – although even here, the arbitrary excesses of communist rule were never far from the surface in the unstable early postwar years. In 1950, for example, the 12th-century Tepl Monastery in Czechoslovakia was stormed by hundreds of secret police during an anti-church campaign. They expelled and imprisoned the monks and turned the monastery into a barracks; it was eventually handed back only in 1990, as a virtually uninhabitable ruin.[28]

Socialist Realism to Modernism: conservation implications of socialism's shifting architectural landscape

In parallel with the repairs and restorations of the early postwar years, new architecture in the socialist bloc also developed along distinctive lines, different from the sharp differentiation of old and new prevalent in the West. Under the influence of Socialist Realism, new buildings were designed in an eclectic historicism supposedly reflecting local tradition. Yet, at the same time, there was a requirement for the opening up of grand classical axes, which on some occasions necessitated radical demolitions of historic fabric and on others left historic buildings isolated in vast open spaces. Sometimes, these grand gestures were prompted by wartime destruction, but their ultimate precedent was Kaganovich's 1935 General Plan for Moscow. It had tried to superimpose a new, Haussmann-like grandeur on the dispersed Russian urban tradition, whose individual elements ranged from kremlin citadels down to individual wooden houses. The 1945 victory further emboldened this programme, expanding its scope to Socialist-Realist skyscrapers and palatial metro lines, usually requiring extensive demolitions of old houses. In contrast to Hitler, Stalin's urban vision had no significant place for Old Towns, and many urban plans across the USSR left elite monuments stranded in *tabula-rasa* redevelopments or vast open spaces, as with the Triumphal Arches of Moscow and Leningrad, the Church of St Simeon of the Pillar in Moscow's Kalinin Avenue and the Golden Gates in Kiev and Vladimir.[29]

Elsewhere in the socialist bloc, the same principle was applied to reconstruction of war-damaged zones, including construction of new arteries outside the old-town cores. In Rostock, a wide boulevard, the Lange Strasse, was built in 1952–5, lined with Socialist-Realist facades in a 'modern Hanseatic

Gothic' style, up to nine storeys high. But, in parallel with this, the town's main commercial street, Kröpeliner Strasse, was partly restored and eventually pedestrianised. In Dresden, a '*magistrale*' axis in a Socialist-Realist style (here, 'Saxon classical') was cut through the Alter Markt area, whose predominantly late-19th-century commercial buildings had conveniently perished in the 1945 bombing. And in East Berlin, the main new axis, the Stalinallee, running eastwards from Alexanderplatz in a neo-Schinkel Socialist-Realist style, mostly replaced little-valued 19th-century tenements. The chief act of deliberate heritage demolition in Berlin, the dynamiting of the dilapidated Baroque Stadtschloss in 1950, was done for ideological rather than town-planning reasons. A similar situation prevailed in Warsaw, where the chief Socialist-Realist set pieces, including a Moscow-style skyscraper, the Palace of Culture and Science, were located away from the Old Town on low-status sites.[30]

From the late 1950s, following Khrushchev's 1954 call for industrialised building and denunciation of Socialist Realism, the socialist bloc converged rapidly with the West in its attitude to the relation of old and new in architecture. This allowed it to play a full part in implementing the principles of the Charter of Venice in the mid-1960s, and from the late 1960s it even saw some mild environmental protests, echoing the Western upheavals we traced in Chapter 9. The 1958 Moscow conference of the International Union of Architects facilitated the rapid Soviet assimilation of Modernist principles, including the functional spacing-out of the city and the differentiation of new and old. This soon spread beyond the USSR: in the GDR, the extension of the Stalinallee (renamed Karl Marx Allee) in the 1960s switched to Modernist towers and slabs. Resource shortages often dictated considerable time-lags in implementation: the prestigious Plac Grunwaldzki project in Wrocław, where a monumental scheme of 1952 was replaced in 1960 by a proposal for Corbusier-style high towers, was only built in the 1970s.[31]

However, the Khrushchev era in the USSR also saw a sharp turn away from the relative strength of heritage conservation in Stalin's more nationalistic later years, above all in the sudden anti-religious campaign launched by Khrushchev in 1959–64, with over half of Russia's 20,000 surviving churches shut down and many destroyed. Soviet politicians and professionals increasingly argued for a concept of rolling obsolescence and for the scientific necessity of demolition of successive epochs of building stock, a notion embraced more energetically in the post-Khrushchev socialist bloc than in the West. A slightly unexpected heritage repercussion of this embrace of modernisation was a reputational and physical rehabilitation of the dilapidated Bauhaus building in Dessau, previously neglected as inconsistent with Socialist Realism, but hailed as 'one of the finest expressions' of Modernism in mid-1960s' German and Soviet books; it was declared a historic monument in 1964, and the first of two GDR restoration schemes began in 1976.[32]

When linked to a gargantuan programme of industrialised housing construction on the outskirts of historic cities, the effect of the new Modernist zeal, especially in the Soviet Union, was to create a sudden and exaggerated version of the new–old polarisation normal in the West since 1945. In Russian cities such as Novgorod, Kaluga, Rostov or Vladimir, modern construction burgeoned across the outskirts, while some old buildings were respectfully assigned cultural/educational functions within the same city-plans, and conservation efforts in historic areas expanded. In some cases, such as Suzdal, the entire urban area was officially exempted from industrial and mass housing development. Through old-town enhancement and establishment of rural open-air museums, or other restoration projects such as Roman temple anastelosis in Armenia, heritage tourism was systematically encouraged, but with the didactic intention 'to apprehend and train aesthetic taste in the best examples of national art', excluding any hint of commercialism.[33]

In Plovdiv, Bulgaria's second city, a similar museological approach was adopted. A 35-hectare area crammed with mid-19th-century Bulgarian Renaissance houses was designated as an 'Ancient Reserve' (effectively, an open-air museum) in the 1950s and converted to cultural, educational and tourist functions, the inhabitants having been decanted to new housing projects.[34] On the other side of the Atlantic, the 1959 Cuban Revolution called a halt to US-style commercial redevelopment of Old Havana; a new State Department for Restoration undertook as its first major project the restoration of the 1580 Castillo de la Real Fuerza, the capital's oldest building. Under the zoning principle, however, there was always a danger that the heritage zones would be too aggressively hemmed in by mass demolition and redevelopment – as we will see shortly – or, as in the Plovdiv case, that they would become neglected and even derelict.[35]

The Baltic States provide an interesting case study of the diversity of post-Khrushchev planning and heritage policy in action: although the final, harsh phase of Stalinism in 1950–3 had seen calls for removal of 'foreign' or 'German' historic monuments, from 1960 there was an almost complete about-turn, with full-blooded Soviet mass housing production on city outskirts combined with increasingly meticulous conservation of historic centres and restoration projects with latent nationalistic overtones. In Lithuania, for example, the lost power of the medieval Grand Duchy was evoked in a Marienburg-like restoration of one of the main power-bases of the Grand Dukes, Trakai Island Castle (1951–61); rehabilitation of the 240-hectare Vilnius Old Town began as early as 1958. In Estonia, vast peripheral housing developments, culminating in the partly realised Lasnamäe project, originally planned in 1973 to house 180,000, were carefully distanced from the Old Town of Tallinn. The latter's recovery and restoration from wartime Soviet bombing was overseen by a new Scientific Restoration Workshop, established in 1950 by the State Architecture and Building Committee, and expanded into a State Restoration Directorate in 1969. Within the Tallinn Old Town, which was declared a protected zone in 1966 (the first in the USSR), an Inspectorate for Protection of Architecture began systematic urban conservation. From 1959, the early-15th-century Town Hall was purged of 19th-century alterations (in a lengthy programme overseen by Teddy Böckler), and the city walls and towers were methodically restored from 1954 onwards in a flamboyantly Viollet-like way, with tall, red-tiled spires. Occasionally, as in the case of the eastern Estonian town of Rakvere, standard housing blocks were mixed in with 19th-century timber houses in a more contextual manner. Estonia was the first republic of the USSR to pass heritage legislation, in 1961; all-Union legislation only followed in 1976 and 1982.[36]

In the Latvian capital, Riga, the historic core was more variegated, combining a conventional *Altstadt* with a remarkable turn-of-century *Jugendstil* heritage. Here, again, systematic restoration began after Stalin's death in 1953, an early set piece being the 'Three Brothers' in Maza Pils Street, a group of three townhouses restored in a gabled Baltic style in 1953–7, partly reusing fragments from elsewhere. Around the Ratslaukums (Town Hall Square), a complex succession of projects evolved from the late 1950s onwards. Here, a similar story unfolded to that in Hildesheim. The former centrepiece, the fantastically gabled Renaissance Schwarzhäupterhaus (Black Heads House), was demolished in 1948 following 1941 war damage. Instead, from 1958 onwards, a number of new buildings in a contextual Socialist-Realist style by architect O Tilmanis were built around the space, and in 1969–70 a new, boxy Modernist museum dedicated to the revolutionary Latvian 'Red Riflemen' was built at right-angles across the space, coupled with a grandiose new statue of a group of Riflemen. Some Latvians speculated that the building was intended to prevent a reconstruction of the prewar square, but as we will see later, a facsimile of the Schwarzhäupterhaus was eventually built in the 1990s.[37]

376

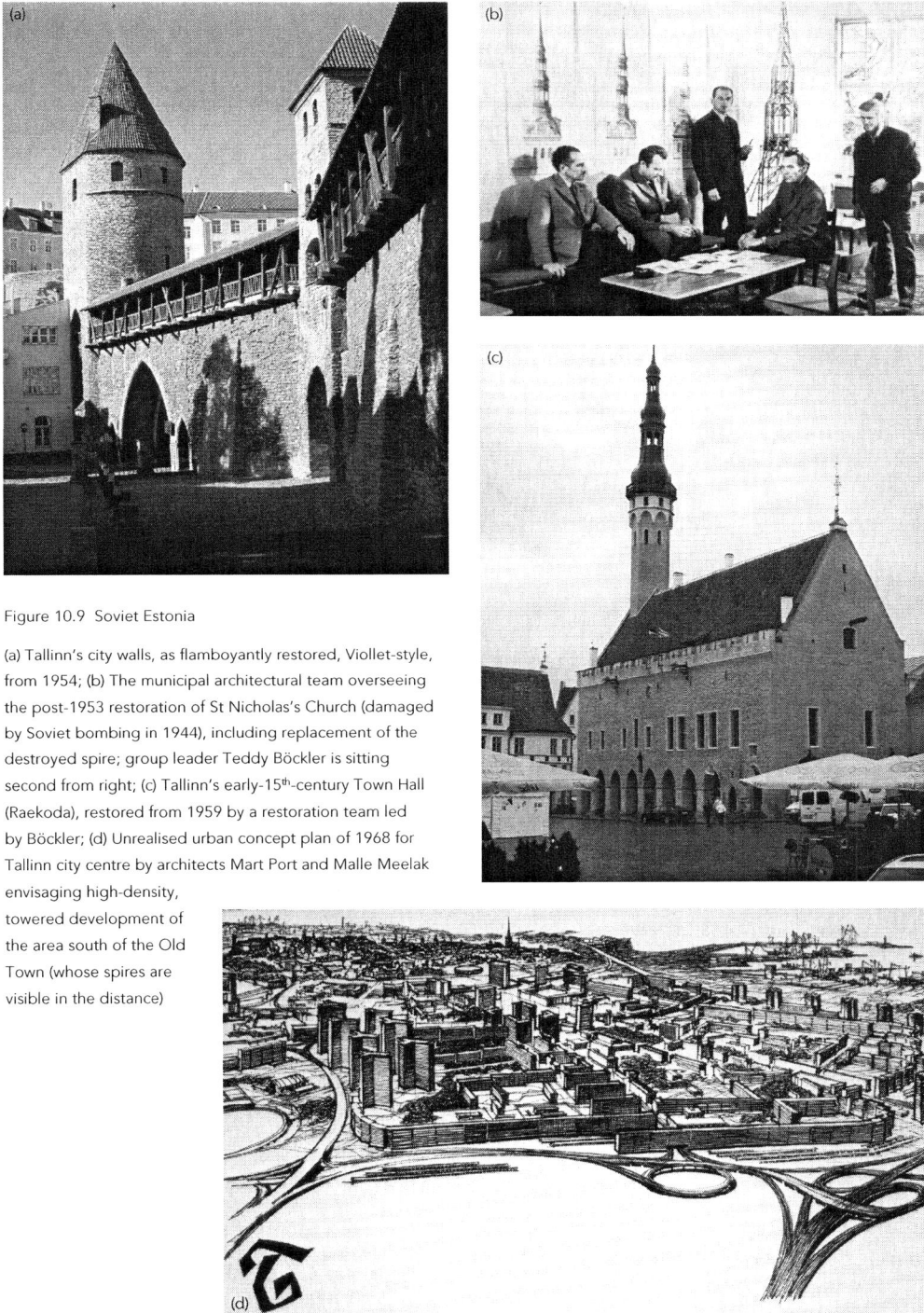

Figure 10.9 Soviet Estonia

(a) Tallinn's city walls, as flamboyantly restored, Viollet-style, from 1954; (b) The municipal architectural team overseeing the post-1953 restoration of St Nicholas's Church (damaged by Soviet bombing in 1944), including replacement of the destroyed spire; group leader Teddy Böckler is sitting second from right; (c) Tallinn's early-15th-century Town Hall (Raekoda), restored from 1959 by a restoration team led by Böckler; (d) Unrealised urban concept plan of 1968 for Tallinn city centre by architects Mart Port and Malle Meelak envisaging high-density, towered development of the area south of the Old Town (whose spires are visible in the distance)

THE NEW BUILDING OF ART EXHIBITION HALL FITS IN WELL WITH THE OLD TOWN

Fig. 10.10 Soviet Latvia and Lithuania

(a) The Three Brothers group in Maza Pils Street, Riga, restored 1953–7; (b) The 1969–70 Red Riflemen monument and museum in Riga's Town Hall Square; (c) Illustration in a 1977 guidebook to Vilnius (Arolfas Medonis, *Vilnius*, Vilnius, 1977), showing the Art Exhibition Hall, a 1967 intervention in the Old Town; (d) Heritage landmark plaque at Pilies 12, Vilnius, commemorating the 1960 restoration of the 16th-century house; the initials 'T.S.' (Soviet Socialist) were carefully removed after 1990 from this and all similar plaques

Plattenbau, protest and international engagement: 1960s' renewal debates in the East

HAVING converged with Western Modernism in the late 1950s, the socialist bloc also echoed the 1960s' Western boom in assertive Modernist redevelopment, in a more exaggerated form: some architects and planners even argued that following rejection of Stalin's Socialist Realism, a proper socialist modernity should involve rejection of everything old.

This forced the Conservation Movement on the defensive for several years and, during the mid-/late 1960s, some city centres became directly overshadowed by prestige tower developments, such as the array of slabs in Moscow's Kalinin Prospekt, planned by Khrushchev across the old Arbat district: the most egregiously intrusive development was probably the 3,200-room Hotel Rossiya (1967–71, by architect Dimitri Chechulin), at the back of the Moscow Kremlin, comprising an ungainly 21-storey slab protruding from a vast podium-like block. In East Germany, whose provincial conservators had been gravely weakened by the 1961 move to a centralised national heritage administration, the mid-1960s saw an increasingly confrontational attitude towards conservation by many contemporary architects and politicians. In 1966, housing rehabilitation was roundly condemned by influential architect Richard Paulick, who argued that seven new prefabricated *Plattenbau* flats could be built for the same cost as four restored ones, while the *Sächische Neueste Nachrichten* attacked the 'scholarly bourgeois' attitudes of Nadler's Dresden conservation office in an article entitled, 'Does a socialist city have to be a museum?' In that year, Nadler nearly gave up his post out of despair at the verbal attacks and the mass demolitions: the years since the destruction of the Berliner Stadtschloss in 1950/1 had seen a succession of further disappearances of individual monuments, including the Potsdam Schloss (to be replaced by a planned gargantuan monument to Karl Liebknecht) and Schinkel's Bauakademie in 1962. But very soon, as in the West, the tide would begin to turn: as early as 1967, Paulick admitted that systematic clearance of tenements would inevitably outstrip the potential pace of new construction and that some old housing would therefore have to be kept. And the following year, as we will see shortly, the demolition of the Paulinerkirche in Leipzig would bring public discontent into the open.[38]

More serious than these headline demolitions, though, were the unprecedented area redevelopments that pockmarked urban East Germany from the 1960s. In some cases, the problem was decay rather than outright clearance, as in Quedlinburg, where the castle and a fraction of the city's timber-framed core were carefully restored and the rest left neglected: many houses were still privately owned, and the owners were reluctant to spend anything on upkeep. More often, especially in the *Bezirksstädte* (regional centres), burgeoning pressure for redevelopment prompted the levelling of swathes of historic fabric and the erection of prestige skyscrapers, as in the centre area of Zwickau, or in Bautzen, where a skyscraper slab was built right next to the *Altstadt*. In 19th-century tenement districts, once the authorities grasped the impracticability of mass demolition, large-scale *Entstückung* programmes were launched instead. The end of the 1960s saw a jarring combination of increasingly ambitious urban conservation projects, including wide-ranging townscape analyses, and increasingly destructive redevelopments, especially in areas relatively untouched by the war, such as Thuringia, where towns such as Gotha were reduced to museum-like cores ringed by modern mass redevelopments.[39]

In the West, the '1968' protests provoked a dramatic policy volte-face, but socialist-bloc policy shifted more cautiously. As in the rejection of Socialist Realism, a decisive role in brokering change

Figure 10.11 Russian heritage in the Brezhnev years

(a) The Hotel Rossiya, Moscow (1967–71, by D Chechulin): 1981 view showing the mid-17th-century St George's Church; (b) Suzdal, view of St Alexander Monastery following its 1960s–70s rebuilding (as a museum of the restoration of the city)

was played by a relatively small elite within the Soviet Union. Here, unlike some satellite states, there was a surprising amount of scope for restrained controversy in areas such as heritage. In the USSR, the scientific branch of conservation saw uncontroversial development throughout the postwar years. The TSNRM mutated in 1968 into an All-Union Production Scientific Restorative Combine (VPNRK), charged with projects ranging in scale up to restoration of the Moscow Kremlin (from 1972) and then in turn transformed in 1981 into an All-Union Specialised Restorative Production Combine (*Soiuzirestavratsiya*) with the same tasks. But the 1960s also saw an upsurge in discontent provoked by Khrushchev's demolition campaigns and aggressively industrialised embrace of Modernism, beginning with a critical 1962 article in the magazine *Moskva* and the foundation in 1964 of a pressure group, '*Rodina*' (Motherland). In July 1965, the year after Khrushchev's dismissal, the USSR Communist Party established an officially backed 'voluntary' preservation organisation on an enormous scale: the All-Russian Society for the Safeguarding of Historical and Cultural Monuments (VOOPIK). VOOPIK was granted massive government aid, including a staff of 1,500 distributed across the country, and subdivided into thematic groupings (including folk art, literary and musical sections, and mass publicity). In the same year, the Politburo also significantly conceded the preservation of a group of tiny Orthodox churches adjoining the planned Hotel Rossiya. Working within the socialist tradition of the mass mobilisation of the whole people, VOOPIK was to harness private citizens' enthusiasm to the Ministry of Culture's programmes of monument inventorisation and conservation workshops, especially by organising student brigades that would make 'broad use of our cultural heritage for the purpose of a patriotic upbringing'. With those resources, unsurprisingly VOOPIK could soon claim a vast membership: by 1975, there were 22 million members in the Russian Federation alone, including 400,000 in the city of Leningrad, and one-third of the entire population of the Armenian Soviet Socialist Republic had joined![40]

This was a very different phenomenon from conservation radicalism in the West, neutralising public agitation while providing a forum for somewhat nationalistic conservation activity on a huge scale; but it formed part of the same story of explosive growth of the Conservation Movement as a whole. And it was followed by further protests against demolitions. In 1967, writer Vladimir Soloukhin bemoaned the litany of destruction since Stalin: 'In place of the unique, rather archaic, typically Russian, matchless city of Moscow, they have built an average European city, not notable

380

for anything special.' Protesters also attacked the ongoing proposals for *tabula-rasa* redevelopment of much of the city centre under the provisions of the city's draft General Plan. This was furiously criticised by *Rodina* member Ilya Glazunov, an architect with close links to party bosses, to whom he submitted in 1971 an album of old Moscow monuments threatened by the General Plan proposals, arguing that the Hotel Rossiya and Kalinin Prospekt were not proud accomplishments but heinous crimes. Although the General Plan was formally approved that year, Viktor Grishin, First Secretary of the Mossoviet (city council), subsequently made a speech criticising the planned demolitions and advocating a more contextually respectful approach. Now the revolt against the clearance policy began snowballing within the elite and spread from Moscow to nationwide decision-making circles.[41]

However, in the USSR, the balance between demolition and urban conservation remained finely balanced right into the 1970s. Protests gained further momentum, as they had earlier in the West, from the provocation of further losses – for instance when, in preparation for the 1972 visit of President Nixon to Moscow, the main road from Vnukovo Airport to the Kremlin was cleared of decrepit houses and churches. In that year, VOOPIK member Sergei Mikhalkov, together with Glazunov and numerous academics, signed a protest letter to the party Central Committee. Favourably impressed, prime minister Alexei Kosygin secured Politburo support for a swing to a conservation-dominated approach, and in 1973 nine conservation zones were created in central Moscow, within which significant demolition was virtually forbidden; in 1974 the Russian Federation government produced a new and radically expanded national monument register, and in 1976–7, new all-Union heritage laws were implemented. There was, however, still no officially recognised Soviet area conservation doctrine. Where demolition continued in old towns, it was often possible to salvage traditional wooden buildings and relocate them in open-air museums; by 1979 there were 17 such museums in the Russian Federation and an 'Ancient Kiev' zone in the Ukrainian capital, and a 100-volume Register of Historical and Cultural Monuments of the Peoples of the USSR was under preparation. In the GDR, controversy burst out slightly earlier, in May 1968, when Leipzig's university church, the 13th–16th-century Paulinerkirche in the Karl Marx Platz (Augustusplatz) was blown up on the orders of Walter Ulbricht, to make way for a university redevelopment designed by Hermann Henselmann and Horst Siegel and built in 1973–5, including a tower in the form of an open book.[42]

In some parts of the socialist bloc, there was less urban destruction from the beginning. In Czechoslovakia, the plentifully staffed State Office for Preservation of Historic Monuments in Prague, headed by Dr Vladimir Novotny, steered the restoration of numerous threatened buildings in the 1950s and '60s and ensured the safeguarding of old towns in the Dvořák urban-conservation tradition. Spared destructive attacks, specialists were able to substantially extend the scope of conservation in Czechoslovakia, including many ensembles of rural folk vernacular, alongside secessionist or early Modernist houses by Loos, Mies and others. In 1975, the president of the Czechoslovak national committee of ICOMOS, Emanuel Hruška, argued that preservation was participating constructively in the 'progressive urban renewal' demanded by socialism, conferring on it an 'organic sense of continuity'.[43]

Although the socialist bloc's gradual convergence with the West partly stemmed from its own internal dynamic, it was also stimulated by an external factor – the emergence of an international conservation network bridging the East–West confrontation and focused (as we will see in Chapter 11) on a succession of conservation charters requiring the endorsement of state governments. Previous East–West collaboration had been episodic: 1956–8 saw a joint project to restore the

Brandenburger Tor, with the West casting the quadriga and the East restoring the building itself. Precedents for a more concerted engagement were provided by the impartiality of the UN and its cultural arm, UNESCO, as well as other built-environment organisations such as the International Union of Architects, in which the Soviet Committee had continued to play a vigorous part.[44] Acknowledging the prestige of the postwar restorations in Poland and the USSR, the 1964 Charter of Venice emphasised there was no single universal framework of restoration theory or political-economic organisation and that each country had to develop conservation policies suited to its own conditions. And significantly, it was in Warsaw, the very next year, that the international conservation organisation, ICOMOS, was founded.

In November 1972, the Soviet volte-face on conservation was followed by the USSR's accession to the UNESCO World Heritage Convention – one of its earliest signatories – and participation of some socialist countries in European Architectural Heritage Year 1975. The entrenchment of détente within the Conservation Movement seemed irreversible and was celebrated in a lavish conference on Central and Eastern European Conservation in New York in November 1975.[45] The new climate of engagement was skilfully exploited by the Poles, whose state conservation workshop service, the PKZ (Pracowanie Konserwacji Zabytków, or Monument Restoration Atelier), became an international restoration consultancy empire rivalled only by its Czech counterpart, the SÚRPMO. Originally established in 1950 by the Ministry of Culture for the Warsaw Old Town project and upgraded in 1960 into a network of national and regional offices, the PKZ built up its illustrious reputation in a succession of projects culminating in the Warsaw Royal Palace reconstruction. Its foreign consultancies were concentrated in socialist countries (especially after 1968, when it received major commissions in cities such as Stralsund (from 1970), Tallinn (from 1978) and Potsdam (from 1977), as part of Poland's reward for its 'fraternal' participation in the invasion of Czechoslovakia, but it also worked in the West, for example in West Germany. Ingeniously, PKZ used a trademark logo based on the blue/white 'distinctive emblem' of the Hague Convention. Within the socialist bloc, a variety of working groups and conferences pursued wider

Figure 10.12 (a), (b) Promotional literature of the Polish PKZ and Czechoslovak SÚRPMO

heritage collaborations, including urban conservation training sessions in Czechoslovakia (1984) and Poland (1986).[46]

1970s and 1980s: Indian summer of socialist urban conservation

IN the socialist countries, the 1970s and early 1980s did not, of course, follow the Western path of welfare-state crisis and early neo-capitalist enterprise in conservation. In the Brezhnev years, change was restricted in scope, with many fits and starts along the way and overt public critiques liable to be repressed at will. In 1970s' Moscow, over 2,200 historic buildings were demolished, whereas 1982 saw largely ineffective demands for a tripling of the Russian Federation heritage list; and in East Germany, the 1970s witnessed a gradual depoliticisation of heritage and disappearance of taboo periods of feudal or bourgeois rule.[47] A popular paperback, *History of Architecture*, published in 1973, and a 1974 architecture guide to East Berlin, included all periods up to and including contemporary Modernism. Impelled by the exemplary effect of the 1975 European Architecture Heritage Year, the GDR introduced a special conservation law of June 1975 – despite the frequent reverses of urban heritage within the country. Increasingly, East German public opinion shifted towards an embrace of conservation, encouraged by educational initiatives and courses from 1968 onwards. As in the USSR, progressive party officials became sympathetic: for example, Nadler's office in Dresden was increasingly protected from Berlin interventions by Hans Modrow (regional party secretary from 1972). Conservation was less often pilloried as anti-socialist.[48]

Despite economic stagnation, the socialist embrace of the Conservation Movement allowed a succession of prestigious restorations of individual monuments to be executed in the GDR from the 1970s, including the Gendarmenmarkt/Schauspielhaus complex in Berlin and the 1870s' Dresden Opera House, sumptuously restored (with a modern auditorium) by Wolfgang Hänsch in 1977–85. Major churches were often restored rather than neglected or demolished from the 1970s, as with the ruined Marienkirche in Prenzlau (from 1970) or Greifswald Cathedral (1981–9).[49] Most significant of all, early 1970s' East Berlin saw a pioneering experiment in tenement conservation, larger than contemporary Western prototypes such as Glasgow's Taransay Street and, in turn, directly influencing Hämer's concept of IBA-Alt. As in Western Europe, the initial rehabilitations were a short-term, stopgap reaction to the out-of-control escalation of the concept of rolling obsolescence (defined in the GDR as 80 years) and redevelopment. A sweeping area-regeneration programme of demolition and rehab began in East Berlin's Prenzlauer Berg district, just north of the centre. It included a courtyard-clearance and rehab project in the Arkonaplatz area (from 1970), along with the better-known Arnimplatz scheme of redevelopment and rehab, planned from 1969 and executed from 1973, including a much-publicised showpiece tenement at Schönfliesser Strasse 5.[50]

This highlighted even more starkly the juggernaut of *Plattenbau* redevelopment, cranking up during the 1970s and early '80s in one old town after another, in a policy announced by new Party leader Erich Honecker in 1973. By the late '60s, isolated and restrained protests were bubbling up, for example in 1968–9 in Jena, where townsfolk organised a civic protest campaign against plans for a towering skyscraper for the VEB Zeiss optical works in the town centre (again by Henselmann, this time shaped like a telescope). Although unsuccessful, the campaign inspired later citizens' protests in other places, such as Halle. But the programme continued regardless: partly, the trouble stemmed from the low rents prescribed by GDR law and concomitant disrepair. Arguably, the internal

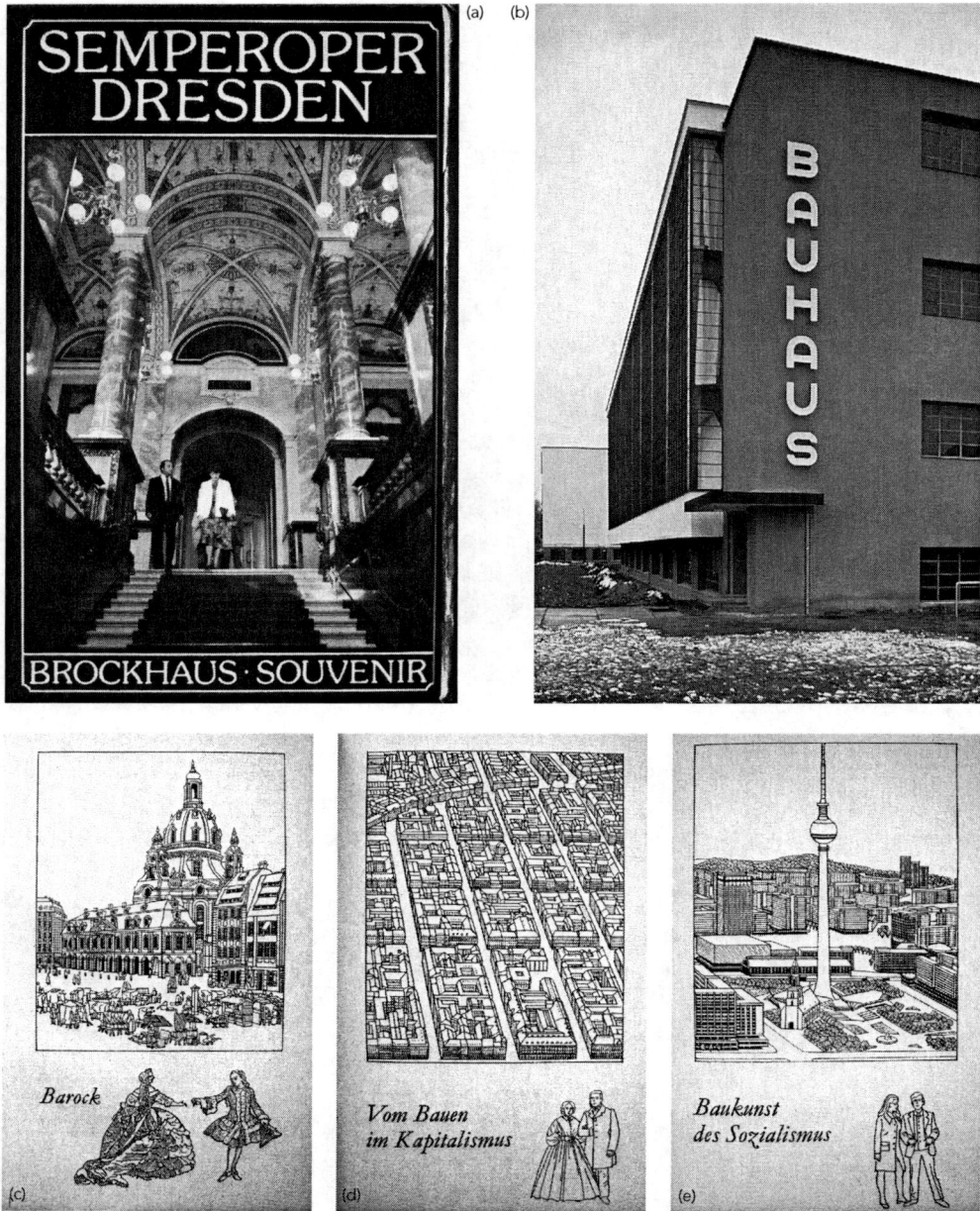

Figure 10.13 GDR heritage in the 1970s

(a) Interior of the restored Dresden Opera House, featured in a 1988 GDR tourist guidebook; (b) 2009 view of the Dessau Bauhaus, as partly restored from 1976 by the GDR authorities and again subsequently; (c), (d), (e) Views of the 'Baroque', 'capitalist', and 'socialist' periods of German architecture, from Georg Piltz's 1973 picture-history

Figure 10.14 Rehab-Ost

(a) The Arkonaplatz rehab scheme in East Berlin, from 1970;
(b), (c) Schönfliesser Strasse 5, showpiece of the Arnimplatz
rehab project (planned from 1969, executed from 1973); (d)
The Nikolaiviertel 'heritage island', planned from 1976,
authorised by the Politburo in 1980 and executed in 1983–7
(masterplan by the collective of architect Günter Stahn)

dynamic of the Conservation Movement itself ultimately fuelled political change: public anger at the demolition juggernaut and at Soviet Modernist regimentation contributed to the impetus of the 1989 revolution against communism.[51]

In the 1980s, GDR policy on historic areas began to converge with the contemporary Western movements of 'Heritage Industry' and facsimile reconstruction. Towns like Greifswald saw attempts to build *Plattenbau* in a 'historic' style, as did Berlin's Nikolaiviertel reconstruction project, planned from 1976, authorised by the Politburo in 1980 and executed in 1983–7. Here, in an old-town zone left virtually empty of buildings by the '60s, a tourist heritage-island was ingeniously created, using Sitte-like groupings of facsimile buildings and generic, neo-traditional *Plattenbau* blocks, masterplanned by architect Günter Stahn's collective bureau. Elsewhere in East Berlin, an almost postmodern sense of neo-historicist image emerged in 1980s' projects for Prenzlauerberg and Friedrichstrasse (the latter including an entertainment centre, the 'Friedrichstadt Palast', built in 1981-4 by architect Manfred Prasser in an Art-Deco-style *Plattenbau*). These areas, after reunification, became hotspots of tourist and capitalist intensification.[52]

In Moscow, too, the 1980s' *perestroika* era saw an upsurge of postmodernism and contextual design – although the 1980 Moscow Olympics prompted a last outburst of demolition of crumbling houses and churches, some replaced with 'fire-proof' copies. In 1976, following *Mossoviet* (city council) approval of a conservation-oriented plan for central Moscow, Viktor Grishin decreed that no new building there could exceed existing heights – a ruling that cut the height of a new TASS headquarters block (1979–81, by G Somov) from 24 to nine storeys and forced the omission of an intended tower from the Izvestiya publishing office. Henceforth the favoured epithet for new buildings in central Moscow was '*spokoiniy*' (tranquil), and a postmodern spirit of decoration and imitation took hold. This was exemplified in a historically styled extension to the KGB Headquarters in Lubyanka Place and the Arbat pedestrian shopping area, where architect Alexei Gutnov honed his contextual (*sredovoi*) philosophy from 1982 onwards. Eventually, heritage advocacy became one of the most conspicuous pre-*perestroika* movements, with ordinary Muscovites blockading threatened buildings and nascent Russian nationalist movements such as *Pamiat* (Memory) and *Otechestvo* (Fatherland) vigorously embracing the cause. During the 1980s, the Soviet state embraced conservation on a massive scale, allocating hundreds of millions of roubles at Union and republic level to rectifying the demolitions and neglect of previous decades. In 1986, at the 27th Soviet Communist congress, Boris Yeltsin, then secretary of the *Mossoviet*, argued that Moscow had lost its historical authenticity and that firm action would be needed to correct the situation: ironically, nine years earlier, as Yekaterinburg party secretary, Yeltsin had ordered the demolition of the Ipatiev villa in which the Romanov royal family had been murdered. In late Soviet Estonia, opposition was carried still further by the 'Tallinn School' of designers, a postmodern grouping led by Leonhard Lapin and others, who boldly attacked establishment architects such as Mart Port.[53]

Black sheep of the bloc: Romania in the 1970s and '80s

THE beneficial interaction between conservation and the peaceful revolution of 1989–91 was highlighted by the very different trajectory of events in the 'black sheep' of communist conservation, Romania. Here, in a demonstration of totalitarian regimes' ability, and occasional tendency, to embark on wild policy reverses concerning the built environment, the government of

Figure 10.15 Moscow contextualism in the 1980s

(a) 1981 article in *Stroitelstvo i Arkhitektura Moskvy* on the newly completed TASS building (by G Somov), headed 'New Building in Historic Context'; (b) 1980s' project for restoration of Old Arbat Street by M Posokhin, A Gutnov and Z Kharitonova

Nicolae Ceauşescu embarked in 1977 on a nationwide programme of mass urban demolition and redevelopment, at a time when most other countries were heading, or beginning to move, in the opposite direction.

Ironically, Ceauşescu's Romania had earlier, in the 1960s, established an enviable international reputation for sensitive urban conservation and for a contextual, regionalist modern architecture, in the early/mid-'60s work of Nicolae Porumbescu and others. Underlying the diversity of the Eastern bloc, Romanian government conservation was orientated to 'French' rather than 'German' traditions and values, just as postwar archaeology in communist Albania was organised around prewar Italian Fascist precedents.[54] Romania's Directorate of Historic Monuments (Directia Monumentelor Istorice, established in 1959, with a predecessor commission of 1951) had, by the mid-1970s, listed nearly 10,000 buildings and carried out over 100 major restorations and was widely regarded as an international leader in urban conservation. In 1966, the directorate's researchers calculated that preservation of historic centres was two/three times cheaper than rebuilding anew (the opposite to Paulick's calculations the same year in East Germany). Accordingly, a programme to rehabilitate the entire historic centre of Bucharest was begun. A 1974 law for 'systematisation' of urban and regional planning seemingly bolstered this trend, by providing in principle for 'reservations of architecture and urbanism'.

In fact, however, this law was the harbinger of a dramatic policy reversal, driven by a retrograde attempt by Ceauşescu to move towards an urbanism combining heavy industrialisation

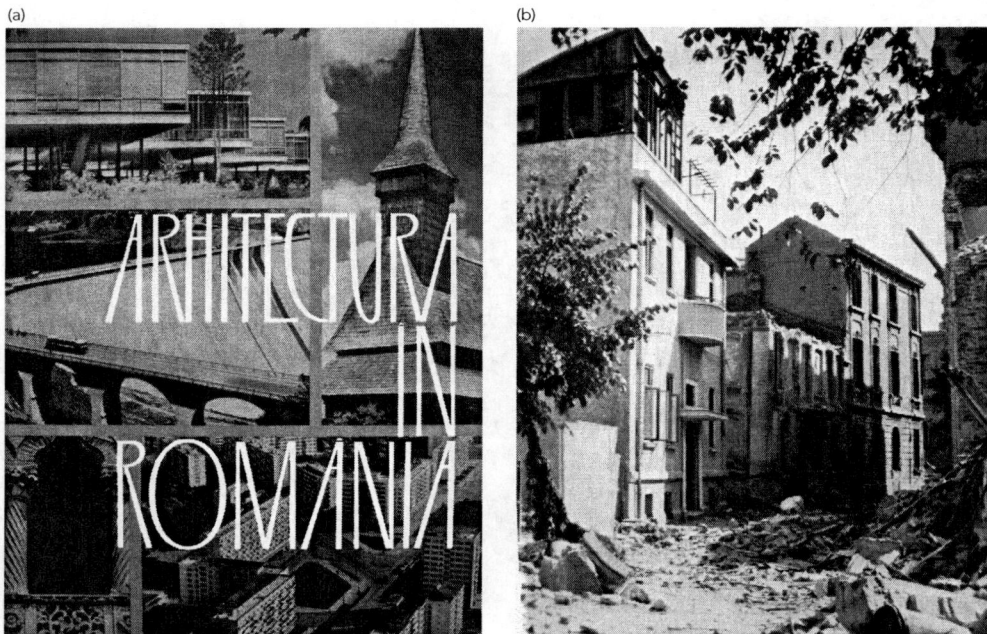

Figure 10.16 Romania – 1960s and 1980s

(a) Front cover of 1965 monograph on Romanian architecture (Prof. arh. Gustav Gusti, *Arhitectura în România*, Bucharest, 1965) showing harmonious coexistence of old and new; (b) Late 1980s' demolitions in Banu Maracine and Negru Voda Streets in Bucharest (published in D C Giurescu, *The Razing of Romania's Past*, 1990)

and Stalinist Socialist Realism. In 1977, citing damage caused by a recent earthquake, the government announced a new nationwide policy of mass redevelopment and demolition of historic urban fabric. Not only Bucharest, but all major towns and villages would be radically 'systematised' with prefabricated apartment blocks, and the number of villages would be slashed from 13,000 to 6,000. In November 1977, the government simply disbanded the Directia, replacing it with a tightly controlled 'Central State Commission for National Patrimony' and, the following year, the demolitions began, further ramped up in scale in a May 1981 decree. Although international protests steadily grew, Ceauşescu relentlessly pushed through the policy, whose centrepiece was the vast *Victoria Socialismului* (Victory of Socialism) reconstruction scheme in Bucharest – a project recalling Hitler's Germania rather than Haussmann's Paris in scale. This boldly carved a grand east–west axis through the urban fabric, heading it with the gigantic, Socialist-Realist-classical *Casa Republicii* (1984–9, by architect Anca Petrescu).[55] In an interview with the *New York Times*, the Bucharest City Architect, Alexandru Budişteanu, hailed this monumental new centre as a 'unique, epoch-making achievement in the most important project of systematisation, construction and architecture ever implemented on Romania's territory'.[56] In December 1989, however, following a violent *coup d'état*, Ceauşescu and his wife were executed, and the far-fetched adventure was over.

From 1989 onwards, as we will see later, the Conservation Movement became more internationally homogeneous, even as its traditional driving impetus fragmented and dissipated. Potential approaches to heritage in different European countries converged to the point where a rogue policy such as Ceauşescu's became impossible, and the epicentre of radical redevelopment moved instead to the rapidly expanding society and economy of China – whose neighbour, North Korea, remained a quirkish last redoubt of Stalin-style heritage attitudes.

(a)

(b)

Figure 10.17 Juche heritage

(a) The Mangyongdae House, near Pyongyang, a thatched-roof group claimed to have been the birthplace (1912) of President Kim Il-Sung and inaugurated as a historic site in 1947; (b) The Korean Film Studio, Pyongyang, inaugurated 1947: early 1980s' view of film-set street in the style of 'the Japanese imperialists' colonial rule'

Charters and conventions

The internationalisation of heritage, 1945–89

Safeguarding an architectural heritage is generally considered one of the fruits of Western
civilization. It has emerged from the Western view of history, from the West's links with
its own past, from the role and significance of Western monuments within a given
environment, from Western artistic sensibility and ethical values.

Raymond Lemaire, 1983[1]

International conservation on the move

OVER the past eight chapters, since the French Revolution and the upheavals triggered by Napoleon, most of the running in the dynamic development of the Conservation Movement was made by the nation-state, whether violently, as in the early 20th century, or peacefully, as in the late 19th and late 20th centuries. Ideals of a universal, or common, international patrimony bubbled to the surface, in the writings of internationalist theorists like Riegl or in international initiatives like the Charter of Athens, only to subside once again. Internationalism could always be trumped and even exploited by nationalism, as we saw in the Franco–German war of words over the damage to Reims Cathedral, or the claims that Anglo-American bombing was an attack on 'European culture'. And the Cold War, although between rival international systems rather than nations, initially kept alive an atmosphere of confrontation that helped impede potential collaboration in the heritage.

Despite all the early hopes invested in the United Nations, by the 1950s the fear of devastating nuclear war was becoming pervasive. When the UN's cultural arm, UNESCO, organised a conference in May 1954 at the Hague Peace Palace to update the 1907 agreement with a new Convention for the Protection of Cultural Property in the Event of Armed Conflict, this seemed likely to be a futile gesture. When the 1954 Convention's preamble referred to the 'cultural heritage of all mankind', it seemed probable that that heritage would soon be reduced to radioactive rubble.[2]

The chief practical outcome of the 1954 Convention was a quirkish system of small blue and white marker shields to be affixed to the outside of significant monuments to indicate that they should be protected in case of war, with explanatory text in the official languages of the Convention (English, French, Russian and Spanish). The diagonal saltire-type pattern of this 'Distinctive Emblem' followed a precise formula specified in article 17 of the Convention – a formula also intended for other relevant contexts, such as the armbands of protection officials or the roofs of lorries evacuating art treasures. Adopted at the discretion of national authorities, generally after a

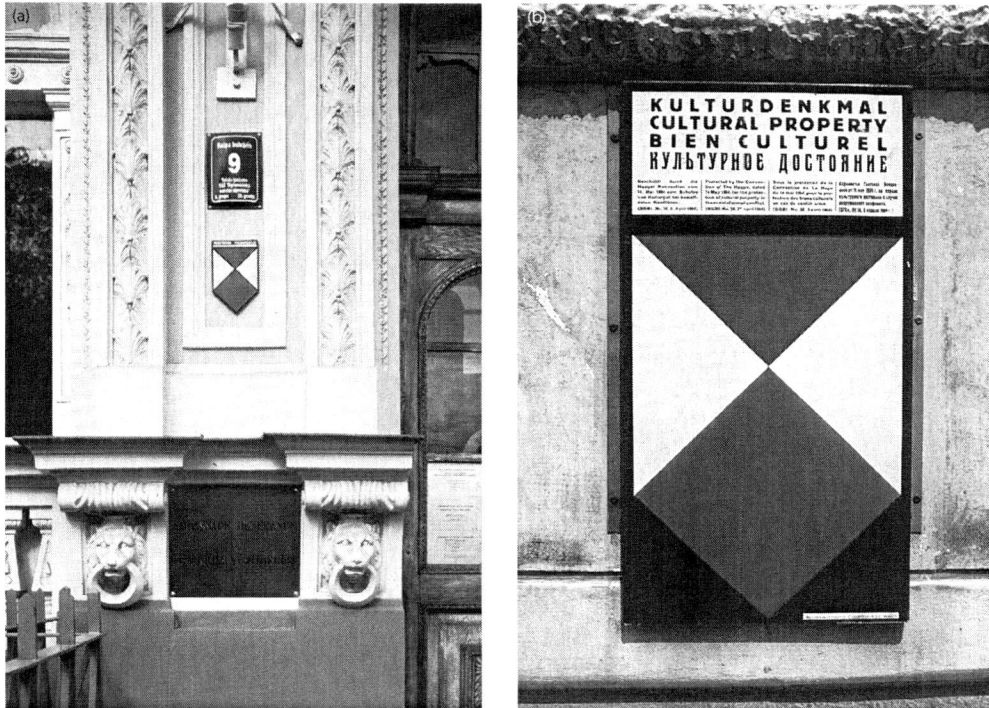

Figure 11.1

(a), (b), (c) The 'Distinctive Emblem' established by the 1954 Hague Convention, as used in Riga, Vienna and Ypres

considerable interval, the little plaques proliferated in some countries, for example in the Low Countries, or in the socialist bloc (where they chimed with the official rhetoric of 'peace'). In a few cases, the Distinctive Emblem was also appropriated by national initiatives, such as the Polish state consultancy PKZ. Many countries largely ignored the system, perhaps doubting its efficacy in protecting against 50-megaton thermo-nuclear missiles.[3]

Yet international conservation was now, slowly but inexorably, on the move. In September 1963, in the shadow of the Cuban crisis, George L Mosse asked at an international conference in Colonial Williamsburg whether the Conservation Movement had successfully thrown off its addiction to nationalism: 'In all our preservation and restoration there has been

391

a political motive implied or expressed ... Can 20th-century conservation transcend its own [nationalist] past?'[4] The self-destruction of the tradition of militant national heritage in 1939–45, and the retreat of nation-based conservation to a lower-key, welfare-state approach, had cleared a wide field of action for internationalism, and the looming menace of global 'Armageddon' added a new note of urgency. Within a few years of that 1963 conference, a remarkable variety of innovative and long-lasting international initiatives had begun to burgeon, initiatives which were openly interconnected, unlike the old divided structure of national heritages: the latter had shared many underlying values, yet promoted themselves as self-contained worlds in opposition to all the others. For the first time, the Conservation Movement openly declared itself to be a single coalition with a potentially common agenda.

Return to Rome: laying the groundwork in the 1950s

BEFORE exploring the diverse facets of this new internationalism within the Conservation Movement, we need to retrace our steps slightly, to tease out its immediate roots in the ideas and practices of the early postwar years. Here, as we saw earlier, a key role was played by a limited group of European conservationists, concentrated in Brandi's Italy. In a world turned upside down by wars and ideological splits and menaced by nuclear annihilation, they tried to ground the Conservation Movement, at an international level, in the reassuring foundation of humanistic idealism. This approach relied initially on the work of professional experts and on the Enlightenment narrative of progress, but also allowed for an eventual shift to a more postmodern relativism and acceptance of cultural diversity: Paul Philippot (Centre for International Conservation in Rome director, 1971–7) argued that 'conservation is a cultural decision'.[5]

The fruitful interplay between humanism and rationalism in the international conservation efforts of the 1960s–'70s – a variant on themes that also dominated the contemporary Modern Movement – was exemplified in the prolific writings of Piero Gazzola, chairman of the Venice Charter drafting committee. Gazzola claimed that international conservation could restore 'harmony' and 'serenity' to environments degraded and alienated by the 'programmed barbarism' of utilitarian mass redevelopment, while remaining grounded in narratives of progress and ethical betterment: conservation was 'above all a moral problem'. He argued in 1971 that 'our defence of historical environment must be conceived in active, modern, scientific terms', exploiting 'the immense progress achieved in science and technology – a progress which, casting off the present disorder, must lead us to a new reality, in a humanistically harmonized world'. Humanistic redemption could be achieved by broad definition of heritage values, while avoiding the totalitarian extremism that had vitiated the all-embracing concept of the *Heimat*. The key word was 'culture': Gazzola argued that 'anything which leads to culture, and anything which results from it, possesses monumental value'.[6]

The initial groundwork for the new conservation internationalism was partly laid by initiatives within individual countries. First among these was, of course, the work in Poland. But by the early/mid-1960s, this had been eclipsed by the more overtly spectacular project to rescue the 1260-BC temples of Abu Simbel in Egypt, threatened by construction of the Aswan Dam and formation of Lake Nasser. This vast, UNESCO-backed project invested the idea of 'international cultural property' with an enduring aura of glamour, and generated a veritable industry of international conservation consultancy. Its Swedish consultant engineers Vattenbyggnadsbyrån, recommended that the temples' great carved figures should be physically relocated, by cutting them up into blocks and moving them 200 yards back and 200 feet upwards; a reinforced concrete store was built to support the temple ceilings. At its height, in 1964–5, when the works were protected by a coffer dam, the Abu Simbel project involved a workforce of 1,900. Eighty were European staff and consultants (including numerous Poles) and the rest were Egyptian technicians and workers, all housed in a small hutted town: the successful relocation was hailed in an early 1966 article as 'a combined wonder of the ancient and modern world'.[7]

At a more general ideological level, collaboration within conservation was encouraged by the postwar proliferation of non-governmental organisations in other areas of the built environment, such as architecture (CIAM, Team 10, UIA) and planning (IFHTP, ISOCARP). The shift of CIAM towards a more 'sensitive' urbanism at its Bergamo meeting (the 'Heart of the City', held in the medieval Palazzo della Ragione) was only one of many international initiatives aiming at a more humanistic urban development.[8] The key transitional steps were concentrated in Italy, especially in Rome, where existing policies and administrative machinery provided a strong foundation for the internationalisation of the Conservation Movement. The ideas of Brandi, which we traced in Chapter 8, deeply affected continental conservationists such as Philippot and Gazzola and boosted the new ideals of humanistic conservation and of the reconciliation of art and science, even if Brandi's preoccupation with 'artistic creation' was far too narrow to serve as a basis for the new international

Figure 11.2 Museum model of the UNESCO-backed relocation of the Abu Simbel temples in Egypt, completed 1966

393

heritage world-outlook. Later, Giovannoni's *diradamento* concept, as modified by Cervellati in Bologna and by Aldo Rossi (through his concept of 'type'), exercised a wider effect on the international development of urban conservation.

Almost inevitably, Rome was chosen in 1956 as the base for the first significant postwar initiative of internationalisation within the Conservation Movement, an initiative that emphasised documentation and education rather than active preservation: the International Centre for the Study of the Preservation and Restoration of Cultural Property. This was established by UNESCO following discussions since 1953 and a decision of the 1956 general conference of UNESCO in New Delhi (proposed by Frédéric G Gysin, representing the Swiss delegation); Brandi, Gysin and Gazzola were included on its provisional council.[9]

The Rome Centre, as it was initially known, was not concerned mainly with architectural conservation, but with 'cultural property' as a whole. Accordingly, its senior staff initially came not from within architecture, but from the well-established field of museum conservation and art history. The first Director, the Dundee-born Harold J Plenderleith, was an eminent museum conservator, founder of the British Museum's renowned conservation laboratory (responsible for the restoration of the Dead Sea Scrolls) and had just published a definitive book on the subject, *The Conservation of Antiquities and Works of Art* (1956). He was supported by Deputy Director Paul Philippot, an art historian with a background in jurisprudence. Initially, the Centre set out to build up a multi-disciplinary network of experts and conservation organisations. Its education strategy piggybacked on existing Italian initiatives. A pioneering course in architectural conservation, started in the Faculty of Architecture in the University of Rome by Guglielmo De Angelis d'Ossat, was taken over around 1965 by the Rome Centre: this 'ARC' course was later augmented by further courses, for example in 1968 on mural paintings ('MPC'); in 1974 Philippot elaborated this experience into a 'typology of training'.[10]

As with the previous advances of conservation at a national level, the rhetoric of threat and loss shaped this international movement, with the Rome Centre playing a key role. The key outrages that helped establish the threat agenda internationally both occurred, predictably, in Italy, when both Florence and Venice suffered severe flooding in November 1966. Of the two, the Florence flood was the more spectacular and shocking, but that in Venice was more long-lasting in its heritage implications.

In Florence, freak weather conditions and landslides on the night of 4 November 1966 caused a wall of water up to 20 feet high to rage through the old town, reaching levels commemorated today by small plaques on the walls of city buildings and inundating in mud countless artistic and cultural treasures, including 14,000 immoveable works of art and 1 million books in the *Biblioteca Nazionale Centrale* alone. By coincidence, Harold Plenderleith was in the city that night, stranded in a hotel near the Ponte Vecchio. After first rescuing its own director, the Rome Centre assumed a focal position in coordinating the unprecedented cultural rescue campaign, along with UNESCO and the Italian government. It marshalled an international army of young volunteers, who carried out emergency inventories and conservation works, both in the immediate aftermath and over the ensuing years.[11] But almost none of this effort fell under the heading of architectural conservation, even broadly defined.

The case of Venice was different as, while the 1966 flood was much less violent and many fewer works of art were affected, it reinforced a well-established sense of crisis, focusing on issues of subsidence and structural water damage. The flood dramatically drew international attention to

Figure 11.3 The 1966 floods

(a) Piazza Santa Croce, Florence, following the 1966 Florence
Flood; (b) View of the 1966 Venice Flood

these longer-term, structural issues: where the English SPAB campaign about St Mark's in the 1870s had been indignantly rejected, this more broadly based external pressure and concern was gratefully received. Previous efforts within Italy had been largely ineffective: the national government had passed a special law in 1956 and set up a Venice initiative, and Italia Nostra had restored a modest row of houses in Calle Lanza. The 1966 flood prompted a range of immediate repair initiatives by specialist NGOs such as the US-based 'Save Venice, Inc'. More significant was the massive campaign of concern and support spearheaded by UNESCO, beginning in 1967 with a study of the longer-term geological and marine causes of decay. In 1968–9, UNESCO organised an inventory of threatened areas of the city and formed an International Advisory Committee for Venice, and in 1970 its director-general sent a confidential note to the Italian administration, urging efforts at the 'highest governmental level' to save Venice. By April 1973, an Italian law allocating special funds of 300 billion lire was in place, but even after this, in 1974, a municipal zoning plan still envisaged redevelopment of 'lesser' areas, stirring up further outrage among conservation bodies across Europe.[12]

The pre-eminence of Italy in these efforts was emphasised by the more limited international impact of initiatives in other traditional centres of heritage antiquity. The nearest rival in status was Athens, where the years from the 1970s witnessed a fresh upsurge in research and restoration work around the Acropolis. But this set piece lacked a broad 'hinterland' of urban heritage in the Italian manner: Greek monument designation was still overwhelmingly dominated by classical archaeology. Even in 1974, only 110 out of some 4,000 protected sites were medieval or later in date: 1979 saw the first comprehensive government review of historic towns and villages (totalling 400). We saw above the way in which, earlier in the century, Balanos's increasingly controversial restoration efforts on the Acropolis had been paralleled by the equally radical American-led campaign of excavation, restoration and landscaping in the Agora, a campaign that intensified under postwar director

Figure 11.4 The Stoa of Attalus

(a), (b) External and internal view (2010) of the rebuilt stoa (1953–6), showing its immediate juxtaposition with the Athens–Piraeus railway

(1946–67) Homer A Thompson: 1953–6 had seen the construction of a replica of the Stoa of Attalus II as a store for Agora excavation finds, on the basis of a largely hypothetical design by John Travlos, longstanding Architect of the School's excavations (between 1940 and 1973).[13]

From the 1970s, however, the spotlight shifted back to the Acropolis, as the emphasis there switched from radical interventive restoration to conservative repair. The initiative lay with UNESCO, which in 1971 appointed an international experts' committee to investigate the rapid deterioration of the temples' fabric and in particular whether the earlier restorations by Balanos and others, including use of iron reinforcement rods and cement drums for the Parthenon colonnades, had worsened their condition. Under the patronage of prime minister Konstantinos Karamanlis, a working group was set up in 1975 and three years later transformed itself into a permanent Committee for the Preservation of the Acropolis Monuments, working under the aegis of the Central Archaeological Council of the Ministry of Culture.[14] The first target for this renewed repair work was the Erechtheion; a 1977 international symposium and published study (*The Study for the Restoration of the Erechtheion*) was followed in 1979–87 by a methodical programme of reconstruction. By 1983, a detailed study for restoration of the Parthenon had appeared, in parallel with a second international meeting. The organisers relentlessly stressed the international significance of the scheme: C Bouras argued in 2007 that the obligation to protect 'monuments of unique value for the entire world' carried a 'heavy responsibility'. In some ways, the new programme was uncannily reminiscent of the 19th-century English Gothic Revival efforts to 'un-restore' the work of previous generations: Balanos's efforts were condemned (and his metal reinforcement bars were removed) even as the strategy of *anastelosis* generally continued apace. Substantial parts of the monuments were repeatedly dismantled and reassembled: the 319 blocks of the temple of Athena Nike were completely taken down twice within a century (in 1935–40 and 2000–9) and the Parthenon changed gradually but inexorably from a fragmentary ruin in the 19th century to a largely reassembled structure.[15]

Conservation issues in another traditional focus of international heritage concern, Jerusalem, were complicated by the ethnic and political conflict of Israel and the Arabs, even though the city had enjoyed a considerable continuity of planning and development policy since the beginning of the British mandate: the consensus that the Old City should be picturesquely isolated in a 'Sacred Park' was accentuated under the post-1967 Israeli planning regime. Critics from both communities joined in furious protests in 1972 against the skyline impact of a series of high-rise hotel projects (within a permissive 1968 master plan), culminating in proposals for a Hyatt Hotel on Mount Scopus and a 22-storey tower annexe to the King David Hotel. An 'Old Jerusalem Committee' of international luminaries, founded by Mayor Teddy Kollek in 1973 (and including Buckminster Fuller, Bruno Zevi, Louis Kahn and Nikolaus Pevsner), condemned high buildings and called for comprehensive urban conservation; and the same year, the city council voted to curb building heights in all new projects. After 2000, however, the Israeli planning concept of the 'Holy Basin' as a national park attracted increasing criticism for its potential role in restricting Palestinian urban growth.[16]

The post-1960s' upsurge in international conservation was largely a boom in administration: in committees, charters and councils. Its character echoed the existing, bureaucracy-led characteristics of conservation at a national level, directed by administrators and planners as much as by ideologues. As we saw earlier, the interwar work of the League of Nations, especially in the Athens Conference and Charter of 1931, had provided a broad initial foundation for international conservation, guiding it towards an ideal of scientific museological accuracy and a strong ethical framework of 'good' and 'bad' practice. Immediately after World War II, the League of Nations's cultural organisations were transferred to the patronage of the United Nations's new cultural, social and economic branch, UNESCO, based in that other traditional cultural focus of the Old World, Paris. UNESCO stimulated a vast range of activities in heritage and environment, including a 1951

Figure 11.5 1970s' view of the safeguarded skyline of Jerusalem

initiative to establish an International Committee on Monuments that could coordinate campaigns such as the Abu Simbel rescue. The year 1956 saw the upgrading and renaming of the well-established International Office for the Protection of Nature, as the International Union for the Conservation of Nature and Natural Resources (IUCN). Later UNESCO initiatives in heritage included a 1962 Paris conference and recommendations on protection of natural landscape.[17]

But the most significant early initiative of UNESCO was undoubtedly the 1956 foundation of the Rome Centre itself. Its budget was significant in scale, at 1 per cent of UNESCO's overall income, and it organised itself like a miniature United Nations, with a 'membership' of individual countries (Austria being the first to join, in 1957). Membership built up to 24 by 1961 and 53 by 1971, in which year Philippot succeeded Plenderleith as director. He shortened the organisation's name to 'Centre for International Conservation in Rome' and moved it to premises provided by the Italian government in the hospice of San Michele, expanding its 'typology' of training courses with a range of additional programmes and an annual newsletter (from 1973). His successor, English conservation architect Bernard Feilden (1977–81), devised a slightly unwieldy acronym for the Centre (ICCROM) and introduced a Modernist architectural theme of interdisciplinary collaboration: the subsequent director, from 1981–8, was Turkish archaeologist Cevat Erder. By then, however, the role of ICCROM had been overshadowed by other initiatives less tied to traditional cultural authority-structures. For example, it ceased to be a unique centre of international conservation training following UNESCO's 1973 establishment of a new Training Centre for Preservation of Cultural and Natural Heritage at Jos, Nigeria.[18]

The role of the Rome Centre was also bypassed by a succession of wider initiatives concerning international harmonisation and doctrine. These began in 1957, the year after its foundation, with a UNESCO-sponsored conference in Paris, whose title echoed that of the interwar Athens conference – the International Conference of Architects and Technicians of Monuments. At this meeting, delegates voted to work towards a permanent international committee to codify and uphold the principles of the Conservation Movement. For this purpose, a Second International Conference of Architects and Technicians of Historic Monuments was convened in 1964 in Venice. Showing once again how the new international initiatives were interwoven with Italian structures, Italy provided the chairs of both the conference itself (De Angelis d'Ossat) and the committee for drafting the new international Charter for the Conservation and Restoration of Monuments and Sites, which would be the conference's main output (Gazzola). The 23-member drafting committee also included Roberto Pane (Italy), Jan Zachwatowicz (Poland) and F Sorlin (France): the rapporteur was Raymond Lemaire (Belgium) and ex-officio representatives included Hiroshi Daifoku (representing UNESCO) and Philippot and Plenderleith of the Rome Centre.[19]

This 'Charter of Venice', with its interweaving of Italian and international elements, was based especially on Brandi's grammar of conservation and on the current Italian official norms – themselves stemming largely from the Charter of Athens. Its text, drafted by Roberto Pane and Piero Gazzola with substantial input from Philippot and others, resoundingly affirmed heritage internationalism: 'People are becoming more and more conscious of the unity of human values and regard ancient monuments as a common heritage. The common responsibility to safeguard them for future generations is recognised.' Reacting against the somewhat cavalier freedom of many postwar restoration projects, the Venice Charter reasserted the moral prohibitive force of the Ruskin and Anti-Scrape tradition of material authenticity, but now in a more up-to-date guise, shaped by the Modern Movement preference for a contrast of old and new. Strengthened by the concept of

Dear Harold,

Congratulations on your 90th birthday and all my best wishes.

Your great book Conservation of Works of Art and Antiquities changed my life. I first saw it on the bookshelves of Lew Majewski's office in New York University and he described it as his "Bible". It was Jim Marston Fitch who arranged for me to lecture to the future conservators in Lew's course about York Minster and I am afraid they were not very interested in my problems, but your book helped me solve them because it introduced me to Relative Humidity. As an architect I dont remember being taught anything about this important question but through your book I came to understand buildings as spatial, structural environmental systems.

And then I wanted other architects to understand this so I began teaching at ICCROM and then became your successor as the Third Director and had the benefit of your wise advice.

You will be glad to know that today we have elected the fifth Director.

Keep well and have a Happy Birthday

Yours sincerely Bernard.

9 May 1988.

Figure 11.6 UNESCO and ICCROM

(a) 2009 view of the UNESCO headquarters in Paris, designed by Marcel Breuer, Pier Luigi Nervi and Bernard Zehrfuss and completed 1958; the complex was nicknamed the 'three-pointed star' owing to its complex layout; (b) Cartoon of the relationship of ICCROM and UNESCO, by J A Flores (participant in ICCROM Architectural Conservation Course), 1971; (c) 1988 letter to H Plenderleith from B Feilden, congratulating him on his 90th birthday

'reversibility' of new alterations, this now became a central shibboleth of conservation ethics. The Venice Charter combined this with a museum-language of strict, scientific protection, dominated by prohibitions of alterations, demolitions or restorations, arguing that it was 'our duty to hand [monuments] on in the full richness of their authenticity'.[20] However, there were also significant markers of the new, broader approach to the values of the monument and to the urban scale of conservation – doubtless inspired here by the urbanist theories of Giovannoni from the Fascist years. The definition of monument included not just single buildings but also:

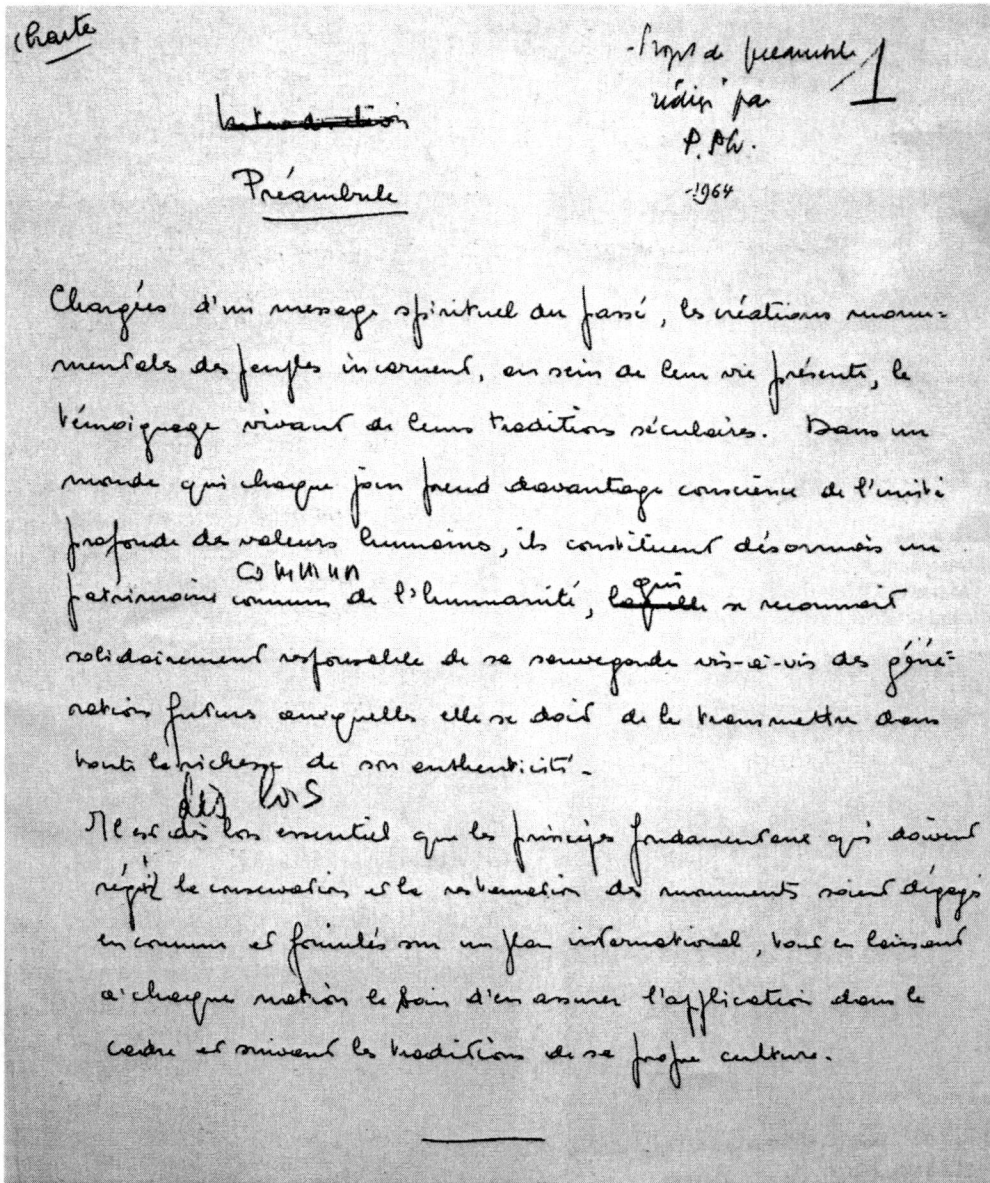

the urban or rural setting, in which is found the evidence of a particular civilisation, a sign of development or a historic event. This applies not only to great works of art but to more modest works of the past, which have acquired cultural significance with the passing of time.

The Giovannoni concept of integrated conservation and planning was also resoundingly echoed. We will trace later the radical spread of the term 'cultural significance' around 2000, but we should bear in mind that it was already clearly present at Venice in 1964.[21]

Global and regional foci: ICOMOS and the Council of Europe

Now that the international Conservation Movement was equipped with a new foundation doctrine, the next step, as foreshadowed in the 1957 conference, was to establish an organisation to act as the custodian of that doctrine: the International Council on Monuments and Sites, or ICOMOS. Here UNESCO took the initiative, boldly selecting Warsaw as the site of ICOMOS's inaugural general assembly. This exploited Poland's unique prestige in the Conservation Movement as a way of involving the socialist bloc in ICOMOS from the beginning (even if the restoration of Warsaw was hardly consistent with the anti-restoration doctrine newly reaffirmed in the Venice Charter!)[22] The emergence of a specifically architectural conservation doctrine and network – compared to the wider concerns or ICCROM – was reflected in the growing dominance of architects within this network, including the early leaders of ICOMOS, Gazzola (its first president) and Lemaire (its first secretary, until 1973, thereafter president until 1981). Lemaire, whose late 1950s' Heverlee chapel had presaged the Venice Charter concepts of separated old and new, and reversibility, was probably the focal figure of the entire network. In addition to his leadership of ICOMOS, he acted as the personal representative of UNESCO's director-general in Jerusalem and carried out missions for the UN, UNESCO and ICCROM at sites such as the Temple of Borobodur, the Athens Acropolis, the Temple of Zeus at Jerash and the painted churches of Moldavia; in 1976 he founded an ICOMOS-inspired conservation course in Bruges (transferred in 1981 to the University of Leuven) and within Belgium he was awarded the *Ordre de Léopold* in 1963 and the *Ordre de la Couronne* in 1973.[23]

ICOMOS was organised on the basis of national committees, concentrated initially in Europe and North America. In some cases, the national committee acted as an intermediary with non-governmental organisations within their countries. Some national committees also undertook supra-national projects and 'propaganda activities': France hosted the organisation's head office. To avoid becoming just a talking-shop, ICOMOS launched a concerted international study of one key theme: the challenges posed by tourism. This continued the Giovannoni philosophy of constructive engagement of urban conservation with wider forces of economic and social development, although some doubted the chances of success: Gazzola expressed extreme suspicion of the risk of degradation of heritage authenticity by interpretative 'valorization' and 'superficial and showy'

facing page
Figure 11.7 Draft text of Venice Charter handwritten by P Philippot, 1964

401

commercialisation.[24] Various international symposia were accordingly held from 1966 onwards on the problems of conservation in historic towns, and the second ICOMOS general assembly, at Oxford in 1969, was dedicated to the theme of cultural tourism. A total of 24 national committees participated, overwhelmingly from Europe, with the largest delegations from Italy (including Gazzola and Pane), West Germany, France, Czechoslovakia and the UK.[25]

To propagate the ideals of ICOMOS, and act as a 'living link' between specialists in the humanistic and scientific aspects of conservation, Lemaire also founded an ICOMOS journal. The idea had taken shape at Venice in 1964: Gazzola argued that it would help promote the 'universal values' of conservation and counter any 'chauvinistic pride'. At Warsaw, the Belgian delegation offered to organise the new journal (to be called *Monumentum*). Its first issue duly appeared in 1967. Implicit in its approach was a presumption that Western norms of conservation, as embodied in the Charter of Venice, were superior to the more overtly politicised Eastern-bloc approaches and should be standardised as the international norm. Lemaire edited *Monumentum* from its foundation until 1981. As a more heavyweight counterpart to *Monumentum*, Bernard Feilden (third head of ICOMOS) published in 1982 the first full-scale practical manual of conservation, which presented the ideas of Brandi and the Charter of Venice (including reversibility) within a framework of modern, scientific, multi-disciplinary conservation, defining it in somewhat Morris-like terms as 'the action taken to prevent decay'.[26]

All in all, the new internationalism expressed fairly exactly the geo-cultural context of the Conservation Movement, overwhelmingly dominated by Europe but with North America as a strong second focus of ideas and activity. And by integrating the socialist bloc into this Western-dominated structure, the new heritage internationalism may also, in a small way, have helped lay the cultural foundations for the political integration that would follow the dramatic events of 1989–91.

This Western dominance was strengthened by the forceful interventions of the Council of Europe (COE), a specifically Western system created in 1949 in response to the outbreak of the Cold War (following an initial 'Congress of Europe' held in May 1948 in the Ridderzaal in the Hague). It acted as the crucible not only for the foundation of the European Union but for the development of a strong European stance on human rights. From the very beginning, the Council of Europe projected a vigorous presence in the cultural and heritage sphere, especially following a 1954 European Cultural Convention. It promoted individual initiatives by figures such as Gazzola, who wrote in the mid-1960s of the 'indivisible unity' of the 'living organism' of European heritage, threatened by war and ideological conflict: 'The whole of Europe is a "historic centre."' In 1963, following a report by Ludwig Weiss, the Consultative Assembly of the Council of Europe set out the first pan-(Western) European programme of heritage protection (Recommendation 366) and authorised a COE-assisted pilot rehab project in Venice, followed by a succession of colloquia and meetings, culminating in a synthesis conference at Avignon in 1968: at the latter, Gazzola stressed the pre-eminence of 'cultural values' and 'social order' in urban conservation.[27] At the same time, the potential of the voluntary sector was acknowledged in the foundation of Europa Nostra, a federation of voluntary societies concerned with built and natural heritage. Enjoying COE consultative status, it held annual conferences from the late 1960s. As we will see shortly, both the COE and Europa Nostra (under the presidency of Duncan Sandys) played a crucial role in the culminating initiative of the first phase of internationalisation of the heritage: European Architectural Heritage Year 1975 (EAHY).

402

Stockholm 1972: gateway to global environmentalism

THE years when the Conservation Movement seized control of the urban planning system in Western countries, in the early and mid-1970s, were also years of triumph for its burgeoning international network. In both cases, environmentalist activism played a prominent role – something that sat slightly incongruously alongside the role of the expert in organisations such as ICOMOS, but which foreshadowed the shift in the general values of conservation towards looser and more emotive frameworks, from art-historical to cultural, from tangible to intangible, from object-focused to process-orientated, from objective to relative and subjective. The emergence of the often impassioned environmentalist movement, in conjunction with the energy crisis of the early 1970s, for the first time allowed international conservation to legitimately raise its emotional and political temperature in a wide range of areas.

Internationally, a key legitimising role was played not by small-scale activism, but by a succession of unwieldy global conferences, above all the June 1972 UN Conference on the Human Environment in Stockholm – the first of the global 'mega-conferences' which attracted a vast assortment of official and unofficial bodies to participate or contribute from the sidelines. Some national groups devoted substantial efforts to preparation for Stockholm. For example, the UK government commissioned an ambitious survey of public opinion on the environment, to 'fortify' the UK delegation, and a committee chaired by the Countess of Dartmouth prepared a book entitled *How Do You Want to Live? A Report on the Human Habitat*. In this book, once iconoclastic concepts such as 'habitat' and 'participation' were now thoroughly codified and assimilated into the system. The book argued that 'conservationists used to be thought cranks' but now had the power and potential to achieve a social and economic revolution and 'bring about a genuine harmony between people and people as well as between people and their habitat'. Many participants, however, eventually departed Stockholm in frustration, feeling that it had been too general, too partial, or skewed by special interests. It disappointed built-environment activists because of the dominance of nature conservationists and campaigns against 'pollution'. And a follow-up UN Habitat conference in Vancouver in 1976, intended to remedy this bias by focusing on poor housing and urban squalor, was in the event dominated by squabbling between pro-Israeli and pro-Arab factions.[28]

For conservationists, the finite character of natural resources provided a powerful metaphor in favour of permanence and restraint, not only in economic and social policy (where E F Schumacher's *Small is Beautiful* (1973) warned against using up 'natural capital') but also in the heritage. An ecological-social argument emerged especially clearly in the Council of Europe's Amsterdam Charter (1975) which identified architecture as a non-renewable resource and 'irreplaceable capital', and a social asset, helping combat loss of identity and alienation: *Architectural Review* editor Peter Davey asserted simply that 'Old buildings are irreplaceable, like oil.'[29]

But the conferences were less important for their own content and outcomes than for the way they encouraged further initiatives in more specific areas, including heritage. Some of these piggybacked onto the national efforts, and others sprang up within wider international frameworks. In the wake of Stockholm, the international Conservation Movement was energised, as we will see shortly, by two individual initiatives in particular: within Europe, by EAHY 1975, which radically reinvigorated the cause of European urban conservation; and globally, by the 1972 UN Convention on World Heritage, which would lead to the wildly popular designation system of the World Heritage

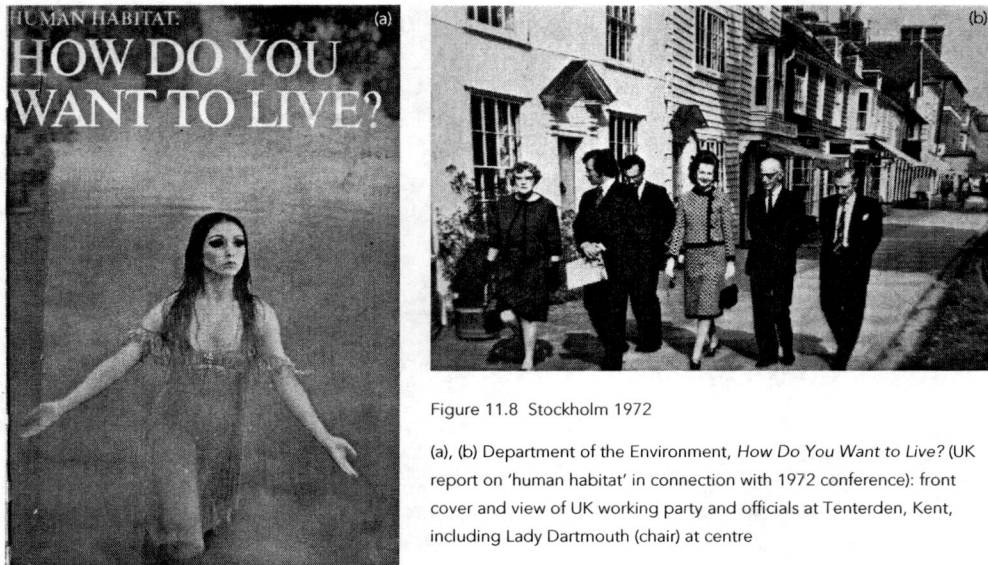

Figure 11.8 Stockholm 1972

(a), (b) Department of the Environment, *How Do You Want to Live?* (UK report on 'human habitat' in connection with 1972 conference): front cover and view of UK working party and officials at Tenterden, Kent, including Lady Dartmouth (chair) at centre

List. Alongside these high-profile developments, a less spectacular thread of consistency was provided by the gradually mutating apparatus of conservation charters. Over the decades following Venice, the charters evolved into a vast, interlocking, cumulative structure of definition, advice, exhortation and regulation – impenetrable to the outsider and sometimes, even, to the insider. Yet this diversity of detail concealed a core corpus of relatively few key ideas: namely, the principle of minimum action or intervention; the principle of 'truthfulness' in any intervention (through its 'legibility'); and the need to define accurately, and respect, the authentic significance of the heritage object.[30] This was, arguably, a remarkably narrow theory base, relying on a few stereotypical dogmas largely handed down from the 19th century. The combination of restricted intellectual scope and vast, relentless output had an inevitable result in the sequence of individual charters themselves. A slow-moving repetitiousness in the development of the overall theoretical corpus was combined with a huge overlap in the detail: each individual charter was largely made up of statements that could also be found in most others, not unlike liturgical formulae within Christianity, such as the Creed, the Magnificat, the Lord's Prayer: each one a little different in emphasis, but all essentially the same.

Thus, one cannot easily construct a narrative of charters, with one leading to the next, or one charter standing for one distinct position and the next for something quite different. But the sequence of charters did reflect some broader trends within the Conservation Movement, including an ever broader definition of heritage and its authenticity, as well as a shift from a museological or archaeological approach, regulated by experts in accordance with supposedly objective artistic or historical values, towards a looser, more diffuse framework, driven by emotive and relativistic ideas of cultural or spiritual significance and shared community meaning. One can see that shift taking effect already by 1975, for example, in the Amsterdam Declaration and Charter, whose rather impassioned language contrasted strongly with the Venice Charter only a decade earlier. And in parallel with this, there were demands for an ever more comprehensive, integrated approach to the *organisation* of international conservation.

404

European Architectural Heritage Year 1975

THE Amsterdam Charter was the culmination of a network of European projects in 1975, for which the umbrella title of European Architectural Heritage Year (and equivalents in other languages, such as *Europäisches Denkmalschutzjahr*, or EDSJ) was devised. The unique success of EAHY stemmed partly from its timing, at the crest of the Western wave of conservation zeal and anti-Modernist revulsion, but also from the way it exploited the interdependence of national and international within the Conservation Movement, by harnessing the spirit of national competitiveness to a loosely defined international project – the same formula that would subsequently, from 1988, be exploited by Modern architecture heritage group DOCOMOMO. EAHY, however, was a strictly time-restricted initiative, originating in an elite conservation network within the Council of Europe and Europa Nostra, many of whose members, like Robert Matthew in Scotland, held pro-Modernist views that were somewhat at variance with the generally anti-Modernist spirit of EAHY.

The general concept of a European Architectural Heritage Year was first conceived by Duncan Sandys in 1969, following the tourism-themed ICOMOS Oxford General Assembly and COE conferences from Palma to Avignon. EAHY was formally proposed in 1971 by an intergovernmental Committee on Monuments and Sites instituted by the COE. It would be the culmination of a wider five-year programme of activity aimed at the preparation of a new international charter, tailored to the specific circumstances of Europe. A previous COE-sponsored 'European Conservation Year' in 1970 had had little impact, as, like Stockholm, it was dominated by nature conservationists. A three-year lead-up to EAHY was formally inaugurated in July 1973 in Zurich, with Sandys chairing the organising committee. The initiative was co-sponsored by ICOMOS and UNESCO as well as the Western-dominated COE and European Union – which allowed the involvement of some socialist countries (Hungary, Poland, USSR). The remaining 20 participant countries were all Western, including the Vatican City. During the three-year lead-up period, the COE sponsored a range of publications and other activities, including a 1973 book on European urban conservation. The year 1975 saw publication of another book, *Modern Interventions in Historic Towns*, sponsored by Cembureau (the European Cement Association).[31]

The main EAHY strategy fell into two stages. First, a range of 50 pilot conservation projects would be organised by national committees, under the coordination of the COE's Committee on Monuments and Sites, to test in practice 'new ideas on the rehabilitation of the cultural heritage as part of regional and urban planning'. This would hopefully show, in implicit contrast to Modern Functionalism's assumed hatred of old districts, the need for their 'useful integration into the functional life of the city'. The lessons of the pilot projects would then be evaluated at an October 1975 congress in Amsterdam and synthesised into a 'Declaration of Amsterdam' and a European Charter of the Architectural Heritage, to be issued (again) by the COE. Owing to the prominence of the COE in the pilot programme, the participating socialist countries avoided the case studies, while attending the Amsterdam Congress itself: there were over 1,000 delegates from 25 nations, including all 18 COE members. Alongside this overall strategy, each national committee could promote any local initiatives it wished, including exhibitions, lectures and preservation campaigns, as well as fully fledged books at a national level (such as a comprehensive survey of Dutch conservation published in early 1975 by the *Monumentenzorg*). A supra-national overview with a participatory slant was presented in Yona Friedman's book, *Votre ville est à vous,* published by the COE in 1975, while the more elitist Europa Nostra produced a 1975 report that stressed the role of conservation in a traditional civic-art manner, as a way of eradicating eyesores. The theme of unity-in-diversity was emphasised to the point of repetitiveness, with well-established local urban conservation traditions trumpeted by the EAHY national committees: in Italy, a Giovannoni-style integration of conservation and planning; in Germany a *Heimatschutz*-style search for wholeness; and in Scotland the Geddes tradition of 'conservative surgery'.[32]

The establishment-dominated organisational formula of EAHY was exemplified in the UK initiative, organised by a committee chaired by the Countess of Dartmouth, with Prince Philip and government ministers as figureheads and Sandys representing the COE; the secretariat was provided by the Civic Trust with government support, and separate committees were set up for Scotland, Wales and Northern Ireland. A major inconsistency in the two-stage pilot project and synthesis strategy was that owing to shortage of time, only well-established projects could credibly be used in the pilot studies. Within the UK, although the official focus of the national EAHY campaign was on expansion and enhancement of conservation areas (along with enhancement of environmental education) and Lady Dartmouth argued that individual prestige projects were 'un-British', in practice the emphasis was on elite projects: the impressive-looking UK EAHY budget of £2.75 million was largely a rebadging of existing grants for the most outstanding conservation areas. The four pilot studies (Chester, Edinburgh New Town, Poole and the National Trust for Scotland's 'Little Houses' rehabilitation programme in small historic towns) were all well established and well financed, combining social regeneration with underlying 'gentrification'.[33]

(a)

(b)

above and facing page

Figure 11.9 European Architectural Heritage Year

(a), (b), (c) *What is our Heritage?* (UK report for EAHY 1975), front cover and pages outlining volunteer activities for EAHY 1975; (d) Dutch symbol for EAHY 1975; (e) Amsterdam 1975 Conference: visit by Prince Claus to Madurodam miniature theme-park; (f) Danish case-study report for EAHY 1975

The October 1975 Amsterdam conference, which comprised a bewildering range of thematic sessions, culminated in the issuing of a Declaration and Charter, focusing on the urban heritage and reiterating themes familiar since Giovannoni and Dvořak: the threat from modernity; the need for socially committed conservation embedded in the planning and education system; and the importance of the everyday, 'lesser' urban heritage. All this was now updated for the modern age of rational management, under the catchphrase of 'integrated conservation' (a term already much-used by Gazzola).[34] In case anyone felt that the

MONUMENTENJAAR 1975

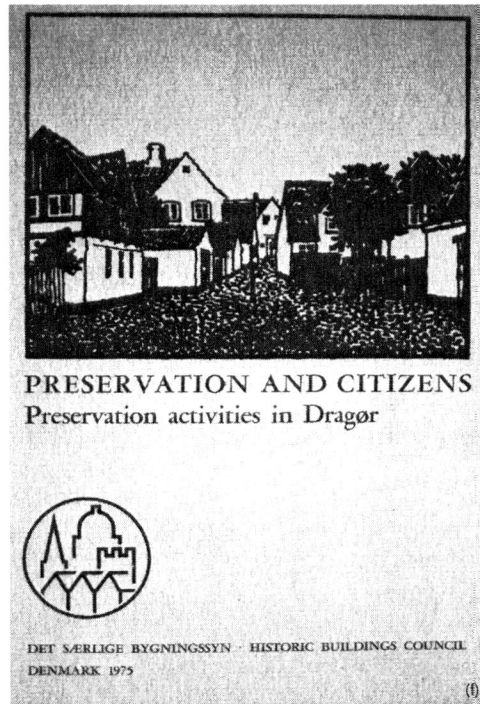

Steek een hartje toe

PRESERVATION AND CITIZENS
Preservation activities in Dragør

DET SÆRLIGE BYGNINGSSYN · HISTORIC BUILDINGS COUNCIL
DENMARK 1975

shift towards managerial corporatism was inappropriate in an age of participation and protest, this new European strategy was reinforced through a geopolitical rhetoric of European integration and a language more impassioned than that of the Charter of Venice 'technicians'. In a post-nationalist version of *Heimat* rhetoric, the Declaration argued that, 'apart from its priceless cultural value, Europe's architectural heritage gives to her peoples the consciousness of their common history and common future. Its preservation is, therefore, a matter of vital importance.'[35]

The chief importance of Amsterdam was in its exhortatory effect on the planning systems of less conservation-minded states and in the further deluge of international initiatives in urban conservation that it unleashed, in Europe and more generally – all agreeing on the need for a holistic approach. Already, ICOMOS had intervened in support of this agenda with a May 1975 symposium on the challenges facing smaller historic towns, especially in relation to tourism, held appropriately in Rothenburg ob der Tauber. Its proceedings, chaired by Lemaire, warned against economic over-exploitation and inspired the October 1975 'Bruges Resolutions on the Rehabilitation of Historic Towns'. UNESCO's 1976 'Recommendation concerning the Safeguarding and Contemporary Role of Historic Areas' had a slightly different emphasis, on reconciling urban conservation with change and modern interventions (also tackled in a UNESCO 1972 intergovernmental conference on European cultural policies in Helsinki, and a subsequent programme of urban pilot studies). Also stemming from ICOMOS, in 1987, was the most important single global charter dedicated to urban issues: the Washington Charter for the Conservation of Historic Towns and Urban Areas. It consolidated everything that had happened in the field since the first tentative steps in the Venice Charter and set out a very broad framework for urban conservation, stressing yet again the need for organisational and social interconnections within urban development (in other words, integrated conservation). In this intensifying focus on 'The City' as opposed to the entire built environment, conservation was also marching closely in step with contemporary developments in postmodern architecture.[36]

The Amsterdam Charter's stress on integrated conservation, as a process rather than an object, was taken up with gusto, both by national governments and in international initiatives. The following year, in 1976, a Resolution of a Committee of European Ministers urged governments to adapt their laws to integrated conservation. A framework for doing precisely that was set out nine years later, in the 1985 Granada Convention for Protection of the Architectural Heritage of Europe (and a follow-up convention in Malta in 1992), which set up a procedure for national governments to establish statutory measures of integrated urban conservation. The idea that urban conservation was as much a process as a set of things, and should be mobilised as part of a new cultural politics, was expressed globally in the Petropolis Charter, a Brazilian initiative of 1987 about preservation of historic town centres. In impassioned language, it stressed the importance of the intangible heritage as part of the culture of the city: 'Urban historical sites may be considered as those spaces where manifold evidences of the city's cultural production accumulate'; preservation was part of a dynamic process of 'expression and consolidation of citizenship', setting the built heritage in a 'wider totality' along with the 'everyday views expressive of their dwellers as well'.[37]

The beginnings of 'World Heritage'

EAHY was typical of the development process of the international Conservation Movement, in the way it flared up out of a complex range of collective efforts and then dispersed once again into a succession of follow-on initiatives. The major exception to this fitful pattern, as an international conservation initiative whose prominence has only grown with time, was what became known as 'World Heritage' – a top-down UNESCO initiative of traditional League of Nations/UN type, set out in a 1972 convention (the Convention concerning the Protection of the World Cultural and Natural Heritage, adopted by UNESCO in November 1972 and implemented in 1975). The only local partners of World Heritage were, strictly, the national governments themselves, or 'states parties'.

The World Heritage system stemmed ultimately from a series of initiatives in the USA, focused chiefly on the natural heritage and inspired by the National Parks experience. The suggestion of 'a trust for the world heritage' was first made by environmentalist Russell Train, at a 1965 White House Conference on International Cooperation convened by President Johnson, and the environmental programme of the Nixon Administration in 1968–71 elaborated this into a fully fledged convention for natural landscapes, bolstered by IUCN proposals to the Stockholm Conference. This natural-heritage initiative was paralleled by a UNESCO convention for 'cultural properties' (originating in a 1960s' UNESCO campaign for 'preservation and strengthening of the cultural heritage of mankind' and expert meetings in 1968–9). After much wrangling between UNESCO and IUCN, these were combined into a single, UNESCO-administered system, presided over by UNESCO's Senegalese Director-General (1974–87), Amadou-Mahtar M'Bow; significantly, the 1972 World Heritage convention coincided with the centenary of the founding of Yellowstone National Park. The need to balance the cultural and natural strands of World Heritage and minimise the bias stemming from state-party pressures, proved a constant running sore in the system's early days. Architect Francois LeBlanc (director of the ICOMOS Secretariat from 1979) recorded in 1984 the difficulties caused by the greater subjectivity of cultural than natural evaluation: 'Bisons have more in common than cathedrals!'[38]

Several years of discord between ICOMOS and states parties over the system of evaluation of cultural properties were only resolved at a 1979 conference in Luxor, when a more centralised system coordinated by French art-historian Leon Pressouyre was established. Here there was no attempt at a general coordination or control of world heritage in the broad sense. Instead, individual 'properties' of 'outstanding universal value' (OUV) were to be nominated by states parties, in accordance with operational guidelines agreed by ICOMOS, IUCN and ICCROM, assessed (in the case of cultural properties) with reference to six cultural criteria of OUV, approved in 1976 by UNESCO. Originally devised by M'Bow's predecessor as Director-General, Rene Maheu (1962–74), the OUV concept emerged gradually and was only conclusively defined in 2005. Designated properties would be inscribed on a list, which would also specify if they were 'endangered'. Justified initially by evocations of the 'duty of the international community', the scheme, because of its simplicity and the possibility of UNESCO grant assistance, proved hugely popular – despite the initial misgivings of influential figures such as Philippot. Originally, in 1975, 20 states parties ratified the convention, but by 2009 this figure would soar to 185, with no less than 878 properties on the World Heritage List (679 cultural, 174 natural, 25 combined: popularly known as 'World Heritage Sites'). Administered by a UNESCO secretariat (known since 1982 as the World Heritage Centre) and a World Heritage Committee of 21 members nominated by states parties, the system also contained a modest element of enforcement, in that properties subsequently 'endangered' by inappropriate development could be publicly branded as such, and ultimately removed from the list.[39] By 2010, only one cultural property had actually been removed (Dresden Elbe Valley, owing to an obtrusive new bridge), but the real effect of the system was hortatory, to encourage or shame states parties to enforce their own planning systems more rigorously.

above and facing page

Figure 11.10 World Heritage

(a), (b), (c), (d), (e), (f) World Heritage marker signs in varying formats: Birkenau (Auschwitz), Speyer, Tallinn, Vilnius (with extra plaque commemorating speech by President G W Bush), Regensburg Altstadt, Mexico City (University); (g) 1981 official World Heritage inscription letter for the Würzburg Residenz

(d)

(e)

(f)

(g)

UNITED NATIONS EDUCATIONAL,
SCIENTIFIC AND
CULTURAL ORGANIZATION

CONVENTION CONCERNING
THE PROTECTION OF THE WORLD
CULTURAL AND NATURAL
HERITAGE

The World Heritage Committee
has inscribed
Wurzburg Residence with the Court Gardens
and Residence Square
on the World Heritage List

Inscription on this List confirms the exceptional
and universal value of a cultural or
natural site which requires protection for the benefit
of all humanity

DATE OF INSCRIPTION
30 October 1981

DIRECTOR-GENERAL
OF UNESCO

Significantly, in view of this system's inclusion of nature conservation, and its growing links with international tourism and capitalist modernity (see Chapter 12), the first country to accede to the Convention, in December 1973, was the USA; from the very beginning of property designation in 1978, the system drew significantly on US National Park Service policies, including a system of buffer zones drawn from 1940s' American practice. Initially, despite the equality of esteem of cultural and natural heritage, most designated properties were of a traditional monumental architectural character and were concentrated in Europe. Some were individual buildings, but more typical was a complex group-property such as Bryggen in Bergen, where 1979 saw designation of the surviving area of wooden merchants' houses, including those 'restored' after 1940s–50s' damage. We will review later challenges to the World Heritage system in the next chapter.[40]

Years of confidence – and complacency?

T HE ascendancy of traditional European and North American values in this foundation phase of World Heritage, and the domination by elite experts, lasted until at least the late 1980s. The climate of triumph, shading into complacency, was vividly expressed in the proceedings of yet another international meeting – the First International Congress on Architectural Conservation, at the University of Basel in March 1983. Co-organised by ICOMOS and ICCROM with the Institute for International Art Festivals, this aimed to create (yet another) platform for 'regular exchange of all the most recent ideas' on conservation: its hardly innovative theme was that of 'Historic buildings, their significance and their role in today's cultural setting'. Attended by a galaxy of the international heritage elite, including Sandys, Lemaire and Feilden, as well as miscellaneous political figures such as the Tunisian prime minister and the ex-president of Senegal, it presented a microcosm of Venice Charter-style heritage values and expert-dominated Eurocentrism – an ethos proudly trumpeted by ICOMOS president Lemaire in his introductory address. On the one hand, Lemaire argued, all the key 'ideas and skills' of conservation had originated in Europe and the West, including the 20th-century modernist interpretation embodied in the Venice Charter: they had 'emerged from the Western view of history, from the West's links with its own past, from the role and significance of Western monuments within a given environment, from Western artistic sensibility and ethical values'.[41] Now, in an age of conservation triumph, this world-view could assert a global authority and universal applicability, albeit qualified by an element of cultural diversity. 'The principles underlying conservation are being accepted by everyone', Lemaire argued: in 'every country in the world', they were recognised as

> a universal aspiration, innate in human nature and in man's awareness of needs associated with the development of his culture . . . [a] framework which nurtures a feeling – basic to man's spiritual equilibrium – that the past abides, that race, culture, intelligence and creativeness continue. The architectural heritage is, in fact, the chief determinant of a society's identity.

With a few changes of wording, this paean to the inevitable and irresistible rise of the Conservation Movement might easily have been a speech by Le Corbusier or Gropius, talking confidently of the universal applicability and socially redemptive power of modern architecture.

But as always in the fluctuating story of the Conservation Movement, this time of elite ascendancy proved short-lived: already, by 1983, the newly established grand narrative was breaking up. Partly, this resulted from the sub-division of heritage into specialist discourses, a trend pioneered by the boom in industrial archaeology, and celebrated in a succession of highly specific international conventions. Some focused on different types of building or construction, like vernacular building, the subject of a conference at Plovdiv, Bulgaria, in 1975 and an ICOMOS charter in 1999; industrial buildings and civil engineering, dealt with in a 1990 resolution of the European Committee of Ministers; modern architectural heritage, addressed in a 1991 Committee of Ministers Resolution on the 20th-century architectural heritage; or mud brick-building, dealt with in two Yazd Symposia in Iran in the 1970s. Others focused on specific professional skills or groups: for example, the resolutions of the European Committee of Ministers on the craft trades. 1981/6; the archaeology charters (ICOMOS in 1990 in Lausanne, Council of Europe in Malta, 1992); the ICOMOS Colombo education and training guidelines of 1993; or the Sofia Principles on recording of buildings and sites, 1996.[42]

Partly, this fragmentation of the conservation grand narrative also reflected the geopolitical shift towards market competition, already seen in Western countries in the 1980s' 'heritage-industry' debates, but increasingly pervasive after 1989. As we will see in Chapter 12, the assumption that capitalist influences could be isolated under the blanket heading of tourism or 'Disneyland', as in the 1969 ICOMOS Oxford proceedings, seemed increasingly unsustainable. Most important of all, however, was the effect of the wider climate of cultural postmodernity, with its corrosive effects on all authoritative discourses and linear Enlightenment concepts of progress – including the Conservation Movement's own historical narrative of perpetual advance.[43]

Within conservation, the effect of this shift was to spotlight the inconsistency between the constant hankering for authoritative, authentic definitions and the unstoppable dynamic of urban conservation towards ever-widening diversity, towards heritage as a social or cultural process rather than physical product. The growing international stress on cultural diversity was showcased in the

Figure 11.11 Images of the 1979 ICOMOS-Australia meeting in the historic mining town of Burra, South Australia, at which the Burra Charter was adopted

renowned Burra Charter, which appeared in 1979 and was updated no less than four times, in 1981, '88, and '99. The Burra Charter, unusually for such a seminal document, was not an international but a national document, drawn up by the Australian committee of ICOMOS. It was catapulted to a leading role within international conservation largely by accident. At the fifth ICOMOS General Assembly in Moscow in 1978 (the last assembly presided over by Lemaire), a draft revision of the Charter of Venice was rejected, following disagreement.[44] The resulting gap was filled by the innovative Burra document the following spring. This strongly reflected the special polarisation within the Australian heritage between European materialistic culture and the intangible values of the aboriginal inhabitants. To bridge this gap, cultural diversity was made into the central theme of the charter and became a new focal criterion of authenticity, eclipsing artistic or historical value, and striking a new general balance between the universal and the local.

Under the Burra framework, in a reprise of Herder, universality and cultural specificity peacefully complemented each other in a stable balance. Universality now expressed the equal validity of different cultures, and authenticity became an expression of international cultural equity, simply resulting from the fidelity of each cultural monument to the internal logic of its own culture – a subtle difference, as we will see, from the more strictly hierarchical World Heritage framework of outstanding universal values. Burra distinguished between 'cultural significance' criteria (including 'aesthetic, historic, scientific, social or spiritual value for past, present of future generations'), which it regarded as legitimate values of conservation, and so-called 'cultural values' (including political, religious, spiritual and moral factors), which were at one remove from conservation and might have to be left unresolved and even in conflict. In other words, there was still a kind of emotional hierarchy, with the more extreme passions kept away from conservation.[45]

By 1989, with a second revision of Burra newly completed, the international colossus of architectural conservation still seemed to be intact and moving forward. But, in some respects, the Burra Charter was arguably a Trojan horse, projecting the values of postmodern relativism into the heart of the authoritative, Western-dominated organisational structure of the Conservation Movement and undermining the conception of conservation as an Enlightenment narrative of progress. And as we will see in the next chapter, fundamental change was now in the air, change that would begin to undermine the imposing structure that had been built up over the previous two centuries, even as it celebrated its newly won ascendancy.

The Contemporary Story

– 12 –

Heritage in the age of globalisation

Post-1989

If all values are equal, then there's no real value any more.

Jukka Jokilehto, 2006[1]

Iɴ this brief concluding chapter, history converges with the present. Covering only the last two decades, it is inevitably more speculative and eclectic in the trends and connections it traces, grouping its subject matter thematically rather than in a tidy narrative. Its subject matter, contrary to Lemaire's confident assertions of final victory in 1983, is a work in progress. Although it is not yet clear whether 1989 will come to seem as decisive a turning-point in the Conservation Movement as 1789, we are undeniably witnessing another fundamental shift of values in the heritage world.

This turning-point in the evolution of conservation, however, has not arrived suddenly or without warning. As we saw in chapters 9, 10 and 11, some elements, like the shift from a nationalistic to an international or global perspective, took several decades to develop. Yet it has only become recognisable as a paradigm-shift since all these elements came together from the 1980s and '90s. Even as late as the 1970s, the newly victorious Conservation Movement still seemed inextricably entwined with ideals of Enlightenment Progress, national identity and historical destiny. Conservation, after all, originally became a Movement under the influence of the overwhelming sense of driving historical force that swept Europe after 1789, and its ideological self-definition as a narrative revolved around constantly self-reinforcing internal concepts of authentic essence and 'life'. It also developed strong external ideological borders, including a highly defined chronological and thematic front-line of perpetual advance.

That sense of authoritative movement and destiny, however, began to weaken during the later 20th century – the very time that the movement seemed to be sweeping to final victory. Now, between 1945 and 1989, the aggressive militancy of the interwar years had been succeeded by milder processes of soft-nationalist spread and protest-led ascendancy. Conservation's victories were made possible by state power, yet they also helped undermine it. And the relativistic ethos of cultural postmodernity, and the general political and economic processes of globalisation, stimulated a further shift in conservation values in the 1990s. Conservation's old expansionist driving force, and its cohesion, both now appeared to be ebbing away, as it became both more static and more fragmented in character.

The far-reaching character of this destabilisation process was highlighted, for example, in a comprehensive German national research project of 2009 onwards, *Denkmal-Werte-Dialog*. This argued that, in the previous two decades, conservation values had been:

thrown into doubt at both a theoretical and practical level. Seemingly secure professional foundations have lost their general acceptance, the power of heritage authorities
has vanished, and cultural heritage has come into head-on collision with commercialism – and all this without any coherent and practical new value systems in prospect,
that could give an up-to-date definition to the place of heritage within globalised
society.[2]

In the rest of this chapter, key aspects of this broad shift in values and content are reviewed
under several thematic headings: topics that substantially overlap in practice. We begin with the way
in which the sharp self-definition of conservation has been challenged at both a geographical and
economic level, by the processes of globalisation. Then we focus more specifically on conservation's
internal theoretical discourses of authenticity and their displacement away from fixed, authoritative
monuments towards the amorphous territory of intangible heritage or memory landscape. And
finally, we review perhaps the most striking shift in the new and uncertain world of post-1989
conservation: the decline in authority of the old, Ruskinian/Modernist emphasis on the distinction
between new and old, and between restoration and original. Increasingly, under the influence of the
postmodern shift from reality to image and spectacle, all these have started to become mixed
together, with disorientating effects.

Of course, there are still unambiguously urgent issues and discourses of progress in today's
globalised built environment, including the driving force of economic expansion and urbanisation
in Asia, or the ever-expanding world-outlook of sustainability, with its escalating claims of threat
from climate change and its habitat discourse focused on combating poor housing and environment
across the world. In all these cases, however, the implications for architectural conservation have
yet to become fully clear – including whether they are likely to reinvigorate traditional Western
conservation values or to further undermine them.

Heritage without borders

WE traced in the previous chapter how a distinct and coherent international heritage network
began to coalesce out of the elite conservation values and forces of the postwar years. Now,
after 1989, that process stepped up a gear, as international became global. In terms of cultural
geography, the most obvious manifestation of this dissolution was the breaking-down of
conservation's old geopolitical boundaries, between West and East, and even between different
national traditions. These intermediate units of heritage organisation were increasingly squeezed
out between the extreme individualisation of market choice on the one hand and the globalising
values of world heritage on the other – although the persistence of 'national heritage' was
demonstrated in Stubbs and Makaš's prodigious 2011 compendium on European and American
conservation, sub-divided into over 70 parallel national narratives.[3] The sharp, highly structured
passions of national *Heimat* were beginning to merge back into the general or universal – yet, at the
same time, there was a new and chaotic upsurge of local cultural narratives and demands. How
would all this affect the Western, Eurocentric values of traditional conservation?

In the event, some of the most dramatic ruptures took place within Europe, as highlighted in
the post-1989 collapse of the socialist bloc and disappearance of ancien-régime 'monuments',

including 'The Wall' and the GDR itself, and the subsequent creation of new, post-communist landscapes. In many parts of Europe, entrenched heritage divisions began to heal, for example in Ireland, where the old Ascendancy landscape of country houses and Georgian Dublin finally became accepted as a legitimate element of the national heritage. Even in the former Yugoslavia, although the 1990s' outbreak of ethnic warfare saw towns and villages in Croatia, Bosnia and Serbia undergo culturally targeted attacks – most notoriously the 1991 shelling of the World Heritage city of Dubrovnik – eventually this ravaged landscape, too, became the focus of concerted international repair efforts, symbolised above all by the reconstruction of the Stari Most of Mostar, destroyed in 1993 and rebuilt in 1996–2004 in a EUR 15m campaign. In some places, however, the politicisation of heritage continued largely undiminished, as in the Middle Eastern conflict, where even the new rhetoric of international charters became appropriated for partisan purposes: 2008, for example, saw the publication of a 'Bethlehem Charter' on the rehabilitation of Palestinian historic towns, all in accordance with up-to-date integrated principles, but justified by the overall aim of affirming 'a practice of cultural claim, set against the forced modernisation of colonial Israeli settlements'.[4]

Figure 12.1 Memory landscape of the Cold War: GDR border-crossing 'highway speed limit' sign at Friedrichstrasse, Berlin, in June 1990, just prior to German reunification

One of the most blatant of the globalising shifts of the 1990s was the growing commercialisation – or, as Choay put it, 'industrialisation' – of the heritage. In Chapter 9, we traced the beginnings of a heritage industry within Western countries in the 1980s, but now, following the collapse of the socialist bloc after 1989, that process became much more pervasive.[5] As recently as the 1970s, the dramatic conservation advances in the West had still been strongly flavoured with utopian and anti-capitalist values. Prior to that, and right back to the 19th century, commercialisation had been widely seen as an external force, sometimes enhancing the heritage but more often threatening it – a menace that provoked the constant rhetoric against unsightly advertisements or 'Disneyland pastiches'. From the 1990s, however, commercialisation ceased to be unambiguously 'the other', and began to infiltrate the values of conservation itself – in the same way that nationalism had in the 19th and 20th centuries.

This shift in dominant ideology had both a positive and negative side. Although some of the most intensively exploited heritage hot spots, such as Rothenburg ob der Tauber, remained outside the World Heritage system, in general World Heritage status became intimately bound up with tourist promotion. For example, Lyon, in the year after its inscription in 1998, saw a 29 per cent increase in tourist visitors. This could act as a benign spur to economic growth and could even (in the opinion of the World Bank) potentially foster the cause of world peace, by neutralising old passions and focusing people's attention on what they had in common, through its stress on universal values. In Europe, the traditional emphasis on the Old Town as a focus of cultural identity and 'life' was transformed from a divisive to a unifying factor. This process began with the 1980s' postmodern embrace of The City as a socially mixed, cosmopolitan ideal and rose to a climax in the 1990s' and 2000s' emergence of a network of 'World Heritage cities' linked by low-cost air travel, competing with, but also complementing, one another through government-sponsored cultural festivals and economic-enterprise promotion. The World Heritage system actively emphasised values of peace and reconciliation, as shown in the exemplary and even-handed Polish presentation of the history of Malbork Castle since its inscription as a World Heritage property in 1997, complete with enthusiastic research by curator Ryszard Rząd into the 19th-century German restorations once reviled by Zachwatowicz.[6]

But that process of unification under the hegemony of the market also had adverse consequences. Just as the pre-1789 world of heritage was dominated by a strong hierarchy, with Rome at its head, so the post-1989 network of competing cities, especially within Europe, became structured and subtly homogenised by a hierarchy of branding, with elite 'iconic' cities at the apex (Florence, Venice, Paris, etc.) and lesser contenders (Berlin, Barcelona, Prague, Munich . . .) struggling for status below. But the more cities competed, in many ways the more they became alike, given their reliance on a standardised recipe of branding, staging and 'facile semanticisation' (Choay).[7] This recipe employed a mix of renowned, long-established 'trademarks' (often radically and repeatedly reshaped, as in the case of Viollet's Notre-Dame), with supposedly more individualistic modern or signature heritage in up-and-coming cities – such as the cults of Gaudí in Barcelona or Mackintosh in Glasgow – as well as a more general common background of 'vibrant, mixed use' hubbub.

A typical example of this process of urban heritage branding was the Temple Bar project in Dublin, a 15-year revitalisation programme that transformed a dilapidated, demolition-threatened inner zone of the Irish capital into a showpiece of mixed-use regeneration. It arose initially from pressure by heritage groups and the desire of Taoiseach (Prime Minister) Charles Haughey to revitalise the area as a 'living' monument to Dublin's year as European City of Culture in 1991,

420

Figure 12.2 Globalisation and commercialisation

(a), (b) Restoration works clad in giant advertisements, Venice, 2011; (c) The Leaning Tower of Pisa, 2008.

exploiting its 'indigenous cultural heritage' to make it 'a city which can stand comparison with any other European capital'. Supported by special legislation and £25 million of EU funding, two government-subsidised development companies mounted a 1987 architectural competition: the winning masterplan, by Group 91 Architects, ingeniously knitted together rehabilitated everyday buildings and new interventions with Townscape-style landscaping. But the eventual result of all this attention to individuality and place looked curiously like its counterparts across Europe. The close interrelationship between Old Town and modern zoning had always presupposed an element of standardisation, but now the pressure of the market was carrying that process rather further.

Neo-modernist architect-critic Rem Koolhaas rhetorically asked in 1995, 'Is the contemporary city like the contemporary airport – "all the same"?' Arguably, heritage environments formed an integral part of that 'generic' standardisation process.[8]

421

(a)

Figure 12.3 'Urban regeneration' in Dublin

(a) Concept drawing for regeneration of the Temple Bar district by Kenneth Browne, published in the *Architectural Review*, November 1974; (b) The winning masterplan for the Temple Bar project, 1991; (c) 2007 street view in Temple Bar

The increased commodification of heritage extended beyond the narrow world of the European Old Town, however. In general, heritage was now expected to 'do' rather more than before, especially in an urban context, whether at a national, European or global level – as exemplified in the role of World Heritage Sites as hot spots of tourism and incoming investment. It was expected not just to act as an architectural, social or cultural catalyst but to play its part as an element in managed strategies of economic regeneration, aimed at reinventing deindustrialised cities through wide-ranging strategies of reconceptualisation and reconfiguring, implemented by a range of governmental and private agencies. For example, the 1980s–'90s' regeneration of Castlefield, an area of decayed 19th-century industrial and transport infrastructure, just west of the city centre of Manchester, Northern England, built on the large-scale preservation designations that followed the boom in industrial archaeology as the stimulus for a two-stage strategy: first, a boom in heritage tourism; second, a programme of cosmopolitan mixed-use urban renewal and property speculation. This strategy, in the event, met with only partial success, owing to sharp economic fluctuations.[9]

Figure 12.4 'Urban regeneration' in Northern England

(a) The Castlefield regeneration zone, Manchester. In the background is the 'iconic' 47-storey Beetham Tower (by Ian Simpson, 2004–6); on the left is the Air and Space Hall of the Manchester Museum of Science and Industry (MOSI), a late-19th-century market hall converted into an aviation museum in 1983 and incorporated in MOSI in 1985; (b) 'Chimney Pot Park', an area of late-19th-century terraces in Langworthy, Salford, converted in 2007–10 by developers Urban Splash and architects 'shedkm' into a 'signature project' – 'your very own, very modern Coronation Street'

In 2010, Rem Koolhaas went so far as to argue that conservation had now become 'an entirely new architectural language of disguised consumerism'. But there were still strong limits to the market impulse within conservation, as shown by the hostile reception often given to any more extreme commercialised initiatives. And, from the 1990s, an increasing range of writers, such as Nezar Al Sayyad or Barry Wellman, highlighted the emergence of hybrid patterns of modernisation, or 'glocalisation', combining elements of global capitalism and material progress with evocations of local cultural and landscape context – in much the same way that interwar nationalism had combined more restricted universal and local elements. These hybridities were often most striking and distinctive in countries beyond the traditional Western heartland of the Conservation Movement. In the planned modernisation of the Saudi city of Medina, for example, an advanced road and rapid transit system, and an array of Islamic-styled multi-storey hotels, were strikingly juxtaposed with a cherished religio-cultural landscape of holy sites.[10]

Dissolution of the 'real monument': intangible heritage, cultural landscape, memory landscape

IF the old national boundaries of heritage were being challenged by the external forces of globalisation, at the same time the coherence of conservation was being eroded from within, by the postmodern idea that nothing was of any greater importance than anything else. Under the old Conservation Movement, one of the key driving forces of heritage was that of its ever-expanding scope, not just chronological but thematic, from the individual commemorative monument to entire buildings, entire towns, entire landscapes. Under nationalism, the need for each country to be unique

423

and the sense that conservation's expansion must automatically be targeted at things of special significance had counterbalanced this levelling-out process. But, under postmodernity, those restraints vanished, and the expansion process was carried much further, to the point where it also began to undermine the principles of special importance, rarity and threat that had given conservation its sense of public urgency and communicability throughout the 19th and 20th centuries. If everything and anything could become heritage, how or why should anything be picked out for special treatment or protection? As Jukka Jokilehto suggested in a 2006 lecture, 'If all values are equal, then there's no real value any more.'[11]

Of course, elite monuments continued to command often intense international attention, as shown by the high-profile campaign to rescue the Leaning Tower of Pisa following its near-collapse in 1995 after botched foundation work – a crisis finally remedied in 1999–2000 by an international project led by Polish geotechnic engineer Michael Jamiolkowski. But elite monuments such as this also now flourished as part of wider cultural movements, including the capitalist-led commodification of heritage environments for consumption as images, as well as more collectivist, socially integrated ideals. As Jokilehto put it in 2010, the 'modern conservation movement started with the recognition of monuments and sites. It has since evolved into a holistic approach to historicised built and natural environment'.[12]

In the broader definition of what could constitute a monument, the gates were now thrown wide open to all-comers, under the influence not only of postmodern relativism, but also of the radically new field of intangible heritage. A century earlier, Riegl had pointed to the importance of reception in some heritage values. But now reception seemed to have become all-important, just as in contemporary art or literature, and the value attributed to any heritage object began to depend entirely on the present-day host culture. Heritage academic Gregory Ashworth wrote in 2006, for example, of 'pluralising the past' and proposed a plethora of new, politically and culturally shaped models for the valorisation of heritage. Leading the charge of the new, polycentric world outlook was the concept of the cultural landscape. This was an idea originally devised by interwar geographers preoccupied with natural rather than built landscapes: Carl Sauer had argued in 1925 that 'The cultural landscape is fashioned from a natural landscape by a cultural group. Culture is the agent, the natural are the medium, the cultural landscape is the result.'[13]

In the heritage context, this concept mutated into a fuzzy postmodern echo of *Heimat*. This was a landscape that encompassed and expressed all aspects of culture, to be sure. But now, it was not just the culture of a single, unitary nation, but a cultural diversity of multiple groups, both at any one time and over time. Psychologists started to talk of landscapes as palimpsests, constantly and fluidly remade, in a kind of extended performance, in the light of changing values and concerns. Within the built heritage, as we saw in Chapter 11, the Burra Charter first decisively opened the door to this concept, in the late 1970s and '80s. Burra was particularly well attuned to this discourse of cultural diversity, because of its strong emphasis on the polarisation between Western materialism and aboriginal landscape spirituality.

But matters did not stop there. Instead, the cultural landscape concept spawned a more extreme offshoot from the 1990s onwards: the 'memory landscape', a landscape laden with multiple and often conflicting political commemorative associations. Partly, this represented a postmodern version of the war-memorial landscapes of the earlier 20th century, recasting commemoration in a more democratic form better attuned to individual choice, dominated by fluid metaphors rather than fixed memorials or elite scholarly histories. With the ending of the real menace of nuclear annihilation, the 20th century's demand for ideological discipline in commemorative matters vanished, and almost unlimited individualism and freedom became possible.[14]

Inevitably, the new stress on subjectivity and flux potentially opened the door to false recollection or simple fiction. As acknowledged by the earlier-20th-century advocates of collective memory, such as Maurice Halbwachs, the terms 'memory' and 'recollection' in this context are really just metaphors and do not correspond to the actual memory of any individual person, just as they do not correspond to the 'facts of history'. Proust had already argued in the 1920s that memory and forgetting were paradoxically intertwined:

> If, owing to the work of oblivion, the returning memory can throw no bridge, form no connecting link between itself and the present minute, if it remains in the context of its own place and date, if it keeps its distance, its isolation in the hollow of a valley or upon the highest peak of a mountain summit, for this very reason it causes us suddenly to breathe a new air, an air which is new precisely because we have breathed it in the past, that purer air which the poets have vainly tried to situate in paradise and which could induce so profound a sensation of renewal only if it had been breathed before, since the true paradises are the paradises that we have lost.[15]

The ever-shifting character of memory landscapes is not a new idea. As we saw in previous chapters, important elements within the Judaeo-Christian cultural tradition insistently denied the permanence of any worldly structures. Arguably, medieval Christendom itself was one single large and ever-shifting memory landscape, with its mixture of pseudo-history, relics and idealised holy places, constantly evolving within a framework of progress towards salvation. As this landscape broke up into individual nations, religious denominations and secular movements in the Renaissance and Enlightenment period, each triumph or disaster inspired its own fleeting mixture of objects, battle scenes or ideologies, some concerned with revolutionary progress and others with traditions and established causes. The Romantic Movement generated a wide variety of metaphoric, politically charged landscape concepts, whose frequent incorporation of ruins powerfully stressed the transience of temporal power. And the close relationship of memory and invention permeated all ideologies of nationalism: Benedict Anderson argued in 1991 that national identity was a matter of politically dictated myth-making and creation of 'imagined communities'.[16] But, on the whole, all these concepts still aimed at promoting authoritative identities, under the control of elite forces.

By contrast, in the extreme postmodern fluidity of the memory landscape concept, all authoritative narratives were comprehensively dissolved. In a memory landscape, everything is provisional, not unlike the designs of digital space in contemporary architecture.[17]

Typically of postmodern culture, memory landscapes are mostly structured through a branded approach, allowing for a combination of elite and popular narratives.[18] In 2009, for example, the august Society of Architectural Historians mounted a study tour of the 'architecture, urbanism and commemorative landscapes' associated with the civil rights movement in Alabama and Georgia, part of which had been officially designated as the 'Selma and Montgomery National Heritage Trail'. In each city, both historians and 'former participants' were on hand to 'help us understand the monuments and landscapes we see, and recall those that are no longer available to see'. The previous year, by contrast, Mississippi governor Haley Barbour announced the inclusion of the birthplace of singer Elvis Presley at Tupelo, as part of the 'Mississippi Blues Trail', to 'honor the work, memory,

425

and music legacy of the King himself'. In a logical extension of this approach, preservation theorist Ned Kaufman argued in his 2009 book, *Place, Race and Story*, for an all-inclusive, community-based concept of 'storyscape'. He claimed that this eclectic populism could help combat the commodification of the historic environment, but the direct overlap with Disneyland seemed even more striking.[19]

Within a memory landscape, the role of authentic architectural heritage, or of buildings in general, is that of an optional extra: in true postmodern fashion, images and discourses seem as 'real' as solid structures. Sometimes, isolated preserved buildings are enlisted to help anchor otherwise reshaped landscapes, rather like the *spolia* or *pars-pro-toto* fragments of early Christian churches. In 2005, for example, a growing movement for affirmative commemoration of the community life of early public housing in Hong Kong under British colonial rule inspired a project to preserve from redevelopment the last surviving 'Mark I Resettlement' block of the pioneering Shek Kip Mei complex (1953–5) – even as the vast built reality of the Resettlement estates disappeared completely.[20]

The most stimulating climate for the burgeoning of complex memory landscapes was that of 'hurtful' rather than affirmative memory. This was exemplified in the post-1989 proliferation of memory landscapes focused on the European genocide of World War II, now generally labelled 'The Holocaust' (following the disappearance of the previously all-consuming threat of the 'nuclear holocaust'). For this, in place of the obsolete, monumental style of World War II commemoration, a new, more fluid postmodern pattern was developed, mingling poetic landscape-traces with the forceful interventions of new, jaggedly Modernist museums by architects such as Daniel Libeskind,

Figure 12.5 Mass housing as heritage

(a) Ground floor flat in Hellersdorfer Strasse 179, Berlin – a typical 'WBS 70' prefabricated housing block, completed 1986. This flat was converted in 2004 into a 'Plattenbau museum' by developers Stadt und Land WoGeHe;
(b) Mei Ho House, Shek Kip Mei, Hong Kong – last survivor of the mid-1950s' 'Mk I Resettlement' blocks, seen here under conversion into a community museum in 2009

catering for so-called 'dark tourism'. Libeskind's stabbing iconic forms were significantly presaged in architect Hugh Hardy's 1978 memorial-style replacement for a terraced townhouse in West 11th Street, Manhattan, destroyed by an accidental explosion of anarchist bombs in 1970: Hardy's design maintained the cornice-line of the classical terrace but erupted into diagonal forms below. In a demonstration of the ephemerality of 'hurtful memory landscapes', the incident and the group involved (the Weathermen) are now largely forgotten, leaving the house-memorial as a baffling visual aberration in the streetscape.[21]

Common to all Holocaust memory landscapes was the flavour of a secular religion of martyrdom, combined with omnipresent touristic exploitation. There were, equally, great divisions, notably between didactic monuments in the commemorative sense identified by Riegl and 'authentic' sites such as remains of concentration camps. Yet these latter were *not* important chiefly for the integrity of buildings and remains, which often freely shifted in arrangement and detail. This battle of interpretations reached a climax at Auschwitz-Birkenau (in Polish, Óswięcim), a scene of incessant conflict between different myths of martyrdom and redemption. Prior to 1989, as we saw earlier, the Polish communist government had focused on commemoration of the camp's political prisoners. After 1989, all this was swamped in a wave of competitive memory landscape-remaking, including a new fashion for 'march of the living' Holocaust tourism by North American Jewish groups. For all of these competing agencies the physical remains are less important than the impassioned, yet ephemeral secular rituals that take place there. These years also saw the first Soviet Gulag museums, notably the 'Perm 36' memorial centre, founded in 1994.[22]

Within post-reunification Germany, there developed a different kind of memory landscape, led by the popular movement for 'digging for traces' of history in the places of genocide planning. Partly inspired by postmodern literary theory, in the relativistic idea that a text, rather than being self-contained, always bears a trace pointing to both past and future, this looked on urban environments as constantly reworked accretions of objects and values. In Berlin's old government quarter, the so-called Topography of Terror zone on the site of the Gestapo headquarters in the Wilhelmstrasse left a contentious interpretation-museum project unfinished as a ragged, open wound in the city.[23] The Reichstag building, too, became the site of intensive 'memory work'. Previously rebuilt in the 1960s by architect Paul Baumgarten to segregate modern interiors and the 19th-century shell, it became an artistic installation in 1995, by being wrapped in plastic by artists Christo and Jeanne-Claude; then, between 1995 and 1999 it was reconstructed to house the reunification parliament by Norman Foster, combining iconic modernism and memory traces (including reverently preserved Soviet soldiers' graffiti).[24] These traces were enriched but also confused by the equally ephemeral traces of the Berlin Wall and the former GDR regime, including yet another rebranding and rearrangement of the Neue Wache, in 1993, as a 'National Memorial to Victims of War and Tyranny', and by more-recent intentional-commemorative monuments such as Peter Eisenman's Monument to the Murdered Jews of Europe, built in 2003–5: a metaphoric built landscape comprising a grid of undulating blocks and paths.[25] In Berlin, we witness something strikingly different from the traditional, self-contained monuments of World War I: a composite memory landscape, spreading out across a major city, encompassing zones as big as some of the subjects of traditional architectural area conservation. And although the hurtful events recalled were undeniably all too real, the reliance on commemoration through metaphorical poetic gestures and minimalist art has a curious obscuring or alienating effect, giving the city something of the atmosphere of an Expo or even a historical theme park.[26]

Figure 12.6 Holocaust as heritage

(a) Visitor information board in the Auschwitz-1 camp museum, 2007; (b) 'March of the living' at Auschwitz-Birkenau museum, 2007; (c) Crumbling stone facing at the Zeppelinfeld Tribune, Nuremberg, 2009; (d) Car rally at the Zeppelinfeld Tribune, 2011; (e) 'Jewish ghetto'-themed zone in the Vilnius Old Town World Heritage Site, completed 2010

The Nara Declaration: dissolving authenticity

THE breaking-open of the narrow definitions and narratives of heritage, especially in the 1990s, naturally had radical implications for the doctrinal certainties around which the international empire of conservation had flourished. The focus of conservation's response, predictably, was yet another charter, or (to be more precise) a Declaration – the 1994 Nara Declaration – which addressed the increasing instability of the concept of authenticity. Definitions of authenticity, after all, had underpinned all doctrinal definitions from the 1964 Charter of Venice through to the outstanding universal values and operational guidelines of the World Heritage Convention.

In 1992, the proposed accession to the Convention by Japan, whose government was worried that its distinctive heritage practices would not be easily accommodated by the Western-dominated corpus of conservation theory, triggered an energetic burst of debates and meetings, mostly under the aegis of ICOMOS. Its Secretary-General (1990–3), Canadian architect-academic Herb Stovel, was determined to use this opportunity to carry out a first-principles reassessment of concepts of authenticity. A round of meetings in 1994 culminated in a conference at Nara hosted (at a cost of US $1 million) by the Japanese governmental Agency for Cultural Affairs – a curious echo of the interwar decision to hold the first international conservation conference in the city where Balanos's *anastelosis* efforts were underway. At the Nara convention, at which ICOMOS was supported by UNESCO and ICCROM, 55 international heritage experts (with Norwegian Knut Einar Larsen as coordinator) debated the issues of authenticity in depth, especially in relation to the World Heritage Convention. At the end, a formal international Declaration on Authenticity was drafted by Stovel and Lemaire. Their joint authorship symbolised the generational shift from the now old-fashioned, dogmatic theory of the mid-century, exemplified by Lemaire, to the greater flexibility and uncertainty of the post-1989 era.[27]

Despite the doubts of Lemaire, with his viscerally elitist instincts, and the worries even of Stovel that the declaration might become a 'shield' for 'anyone determined to do anything they want', the Nara Declaration ended up as a ringing endorsement of postmodern relativism.[28] It bluntly contended that, although 'authenticity appears as the essential qualifying factor concerning values', it was virtually impossible to arrive at any fixed or objective definition of it, as this would involve unacceptable 'mechanistic formulae': 'authenticity is a value judgement'. Instead, the respect due to all cultures required that all 'heritage properties must be considered and judged within the cultural contexts to which they belong', to ensure 'attributed values' were 'truly representative of a culture and the diversity of its interests'; both tangible and intangible expression were equally valid. What was needed was 'global respect and understanding' of heritage diversity and the 'legitimacy of the cultural values of all'. The overarching, collectivist rhetoric previously deployed by the nationalist conservationists had now been appropriated by globalisation and postmodernity: the key role of authentic monuments was 'to clarify and illuminate the collective memory of humanity'.[29]

The effects of the new relativism were felt directly within the doctrinal corpus of conservation itself, especially in a growing tension between the authenticity implications of tangible and intangible heritage. The former retained the traditional conservation requirement for historical or material authenticity, the latter presupposed constant remaking. As Australian heritage architect-planner Joan Domicelj noted, 'Nara told us authenticity did not require any significant place to stay frozen as is – that the outstanding values of a place could be sustained dynamically, so long as the stories remained credible and truthful.'[30] Conservation had always been about creating and enhancing

429

'stories' – a process that had involved constant remaking of built fabric, not least through restorations – but its activities had presupposed at least an enduring foundation. Ruskin had pleaded, 'When we build, let us think that we build forever.' Now, all that certainty had been jettisoned.

Within the recently formed and still-growing organisational structure of international conservation, and its vast array of charters and other definition-documents, the new doctrinal relativism led to a growing sense of instability. One writer argued in 2010 that 'the existence of the many documents leads to devaluation of their contents, weakens their position or undermines their sense altogether', so that 'any conservatory interference can be legitimized by any chosen text'. Least affected among the international bodies, owing to its relative autonomy and flexibility, was ICCROM, especially because it enjoyed a substantial and stable annual budget (by 2010, nearly EUR 5 million), 75 per cent of which was contributed by its member states, allowing it to run substantial surpluses on its annual expenditure. In a 1994 programme redefinition, it agilely shifted its operational emphasis from direct teaching and authoritative coordination to conflict-solving. A new programme of 'Integrated Territorial and Urban Conservation' applied the now long-established formula of integrated conservation (popularised in the Amsterdam Declaration) to the fragmented urban landscapes of the globalisation age, focusing initially on the post-socialist societies of Eastern Europe.[31]

Figure 12.7 The entrance of ICCROM, Via di San Michele, Rome, in 2009

ICOMOS was in a far more uncertain position, being both excessively taxed with external pressures and ambitious aspirations, and excessively small in relation to those demands. By comparison with IUCN's several hundred thousand members and 150-strong Swiss-based headquarters, ICOMOS's 10,000 members and tiny Paris secretariat (usually comprising around five staff), although adequate to service its own internal structure of national committees, international scientific committees and three-yearly general assemblies, was not large enough to adequately cover the external demands on it as an international doctrinal reference point and arbiter of World Heritage nominations. Increasingly, this system began to split at the seams, with many eligible cases (within the ever-spreading categories of significance) often simply 'falling through the cracks'.[32] Most serious of all, the UNESCO World Heritage organisation itself was becoming increasingly challenged, not least through the mismatch between demands and resources. Here, the inevitable shift from the authoritative Eurocentricity of the Lemaire generation towards a polycentric, competitive globalisation was an important contributory factor, with developing countries exploiting the far wider criteria of cultural landscape to make forceful inscription bids. Jukka Jokilehto, working as a full-time World Heritage designation adviser after his retirement from ICCROM (latterly as Assistant Director-General) in 1998, recalled that the years after 2000 saw a sudden boom in 'cultural landscape proposals coming simply out of the blue!'[33]

One way of trying to keep ahead of these developments was to try to make the definitions of value and authenticity wider and more flexible, shifting away from the old European elitism. This approach was first advocated in a 1998 World Heritage strategy report, which attempted to differentiate, Brandi-style, between universal core values and authentic responses to these in each society – a balance characterised by the concept of 'creative diversity'. In 2004–5, ICOMOS developed the theme further, issuing a report that tried to correlate Outstanding Universal Values with local or place-specific expressions and responses, including cultural associations, creativity, spatial, natural resources, movement of peoples ('roots') and development of technologies. A revised set of World Heritage operational guidelines of 2005 again differentiated between universal values and diverse expressions. Jokilehto argued optimistically that all these specific manifestations were linked by a common concern 'about truth, an ideal common to all countries'.[34]

In the issue of how to reconcile World Heritage and intangible heritage, it seemed initially as if the latter might become a new category of cultural World Heritage, paralleling the existing system. Cultural landscapes were accepted as eligible for World Heritage status in 1992 and, gradually, it became clear that this concept would become especially important outside Europe. Four years after Nara, in 1998, UNESCO issued a 'Proclamation of Masterpieces of the Oral and Intangible Heritage of Humanity', together with an initial draft list of potential intangible heritage sites. In 2003, this was elaborated into a full-blown UNESCO Convention for Safeguarding the Intangible Cultural Heritage, which defined the intangible heritage through the sweeping rhetoric of opposites and 'life' once used by *Heimat* propagandists. It was 'traditional and living at the same time' and could include just about any cultural practice, albeit with a few arbitrary political constraints: for example, nothing could qualify as intangible heritage that violated 'mutual respect among communities' or impeded 'sustainable development'. This was followed the next year by the even more pithily named 'Yamato Declaration on Integrated Approaches for Safeguarding Tangible and Intangible Cultural Heritage'. This argued that, 'as intangible cultural heritage is constantly recreated, the term authenticity as applied to tangible cultural heritage is not relevant when identifying and safeguarding intangible cultural heritage'.[35] And in 2005, a revised set of World Heritage Operational Guidelines was issued, engulfing the old four simple criteria of materially based authenticity – design, material, work-manship, setting – within a proliferation of new criteria: traditions, technology, language, even 'spirit and feeling' (paragraph 82). In the process, it seemed that intangible heritage had been decisively reclaimed by the mainstream World Heritage system.[36]

However, all these theoretical debates about values and definitions, so dear to the international heritage leadership, were increasingly overshadowed by the practical and often harsh reality of the politicisation of the World Heritage Committee, with countries often no longer represented by experts but by ambassadors tasked with 'horse-trading' and manipulation of the nominations process. In 2010, Mechtild Rössler, head of World Heritage policy, argued that the system was above all aimed at saving 'the most outstanding places on earth'.[37] But the outcomes on the ground frequently seemed to undermine any objective assessment of the 'outstanding universal values' of nominations. For instance, at an August 2010 UNESCO meeting in Brasilia, 21 new properties were added to the list, even though the expert advice had suggested only ten were eligible. The additional designations included a village near Riyadh associated with the Saudi royal family and an imperial palace in Vietnam earmarked for a forthcoming millennium festival. At the same time, political pressures mounted on the system of endangered properties, for example with claims that the removal of the Galapagos Islands from the endangered list in 2010 was influenced by political lobbying from the Ecuadorean government. The first delisting of a cultural heritage property, that of the Dresden Elbe Valley in 2009 because of an intrusive road bridge, had little deterrent effect on more audacious challenges, such as the 2010 proposal to rebuild the ruined 11th-century Bagrati cathedral in Georgia, backed both by President Mikheil Saakashvili and religious leader Patriarch Ilia II, in a project of near-facsimile character.[38] The maintenance of a hard line in a case such as this was undermined by the fact that already, as we will see later, radical facsimile schemes were proliferating near other World Heritage properties, as in central Moscow. Little of this was directly to do with architectural conservation in the traditional sense: but that only underlined the pervasiveness of the postmodern destabilisation of 'traditional values' and of any definitions of authenticity.

End of the Movement: conflating the old and new

PERHAPS the most fundamental of the special characteristics of the Conservation Movement, the idea that really made it a 'movement', had been its insistence on a sharp distinction between old and new, an idea stemming both from the general Western concept of the inexorable march of History, and from the specifically architectural, Ruskinian and Modern Movement insistence on 'honesty' and condemnation of 'pastiche'.

At first glance, the post-1989 years simply seemed to continue that pattern: indeed, the replacement of heritage-friendly postmodernism in the mid-1990s by a new and flamboyant iconic modernism seemed to make the contrast sharper than ever before. In Britain, continuity was accentuated by the continuing power of the Anti-Scrape ideology, with SPAB still projecting a vociferously anti-restorationist stance in English cases such as the IRA bomb-damaged St Ethelburga's Church in London and in Scottish castle-rebuilding projects such as the long-running (1997 onwards) saga of Caisteal Tioram. Even the restoration of a fire-damaged royal palace, Windsor Castle, stirred up fierce controversy in England, as did the National Trust's £20 million rebuilding of the fire-gutted 18th-century country house of Uppark in 1989–94.[39] The 1990s and 2000s' neo-Modernist vogue for

432

Figure 12.8 Distant skyline view (from north-west) of central Cologne, showing on the left the LVR Tower at Köln-Deutz, across the Rhine from the Cathedral and the epicentre of a conflict over 'threats' to World Heritage Sites from skyscraper proposals. In this instance, the city was placed on the UNESCO 'red list' following completion of the tower in 2005: the dispute was only resolved when the city authorities cancelled further towers planned nearby

aggressively iconic tall buildings also seemed to continue the sharp old–new separation, provoking fierce confrontations between modernising developers and conservation protesters over proposed skyscrapers in World Heritage buffer zones in Vienna, London and Cologne.[40]

But if we look closer at these relationships, the confrontation of new and old starts to seem more a matter of a contrast of images than of realities – as so often in postmodern culture. On the one hand, the new iconic modernism of the 1990s and 2000s seemed to exemplify what Guy Debord, as early as 1967, had dubbed *la societé du spectacle*, a culture whose authenticity rested in images and 'empty signifiers' rather than (as with the original Modernism) in the grand narratives of social progress. Cultural critic Fredric Jamieson argued compellingly in 1991 that the postmodern era was permeated by a crisis of historicity, with everything reduced to 'pastiche'.[41]

At the same time, conservation moved to finally close off its own internal narrative of Progress, by a bold move from around 1990 towards recognising the original Modern Movement as heritage. This movement was led by DOCOMOMO, a vigorous global confederation of individual amateur historians and architects, rather in the old anglophone volunteering tradition. Formed in 1988 by a largely European core grouping led by Dutch architects Hubert-Jan Henket and Wessel de Jonge, by 2010 DOCOMOMO had expanded into a worldwide network of over 60 national and regional working parties. Its interests included canonical or heroic works alongside many other aspects of Modernism, including postwar mass housing and commercial building, and modernisms divergent from the Western canon. The achievement of DOCOMOMO was precisely to set the Modern Movement *in* history.[42]

The new stress on Modernist heritage was also, however, a natural ally of image-based commodification and branding, not least in the intrinsic tendency of Modernist restorations towards a highly polished state of visual perfection dictated by iconic photographs. In 2008, Barry Bergdoll of MOMA indicted the 1983–6 facsimile-rebuilding of Mies van der Rohe's Barcelona Pavilion as evidence of 'the commodification of a highly selective view of Modern Movement history'; and Italian DOCOMOMO historian Andrea Canzani warned of a 'Dorian Gray syndrome' within Modernist historiography, 'where two-dimensional images dominate over the three-dimensional shapes and erase the dimension of the use and material authenticity', substituting instead an 'iconic authenticity' of images.[43] MoMo iconic authenticity and the increasingly image-preoccupied World Heritage system seemed to be natural allies, as emphasised in the case of Le Corbusier's 1960–5 design for St Pierre Firminy Vert, posthumously constructed by his disciple, José Oubrerie, in

(a)

Eine Schloßführung im Jahre 1979

Zeichnung von Karl Holtz

„. . . und dies hier ist die berühmte alte ‚neue' Sachlichkeit· . . .!"

(Oktober 1929)

261

(b)

Figure 12.9 Die alte 'neue' *Sachlichkeit*

(a) 1929 cartoon by Karl Holtz, 'Stately Home Tour in 1979'. The caption reads, 'And here we have an example of the famous old "Modern" Movement'; (b) 'Berlin Modernism Housing Estates' World Heritage Site (inscribed in 2008, and comprising six separate estates): information board at the Britz Siedlung, 2011

1968–2006, and then immediately included in 2006 as part of a proposed World Heritage cluster-nomination of Corbusier works.[44]

In combination with contemporary architecture's confusing revival of modernist forms, the move to embrace Modernism as heritage gave a final deathblow to the old Lemaire-style separation of old and new and effectively blocked off any further dynamic chronological evolution of conservation: there was no longer a significant campaigning front line towards which the Movement could push. Previously, the monument-designation system in Finland had been exceptional in refusing any chronological cut-off date for eligibility and in embracing Modernism as an integral part of the heritage, but now any old–new segregation segregation was widely questioned. In a *reductio ad absurdum* of this position, Rem Koolhaas argued in 2004 that city zones in Beijing should be designated as heritage even before they were built![45]

The disorientating effects of the mixture of old and new were seen at their most convoluted in the issue of how to reconcile modern development with historic urban fabric, with the advocates of a clear separation increasingly forced on the defensive. In an autumn 2008 English Heritage bulletin on the issue, *The Old and the New*, chief executive Simon Thurley and architect Richard McCormac argued not for a contrast but for a harmony of differentiated new and old – an artistic modern contextualism. The Lemaire philosophy of refined differentiation was subtly appropriated, and made unrecognisable, by iconic modernism. Its new projects in historic settings usually aimed at an integrated dialogue of old and new elements within intensely poetic visions. As an inspiration, neo-modernist architects often cited 1950s–'70s' museum or gallery conversion projects by Carlo

434

Figure 12.10 DOCOMOMO

(a) The June 2002 special retrospective edition of the DOCOMOMO International Journal: the front cover features a mosaic of 'DOCOMOMO Friends' past and present; (b) 2010 DOCOMOMO-International Conference, tour of the Universidad Nacional Autónoma de México campus in Mexico City: on the left (taking photo) is Hubert-Jan Henket, founding chairman of DOCOMOMO

Scarpa (see Chapter 8). Old and new were still clearly differentiated in a project such as I M Pei's Grand Louvre of 1983–9, with its giant glass pyramid, but the multi-layered complexity increased as the years went on. For example, Peter Zumthor's Kolumba museum in Cologne (2001–7) encased an accumulation of Roman, medieval and 20th-century fragments in a restrained, but highly individual box-like shell, with the aim of presenting a 'memory landscape of the two-millennium history of the city'; while David Chipperfield and Julian Harrap's restoration of Berlin's Neues Museum (1998–2009) consciously evoked the precedent of Döllgast in Munich – as well as the English Ruskinian and Anti-Scrape world-outlook – in its dense web of juxtaposed new interventions and repaired, ruined fragments. Carried to extremes, however, this poetic mixture of preserved fragments and iconic interventions could become incoherent or incomprehensible, as in Enric Miralles's Scottish Parliament project of 1998–2004, or Herzog & de Meuron's Caixa Forum art gallery in Madrid (completed in 2008): a giant brown metal box planted on the shell of an 1899 power station, whose base had been bizarrely chopped away.[46]

Facsimile monuments: the image of heritage

INCREASINGLY, though, a more radical heritage alternative gained in popularity: the building not of contextual new 'interventions' but of facsimiles of vanished individual buildings or even entire streets or districts. Architecturally, this represented a revival of Viollet's philosophy of restoration of ideal forms, as subsequently extended to an area scale in post-1918 Ypres and post-1945 Warsaw. In Western Europe, where facsimile building had been previously stigmatised as 'pastiche', or as 'Disney', the late 20th century had seen the beginnings of a modest revival of popularity, often on the borderline between restoration and recreation. The post-1970s expansion of the Acropolis *anastelosis*

435

Figure 12.12 Strategy for the Acropolis

(a) The Propylaea in 2010, showing reconstruction of portico roof; (b), (c) Dismantling and re-erection of the Temple of Athena Nike, 2000–10 (views of 2008 and 2010); (d) The stone-yard on the south side of the Parthenon, 2010; (e) 2010 view of the new Akropolis Museum (Bernard Tschumi architects, opened 2009) and demolition-threatened neo-classical house, subsequently reprieved

facing page

Figure 12.11 European interventions

(a) Kolumba Museum, Cologne (Peter Zumthor, 2001–7); (b) Neues Museum, Berlin (David Chipperfield and Julian Harrap, 1998–2009); (c) Caixa Forum, Madrid (Herzog & de Meuron, completed 2008); (d) Hotel Fouquet Barrière, Paris (Edouard François, 2003–6: a 'facsimile' of Haussmann boulevard facades in grey concrete, punctured randomly by new windows)

437

programme in Athens, for example, strayed significantly into the area of recreation, and the decade following 2000 saw the Athena Nike temple dismantled and re-erected yet again, the Propylaea partly reroofed and substantial enhancement works on the colonnades and entablature of the Parthenon: the official reconstruction newsletter argued in Viollet-like terms that, in this way, 'some of the structural and formal authenticity of the monuments is regained'.[47]

Elsewhere, things soon went further. In Scotland, for example, the 1990s saw the start of a rolling programme of facsimile reconstruction at Stirling Castle by government agency Historic Scotland, including the reconstruction of the Renaissance Great Hall in 1997–9 and the start of internal reconstruction of the Palace Block in 2008: the project was refused a Europa Nostra Heritage Award in 2001 on grounds that it was a 're-creation' rather than conservation.[48] In Britain and the USA, the 'New Urbanism' architectural movement headed by Andres Duany and Elizabeth Plater-Zyberk proselytised a related philosophy of new architecture in quasi-facsimile styles, in strong opposition to the Ruskin-Charter of Venice concepts of new/old differentiation. Their ally, Prince Charles, resigned as the SPAB's patron in 2009 over this issue, after its secretary, Philip Venning, vetoed the publication of a journal foreword by the prince. The latter had criticised the Morrisian doctrine of 'honesty' in new additions as having been 'used too often to justify unsatisfactory alterations and ugly additions'. And, in 2008, in *The Old and the New*, Hank Dittmar, head of the Prince's Foundation for the Built Environment, condemned outright the Venice Charter demand for 'difference rather than continuity', as being an outmoded relic of the Cold War years.[49]

All these, though, could still be seen as a continuation of the Viollet agenda of recovery of an 'ideal form'. Under globalisation, however, the new movement of city-branding, combined with elements of memory-landscape emotionalism, gave facsimile-building a far more powerful new motive. Within the European heartland of the Conservation Movement, this shift emerged most clearly in the former East, in the radical recycling processes of post-socialist and post-Cold War cities. In Moscow, for example, the modestly postmodern *sredovoi* contextualism of Alexei Gutnov in the late Soviet 1980s, combined with the lack of any legal definition of area-based conservation, provided the seed-bed from which erupted, after 1991, a startling proliferation of projects for recreation of communist-destroyed buildings, celebrating the new capitalist and religious Russia. Many of these projects were conceived by Gutnov's pupil Boris Yeremin, and some were carried through to rapid and spectacular fruition: for example, the rebuilding of copies of the Kazan Cathedral in Red Square and part of the wall around the Kitai-Gorod (both 1993), and of the Cathedral of Christ the Saviour (1994-2000).[50] These all formed part of the construction boom initiated by Moscow's powerful mayor from 1992 to 2010, Yuri Luzhkov – a boom that involved a vigorous but anarchic mixture of extensive demolitions, eclectic new constructions in both classical palace and iconic modern styles and commemorative monuments of a controversially flamboyant character, such as sculptor Zurab Tsereteli's gargantuan, 300-foot-high statue of Peter the Great (1997). A prominent role in this campaign was played by a more aggressive variant of facsimile-building, involving demolition and replacement of genuine old buildings, even as Moscow's annual municipal conservation budget was cut (from £150 million in 1989 to £8 million in 2004). This programme initially focused on relatively modest replacements of 19th-century mansions, but by 2004–6 it had expanded to include a structure as massive as Alexei Shchusev's Moskva Hotel (1932–8), demolished and rebuilt complete with its curious asymmetrical wings. Critics argued that Luzhkov was responsible for 'bulldozing Moscow's architectural heritage and replacing it with mock-palaces', but he boldly riposted that 'in Moscow culture, the idea of a replica sometimes has no lesser meaning than the original had. The meaningful

(a)

Figure 12.13 Stirling: barracks, castle, palace?

(a) The Great Hall (reconstructed 1997–9); (b) the Palace interiors (reconstructed 2009–11); (c) 2011 promotional poster on display at the castle

(c)

(b)

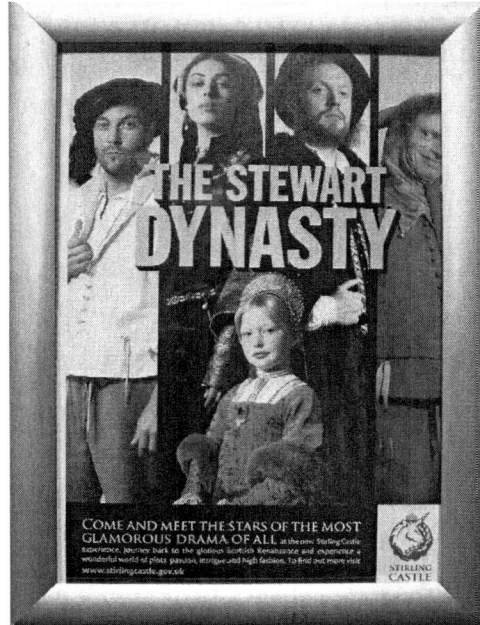

historical and cultural "load" carried by the replica is frequently richer and wider than the original architect's solution.'[51]

In the subordinate countries released from Soviet rule in 1989–91, the picture was fairly similar – for example in Hungary, where the medieval castle of Visegrad (previously restored extensively in 1927–44 by Janos Schulek) was targeted for a Viollet-style fortified embellishment in the 1990s, or in Kiev, where the 11th-/17th-century Cathedral of the Dormition (demolished in 1941) was rebuilt in 1998; or in the Latvian capital, Riga, where the town hall square became ringed in the 1990s and early 2000s by an eclectic array of facsimile buildings. The most spectacular of these, a

439

Figure 12.14 The Cathedral of Christ the Saviour, Moscow, rebuilt in facsimile in 1994–2000

copy of the gabled 14th-century 'Schwarzhäupterhaus' demolished in 1948, was built in 1995–9 by architects V Bratuškis and others and was later joined by other buildings, including the classical Town Hall of 2000–1 (a replica of a 1750–65 original, razed in 1954 and replaced by university laboratories, themselves demolished in 2000). Far more speculative, and rather larger, was the facsimile of the 16th-century Palace of the Grand Dukes of Lithuania in Vilnius, built in 2001–9 at the instigation of prime minister Algirdas Brazauskas: the 'original' had been demolished 200 years earlier, under the Russian empire.[52]

440

(a)

Figure 12.15 Baltic facsimiles

(a) Riga, facsimile of the 14th-century 'Schwarzhäupterhaus' (demolished 1948), built 1995–9 by architects V Bratuškis and others; (b) Vilnius, the rebuilt Palace of the Grand Dukes of Lithuania (2001–9)

(b)

The foremost set piece of the new-old, mixed-up and mixed-together world of European urban conservation was located in central Germany: the vast project for the reconstitution of the historic centre of Dresden, still at reunification in 1989 a disjointed hotchpotch, despite all the efforts of Nadler and others under the GDR. Elsewhere in post-1989 Germany, vigorous recreation initiatives had included the proposals to rebuild large chunks of the Frankfurt *Altstadt*, the 'Dom-Römer-Areal' (where a 2011 competition proposed a combination of facsimiles and traditional-style new designs on 35 historic plots); the citizens' campaign of 2006 onwards to reconstruct the ruined courtyard behind the Nuremberg Pellerhaus; the long-running Berlin controversy about the proposed demolition of the 1970s' Palast der Republik (finally achieved in 2009) and its replacement by a facsimile *Stadtschloss* fronting a shopping complex; and the rebuilding project for Schinkel's nearby Bauakademie, demolished by the GDR in 1962.[53] In Dresden, the focus of reconstruction efforts was the area between the Frauenkirche and the still partly ruined Schloss. Even under the GDR, work had begun in 1986 on the historically styled Hotel Dresdner Hof (by architect Wolfgang Levin of Berlin) but, during the 1990s, the focus of work shifted decisively to the ruin of the Frauenkirche. As early as February 1990, a newly established reconstruction association issued a 'Call from Dresden', and 1993 saw the full-scale start of work: nearly EUR 100 million of the (eventually) EUR 179 million cost of reconstruction was raised by a massive public donation and subscription campaign.[54]

The Frauenkirche project was first and foremost a straightforward restoration of a war-devastated great public building, in the Ypres Cloth Hall tradition. In that sense, it differed little from the Dresden restoration projects of the GDR years. And the surviving remains – the ruined stump and other salvaged masonry blocks – were more substantial than in some cases such as the Warsaw Royal Palace, which had been completely demolished. This allowed the completed exterior, in its mixture of bright new stone and blackened reused fragments, to become itself a graphic memory landscape, whereas the interior seemed more comprehensively 'new', in its garish mixture of white, pink and green colouring and its shiny new woodwork. What differentiated this from the scientific expert restorations of the GDR was the populist political-commercial instrumentalisation that attended the project.

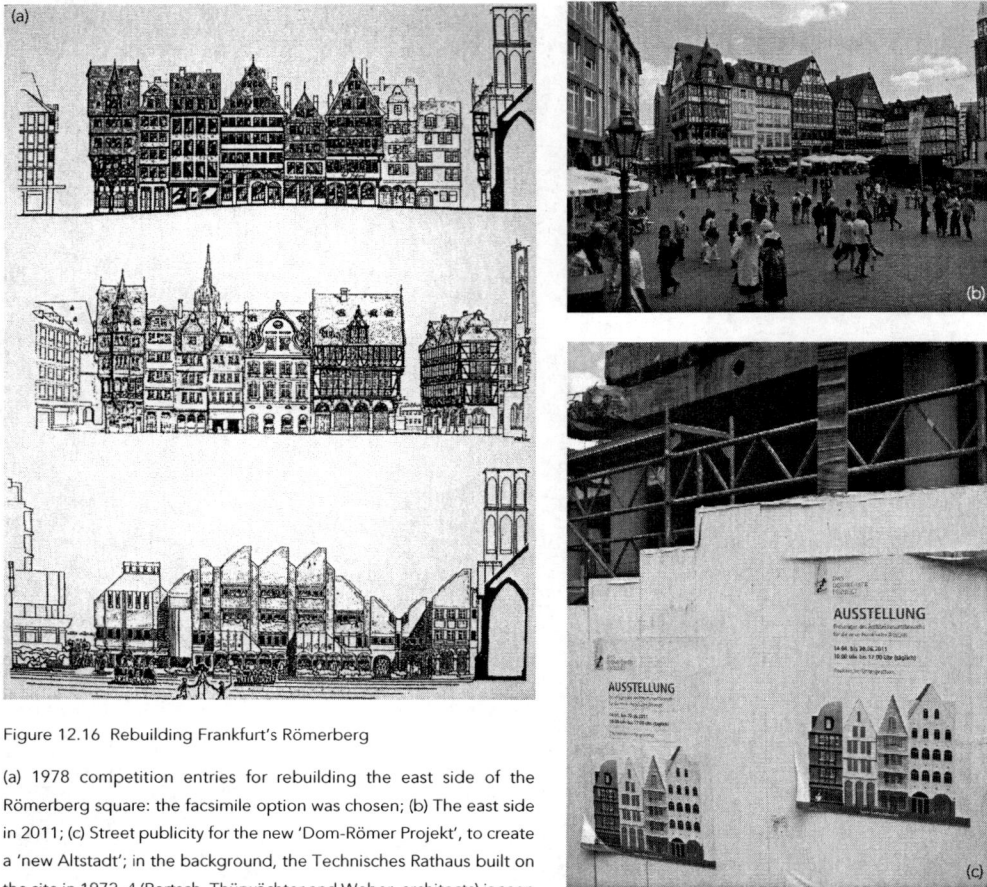

Figure 12.16 Rebuilding Frankfurt's Römerberg

(a) 1978 competition entries for rebuilding the east side of the Römerberg square: the facsimile option was chosen; (b) The east side in 2011; (c) Street publicity for the new 'Dom-Römer Projekt', to create a 'new Altstadt'; in the background, the Technisches Rathaus built on the site in 1972–4 (Bartsch, Thürwächter and Weber, architects) is seen under demolition

The consecration of the 'restored' Frauenkirche in October 2005 saw a vast outpouring of symbolism and emotion, including the carrying of a 'cross of nails' from Coventry Cathedral in the inaugural procession. In a more restrained echo of Warsaw half a century earlier, the project was presented as a unifying symbol of pride in the reunified Germany as a beacon of peace and internationalism. Federal president Horst Köhler proclaimed at the inauguration that the rebuilt Frauenkirche:

> connects together people all over the world. Anyone who wants to explain to school-children what a peaceful Europe means for us, should take a class trip to Dresden and Coventry! What's been achieved here in Dresden should give encouragement to everyone in Germany: the rebuilding of the Frauenkirche is an achievement of the entire nation.[55]

But the Frauenkirche was only the spearhead of a much larger reconstruction project in Dresden, in which the new forces of image-led globalisation were seen more openly at work. This

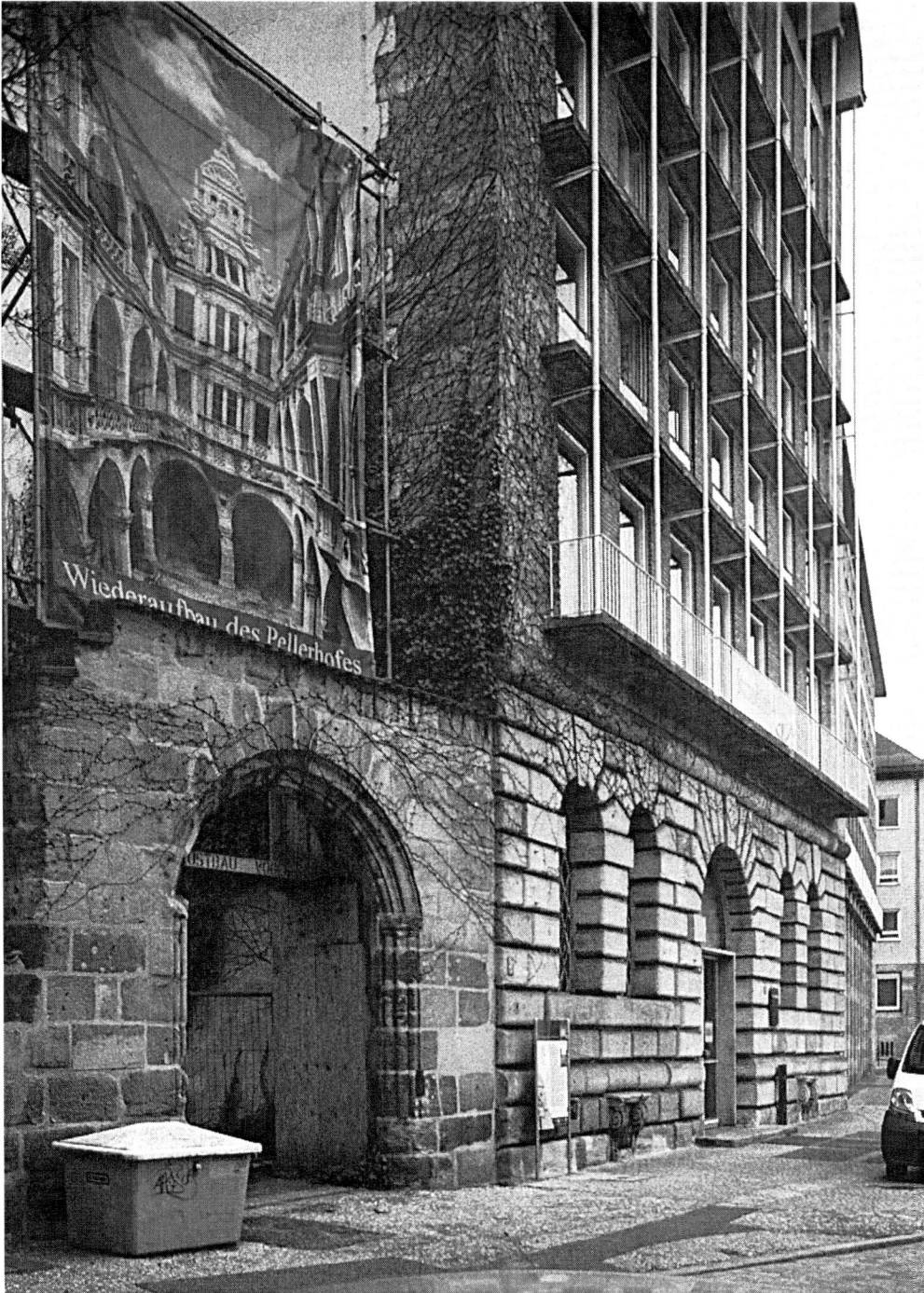

Figure 12.17 The Nuremberg Pellerhaus seen in 2009, adjoined by a giant information board about the proposed courtyard restoration scheme

Figure 12.18 Berlin's Palast der Republik: for and against

(a) 'Protect the Palast!' Leaflet by 'Pro Palast' campaign, 1995; (b) The last remnants of the Palast under demolition in November 2008

project aimed at restoring the historic silhouette of the 'Florence of the Elbe', an ideal image of pre-industrial harmony defined by the 1740s' paintings of Bellotto. This would require the vast spaces between the surviving public monuments, including the demolished Neumarkt area, to be filled with an instant *Altstadt*, here comprising not medieval gabled buildings (as in Frankfurt) but Baroque-style facsimiles of the pre-1945 streets, built quickly with rough brick carcasses and imitation plaster facades. At first glance, all this seemed an uncanny echo of the 1940s/50s' reconstruction of Warsaw. But almost opposite economic forces were at work, forces of city-marketing and capitalist intensification that had identified Dresden as a potentially affluent hot spot in the depressed east. The year 1999 saw the foundation of the Gesellschaft Historischer Neumarkt Dresden (GHND), a private restoration trust that linked up property developers with the campaigns of the new generation of pro-capitalist *Bürgerinitiative*. Unlike, for instance, Gdańsk, the rebuilt *Altstadt* would contain not ordinary homes but smart shops, apartments and hotels. Although some conservationists expressed unease that the 'new old town' would amount essentially to a market-driven 'Disneyland', the new streets of the recreated Neumarkt quarter began sprouting rapidly, aided by slick publicity material and public support.[56] In 2007, there was an unexpected twist, when a competition for a new cultural centre on one of these sites was won by Stuttgart firm Cheret & Bozik with an assertively iconic, neo-modernist design, bristling with jabbing forms in the deconstruction style of Zaha Hadid or Frank Gehry. Immediately, a storm of protest erupted against this Modernist 'desecration' of the 'historic' Neumarkt vision, even although much of this threatened *Altstadt* was still as yet unbuilt. The GHND, naturally, led the opposition and, eventually, in 2010, the menace to Dresden's new *Altstadt* was averted by a city council vote.[57]

The coincidence that the 2005 completion of the Frauenkirche was the centenary of Dehio's Strasbourg speech prompted an increasingly stormy debate in Germany about the ethics of facsimile restoration. Pro-reconstructionists like Winfried Nerdinger and Uta Hassler argued that a building's design was like a music score and could be repeatedly 'performed', while opponents protested that

(a)

(b)

(c)

Figure 12.19
The Dresden
Frauenkirche

(a) Ruins of the
Frauenkirche seen on
'currency reunification
day', 1 July 1990;
(b) The consecration
of the rebuilt
Frauenkirche,
30 October 2005:
procession including
'cross of nails' donated
by Coventry Cathedral;
(c) Commemorative
Frauenkirche edition
of 'Super Illu' popular
magazine, 2005

the undermining of buildings' 'irrevocability' would turn the entire built environment into a glorified Expo park. In 1998, ETH Zürich conservation director Georg Mörsch had already flatly claimed that '*Rekonstrution zerstört*' ('Reconstruction destroys') and in 2011 he and other historians vehemently condemned facsimile-building as an *Attrappenkult* (dummy cult).[58] Although this controversy in some ways echoed the restoration debates of 100–150 years previously, it also underlined the growing uncertainty concerning the traditional values of the Conservation Movement, in its European heartland.

Beyond the heartland: towards new authenticities?

THESE processes of destabilisation of authenticity, especially through facsimile urbanism, became far stronger with the globalisation of architectural heritage. These burst open the European and Western value-system of the Conservation Movement, spotlighting parts of the world which were undergoing development as rampant as any in 19th-century Europe, but which did not necessarily share the Western preoccupation with material permanence.

In Singapore, for example, a formula of ceaseless, state-planned renewal increasingly provided, from the 1980s, for preservation of heritage-islands embedded in modern developments. The redevelopment of early postwar phases of planned development in 'heartland' communities such as Queenstown also prompted the beginnings in Singapore of a distinctive Modernist heritage nostalgia. In the booming cities of Mainland China, the heritage-island approach was carried to far greater extremes. Prompted by the competing but separate agendas of the Ministry of Construction (for output) and the Ministry of Construction (for preservation), the vast steamroller of state-sponsored redevelopment and urban modernisation increasingly made provision for numerous pockets of carefully restored monuments – an approach reminiscent of the 19th-/early-20th-century narratives of Haussmann or even Mussolini, but on a far vaster scale. In Beijing, for example, following 40 years of socialist neglect or destruction of old areas and a decade of rampant capitalist development and clearance in the 1990s, 2002 saw the designation of 25 historical areas in the inner city. But the conservation work here was sometimes of a radical, invasive kind – a 21st-century equivalent to Wyatt's improvements and motivated by city-branding agendas and events such as the 2008 Olympic Games. For example, an improvement scheme of that year for the affluent Qianmen historic shopping zone south of Tiananmen Square, by a design team from Tsinghua University (under Prof. Wu Liangyong), saw the regularisation, pedestrianisation and heritage branding of the half-mile-long main thoroughfare (complete with 'period' tramway) and the clearance of dilapidated original shops and their replacement by 1920s-style replicas. The 16th-century Yongding Gate, destroyed in 1957 under Mao during the redevelopment of the city walls with ring roads, was rebuilt in facsimile in 2004.[59]

Elsewhere, more radical challenges to Western-style authenticity gained ground. The leading Indian conservation architect, A G Krishna Menon (co-drafter of the 2004 INTACH Charter for Conservation of Unprotected Architectural Heritage and Sites in India), argued that,

> in the West, it [authenticity] is determined by the awareness of time's irreversibility which emphasizes the temporal qualities of objects and events – 'the golden stain of time' – but in India, the cyclical perception of time places no critical temporal value on man-made objects but transfers the quality of authenticity to the site in which the object exists.

Figure 12.20 Singapore's heritage districts

(a), (b) Two views of Singapore's Chinatown heritage district, showing the juxtaposition with modern public housing and with the Chinatown metro station in Pagoda Street (opened 2003); (c) Heritage trail display (celebrating the modernist legacy of the 1970s–80s) in Ang Mo Kio New Town, Singapore

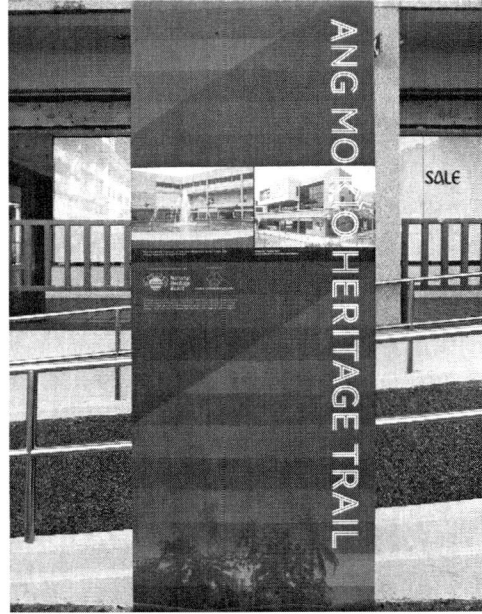

(a)

(b)

(c)

The new emphasis on pluralism within international conservation ensured there would be no doctrinaire obstruction from organisations such as UNESCO. For example, in Bhaktapur, Nepal – a long-standing World Heritage Site, designated in 1979 – the god-house of the goddess Tripurasundari (a structure largely built in 1934 and incorporating 17th-century fragments in *spolia* style) was replaced in 2006–8 by a semi-facsimile constructed by the local municipal heritage office. The new building, four feet higher and more elaborate than its predecessor, was financed by tourist entrance fees and justified by reference to the authenticity criteria of 'spirit and feeling' in the 2005 World Heritage Operational Guidelines.[60]

Figure 12.21

(a), (b) The 1995–2009 reconstruction of Gyeongbokgung Palace in Seoul, a late-14th-century courtyard complex twice largely destroyed by Japanese invasion and occupation (in 1592–8 and from 1911). The project involved demolition of the domed, classical Japanese Government-General Building, erected in the middle of the site in 1916–26 and shown in image (a). In its place, facsimiles of predecessor buildings were constructed, including the Gwanghwa and Heungye Gates – visible in the foreground and background of image (b), taken in 2012

Some of the most radical recreation projects in Asia, vigorously intermingling facsimile-building, globalisation, city-marketing and correction of hurtful memory, have sprung up in South Korea, where the demands of tourist marketing have profitably interacted with nationalist political reactions against previous centuries of external oppression. The 1995–2009 project to reconstruct the Gyeongbokgung Palace in Seoul, a late-14th-century courtyard complex twice largely destroyed by Japanese invasion and occupation (in 1592–8 and 1911), required demolition of a massive, classical Japanese government building that had occupied the site since the 1920s (latterly used, ironically, as the National Museum of Korea). Despite vigorous debate as to whether this should be preserved, either for its architectural value or as a testimony of hurtful memory, the building was finally demolished in 1995 and a recreation of the palace courtyards took its place, under the direction of craftsman Shin Eung-Su, himself officially registered as an 'important intangible cultural property'. More extreme still was the protracted (1996–2010) and costly (US $600 million) effort by South Chungcheong Province to build a 'Baekje Historical Village' that would recreate a palace destroyed by invaders virtually without trace in AD 660, as part of a major tourist-promotion effort by a backwater region. The reconstructed palace, complete with a five-storey pagoda built by master carpenter Choe Gi-Yeong, formed part of a Lotte World resort, including an 18-hole golf course, shopping mall, hotels and cultural institutes. Following its completion, moves began to have the complex inscribed on the World Heritage List in 2015.[61]

In such places as this, we see history still very much on the march, but it is unclear what role – if any – can be played in that process by the traditional Conservation Movement, with its baggage of Western ideologies and dogmas. In the Epilogue, we end our account by briefly considering that very question.

– Epilogue –

Once history reaches a certain level of excess, life itself starts to crumble and degenerate –
and this degeneration, in turn, eventually corrupts history itself.

F Nietzsche, 1874[1]

OVER a century ago, Georg Dehio asserted confidently before Emperor Wilhelm II that 'to protect monuments is not a matter of enjoyment but of piety. Aesthetic and even art-historical judgments fluctuate; here, we have found an unchanging mark of worth.' The roots of the modern Conservation Movement may partly have sprung from the chaos of the French Revolution and the Industrial Revolution, but now, Dehio believed, all was firmly and authoritatively anchored in the strong nation-state.[2] A decade later, in the vitriolic exchanges over Reims, his claims were already looking threadbare, and a further 30 years on, the entire Heritage Militant lay literally and metaphorically in ruins. Yet, in its new, more peaceful and internationalist guise, the Conservation Movement only seemed to go from strength to strength.

What, then, of the position today, over half a century on again? Has the Conservation Movement been able to successfully reinvent and reinvigorate itself again in the new age of globalisation and postmodern relativism? If it has always (despite its own frequently 'anti-modern' propaganda) been an offshoot of Progress and Western modernity, a highly defined, assertive movement dedicated to using the past for useful modern purposes, then where does it stand in today's age of postmodern relativism and 'deconstruction' of all grand narratives and norms? One considered response came from Jokilehto in 1999:

> Against this new background, one can well ask if the conservation movement, as it evolved from the eighteenth century, cannot be considered as concluded, and whether modern conservation should not be redefined in reference to the environmental sustainability of social and economic development within the overall cultural and ecological situation on earth.[3]

In other words: conservation might only be able to advance effectively into the new age by surrendering its core identity as a specific, unique phenomenon and its primary concern with the historic built environment, and by letting itself be swallowed up into nebulous overarching causes. And indeed, as we saw above, the Conservation Movement over the past two decades has begun to display signs of a movement in a state of disorientation, both in its ideas and in its built forms – as in the bewildered debates stirred up by the boom in facsimile building.

In 1903, Alois Riegl argued that old buildings were

> a mere perceptible substratum whose task is to stimulate that feeling which is evoked
> in modern people by the inexorable ritual of being and passing away, of the arising of
> the particular out of the general and its inevitable reabsorption back in the general.

Could what Riegl said of *Alterswert* now be applicable equally to the entire Conservation Movement? Could the latter's highly defined, assertive character be beginning to blur and fade back into the 'general'? And might that dissolution be an echo, or a reflection, of a wider 'fall of the West', in which innovations originally devised in Europe and America are increasingly now being appropriated and exploited, or discarded, by China, India and others? To be sure, in some non-Western cultures, the sense of irreversible, driving Time and History, that powered the story of conservation, is lacking, and more fluid, cyclical concepts of time are prevalent. The outcome in conservation is a value-system in which the strong distinction between original and copy become almost meaningless – with results that we touched on in Chapter 12.[4] To engage with grandiose geopolitical speculations and theories, however, is emphatically outside the scope of this book, not least because the likely future pattern and character of the spread of conservation theory and practice in the developing world is so uncertain. Nor can one hope at this stage to credibly predict the future historical trajectory of conservation. We cannot know yet whether the story of conservation will turn out to be one of rise, ascendancy and fall: whether what we have traced in this book has been the narrative of a historical movement driven inexorably forward by circumstances to flourish, expand and then begin its decay and dissolution.

Will the Conservation Movement prove to have been just a fleeting moment in the broad history of the built environment, or do its ideas, achievements and progress over the past two centuries represent an irreversible shift in values? Perhaps the present period of apparent disorientation is only a temporary interlude, until a new phase of modernity and historical crisis, centred outside conservation's traditional European-American heartland, can provide a sufficient external threat to reinvigorate conservation as a real, urgent movement and provide it with a renewed grand narrative. In longer-term perspective, are we, today, confronting a fundamental shift in heritage values, or only a temporary blip in a long and ongoing story?

In the end, we can only conclude that to all these questions there are no clear or simple answers. Certainly, this book offers none. Readers will have to weigh up the accumulated evidence and make up their own minds. But even if conservation eventually ceases to exist and to thrive in its present-day form, as a vast global movement, one thing is certain: the aura and fascination of 'the monument' will endure in some way or another, whether in tenacious survival or in gradual but inexorable dissolution.

> We are not impotent – we pallid stones.
> Not all our power is gone – not all our fame –
> Not all the magic of our high renown –
> Not all the wonder that encircles us –
> Not all the mysteries that in us lie –
> Not all the memories that hang upon
> And cling around about us as a garment,
> Clothing us in a robe of more than glory.
> Edgar Allan Poe, 1833[5]

– Notes –

Introduction

1 Anthony Trollope, *Barchester Towers*, 1857 (Penguin edition, 2003, 183).
2 Thucydides, *History*, 1.1.22.
3 J Jokilehto, *A History of Architectural Conservation*, Oxford, 1999/2004.

Chapter 1

1 Luke 21, 5–7.
2 D Arnold, *The Monuments of Egypt*, London, 2009; *Conservation*, Getty Conservation Institute Newsletter, vol 23, 2, 2008; M McArthur, *Confucius*, London, 2010.
3 P Siewert, 'The Ephebic Oath in Fifth-century Athens', *Journal of Hellenic Studies*, vol 97, 1977, 102–11; information from Wolfgang Sonne.
4 Jokilehto, *History*, 1999 edition, 4.
5 Lycurgus, *Against Leocrates*, 81, quoted in C Cunningham, *The Akropolis*, Milton Keynes, 1979, 8; L V Barrelli, in *Geschichte der Restaurierung in Europa*, vol 1, Worms, 1991, 122; T Spawforth, *The Complete Greek Temples*, London, 2006.
6 Olympia: W Nerdinger (ed), *Geschichte der Rekonstruktion: Konstruktion der Geschichte* (hereafter *GRKG*), Munich, 2010; Spawforth, *Greek Temples*, 152–4. Theseus: C Erder, *Our Architectural Heritage*, UNESCO, Paris, 1986, 28; J Rykwert, *On Adam's House in Paradise*, New York, 1972, 106.
7 Erder, *Heritage*, 31.
8 N Wibiral, 'Ausgewählte Beispiele', *Österreichische Zeitschrift für Kunst und Denkmalpflege* (hereafter *ÖZKD*), 36, 1982, 93–8; E Thomas, *Monumentality and the Roman Empire: Architecture in the Antonine Age*, Oxford, 2007.
9 Exploitation of *pietas*, *virtus* and visual imagery in the construction of Augustan power: see P Zanker, *The Power of Images in the Age of Augustus*, Ann Arbor, MI, 1990, 120–276; Virgil, *Aeneid*, vi 759, viii 715–31.
10 Plutarch, *Moralia* 559a; E Povoledo, *New York Times*, 13 November 2008. Rykwert, *On Adam's House*, 175; Favro, 'The Iconic City of Ancient Rome', *Urban History*, 33, 1, 2006, 20–38; B Ward-Perkins, *From Classical Antiquity to the Middle Ages*, Oxford, 1984, 38–48; F Castignoli *et al*, *Topografia e urbanistica di Roma*, Bologna, 1958.
11 Rykwert, *On Adam's House*, 175; Erder, *Heritage*, 45.
12 Rykwert, *On Adam's House*, 174–5.
13 S Sorek, *The Emperors' Needles*, Exeter, 2010.
14 *GRKG*, 53.
15 Ward-Perkins, *From Classical Antiquity*, 12.
16 J H Newman, *An Essay in the Development of Christian Doctrine*, 1845 (revised 1878).
17 R Krautheimer, *Rome: Profile of a City*, Princeton, NJ, 1980; R Lanciani, *The Destruction of Ancient Rome*, London, 1899/1906.
18 1 Moses 11.4; Revelations 21.2.
19 Information from Ian Campbell.
20 Ward-Perkins, *From Classical Antiquity*, 89; F Choay, *The Invention of the Historic Monument*, Cambridge, 2001 (first edition, 1992), 20–7.
21 M Greenhalgh, *The Survival of Roman Antiquities in the Middle Ages*, London, 1989; C Pharr, *The Theodosian Code* (hereafter *TC*), Princeton, NJ, 1952, 424–5 and 554, courtesy of Ian Campbell.

22 Ward-Perkins, *From Classical Antiquity*, 34–7; Krautheimer, *Rome*; Lanciani, *Destruction*; M Campbell, *Ecce Roma*, Edinburgh, 1992.

23 *TC*, 553–4.

24 Erder, *Heritage*, 57.

25 Cassiodorus, *Variae*, Book 7, 13, 'Formula of the Count of Rome' (see www.gutenberg.org/ebooks/18590); see also Cassiodorus *Variae*, in T Mommsen (ed), *Germaniae Historica, Auctores Antiquissimi*, xii, 1894; C J Goodson, 'Roman Archaeology in Medieval Rome', in D Caldwell and L Caldwell (eds), *Rome: Continuing Encounters between Past and Present*, Farnham, 2011, 17–34; Ward-Perkins, *From Classical Antiquity*, 208 and 212; Erder, *Heritage*, 55.

26 E O'Carragain and C N de Vegvar (eds), *Roma Felix*, Farnham, 2008; Ward-Perkins, *From Classical Antiquity*, 199 and 204; Goodson, 'Roman Archaeology', 23–7.

27 Ward-Perkins, *From Classical Antiquity*, 203, 224, 211; Sopritendenza Archaeologica di Roma, *The Roman Forum*, nd, 7–8; Goodson, 'Roman Archaeology', 27–30.

28 A Hubel, *Denkmalpflege*, Stuttgart, 2006, 18; Ward-Perkins, *From Classical Antiquity*, 215–17; Erder, *Heritage*, 70; J H Stubbs and E G Makaš (eds), *Architectural Conservation in Europe and the Americas*, Hoboken, NJ, 2011, 336.

29 Ward-Perkins, *From Classical Antiquity*, 181–5, 218–19. Syracuse: *GRKG*, 27; Spawforth, *Greek Temples*, 122–3. Arles: C Woodward, *In Ruins*, London, 2001, 33.

30 A Kaldellis, *The Christian Parthenon*, Cambridge, 2009, 164.

31 Hubel, *Denkmalpflege*, 18.

32 Erder, *Heritage*, 71.

33 D Baker, *Living with the Past*, Bradford, 1983, 31.

34 M F Hansen, *The Eloquence of Appropriation*, Rome, 2003; J Elsner, *Papers of the British School at Rome*, 68, 2000, 149–64; Barrelli, *Geschichte der Restaurierung*, 174–5.

35 Castignoli *et al*, *Topografia e urbanistica di Roma*, 389–90; Prince Charles, *A Vision of Britain: A Personal Vision of Architecture*, London, 1989, 147–53; D M Lasansky, *The Renaissance Perfected*, University Park, PA, 2004, 326, 340.

36 Rykwert, *On Adam's House*; Greenhalgh, *The Survival of Roman Antiquities*, London, 1989. Chapel Royal of 1594: see A MacKechnie, 'Sir James Murray of Kilbaberton: King's Master of Works 1607–1634', paper at 22 May 2010 conference, Paul Mellon centre, London.

37 M Trachtenberg, *Building in Time from Giotto to Alberti and Modern Oblivion*, New Haven, CT and London, 2010 ; *Journal of the Society of Architectural Historians* (hereafter *JSAH*), March 2005, 74–99; Choay, *Invention*, 181; M S Briggs, *Goths and Vandals*, London, 1952, 11.

38 N Boulting, 'The Law's Delays', in J Fawcett (ed), *The Future of the Past*, London, 1976, 10.

39 Boulting, 'The Law's Delays', 10.

40 Hubel, *Denkmalpflege*, 18–19.

41 *GRKG*, 85. P Kurmann, 'Restaurierung', in *Geschichte der Restaurierung in Europa*, vol 1, Worms, 1991, 15.

42 P N Balchin, *Urban Development in Renaissance Italy*, Hoboken, NJ, 2008; H Hyde Minor, *The Culture of Architecture in Enlightenment Rome*, University Park, PA, 2010.

43 Choay, *Invention*, 15–17; D Lowenthal, *The Past is a Foreign Country*, Cambridge, 1985, 1, 4.

44 Jokilehto, *History*, 15; M Pollak, *Vom Erinnerungsort zur Denkmalpflege*, 2009, Bohlau Verlag, Cologne; Erder, *Heritage*, 74; I Campbell, 'Rescue Archaeology in the Renaissance', in I Bignamini, *Archives and Excavations*, British School at Rome, 2004, 13.

45 Jokilehto, *History*, 16; see also R E Spear and P Sohm (eds), *Painting for Profit, The Economic Lives of 17th Century Italian Painters*, New Haven, CT and London, 2010; Choay, *Invention*, 27, 33, 38, 181–4; G Clarke, *Roman House, Renaissance Palaces: Inventing Antiquity in 15th Century Italy*, Cambridge, 2003.

46 Jokilehto, *History*, 23; M Campbell, *Ecco Roma: European Artists in the Eternal City*, Edinburgh, 1992; A Day and V Hart, 'The Architectural Guidebook', *ARQ*, vol 11, 2, 2007, 151; V Hart and P Hicks (eds), *Palladio's Rome*, New Haven, CT and London, 2009; J Maier, 'Roma Renascens: Sixteenth-century Maps of the Eternal City', in D Caldwell and L Caldwell (eds), *Rome: Continuing Encounters between Past and Present*, Farnham, 2011, 35–55; Campbell, 'Rescue Archaeology', 14 and 21.

47 Barrelli, *Geschichte der Restaurierung*, 126–9.

48 Jokilehto, *History*, 26–8; Society of Architectural Historians, *Newsletter*, April 2008, 15.

49 K W Christian, *Empire Without End, Antiquities Collections in Renaissance Rome, c.1350–1527*, New Haven, CT and London, 2010; Jokilehto, *History*, 29; Campbell, 'Rescue Archaeology', 14. Information throughout this section more generally from Ian Campbell.

50 Erder, *Heritage*, 74; Jokilehto, *History*, 30–2; P Quennell, *The Colosseum*, New York, 1971; D Karman, *Future Anterior*, Winter 2005, 1–7; Choay, *Invention*, 36–9.

51 Erder, *Heritage*, 76; Karman, *Future Anterior*, Winter 2005, 6; Ward Perkins, *From Classical Antiquity*, 38–40.

52 Jokilehto, *History*, 33; Maier, 'Roma Renascens', 35–55; Campbell, 'Rescue Archaeology', 15–16.

53 Erder, *Heritage*, 82.

54 Campbell, 'Rescue Archaeology', 15.

55 Comune di Roma, *The Capitoline Museums Guide*, Rome, 2006, 4–10.

56 Erder, *Heritage*, 80–1; for 1550–68 excavations at Hadrian's Villa, Tivoli, see Campbell, 'Rescue Archaeology', 18.

57 Jokilehto, *History*, 34, 39.

Chapter 2

1 Erder, *Heritage*, 85.

2 Germanisches Nationalmuseum, *Mythos Burg*, Nuremberg 2006, 294–5.

3 Erder, *Heritage*, 114; S Baldrick and R Clay, *Iconoclasm*, Farnham, 2007, 83.

4 Kurmann, 'Restaurierung', in *Geschichte der Restaurierung*, vol 1, 19–20; *GRKG*, 8, 227–8; J Nivet (Société Archaeologique et Historique de l'Orleanais), *Sainte Croix d'Orléans*, Orleans, 1984, 93; *GRKG*, 224–6.

5 Nivet, *Sainte Croix*, 93.

6 Nivet, *Sainte Croix*, 1.

7 Erder, *Heritage*, 120.

8 Hubel, *Denkmalpflege*, 28; Kurmann, 'Restaurierung', 18, 26–9; W Haas, *Der Dom zu Speyer*, Königstein, 1988, 15–21.

9 A Smolar-Meynart (ed), *Autour du bombardement de Bruxelles de 1695*, Brussels, 1995; C Damas, *Le bombardement de Bruxelles à 1695*, Bruxelles 1951, 51–5; M Culot *et al*, *Le bombardement de Bruxelles par Louis XIV*, Brussels, 1992.

10 Wibiral, 'Ausgewählte Beispiele'.

11 Choay, *Invention*, 184–5; Jokilehto, *History*, 44.

12 D Cressy, 'National Memory in Early Modern England', in J R Gillis (ed), *Commemorations*, Princeton, NJ, 1994; W H G Armytage, *Heavens Below*, London, 1967, 5.

13 *Cornerstone*, vol 28, 4, 2007, 41–3; Hugh Pearce, in D Lowenthal and M Binney, *Our Past Before Us*, Oxford, 1981, 36.

14 Boulting, 'The Law's Delays', 11; M S Briggs, *Goths and Vandals*, 1952, 16.

15 T Cocke, 'Thou Who Changest Not', in *Geschichte der Restaurierung*, vol 1, 34–7; Jokilehto, *History*, 42; Boulting, 'The Law's Delays', 12; Briggs, *Goths and Vandals*, 51–2, 82–8.

16 Fawcett, *Future of the Past*, 13, 43; Baldrick and Clay (eds), *Iconoclasm*, 5; M S Briggs, *Goths and Vandals*, 1952; Cocke, 'Thou Who Changest Not', 34–9.

17 R Moss, 'Appropriating the Past', *Architectural History*, 51, 2008, 63–86 (for example, at the Priests' Church, Glendalough, Co. Wicklow, reconstituted in the late 17th or early 18th century to incorporate an incomplete Romanesque arch).

18 E Impey, G Parnell, *The Tower of London*, London, 2000.

19 Choay, *Invention*, 184–5; Buchanan, *Iconoclasm*, 139.

20 Boulting, 'The Law's Delays', 12; Choay, *Invention*, 187.

21 Choay, *Invention*, 44–5; D Baker, *Living with the Past*; Armytage, *Heavens Below*; O H Turner, 'The Windows of this Church are of Several Fashions', *Architectural History*, 54, 2011, 171–93.

22 *Mythos Burg*, 322. F Russo, 'The Printed Illustration of Medieval Architecture in Pre-Enlightenment Europe', *Architectural History*, 54, 2011, 119–70. For early antiquarianism in Spain, see M M Turina, *La memoria de los piedras. Anticuarios, arqueologos y coleccionistas de antiguedades en la España de los Austrias*, Madrid, 2010.

23 Boulting, 'The Law's Delays', 13–14; R Hill, *God's Architect*, London, 2007, 123; 'Making History: Antiquaries in Britain', 2 February–27 May 2012 (travelling exhibition, Yale Center for British Art).

24 H Prince, in Lowenthal and Binney, *Our Past Before Us*, 36.

25 Haas, *Dom zu Speyer*, 21; *GRKG*, 228–32; Hubel, *Denkmalpflege*, 28; F I Neumann also rebuilt Mainz Cathedral in 1771–7 following a 1769 fire, including a new, Gothic-cum-Baroque pyramidal tower: *GRKG*, 228; R Bernheimer, 'Gothic Survival and Revival in Bologna', *Art Bulletin*, December 1954, 263–84.

26 Cocke, 'Thou Who Changest Not', 34–7.

27 Cocke, 'Thou Who Changest Not', 34–7.

28 Jokilehto, *History*, 42–3; Briggs, *Goths and Vandals*, 99–108.

29 Erder, *Heritage*, 85; Lowenthal, *Past*, 1–4; Rykwert, *On Adam's House*, 58 and 80.

30 Choay, *Invention*, 48; Jokilehto, *History*, 48; Bernheimer, 'Gothic Survival', 272.

31 Erder, *Heritage*, 85.

32 H Cain, *Gipsabgüsse*, Nuremberg, 1995, 200ff; Campbell, *Ecce Roma*; Museum für Abgüsse Klassischer Bildwerke, *Wozu Gipsabgüsse?*, Munich, nd; Jokilehto, *History*, 48–50; Erder, *Heritage*, 89; I Bignamini and C Hornsby, *Digging and Dealing in 18th Century Rome*, New Haven, CT and London, 2010.

33 J M Kelly, *The Society of Dilettanti*, New Haven, CT and London, 2010; Jokilehto, *History*, 48–50; J Ingamells, *A Dictionary of British and Irish Travellers in Italy*, New Haven, CT, 1997; Erder, *Heritage*, 85.

34 Rykwert, *On Adam's House*, 80.

35 M Giambruno (ed), *Per una storia del restauro urbano*, Novara, 2007, xxxix–xxx.

36 P M Harman, *The Culture of Nature in Britain 1680–1860*, New Haven, CT and London, 2009; Diderot, *L'Encyclopédie*, 32 vols, Paris, 1751–77; S Muthesius, *Art History*, June 1988, 235.

37 Abbé Laugier, *Essai sur l'architecture*, 1753; Jokilehto, *History*, 112; Goethe: P Leisching, *ÖZKD*, 1971, 129; Hubel, *Denkmalpflege*, 24–5; A Gebessler (ed), *Schutz und Pflege von Baudenkmälern in der BRD*, Cologne, 1980.

38 E Burke, *A Philosophical Enquiry into the Origin of our Ideas of the Sublime and the Beautiful*, 1757; Choay, *Invention*, 56, 59; Jokilehto, *History*, 50–2; S Muthesius, *Poland*, Königstein, 1994, 53. For example, in late 18th-century Poland, the Dresden-born architect, Szymon Bogomil Zug, designed numerous picturesque English-style gardens, including Arkadia, near Nieborów (1777–98), and contributed an article to Hirschfeld's *Theorie der Gartenkunst* (1785), a key book on late-18th-century gardens in Europe.

39 Ruins: Choay, *Invention*, 88–9; Jokilehto, *History*, 112. Goethe: N Huse, *Denkmalpflege: deutsche Texte aus drei Jahrhunderten*, Munich, 1984, 17. See also Wu Hung, *Ruins in Chinese Art and Visual Culture*, London, 2011.

40 G Coppick, *Fountains Abbey*, Stroud, 2009; M Mauchline and L Greeves, *Fountains Abbey and Studley Royal*, Swindon, 1988, 42–8; Woodward, *In Ruins*, 124–5.

41 Erder, *Heritage*, 109; C C Parslow, *Rediscovering Antiquity*, Cambridge, 1995; Borrelli, 'Restauro e conservazione', 128; N Shrady, *The Last Day*, Viking, New York, 2003; *Economist*, 5 April 2008, 98; Jokilehto, *History*, 56–9.

42 R Cabrera Castro, *Teotihuacan*, INAH, Mexico City, 2003; L Boturini, *Historia de la America septentrional*, 1746.

43 A Potts, *Art History*, December 1982, 377ff; Jokilehto, *History*, 59–65; Wibiral, *Beispiele*, 98; Borrelli, 'Restauro e conservazione', 128.

44 J J Winckelmann, *Geschichte der Kunst des Alterthums*, Dresden, 1764; P Tournikiotis (ed), *The Parthenon and its Impact in Modern Times*, Athens, 1994, especially 26–53; J Stuart and N Revett, *The Antiquities of Athens*, published 1762–87.

45 Potts, *Art History*, December 1982.

46 T Lemain, *Forgery, The History of a Modern Obsession*, London, 2011; C Woodward, *In Ruins*, London, 2001, 152–6 and 162; J de Cayeux, *Hubert Robert*, Paris, 1989; M Bevilacqua, H Hyde Minor and F Barry (eds), *The Serpent and the Stylus*, Ann Arbor, MI, 2006; S E Lawrence (ed), *Piranesi as Designer*, New York, 2007; W B Lundberg and J A Pinto (eds), *Steps off the Beaten Path*, Milan, 2007.

47 J Musson, 'Vanbrugh', *Cornerstone*, vol 29, 4, 2008, 34–7; G Beard, *The Work of John Vanbrugh*, London, 1986, 37–50; G Woodward, 'A Pre-history of Conservation', in *Transactions of the Society for Studies in the*

Conservation of Historic Buildings, 1995; G Webb, *The Complete Works of Sir John Vanbrugh*, vol 4, London, 1928, 29.

48 N Pevsner and E Hubbard, *Cheshire*, 1971/2003, *passim*.

49 M Snodin, *Horace Walpole's Strawberry Hill*, New Haven, CT and London, 2009; M Harney, 'Strawberry Hill', *Architectural Review* (hereafter *AR*), November 2008, 72–7; M Charlesworth (ed), *The Gothic Revival 1720–1870, Literary Sources and Documents*, (3 vols), Hastings, 2002; Briggs, *Goths and Vandals*, 122–6.

50 T Cocke, 'James Essex, Cathedral Restorer', *Architectural History*, vol 18, 1975, 12–22. Jokilehto, *History*, 102–4; Cocke, 'Thou Who Changest Not', 40; A Rowan, *Architectural History*, 1972, 23–33.

51 Choay, *Invention*, 61–2.

52 A Buchanan, 'Wyatt the Destroyer?', in Baldrick and Clay (eds), *Iconoclasm*, 123–143; Cocke, 'Thou Who Changest Not', 34–48; J Frew, 'James Wyatt's Choir Screen at Salisbury Cathedral Reconsidered', *Architectural History*, vol 27, 1984, 481–7; Jokilehto, *History*, 104–6; P Aspin. 'Our Ancient Architecture', *Architectural History*, 54, 2011, 213–32.

53 Fawcett, *Future of the Past*, 6; Buchanan, 'Wyatt'.

54 J M Frew, 'Richard Gough', *JSAH*, 1979, 366–74; Choay, *Invention*, 195–6.

55 Frew, 'Richard Gough', 372; Jokilehto, *History*, 106–8.

56 Buchanan, 'Wyatt'; Jokilehto, *History*, 106–7; Frew, 'Richard Gough', 373–4

57 Hubel, *Denkmalpflege*, 31.

58 I Gow, *Country Life*, 23 November 1995; M Glendinning, R MacInnes and A MacKechnie, *A History of Scottish Architecture*, Edinburgh, 1996, ch 3; D Lockhart, *Scottish Local History*, Autumn, 1997, 7–9; C McWilliam, *Scottish Townscape*; M Glendinning and S Wade-Martins, *Buildings of the Land*, Edinburgh, 2009.

59 Glendinning *et al*, *Scottish Architecture*, ch 3; D Walker, in R Clow (ed), *Restoring Scotland's Castles*, Glasgow, 2001; J Gifford, 'From Blair Castle to Atholl House to Blair Castle', in A Dakin, M Glendinning and A MacKechnie (eds), *Scotland's Castle Culture*, Edinburgh, 2011, 240–54.

60 Glendinning *et al*, *Scottish Architecture*, 130, 96 (and chs 3 and 4); A MacKechnie, 'Caisteal Inbhir-Aora/Inveraray Castle', in Dakin, Glendinning and MacKechnie, *Scotland's Castle Culture*, 213–27; *Mythos Burg*, 337–9; I Lindsay and M Cosh, *Inveraray and the Dukes of Argyll*, Edinburgh,1973.

61 R G Cant, in St Andrews Preservation Trust, *Three Decades of Historic Notes*, St Andrews, 1991, 167; I MacIvor and R Fawcett, 'One Hundred Years On', *Popular Archaeology*, November 1982.

62 A McKechnie, 'Scottish Historical Landscapes', in *Studies in the History of Gardens*, July–September 2002, 214–39; M Stewart, 'The Metaphysics of Place in the Scottish Historical Landscape: Patriotic and Virgilian Themes, *c*.1700 to the Early Nineteenth Century', *Studies in the History of Gardens and Designed Landscapes*, vol 22 (3), 2002. Glendinning *et al*, *Scottish Architecture*, 94–6; 1782 saw reconstruction of the landscaped Hermitage at Dunkeld with a new 'Ossian's Hall'.

63 I B Whyte, 'Schinkel in Scotland', lecture to Architectural Heritage Society of Scotland, 13 February 1992; A S Bell (ed), *The Scottish Antiquarian Tradition*, Edinburgh, 1981; A Ross, *Innes Review*, vol 15, 1964, 122–39.

Chapter 3

1 K F Schinkel, *Concerning the Preservation of the Monuments and Antiquities of Our Country*, 1815 (translated by/courtesy of Jonathan Blower, 2012); Karl Marx, *The 18ᵗʰ Brumaire of Louis Napoleon*, 1852.

2 C Dellheim, *The Face of the Past*, Cambridge, 2004, 45; G Iggers and J M Powell (eds), *Leopold von Ranke and the Shaping of the Historical Discipline*, Syracuse (NY), 1990; L Krieger, *Ranke: the Meaning of History*, Chicago, 1973.

3 Jokilehto, *History*, 281–2.

4 F Tönnies, *Gemeinschaft und Gesellschaft*, Leipzig, 1887 (translation: *Community and Civil Society*, Cambridge, 2001).

5 Marx, *18ᵗʰ Brumaire*.

6 Choay, *Invention*, 63ff.

7 Choay, *Invention*, 66; Jokilehto, *History*, 69–71.

8 F Berce, *Des monuments historiques au patrimoine du dix-huitieme siècle*, Paris, 2005; Choay, *Invention*, 66–73.

9   Choay, *Invention*, 72; G Dehio, *Denkmalschutz und Denkmalpflege in Neunzehnten Jahrhundert*, Strasbourg, 1905 (by courtesy of Jonathan Blower).

10 Choay, *Invention*, 73.

11 A Sutcliffe, *The Autumn of Central Paris*, London, 1970.

12 Choay, *Invention*, 76.

13 Jokilehto, *History*, 72; P Plagnieux, *The Basilica Cathedral of St Denis*, Paris, 2008, 32.

14 L M O'Connell, 'Afterlives of the Tour St Jacques', *JSAH*, vol 60, 4, December 2001, 454–8.

15 Choay, *Invention*, 201; Erder, *Heritage*, 114; Jokilehto, *History*, 73–4; Quatremère de Quincy, *Lettres au Général Miranda*, 1796 (Letter 3).

16 Erder, *Heritage*, 175; T Russell, *The Discovery of Egypt*, Stroud, 2005.

17 K E Meyer, *The Plundered Past*, New York, 1977, 171–5, 179, 182.

18 Spawforth, *Greek Temples*, 42–5 and 148–9; A Mallwitz, *Olympia und seine Bauten*, Darmstadt, 1972.

19 C M Johns, *Antonio Canova and the Politics of Patronage in Revolutionary and Napoleonic Europe*, Berkeley, 1998; G Pavanello and G Romanelli (eds), *Canova*, Venice, 1992.

20 N Gramaccini, 'Trèves sous Napoleon', *Les Monuments Historiques* (hereafter MH), March/April 1992, 30; J van Elbe, *Roman Germany*, Mainz, 1977, 406; Haas, *Dom zu Speyer*, 21; Giambruno, *Per una storia*, xxix–xxx.

21 *GRKG*, 408–10; Jokilehto, *History*, 77–87.

22 M Beard, 'Pompeii', *AA Files*, 58, 2008, 5.

23 M Falser, *Zwischen Identität und Authentizität*, Dresden, 2008, 21–30; *GRKG*, 302; M Hall (ed), *Towards World Heritage*, Farnham, 2011, 8–9.

24 F C Beiser, *Hegel*, London, 2005; T Rockmore, *Before and After Hegel: A Historical Introduction to Hegel's Thought*, Indianapolis, 1993; T P Pinkard, *Hegel: A Biography*, Cambridge, 2000.

25 T F Barnard, *Herder's Social and Political Thought*, Oxford, 1965, 117; Herder, *Ideen zur Philosophie der Geschichte der Menschheit*, 1, 784–91.

26 Fichte, *Reden an die deutsche Nation*, 1808; Fichte, *Werke*, vol 7, Leipzig, 1846, 455–6; G Mazzini, *Life and Writings*, vol 3, London, 1890–1, 31–3; H Kohn, *Pan-Slavism*, Indiana, 1953; A Hastings, *The Construction of Nationhood, Ethnicity, Religion and Nationalism*, Cambridge, 1997.

27 Jokilehto, *History*, 151; M F Hearn, *The Architectural Theory of Viollet-le-Duc*, Cambridge, MA, 1990.

28 'A Window to the Universe': Falser, *Zwischen Identität und Authentizität*, 21–5.

29 R Chapman, *The Sense of the Past in Victorian Literature*, London, 1986, 21–31; G P Gooch, *History and Historians in the 19th Century*, London, 1913.

30 P Watson, *The German Genius*, New York, 2010.

31 Falser, *Zwischen Identität und Authentizität*, 21–5; Hubel, *Denkmalpflege*, 33–5; *GRKG*, 255–7; Jokilehto, *History*, 119–20; Huse, *Denkmalpflege*, 23–5, 35–8.

32 Falser, *Zwischen Identität und Authentizität*, 30; *GRKG* 255–7; Jokilehto, *History*, 119–25; Huse, *Denkmalpflege*, 48.

33 *AR*, 5–2009; Huse, *Denkmalpflege*, 62–6. Although originally a topographic painter, Schinkel had become an architect partly through the influence of a dramatic project of 1797 by Friedrich Gilly for a monument to Frederick the Great in the form of a Grecian temple.

34 Schinkel, *Concerning the Preservation of the Monuments and Antiquities of our Country*, 1815 (courtesy of Jonathan Blower); Jokilehto, *History*, 114–16; Hubel, *Denkmalpflege*, 33–7; Falser, *Zwischen Identität und Authentizität*, 21–39; D Karg, in Scheurmann, *Zeitschriften*, 242–7; L Schreiner, *K F Schinkel und die erste westfälische Denkmäler-Inventarisation*, Münster, 1968; Huse, *Denkmalpflege*, 29, 47–51, 70–6.

35 Huse, *Denkmalpflege*, 32–3; G Moller, *Denkmäler der deutschen Baukunst*, 2nd ed, Leipzig, nd, 9.

36 *Die Denkmalpflege* (hereafter *DD*), February 2007, 157–63; Jokilehto, *History*, 125–7; Huse, *Denkmalpflege*, 69–70, 78–9; R Wagner-Rieger and W Krause (eds), *Historismus und Schlossbau*, Munich, 1975; D Karg, *Zum 200. Geburtstag von Ferdinand von Quast*, Berlin, 2008.

37 *Mythos Burg*, 54, 84, 343, 387; *GRKG*, 255–7, 259–60.

38 G Fliedl, 'Der Kölner Dom als Nationaldenkmal', in *Beiträge zur historischen Sozialkunde*, April 1981, 129–35.

39 Scheurmann, *Zeitschichten*, 88–93; Jokilehto, *History*, 116–19; Hubel, *Denkmalpflege*, 55–7; Huse, *Denkmalpflege*, 39–47, 52–60.

40 M J Lewis, *The Politics of the German Gothic Revival: August Reichensperger*, Cambridge, MA, 1994; Jokilehto, *History*, 117.

41 Jokilehto, *History*, 112–15; Nicola Borger-Keweloh, *Die mittelalterlichen Dome im 19. Jahrhundert*, Munich, 1986.

42 Hubel, *Denkmalpflege*, 49–51; Haas, *Dom zu Speyer*, 15–21.

43 Wagner-Rieger and Krause, *Historismus*; S Muthesius, *Das englische Vorbild*, Munich, 1974, 37.

44 S Brockmann, *Nuremberg, the Imaginary Capital*, New York, 2006, 32–40.

45 Brockmann, *Nuremberg*, 34–5.

46 Woodward, *In Ruins*, 179; J Hagen, *Preservation, Tourism and Nationalism*, Aldershot, 2006. Hubel, *Denkmalpflege*, 49; Brockmann, *Nuremberg*, 34–5. Prague: J Stulc, 'The Birth of the Idea of Protection of Historic Quarters in 19th Century Prague', in A Tomaszewski (ed), *Values and Criteria in Heritage Conservation*, Florence, 2008, 145–6.

47 P Plagnieux, *The Basilica Cathedral of St Denis*, Paris, 2008, 22, 32, 35.

48 V Hugo, 'Guerre aux demolisseurs, 1825–1832', in V Hugo, *Oeuvres Complètes*, vol 2, Brussels, 1837 edition; G Baldwin Brown, *The Care of Ancient Monuments*, Cambridge, 1905, 74; Plagnieux, *St Denis*, 22–3; Berce, *Des monuments historiques*.

49 Montalembert, *Du vandalisme et du catholicisme dans l'art*, Paris 1839 (essay of 1833); Choay, *Invention*, 90–2, 205, 204.

50 Choay, *Invention*, 70; D Repellin, 'Restoration in Old Lyon', in P Harrison (ed), *Civilising the City*, Edinburgh, 1990.

51 Erder, *Heritage*, 127.

52 Jokilehto, *History*, 127–32; M L Catoni (ed), *Il patrimonio culturale in Francia*, Milan, 2007; A Sutcliffe, *The Autumn of Central Paris*, London, 1970; Berce, *Des monuments historiques*.

53 Jokilehto, *History*, 132; F Berce, *Les premiers travaux de la Commission des Monuments Historiques*, Paris, 1979.

54 Berce, *Premiers travaux*, 4; Catoni, *Patrimonio cultural*.

55 Choay, *Invention*, 96–7; Berce, *Premiers travaux*.

56 Berce, *Premiers travaux*.

57 Baldwin Brown, *Care of Ancient Monuments*, 74, 90.

58 Sutcliffe, *Autumn of Central Paris*; L O'Connell, 'Afterlives', 461.

59 Jokilehto, *History*, 141.

60 Huse, *Denkmalpflege*, 85–8; E E Viollet-le-Duc, *Dictionnaire raisonné de l'architecture française*, Paris, 1854–69; M F Hearn, *The Architectural Theory of Viollet-le-Duc*, Cambridge, MA, 1990; J Dupont, 'Viollet-le-Duc and Restoration in France', in *Historic Preservation Today: Essays Presented to the Seminar on Preservation and Restoration, Williamsburg, Virginia, September 8–11, 1963*, Charlottesville, 1966; Jokilehto, *History*, 145, 151, 154–5.

61 Jokilehto, *History*, 141–4.

62 Jokilehto, *History*, 145–7; D D Reiff, *JSAH*, 1971, 17–30; Choay, *Invention*, 104–5, 127, 211–17.

63 M Delon, *The Conciergerie*, Paris, 2000, 35–7; L de Finance, *The Sainte Chapelle*, Paris,1999, 12.

64 Jokilehto, *History*, 147–9; *GRKG* 262–5; J-P Panouille, *The City of Carcassonne* (Editions du Patrimoine, CNMHS), Paris, 2001; CNMHS, *De la place forte au monument*, Paris, 2000; C Amiel and J Pinies, *La cité des images*, Carcassonne, 1999.

65 C R Dulau, *Pierrefonds Castle*, Paris, 1999; Jokilehto, *History*, 154–5. See also G Lescuyer (ed), *Precis historique du chateau de Pierrefonds*, Paris, 1842 (and later editions).

66 *Monumental*, Direction du Patrimoine, 1993; F Enaud, in *Geschichte der Restaurierung*, vol 1, Worms, 1991, 49–64.

67 Jokilehto, *History*, 154–5; Berce, *Des monuments historiques*; L Vitet, *Histoire des anciennes villes de France*, Dieppe and Paris 1833, vol 1, viii; J Jokilehto, in Falser *et al*, *Conservation and Restoration*, Florence, 2008, 29.

68 Jokilehto, *History*, 138; Erder, *Heritage*, 139; P Kurmann, in *Geschichte der Restaurierung*, vol 2, Worms, 1993, 53–62.

69 J R Blanco, 'The Cathedral of Leon', in N Stanley-Price and J Kay, *Conserving the Authentic*, Rome, 2009, 107ff; A S Shchenkov *et al*, (eds) *Pamyatniki arkhitektury v dorevolutsionnoi Rossii*, Moscow, 2002, 44–6, 89–97, 157–8, 176–80, 194–8, 290–6; Stubbs and Makaš, *Architectural Conservation*, 271–2. Key early Austrian restorations included the Kiefermarkt altar (in 1853, by antiquarian novelist Adalbert Stifter and artist Rudolf Eitelberger).

70 *GRKG*, 234; Martin and Krautzberger, *Handbuch*, 2004, 15.

71 *GRKG*, 410–13; L von Klenze, 'Address at the Inauguration of Restoration Work on the Acropolis', 1834 (source: Klenze, *Aphoristische Bemerkungen gesammelt auf seiner Reise nach Griechenland*, Berlin, 1838, 384–7; translated by, and quotation by courtesy of, Jonathan Blower, 2012); Baldwin Brown, *Care of Ancient Monuments*, 217; C Cunningham, *The Acropolis*, Milton Keynes, 1979; Jokilehto, *History*, 89–96; Lowenthal and Binney, *Our Past Before Us*, 24. List of dates of first conservation laws in various countries: Hall, *Towards World Heritage*, 140.

72 *The Victorian*, July 2009, 24.

73 Choay, *Invention*, 82.

74 M Glendinning and S Wade-Martins, *Buildings of the Land*, Edinburgh, 2009.

75 C Dellheim, *The Face of the Past*, Cambridge, 1982, 182; W Bagehot, *The English Constitution*, London, 1867; see also E Freeman, *The Growth of the English Constitution*, London, 1870, 26.

76 I Flour, *Architectural History*, 51, 2008, 211–38.

77 J Elmes, *Metropolitan Improvements, or London in the 19th Century*, London, 1827.

78 R Southey, cited in A Briggs, *Victorian Cities*, London, 1963, 361.

79 The new was 'false pride', while 'all the old's laid low': 'The Crow's Account of Newcastle', Newcastle, 1812; W J Glaser, *Making Lahore Modern, Constructing and Imagining a Colonial City*, Minneapolis, 2008.

80 Glendinning *et al*, *Scottish Architecture*, 240; S Kelly, *Scott-land, the Man Who Invented a Nation*, Edinburgh, 2010.

81 M Girouard, *The Return to Camelot*, New Haven, CT and London, 1981; R Chapman, *The Sense of the Past in Victorian Literature*, London, 1986, 35.

82 Chapman, *Sense of the Past*, 49–57.

83 R Hill, *God's Architect*, London, 2007, 247, 314, 456; S Muthesius, *Art History*, June 1988, 231–54; Dellheim, *Face of the Past*, 26.

84 Hill, *God's Architect*, 123; *Transactions* of the Ancient Monuments Society, new series, 30, 1986, 221; *Heritage Outlook*, July–August 1982, 89.

85 Dellheim, *Face of the Past*, 33–6.

86 Hall, *Towards World Heritage*, 163–76; Dellheim, *Face of the Past*, 83.

87 S Muthesius, *ÖZKD*, 3/4, 1989, 170; N Pevsner, 'Scrape and Anti-Scrape', in Fawcett, *Future of the Past*, 43.

88 *An Essay on the Development of Christian Doctrine*, 1845, revised 1878.

89 J Elmes, *Metropolitan Improvements*; J M Crook, *Architectural History*, 1965; *Ecclesiologist*, 1847, I, 24, 41, 48; Dellheim, *Face of the Past*, 82–4; C Miele, 'Gothic Sign, Protestant Realia', *Architectural History*, 53, 2010, 191–213; Frew, 'James Wyatt's Choir Screen', *Architectural History*, 1984.

90 *Transactions of the Leicestershire Archaeological and Historical Society*, 64, 1990; *Architect*, 10, 1873, 38.

91 Pevsner, 'Scrape and Anti-Scrape', in Fawcett, *Future of the Past*, 41–2.

92 Pevsner, 'Scrape and Anti-Scrape', 46; G Stamp, *AA Files*, 91, 1981–2.

93 C Marx, 'A Conservative Cathedral Restorer', *The Victorian*, July 2011, 8–9; Pevsner, 'Scrape and Anti-Scrape', 82–4 and 88.

94 See e.g. J M Crook and M H Port, *The History of the King's Works*, vol 6, 1782–1851, London, 1973, 641–5; J Allibone, *George Devey*, Cambridge, 1991, 23, 37, 60; E Whittaker, in P Guillery (ed), *Built from Below*, London, 2010. The Lake District was widely known as 'Wordsworthshire' in 19th-century America: Hall, *Towards World Heritage*, 28.

95 Briggs, *Victorian Cities*, 361–79.

96 N Taylor, 'The Awful Sublimity of the Victorian City', in H Dyos and M Wolff (eds), *The Victorian City*, vol 2, London, 1973, 434; *Transactions of the Ancient Monuments Society*, new series, 30, 1986, 221; Dellheim, *Face of the Past*, 60, 67–9; *Heritage Outlook*, July–August 1982, 89.

97 *Scottish Historical Landscapes*, London, 2002; A N Wilson, *The Laird of Abbotsford, A Life of Sir Walter Scott*, Oxford, 1980.

98 R Fawcett, 'Robert Reid and the Early Involvement of the State in the Care of Scottish Ecclesiastical Buildings and Sites', *Antiquaries Journal*. 82, 2002, 269–84.

99 Walker, 'Adaptation and Restoration', 7; J Gifford, 26 April 2008 lecture at Edinburgh College of Art ('Castle Culture' conference).

100 M Dorrian, 'The King and the City', *Edinburgh Architectural Research*, 30, 2006, 32ff; C McWilliam, *Scottish Townscape*, London, 1975, 194.

101 M Williams, 'Planning for the Picturesque', *Architectural Heritage, AHSS Journal*, 2009, 33–53; S Ferrier, *Marriage*, 226–323; *Provincial Antiquities and Picturesque Survey of Scotland*, vol 1, London, 1826.

102 A Bell (ed), *Lord Cockburn, 1779–1979*, Edinburgh, 1979.

103 *Archaeologia Scotica*, 54, Edinburgh 1857; D Laing, *Proceedings of the Society of Antiquaries of Scotland*, 1864, vol 1v, 1864.

104 Cockburn St 'plain but marked character': Glendinning *et al*, *Scottish Architecture*, ch 6.

105 *Monumentum*, vol 13, 1976; vol 2, 1968; C B Hosmer, *Presence of the Past*, New York, 1965; *JSAH*, 1969, 224; also *GRKG*, 254, on Capitol rebuilding after 1815 burning; W Murtagh, *Keeping Time: the History and Theory of Preservation in America*, NewYork, 1997, 25–7, www.globalmountainsummit.org/independence-hall.html accessed 2 January 11.

106 J M Lindgren, 'A Spirit that Fires the Imagination', in M Page and R Mason, *Giving Preservation a History*, New York, 2004, 108ff; *GRKG*, 128–9; Murtagh, *Keeping Time*, 29–31.

107 *GRKG*, 430.

108 S Winghart, 'Die Wartburg bei Eisenach', in Scheurmann, *Zeitschichten*, 82–7; E Badstuebner, *Wartburg*, Regensburg, 1995, 37; *Mythos Burg*, 346–52.

109 Museen der Stadt Nürnberg, *Das Albrecht Dürer Museum*, Nuremberg, 2006, 26–7, 41–5; Goethehaus Museum, *Revolution und Tradition: 150 Jahre Freies Deutsches Hochstift*, November 2009–February 2010 exhibition catalogue. For a later debate about the balance of modernity and authenticity in an 'artist's house restoration' (of the Amsterdam Rembrandthuis by Dutch architect Karel de Bazel in 1908), see B Laan, 'Het interieur van het Rembrandthuis: ontwerp voor een "levende bestemming"', *Amstelodamum*, 85, 1998, 73–80.

110 J Taylor, *A Dream of England*, Manchester, 1994, 73; J L Williams, *Home and Haunts of Shakespeare*, 1893; R Pringle, *The Shakespeare Houses*, Jarrold, Norwich, 1999.

111 Taylor, *A Dream of England*, 73; Pringle, *The Shakespeare Houses*; each 'revival' initiative fuelled the next; the 'restored' house became the centrepiece of a tricentennial celebration in 1864, at which the idea of building a Shakespeare Memorial Theatre was broached (eventually built in 1875–9).

Chapter 4

1 J Ruskin, *The Seven Lamps of Architecture*, London, 1849 (second edition, 1855); Getty Institute, *Historical and Philosophical Issues in the Conservation of Cultural Heritage*, Los Angeles, CA, 1996, 178; Dellheim, *Face of the Past*, 91.

2 Tönnies, *Gemeinschaft und Gesellschaft*.

3 Jokilehto, *History*, 111–12; Pevsner, 'Scrape and Anti-Scrape', 77; Hill, *God's Architect*; Briggs, *Goths and Vandals*, 156.

4 S Muthesius, *The Poetic Home: Designing the 19th-Century Domestic Interior*, London, 2009; Hill, *God's Architect*.

5 Chapman, *Sense of the Past*, 61.

6 Chapman, *Sense of the Past*, 63–6; R Hewison, *Ruskin on Venice*, New Haven, CT and London, 2010; *Transactions of the Ancient Monuments Society*, new series, 30, 1986, 222.

7 J Ruskin, *Seven Lamps*; Cornelis J. Baljon, 'Interpreting Ruskin: The Argument of *The Seven Lamps of Architecture* and *The Stones of Venice*', *Journal of Aesthetics and Art Criticism*, Autumn 1997, 401–14; M Swenarton, *Artisans and Architects*, London, 1989, 4–7.

8 Ruskin, *Seven Lamps*.

9 M H Beyle, *Promenades dans Rome*, vol 1, Paris, 1829; W Denslagen, *Architectural Restoration in Western Europe*, Amsterdam, 1994.

10 John Ruskin, *The Seven Lamps of Architecture*, 1849 (Orpington, 1883 ed), 186; Jokilehto, *History*, 174–81.

11 M Wheeler and N Whiteley (eds), *The Lamp of Memory: Ruskin, Truth and Architecture*, Manchester, 1992.

12 Pevsner, 'Scrape and Anti-Scrape', 47; Marx, 'Conservative Cathedral Restorer', 8–9.

13 Pevsner, 'Scrape and Anti-Scrape', 50.

14 P Thompson, *The Work of William Morris*, Oxford, 1991 (first publ 1967), 67. Jokilehto, *History*, 156–63, 183–4; J Fawcett, 'A Restoration Tragedy', in Fawcett, *Future of the Past*, 80; RCAHMS, *St Mary's Episcopal Cathedral, Edinburgh: A Short History and Guide, RCAHMS Broadsheet 13*, Edinburgh, 2003; Pevsner, 'Scrape and Anti-Scrape', 49; Briggs, *Goths and Vandals*, 170–6, 193.

15 Armytage, *Heavens Below*, C Miele (ed), *From William Morris*, New Haven, CT and London, 2005; P Davey, *Arts and Crafts Architecture*, London, 1980, 59; Thompson, *Morris*, 235; Chapman, *Sense of the Past*, 59 and 174; R Samuel, 'The Vision Splendid', *New Socialist*, May 1985.

16 M Swenarton, *Artisans and Architects*, 16; Davey, *Arts and Crafts*, 9.

17 *Old House Journal*, September–October 1996, 26.

18 Dellheim, *Face of the Past*, 90; Choay, *Invention*, 205–6, 212; Pevsner, 'Scrape and Anti-Scrape', 50–1; Briggs, *Goths and Vandals*, 203–11; Jokilehto, *History*, 184–6. Tewkesbury: www.marxists.org/archive/morris/works/1877/tewkesby.htm.

19 O Hill, *Our Common Land*, London, 1877; *Transactions of the Ancient Monuments Society*, new series, 30, 1986, 17–18; Miele, *From Morris*, 65, 135–7.

20 Pevsner, 'Scrape and Anti-Scrape', 53. Thompson, *Morris*; Jokilehto, *History*, 185–6.

21 Jokilehto, *History*, 183–4; Fawcett, *Future of the Past*, 97–8.

22 Miele, *From William Morris*, 366. See e.g. Cherry Hinton Church, from 1877: P Venning, 'Back to the Battle', *Cornerstone*, vol 32, 3, 2011, 32–3.

23 Miele, *From William Morris*, 59.

24 R Judge, 'May Day and Merrie England', *Folklore*, vol 102 (ii), 1991, 131ff.

25 Chapman, *Sense of the Past*, 81, 97, 124, 149–161, 199–201; S Omoto, 'Thackeray and Architectural Taste', *JSAH*, 1967, 410–17.

26 S Muthesius, 'Why Do We Buy Old Furniture?', 231–45; Muthesius, *Poetic Home*.

27 Muthesius, 'Die Diskussion über Restoration und Preservation', *ÖZKD*, 3/4, 1989, 170–3; J Allibone, *George Devey, Architect 1820–1886*, Cambridge, 1991.

28 J M Dick Peddie, *Proceedings of the Society of Antiquaries of Scotland*, new series, vi, 1884, 465ff; *Transactions of the Ancient Monuments Society*, new series, 30, 1986. In 1877–8 architect R Rowand Anderson tried to preserve an 18th-century classical house in a threatened South Side street: *Scotsman*, 22 November 1877; *Builder*, 5 January 1878, 5.

29 Dellheim, *Face of the Past*, 78–9, 88–92; Society for Protection of Ancient Buildings (hereafter SPAB), *Report*, 1887, 65–77.

30 Dellheim, *Face of the Past*, 79, 85–130; Havelock Ellis, *The New Spirit*, 1890, 24; *SPAB News*, vol 17, 2, 1996, 10–14.

31 Thompson, *Work of William Morris*, 239; Choay, *Invention*, 102–3.

32 Borger-Keweloh, *Die mittelalterlichen Dome*; Hubel, *Denkmalpflege*, 59–63; A Hubel and M Schuller, *Der Dom zu Regensburg*, vol 1, Regensburg, 2011.

33 *Mythos Burg*, 355–7.

34 J M Crook, *William Burges and the High Victorian Dream*, London, 1981; *GRKG*, 431–3; Jokilehto, *History*, 250–2; P J H Cuypers and F Luyten, *Le Chateau de Haar a Haarzuylen*, Utrecht, 1910; M Kok, *Het Kasteel van de Haar*, Utrecht, PhD, 1994.

35 *The Acropolis at Athens*, Athens, 1983, 24; Jokilehto, *History*, 187–91; Choay, *Invention*, 94.

36 M D Bardeschi (ed), *Il Monumento e suo doppio*, Florence, 1981; Lasansky, *Renaissance Perfected*; B Roeck, *Florence 1900: the Quest for Arcadia*, New Haven, CT and London, 2009.

37 Jokilehto, *History*, 246–7; Blanco, 'Cathedral of Leon', 107ff; J Rivera, *Historia de las restauraciones de la Catedral de Leon*, Valladolid, 1903; M V Fernandez, *La Catedral de Leon*, Madrid, 1993. Ridderzaal: Denslagen, *Architectural Restoration*, 188–204.

38 C Delon, *Notre Capitale*, Paris, 1888, 254; Sutcliffe, *Autumn of Central Paris*, 181–8.

39 Hall, *Towards World Heritage*, 205–7; Jokilehto, *History*, 164–5; I Hlobil and R Svacha, *SAHGB Newsletter*, 59, Autumn 1996, 1–3; *GRKG*, 267–270. Diocletian's Palace debates: see the very full account in J Blower, 'The Monument Question in Late Habsburg Austria: a Critical Introduction to Max Dvořák's Denkmalpflege', PhD thesis, University of Edinburgh, 2012.

40 Choay, *Invention*, 94.

41 Swenarton, *Artisans and Architects*, 195; S Muthesius, 'The Origins of the German Conservation Movement', in R Kain (ed), *Planning for Conservation*, London, 1981, 28.

42 Hewison, *Ruskin in Venice*; J Unrau, *Ruskin and St Mark's*, London, 1984, 197ff.; Dellheim, *Face of the Past*, 79–80.

43 Miele, *From William Morris*, 187–97; Unrau, *Ruskin*; Bardeschi, *Monumento*, 51; A Fasolo, *Palaces of Venice*, 2006, 20.

44   E Lemire (ed), *The Unpublished Lectures of William Morris*, Detroit, 1969, 292; A Vance, *William Morris*, London, 1897, 261.

Chapter 5

1   G Dehio, *Was wird aus dem Heidelberger Schloss werden?* Strasbourg, 1901, 41; H Sumner, 'Protection and Production', 1895, cited Dellheim, *Face of the Past*, 77.

2   *Mythos Burg*, 329–33; *GRKG*, 121–3, 271–4; R Koshar, 'Building Pasts', in Gillis, *Commemorations*; W Speitkamp, 'Die Hohkönigsburg und die Denkmalpflege im Deutschen Kaiserreich', *Neue Museumskunde*, 34, 1991, 121–30.

3   G K Chesterton, *G F Watts*, 1904, 3; J Summerson, *Victorian Architecture, Four Studies in Evaluation*, New York, 1970, 2.

4   Jokilehto, *History*, 213–4; V Welter and J Lawson (eds), *The City after Patrick Geddes*, Bern, 2000; G Nietzsche, *Die Fröhliche Wirtschaft*, 1887; Bernd Behrendt, 'August Julius Langbehn, der „Rembrandtdeutsche"', in Uwe Puschner, Walter Schmitz and Justus H Ulbricht (eds), *Handbuch zur 'Völkischen Bewegung' 1871–1918*, Munich, 1999, 94–113.

5   H F Mallgrave and E Ikonomou (eds) *Empathy, Form and Space*, Chicago, 1994; Muthesius, *Englische Vorbild*, 166–73.

6   Armytage, *Heavens Below*, 304, 380.

7   Alois Riegl, *Der moderne Denkmalskultus. Sein Wesen, seine Entstehung*, Vienna, 1903 (published 1929 in collected works); Huse, *Denkmalpflege*, 124–49; R Bentmann, *Hessische Blätter für Volks- und Kulturforschung*, 2/3, Giessen, 1976, 213–46; H Evers, *Tod, Macht und Raum als Bereiche der Architektur*, Munich, 1939, 283–303; M Wohlleben, *Konservieren oder Restaurieren?* (thesis), Munich, 1979.

8   Huse, *Denkmalpflege*, 124–49; Jokilehto, *History*, 214–17; M R Olin, 'The Cult of Monuments', *Wiener Jahrbuch für Kunstgeschichte*, 1985.

9   Olin, 'Cult of Monuments'.

10   Riegl, *Denkmalskultus*.

11   C Applegate, *A Nation of Provincials, The German Idea of Heimat*, Berkeley, CA, 1990; A Riegl, 'Neue Stroemungen in der Denkmalpflege', in *Mitteilungen der K. K. Zentralkommission*, III (iv), 1905, 92ff.

12   Olin, 'The Cult of Monuments'.

13   M Iversen, *Alois Riegl, Art History and Theory*, Cambridge (MA), 1993; R Kain (trans), *Alois Riegl, Problems of Style*, Princeton, NJ, 1992; W Kemp, 'Alois Riegl', in H Dilly (ed), *Altmeister moderner Kunstgeschichte*, Berlin, 1990; Falser, Lipp and Tomaszewski, *Conservation and Preservation*, 289–92; M Iversen, 'Style as Structure', *Art History*, March 1979; Falser, *Zwischen Identität und Authentizität*, 59; Huse, *Denkmalpflege*, 129–30, 146–9; Arrhenius, in Choay, *Invention*, 111–14, 213–14; Scheurmann, *Zeitschichten*, 4.

14   P Paszkiewicz, in K Murawska-Muthesius (ed), *Borders in Art: Revisiting Kunstgeographie*, Warsaw, 2000; A Sutcliffe, *Towards the Planned City*, Oxford, 1981.

15   H James, *Italian Hours*, Boston, 1909; Keats-Shelley Memorial Association, *Spellbound by Rome: The Anglo-American Connection*, Rome, 1890–1914; Hansi (J J Waltz), *Mon Village*, 1913, 16–17.

16   Baldwin Brown, *Care of Ancient Monuments*, 99; F Meinecke, *Entstehung des Historismus*; S Muthesius, *Art, Architecture and Design in Poland*, Königstein, 1994.

17   Muthesius, *Poland*, 10; Tönnies, *Gemeinschaft und Gesellschaft*.

18   J von Altenbockum, *Wilhelm Heinrich Riehl*, 1994, Cologne; W Sonne, 'Heimat in der Architektur', in I Scheurmann and H R Meier (eds), *Echt Alt Schön Wahr*, Munich, 2006, 58–67; S Michalski, *Public Monuments, Art in Political Bondage*, London, 1998; M Falser, *Zwischen Identität und Authentizität*, 44; G R Gillis (ed), *Commemorations*, 1994, Princeton, NJ, 10.

19   Hubel, *Denkmalpflege*, 67–73; E Troeltsch, *Deutscher Geist und Westeuropa*, Tübingen, 1925, 18; W Hartung, *Denkmalpflege und Heimatschutz in wilhelminischen Deutschland*, ÖZKD, 3–4, 1989, 173–81.

20   Huse, *Denkmalpflege*, 150–66; Hubel, *Denkmalpflege*, 97; G Hermann, *Das Biedermeier im Spiegel seiner Zeit*, Berlin, 1913; S Muthesius, *Design History*, vol 16, 4, 2003; A Wentscher, *Das Landhaus*, Bielefeld, 1913; F Hoermann, *Heimkunst und Heimatkunst*, Dresden, 1913; Muthesius, *Das englische Vorbild*, Munich, 1974, 166–73; S Muthesius, 'The Origins of the German Conservation Movement', in R Kain (ed), Planning for Conservation, London, 1981, 37–48.

21   Hubel, *Denkmalpflege*, 97–9; P Schultze-Naumburg, *Kulturarbeiten*, vol 1, 1906; A Achleitner, *MH*, 1993, 159.

22 I Scheurmann, 'Reden über Denkmalpflege', in *DD*, 1/2011, 17–24; Falser, *Zwischen Identität und Authentizität*, 59–66; Blower, 'Monument Question in Late Habsburg Austria'; Olin, 'Cult of Monuments'; Huse, *Denkmalpflege*, 139–45; Jokilehto, *History*, 217; Scheurmann, *Zeitschichten*, Munich, 2005, 13–17, 48–81; E Hubala, 'Georg Dehio', in *Zeitschrift für Kunstgeschichte*, 46, 1983, 1–14; Dehio, *Denkmalschutz und Denkmalpflege in 19. Jahrhundert*, 1905, full translation of speech: courtesy of Jonathan Blower, 2012; A Lehne, in M S Falser, W Lipp and A Tomaszewski (eds), *Conservation and Preservation*, Florence, 2010, 72–3. Tallinn was at that stage, with its then chiefly German inhabitants, called Reval; Riegl was born in Linz. Dehio's famous phrase, 'Konservierien, nicht restaurieren', was in fact first used by someone else at the 1873 1ˢᵗ International Congress for Art History in Vienna: Georg Dehio, *Geschichte der deutschen Kunst*, vol 1, Berlin, 1930, 7; M Wohlleben (ed), *Georg Dehio und Alois Riegl, Konservieren nicht Restaurieren*, Zürich, 1988.

23 Huse, *Denkmalpflege*, 124–49.

24 *GRKG*, 148; W Denslagen, *Architectural Restoration*, frontispiece 1 and 2; Hubel, *Denkmalpflege*, 64–6; *GRKG*, 268–70.

25 Falser, *Zwischen Identität und Authentizität*, 33–5.

26 C Hellbrügge, *Kunsttexte.de*, 2002/2, 3; Hubel, *Denkmalpflege*, 75; Jokilehto, *History*, 197–8, 217; E Rusch, *DD*, 1/1999, 81–5; A von Buttlar, G Dolff-Bonekaemper, M Falser, A Hubel and G Moersch, *Denkmalpflege Statt Attrappenkult*, Basel, 2011, 46.

27 Hartung, 'Denkmalpflege', 178–81; Miele, *From William Morris*, 193; Falser, *Zwischen Identität und Authentizität*, 43–66; Jokilehto, *History*, 195–6; H H Müller, *DD*, 2/1999, 112–15.

28 Olin, 'Cult of Monuments'; Martin and Krautzberger, *Handbuch Denkmalschutz*, 15–16; L Hammer, *Die Geschichtliche Entwicklung des Denkmaltrechts in Deutschland*, 1995; J F Hanselmann, *Die Denkmalpflege in Deutschland um 1900*, 1996; W Speitkamp, *Die Verwaltung der Geschichte*, 1996; Hall, *Towards World Heritage*, 147.

29 Hubel, *Denkmalpflege*, 77–84; Hellbrügge, *Konservieren*.

30 Huse, *Denkmalpflege*, 93–4, 105–11; *Verhandlungen der Heidelberger Schloss-Konferenz*, Karlsruhe 1902; Scheurmann, *Zeitschichten*, 108–13; S Gensichen, in Scheurmann and Meier, *Echt Alt Schoen Wahr*, 96–121; G Skalecki, in Scheurmann, *Zeitschichten*, 102–7; Falser, *Zwischen Identität und Authentizität*, 44–59; Landesamt für Denkmalpflege im Regierungspräsidium Stuttgart, *Traum und Wirklichkeit, Vergangenheit und Zukunft der Heidelberger Schlossruine*, 2005.

31 *MH*, 189, 1993, 26ff; Olin, 'Cult of Monuments'; A Lehne, *ÖZKD*, xliii, 3–4, 1989, 145–81.

32 Falser, *Zwischen Identität und Authentizität*, 65; Blower, 'Monument Question in Late Habsburg Austria'; Olin, 'Cult of Monuments'; Hubel, *Denkmalpflege*, 89–93, 157; Jokilehto, *History*, 218–19; Huse, *Denkmalpflege*, 158–60, 171–9.

33 *SAHGB Newsletter*, 59, Autumn 1996, 1–3; Huse, *Denkmalpflege*, 177–82; M Dvořák, *Katechismus der Denkmalpflege*, 1918; Projektgruppe Denkmalpflege, Universität Kiel, *Materialen zur Denkmalpflege*, Kiel, 1972, 7; Olin, 'Cult of Monuments'; S Casiello (ed), *La cultura del restauro*, Venice, 2005, 239–68; *Restauro*, 3–1972; Kain, *Planning for Conservation*, 49–62; Blower, 'Monument Question in Late Habsburg Austria'; L Frodl, *Idee und Vertwirklichung*, 1988; M Koller, in *Geschichte der Restaurierung in Europa*, vol 1, 65–83.

34 D Borutova, *MH*, 189, 1993, 28; D Crowley, *National Style and Nation State*, Manchester, 1992; Muthesius, *Poland*, 79.

35 Zamek Krolewski na Wawelu, *Wawel Narodowi Przywrocony*, Krakow 2005.

36 *JSAH*, 1971, 228–9; *GRKG*, 336–7 on Fantoft Stave Church, Norway, built *c*.1200, rebuilt/moved 1883 and rebuilt 1993–7 after fire destruction. Catalan conservation: http://arquicatalana.blogspot.com/2010/04/jeroni-martorell-i-terrats.html.

37 Riall, *Garibaldi*, New Haven, CT, 2007, 388–92; Jokilehto, *History*, 198–200; *The Roman Forum* (guide), 9.

38 G Martelli, *Transactions of the Ancient Monuments Society*, new series, vol 19, 1972, 73–8; Northern Arts Council, *Futurism 1909–1919* (catalogue), 1972, 58.

39 *Future Anterior*, vol vi, 1, summer 2009, 70–1.

40 S Casiello (ed) *La cultura del restauro*, Venice 2005, 141–58; Jokilehto, *History*, 198–209; C di Biase, 'Camillo Boito', in Casiello, *Cultura del restauro*, 159–82; M Giambruno (ed) *Per una storia*, 2007; R Etlin, *Modernism in Italian Architecture*, Cambridge, MA, 1991, 124. Boito also ridiculed Ruskin's defence of the external discolouration of St Mark's Venice, arguing it was due to soot: J Otero-Pailos, 'The Ambivalence of Smoke', *Grey Room*, 44, Summer 2011, 102.

41 Casiello, *Cultura del restauro*, 183–238; Jokilehto, *History*, 203–6; F B Cavalieri, in Giambruno, *Per una storia*, 13–20; A M Shanken, 'Preservation and Creation: Alfonso Rubbiani and Bologna', *Future Anterior*, Summer 2010, 61–82; A Rubbiani, 'The Houses of the Bourgeoisie', reproduced in *Future Anterior*, Summer 2010, 83–95; A Rubbiani, *Il Palazzo di Re Enzo in Bologna*, Bologna, 1906. See also the work of G Castellucci around Florence, Lasansky, *Renaissance Perfected*, 46.

42 Jokilehto, 207–9, 220–1; Casiello, *Cultura del restauro*, 269–92.

43 Sutcliffe, *Autumn of Central Paris*, 200–3; Haussmann, *Memoires*, 1890–3, iii, 28 and 501; L Grodecki, *Deutsche Kunst und Denkmalpflege* (hereafter *DKD*), 1971 13–14; P Clemen, in *Zeitschrift für Bauwesen*, 48, 1989, 490–538 and 594–630; Catoni, *Patrimonio culturale*; *Monumental*, December 1993; M Chatenet, 'Inventory and Protection', *Transactions of the Ancient Monuments Society*, 39, 1995, 41–50; P Leon, *Les Monuments Historiques*, Paris, 1913; *MH*, 189, 1993, 5; Berce, *Des monuments historiques*. Terroir was expressed in new architecture, in the 'neo-Normand' villas of the late 19th century, elaborated into a timber regional style in towns like Trouville-Deauville. Toulouse: R Wakeman, *Modernising the Provincial City*, Cambridge, MA, 1997, 51.

44 Japan: N Inaba, 'Authenticity and heritage concepts', in *Conserving the Authentic*, 153–62; Y I Akieda, 'What has been valued in Japanese Architectural Heritage?', in A Tomaszewski (ed), *Values and Criteria in Heritage Conservation*, Florence, 2008, 125–7. Ise: *GKRG*, 34. China: Zhao Xiaofeng, 'Chinese Conservation in Action', lecture at ECA, 24 March 2011.

45 Kain, *Planning for Conservation*, 37–48; D Walker, 'The Restoration of Ruined Castles' (unpublished government paper), 1982 (revised 1985); *AHSS Magazine*, Spring 1997, 17; D Robinson, *Heritage in Wales*, London, 1989: in Wales, likewise, the turn-of-century years saw an ambitious restoration programme at the previously neglected Caernarfon Castle.

46 R Pringle, *The Shakespeare Houses*, Jarrold, Norwich, 1999; J L Wilson, *Home and Haunts of Shakespeare*, 1892; J Taylor, *A Dream of England*, Manchester 1994, 73.

47 C Chappindale, *Journal of the British Archaeological Association*, vol 136, 1983, 1–55. Expansion of archaeology: Raymond Lemaire International Centre for Conservation, *Conservation in Changing Societies*, Leuven, 2006; Briggs, *Goths and Vandals*, 230.

48 N Boulting, 'The Law's Delays', 18–19.

49 J Popham, 'The Relict Countryside', in Lowenthal and Binney, *Our Past Before Us*, London, 1981, 159; A Hale and R Sands, RCAHMS, *Controversy on the Clyde*, Edinburgh 2005; R Munro, *Archaeology and False Antiquities*, London, 1905.

50 Taylor, *Dream of England*, 56–7; P James and W J Harris, *Sir Benjamin Stone*, MA diss., Birmingham Polytechnic, 1989; F N Bohrer, *Photography and Archaeology*, Reaktion, 2010.

51 B Skinner, *A Place in Trust*, Edinburgh, 1981; also *The National Trust, an Introduction*, London, 1980; Miele, *From William Morris*, 65, 129–53.

52 M Hall, in Miele, *From William Morris*, 129–57; Hall, *Towards World Heritage*, 42; M Hall, *Transactions of the Royal Historical Society*, 2003, 13, 345–57; Boulting, 'The Law's Delays', 18; *The National Trust, an Introduction*, London, 1981.

53 Henry Moore Institute, *Angkor Wat: from Temple to Text*, Leeds, 2010; P V Lottin de Laval, *Manuel complet de Lottinoplastique*, Paris, 1857; G Coedes, *Pour mieux comprendre Angkor*, Paris, 1947; E Aymonier, *Le Cambodge*, Paris, 1901–4; G B Brown, *Care of Ancient Monuments*. For a late example of 'cultural looting', see the depradations in Cyprus of the US consul Luigi Palma di Cesnola and the New York Metropolitan Museum in the 1860s/70s: Stubbs and Makaš, *Architectural Conservation*, 350.

54 Hall, *Towards World Heritage*, 65–9; W G Glover, *Making Lahore Modern*, Minneapolis, 2008.

55 *Apollo*, August 1970, 144–5; Miele, *From William Morris*, 202–4. Jokilehto, *History*, 274–7; Curzon had no sense of a schism between 'good old' and 'bad new' architecture; having been brought up in Adam's Kedleston, he greatly admired the colonial classicism of Calcutta and aimed to enhance it by the vast new Victoria Memorial. Contested status of pre–Reformation sites and archaeology in 19th-century Ireland: L Kealy, in N Stanley Price and J King, *Conserving the Authentic*, Rome, 2009, 101–2.

56 Miele, *After William Morris*, 205–6; *Cornerstone*, vol 25, 3, 2004: Prince Frederick Duleep Singh was a prominent member of SPAB, especially in Norfolk and Suffolk, for nearly 40 years between 1891 and 1928. SPAB itself intervened in some demolition controversies within the British Empire, notably in its successful campaign against harbour redevelopment in Famagusta, Cyprus, in 1900–4. See also V Hastaoglou-Martinidis, 'Urban Aesthetics and National Identity: The Refashioning of Eastern Mediterranean Cities between 1900 and 1940', *Planning Perspectives*, April 2011, 153–82.

57 Exceptionalism: for example, William Murtagh argued that the pioneering preservation of Washington's home by the Mount Vernon Ladies' Association in 1858 helped establish a unique American focus on private action, and on monuments associated with national heroes: W Murtagh, *Keeping Time: the History and Theory of Preservation in America*, New York, 1988 (1997 edition), 30. 'Confederate nationalism': P Quigley, *Shifting Grounds: Nationalism and the American South, 1848–1865*, Oxford, 2011; C B Hosmer, *Presence of the Past*, New York, 1965; W B Rhoads, 'The Colonial Revival', *JSAH*, 1976, 239ff; A Downing, *JSAH*, 1975, 290.

58 T Hild, *Journal of Illinois History*, 1995, 79–100 and Autumn 2001, 178–92; D Bluestone, *Journal of Architectural Education*, May 1994, 210–23; Museum of the City of New York, *The American Style: Colonial Revival and the Modern Metropolis*, exhibition, 2011; Hall, *Towards World Heritage*, 130.

59 Alfred Runte, *National Parks, the American Experience*, Lincoln, Nebraska, 1997; Antiquities Act: www.nps.gov/history/local–law/anti1906.htm; *Monumentum*, vol 2, 1968, and vol 13, 1976; S Schama, *Landscape and Memory*, London, 1995, 7–15. Mexico: Stubbs and Makaš, *Architectural Conservation*, 511–13.

60 Hall, *Towards World Heritage*, 23–44; Stubbs and Makaš, *Architectural Conservation*, 485–7; C J Taylor, *Negotiating the Past*, Montreal, 1990; A W Gilg, in Kain, *Planning for Conservation*, 97ff; L Bella, 'The Politics of Preservation', *Planning Perspectives*, 1, 1986, 189–206; H Kalman, *A History of Canadian Architecture*, vol 2, Oxford, 1994, 659, 856; Parks Canada, *Guide to Canada's National Historic Parks*, Ottawa, 1985.

61 Hall, *Towards World Heritage*, 182–94; M Holleran, 'Roots in Boston' and Lindgren, 'A Spirit that Fires the Imagination', in Page and Mason, *Giving Preservation a History*, 81–121; www.paulreverehouse.org/about/paulreverehouse.html; www.encyclopediavirginia.org/Association_for_the_Preservation_of_Virginia_Antiquities; Murtagh, *Keeping Time*, 31, 54–5; Lowenthal, *Past is a Foreign Country*, 2; *Historic Preservation Today*, Charlottesville, 1966.

62 *GRKG*, 128.

63 Rhoads, 'Colonial Revival'; A Downing, *JSAH*, 1975, 290; W K Sturges, 'Arthur Little and the Colonial Revival', *JSAH*, 1973; A G Robinson, *Old New England Doorways*, New York, 1919; Miele, *From William Morris*, 213–35.

64 R Mason, *The Once and Future New York*, Minneapolis, 2009; R Mason, 'Historic Preservation, Public Memory, and the Making of New York City', in Page and Mason (eds), *Giving Preservation a History*, 131–159.

65 K Hafertepe, 'Restoration, Reconstruction, or Romance?', *JSAH*, September 2008, 412–33.

66 G Vinken, *Zone Heimat*, Munich, 2010.

67 Choay, *Invention*, 94, 102–3, 139, 205–15; R Kaltenbrunner, 'Sonderzone Heimat', *Süddeutsche Zeitung*, 23 August 2010, 12.

68 R Home, *Of Planting and Planning*, London, 1997, 170ff.

69 J Stübben, 'Der Baulinien – Plan für die Altstadt Wien', *Deutsche Bauzeitung*, 1896, xxx; J Grimm, *Deutsches Wörterbuch*, vol 1, Leipzig, 1854, 274; F Kaspar, *Die Alte Stadt*, 3–1999, 300; Vinken, *Zone Heimat*; Baedeker, *Denmark*, 1913; Vinken, in Scheurmann and Meier, *Echt Alt Schön Wahr*, 190–201.

70 Home, *Of Planting and Planning*, 141–8; Glendinning *et al*, *Scottish Architecture*, ch 7 (criticism of Haussmann).

71 Sutcliffe, *Autumn*, 190–1; *Bulletin*, vol 1, 29; Kain, *Planning for Conservation*, 177–88.

72 Hubel, *Denkmalpflege*, 89. However, there were also prominent instances of avoidance of *Freilegung* in Germany, as in the case of the long-established tourist hot spot of Bacharach am Rhein, whose medieval walls remained densely embedded in later housing (picturesquely interpenetrated by a public wall-top promenade), as well as directly abutted by a main-line railway viaduct.

73 Giambruno, *Per una storia*, 5–20; Home, *Of Planning and Planning*, 123; Jokilehto, *History*, 220–1.

74 J Stübben, *Der Städtebau*, 1890; Welter and Lawson, *The City after Patrick Geddes*; Sutcliffe, *Towards the Planned City*; S Collini, *Past and Present*, 1976, 72, 86–111.

75 Sutcliffe, *Towards the Planned City*; Hubel, *Denkmalpflege*, 93.

76 Jokilehto, *History*, 220.

77 Muthesius, *Das englische Vorbild*, 168–70; Sitte, *Der Städte-bau nach seinen künstlerischen Grundsätzen*, Vienna, 1889, 49 and 120; Choay, *Invention*, 123–9 and 215–16.

78 J Düwel and N Gutschow, *Städtebau in Deutschland (im 20. Jh)*, Stuttgart, 2001; G R and C C Collins in Kain, *Planning for Conservation*, 63–74.

79 Jokilehto, *History*, 220; C C Collins and G R Collins, *Camillo Sitte and the Birth of Modern City Planning* (2 vols), New York, 1965.

80 Collins and Collins, *Sitte*, vol 2, 189; Düwel and Gutschow, *Städtebau*, 52–3; A Herscher, *JSAH*, 62, 2, June 2003, 212–27; *AR*, March 1980, 16ff.

81 Muthesius, 'Altdeutsche Zimmer'; G Hajós, 'Die Denkmalpflege', *Beiträge zur historischen Sozialkunde*, 4–1981, 140.

82 Sutcliffe, *Autumn*, 209; W Kos and C Rapp (eds), *Alt-Wien, die Stadt die niemals war*, Vienna, 2004; *GRKG*, 450–1 (1897 plaster copy of Parthenon in Nashville, rebuilt 1920–5 in steel/stone).

83 Miele, *From William Morris*, 131–2; Hall, *Towards World Heritage*, 207–8; J R Armstrong, *Transactions of the Ancient Monuments Society*, 1975; K Blent *et al*, *Skansen*, Stockholm, 2005; E A Chappell, *JSAH*, vol 58, 3; L Young, *International Journal of Heritage Studies*, July 2006, 321–38; M Hunter (ed), *Preserving the Past*, 1996, 156–9.

84 Choay, *Invention*, 123–5; Sonne, 'Heimat in der Architektur', 58–67; R Unwin, *Town Planning in Practice*, London, 1909, 135 and 215.

85 B Ladd, *Urban Planning and Civic Order in Germany*, 1860–1914, Cambridge, MA, 1990.

86 S Haps, 'Erhaltung und Gestaltung von Altstadtbereichen', *DD*, 1/2011, 25–31; Hubel, *Denkmalpflege*, 89–99; Muthesius, *Englische Vorbild*, 168.

87 C Buls, *L'Esthetique des villes*, Brussels, 1893; F Strauven, in D Appleyard (ed), *The Conservation of European Cities*, Cambridge, MA, 1979, 104; *Architectural History*, 51, 2008, 172. Buls translation/abridgement of Sitte: *La construction des villes*, Brussels, 1895.

88 Muthesius, *Englische Vorbild*, 168; G Bruce, *Some Material Good*, Edinburgh, 1975, 53–4; J H A Macdonald, *Incongruity and Defacement in Edinburgh and Elsewhere*, Edinburgh, 1907. Cockburn and SCAPA: Miele, *From William Morris*, 62–5 and 133–4.

89 Museen der Stadt Nürnberg, *Albrecht Dürer Museum*, 42–5.

90 *New York Times*, 2 Dec 1894.

91 *New York Times*, 21 February 1893.

92 Brockmann, *Nuremberg*, 116–19; *DD*, 1, 1899.

93 Hagen, *Preservation, Tourism and Nationalism*, 44, 123–5; *DD*, 1/2010, 80.

94 Vinken, *Zone Heimat*, 61–122; Vinken, 'Gegenbild – Traditionsinsel – Sonderzone'; J Stulc, '19th Century Prague', 146–8; J Pavel, *Dejiny Pamatkove Pece Vceskych Zemich*, Prague, 1975; R Josephson, *Borgarhus i Gamla Stan*, Stockholm, 1916.

95 Sutcliffe, *Towards the Planned City*, 142; Sutcliffe, *Autumn*, 188–211; C Delon, *Notre capitale Paris*, Paris, 1888, 254; C Magny, *La beauté de Paris*, Paris, 1911, 122; Chatenet, 'Inventory and Protection', 47; F Roussel, *MH*, 189, 1993, 43ff.

96 Futurists: Choay, *Invention*, 110–11; Etlin, *Modernism*, 124; Giambruno (ed) *Per una storia*.

97 M Vitello, in Giambruno, *Per una storia*, 5–12.

98 D M Lasansky, *The Renaissance Perfected*, University Park, PA, 2004, 31–7; Jokilehto, *History*, 220–1; G Giovannoni, 'Vecchie città ed edilizia nuova' and 'Il diradamento edilizio', in *Nuova Antologia*, 48, 1913, 53–76 and 449–72. He argued that the 'quartiere Rinascimento' should be generally conserved, but thinned out and traversed by a new artery, the Corso Vittorio Emmanuele II.

99 Fulvio Irace, 'Le due città: Piacentini e Angelini', in *Bergamo e il suo territorio*, Milan, 1997, 161–97; M Lupano, *Marcello Piacentini*, Rome, 1991; M Pisani, *Architetture di Marcello Piacentini. Le opere maestre*, Rome, 2004.

100 A Ballantyne and A Law, *Tudoresque*, London, 2011.

101 N Pevsner and Edward Hubbard, *Cheshire*, London 1971 (2003 edition), 38, 162, 169; D Insall & Partners, Chester, *A Study in Conservation*, London, 1968, 27.

102 J Dougill, *Oxford in English Literature*, Ann Arbor, MI, 1998.

103 J Baldwin Brown, *Care of Ancient Monuments*; T Hunt, *Building Jerusalem, the Rise and Fall of the Victorian City*, London, 2004, 232–65; *Survey of London, Monograph 1, Trinity Hospital, Mile End* (by C R Ashbee), London, 1896.

104 Home, *Of Planting and Planning*, 123; Sutcliffe, *Towards the Planned City*.

105 J Bone, *The Perambulator in Edinburgh*, London, 1911; R L Stevenson, *Edinburgh, Picturesque Notes*, London, 1903, 13, 25, 70: Stevenson also argued that 'To be old is not the same thing as to be picturesque: not because the Old Town bears a strange physiognomy, does it at all follow that the New Town shall look

commonplace. Indeed, apart from antique houses, it is curious how much description could apply commonly to either'.

106 Welter and Lawson, *The City after Patrick Geddes* (Geddes on reconciliation of Art and Science, and influence of Darwin, Le Play and Huxley); M Hall, in Miele, *From William Morris*, 133–4, 145–51.

107 P Geddes, *City Development*, Birmingham, 1904; Armytage, *Heavens Below*, 380; P Geddes, *Scottish Geographical Magazine*, August/October 1919, 294–7.

108 Geddes, *City Development*, 216; Geddes, *Scottish Geographical Magazine*, 1902, 302–12.

109 Geddes, *Scottish Geographical Magazine*, 1919, 281–98; Geddes, *Scottish Geographical Magazine*, 1902, 293; Geddes, *City Development*.

110 J Johnson, L Rosenburg, *Renewing Old Edinburgh*, Glendaruel, 2010; Geddes, *City Development*; J Tyrwhitt, *Patrick Geddes in India*, London, 1947; H Meller, *Planning History Bulletin*, 1987, vol 9, 2; Home, *Planting and Planning*, 147–8; G A S Purves, *An Introduction to the Work of Sir Frank Mears*, Edinburgh College of Art/Heriot-Watt University Research Paper no 4, 1983.

111 Geddes, *City Development*.

Chapter 6

1 E Gentile, *Fascismo di Pietra*, Rome, 2007, 68.

2 L Vale and T Campanella, *The Resilient City*, Oxford, 2005; M J Bowden, 'Reconstruction Following Catastrophe', *Proceedings of the Association of American Geographers*, 2, 1970, 22–6.

3 Karl Woermann, in M F Fischer (ed), *Wie lange dauern die Werke?* Munich, 1990, 36.

4 C Forrest, *International Law and the Protection of Cultural Heritage*, London, 2010, 63–73; D Schindler and J Toman, *The Laws of Armed Conflicts*, Leiden, 2004, 21 and 41; Jokilehto, *History*, 282; Lambourne, *War Damage*, 28–36; Martin and Krautzberger, *Handbuch Denkmalschutz*, 18–19; Sutcliffe, *Towards the Planned City*, 170; *Bulletin de la Societé des Amis des Monuments Parisiens*, 1889, 53–4, 91; J Blower, 'Max Dvořák and Austrian Denkmalpflege at War', *Journal of Art Historiography*, December 2009, 2.

5 K Baedeker, *Belgium and Holland*, Leipzig, 1910, 43–4. The 'Eastern Front' experienced far less architectural destruction than in World War II, with only isolated set-piece reconstructions required, such as the 1915 Polish plan for the rebuilding of Kalisz: Ewa Zarębska, 'Kalisz', in (ed.) Wojciech Kalinowski, *Zabytki i Urbanistyky i Architektury. Odbudowa i Konserwacja*, vol. 1, Warsaw, 1986.

6 H Stynen, G Charlier, A Buellens, *Het verwoeste gewest 15/18: Mission Dhuicque*, Bruges, 1985. Leuven university library destructions in World War I and II and controversy over responsibility: N Lambourne, *War Damage in Western Europe*, Edinburgh, 2001, 24–7.

7 D Dendooven and J Dewilde, *The Reconstruction of Ieper*, 1999, 7–9.

8 R Stummer, *Cornerstone*, vol 28, 4, 2007, 34–9.

9 H Clout, 'The Great Reconstruction', *Planning Perspectives*, January 2005, 1–33; O Rigaud, *Reims à l'époque de l'Art Déco*, Paris, 2006; Kurmann, *Restaurierung*, 16.

10 N Lambourne, *Art History*, vol 22, September 1999, 347–63; C Corvisier, *Le Chateau de Coucy*, Paris, 2009; S Garnero, *Conservazione e restauro in Francia 1919–39*, Florence, 2006, 110.

11 S Muthesius, *Art, Architecture and Design in Poland*, Königstein, 1994, 8; M Briggs, *Goths and Vandals*, London, 1952, 231; P Clemen, *zur 125. Wiederkehr seines Geburtstages*, Cologne, 1991; 'Der Rhein ist mein Schicksal geworden', *Paul Clemen 1866–1947*, Cologne, 1991; *Denkmalpflege in Rheinland-Pfalz*, Mainz, 1991; Clemen also presciently warned that in future wars it might be German cities that suffered destruction from an 'armada' of aeroplanes; Paul Clemen, *Die deutsche Kunst und die Denkmalpflege*, Berlin, 1933.

12 G Goodall, *Whitby Abbey*, London, 2002.

13 J Genet (ed), *The Big House in Ireland, Reality and Representation*, Dingle, 1991, 38–41; M A Bence-Jones, *Guide to Irish Country Houses*, 1988; J Boyer and J Bell, *The Secret Army: The IRA 1916–1979*, Dublin, 1979; D Fitzpatrick, 'The Geography of Irish Nationalism, 1910–1921', in *Past and Present*, 78, 1978, 113–44; P Hart, 'The Geography of Irish Revolution in Ireland, 1917–1923', *Past and Present*, 1997; P Hart, *The IRA and its Enemies: Violence and Community in Cork, 1916–1923*, Oxford, 1999; C Townshend, 'The Irish Republican Army and the Development of Guerrilla Warfare, 1916–21', *English Historical Review*, 94, 1979, 371; C Townshend, *The British Campaign in Ireland, 1919–1921*, Oxford, 1975; C Younger, *Ireland's Civil War*, London, 1970; C Maxwell, *Dublin under the Georges*, 1936; F Aalen, *Man and the Landscape in Ireland*, 1978; Blower, 'Max Dvořák'.

14 Glendinning *et al*, *Scottish Architecture*, 388.

15 Düwel and Gutschow, *Städtebau in Deutschland im 20. Jahrhundert*, Stuttgart, 2001, 64–5.

16 Jokilehto, *History*, 205–7; *Daily Telegraph*, 16 July 1902.

17 V Konerding, in Scheurmann (ed), *Zeitschichten*, 114–19.

18 Dendooven and Dewilde, *Reconstruction of Ieper*, 9–14. Avoidance of individualistic self-projection in conservation: see e.g. H Hesse on 'individuality . . . almost perfectly absorbed in its hierarchic function', *Das Glasperlenspiel*, 1943 (English translation, 1969), 13.

19 Ypres: *GRKG*, 310.

20 Dendooven and Dewilde, *Reconstruction of Ieper*, 9–17; E Vandeweghe, 'Between Public and Private', *Planning Perspectives*, January 2011, 105–18.

21 Ellis, *AR*, January 1993; J Favier (ed), *Reconstructions et Modernisations*, Paris, 1991; Garnero, *Conservazione*, 61–2; A Duménil and P Nivet (eds) *Les reconstructions en Picardie*, Amiens, 2003; P Leon, *La guerre at l'architecture*, Paris, 1918.

22 *GRKG*, 108–10; Garnero, *Conservazione*, 65; Clout, 'Great Reconstruction'; *L'Architecture*, 25 October 1924, 276–82.

23 Clout, 'Great Reconstruction', Garnero, *Conservazione*, 28, 70, 283; F Bercé, *MH*, 189, 1993, 85.

24 Garnero, *Conservazione*, 242.

25 O Rigaud, M Bedarida, *Reims, reconstructions 1920–1930*, Reims, 1988; Clout, 'Great Reconstruction'; Garnero, *Conservazione*, 68.

26 Hall, *Towards World Heritage*, 95–107; A Carmel and A J Eisler, *Der Kaiser reist ins Heilige Land*, Stuttgart, 1999.

27 Paszkiewicz, in *Borders*; Armytage, *Heavens Below*, 402; Welter and Lawson, *The City after Patrick Geddes*; H Meller, *Planning History Bulletin*, 1987, vol 9, 2; H Kendall, *Jerusalem, the City Plan*, London, 1948; W Pullan and M Sternberg, 'The Making of Jerusalem's "Holy Basin"', *Planning Perspectives*, April 2012, 225–48.

28 Forrest, *International Law and the Protection of Cultural Heritage*, 71; S M Titchen, *On the Construction of Outstanding Universal Value (*unpublished PhD thesis, Australian National University), Canberra, 1995, 35.

29 Jokilehto, *History*, 187–91; Institut de cooperation intellectuelle (ICI), *Le conservation des monuments d'art et d'histoire*, Paris, 1933; Garnero, *Conservazione*, 21.

30 Jokilehto, *History*, 187–91; Ministry of Culture, *The Acropolis at Athens*, Athens, 1983, 24; F Choay, *La conference d'Athènes*, Paris, 2002; Falser, Lipp and Tomaszsewski, *Conservation and Preservation*, 111.

31 Craig A Mauzy, *Agora Excavations 1931–2006* (The American School of Classical Studies at Athens), Athens, 2006.

32 Jokilehto, *History*, 284–5; Choay, *Conference d'Athènes*.

33 Vinken, *Zone Heimat*, Berlin, 2010, 119; T Arrhenius, 'Restoration in the Machine Age', *AA Files*, 38, Spring 2000, 10–22; *Le Corbusier: La ville, l'urbanisme*, Paris, 1995, 81–8; O'Connell, 'Afterlives of the Tour St Jacques', 464–8; Choay, *Conference d'Athènes*.

34 Arrhenius, 'Restoration', 10–22.

35 Choay, *Conference d'Athènes*; Garnero, *Conservazione*, 176.

36 London and North-Eastern Railway, 'York, Walled City of Ancient Days', 1930 (poster, National Railway Museum, York).

37 T J Colton, *Moscow, Governing the Socialist City*, London, 1995, 228–33, 260–70, 306; Schmidt (ed), *The Impact of Perestroika on Soviet Law*, Dordrecht, 1990, 348–52; Stubbs and Makaš, *Architectural Preservation*, 272–3; L M Kaganovich, *The Construction of the Subway and the Plan of the City of Moscow*, Moscow, 1934; *Generalnyi plan rekonstruktsii goroda Moskvy*, Moscow, 1936; P Stronski, *Tashkent, Forging a Soviet City*, Pittsburgh, 2010, 41 and 47; *GRKG*, 291–4; P Kann, *Leningrad, a Guide*, Moscow, 1988, 279; *Istoriia SSSR*, no 2, 1966, 205–26; Jokilehto, *History*, 258; J Cracraft, *The Petrine Revolution in Russian Architecture*, Chicago, 1988, 9, 174 ; J H Bater, *The Soviet City: Ideal and Reality*, London, 1980; R A French and F Hamilton (eds), *The Socialist City*, New York, 1979; V Stadnikov and O Fedorov, *Samara, Guide to Modern Architecture*, Moscow, 2006, 31; B A Ruble, *Leningrad: Shaping a Soviet City*, Berkeley, 1990; http://oiru.arceologoa.ru/history30.htm.

38 Lasansky, *Renaissance Perfected*, xxxiii, 3–14.

39 G Giovannoni, in ICI, *Le conservation des monuments d'art*, 60–6; Etlin, *Modernism*, 158, 249; D M Lasansky, 'Urban Editing, Historic Preservation, and Political Rhetoric: The Fascist Redesign of San Gimignano', *JSAH*, September 2004, 340–1; Giambruno, *Per una storia*; E Gentile, *Fascismo di pietra*.

40 Gentile, *Fascismo di pietra*, 29 and 52.

41 D Caldwell and E Caldwell (eds), *Rome: Continuing Encounters between Past and Present*, Farnham, 2011, 101–52; Fried, *Planning the Eternal City*, 31–2. Gentile, *Fascismo di pietra*, 68.

42 A Cederna, *Mussolini urbanista*, Rome, 1979, 23, and illustrations 61–6.

43 Cederna, *Mussolini urbanista*, 23, and illustrations 61–6; S Kostof, *The Third Rome*, Berkeley, CA, 1973; Comune di Roma, *L'invenzione dei Fori imperiali*, Rome, 2008; Gentile, *Fascismo di pietra*, 107; F Castignoli, C Cecchelli, G Giovannoni, M Zucca, *Topografia e urbanistica di Roma*, Rome, 1958, 667, 672; *GRKG*, 276–8.

44 Comune di Roma, *L'invenzione*.

45 Fried, *Planning the Eternal City*, 31; *Town Planning Review*, May 1931, 145–62; Gentile, *Fascismo di pietra*; Woodward, *In Ruins*, 28–31; Lasansky, *Renaissance Perfected*, 83–98; Meyer, *Plundered Past*, 181; D Lowenthal, 'The Heritage Crusade and its Contradictions', in Page and Mason, *Giving Preservation a History*, 37; A Elon, *The Israelis: Founders and Sons*, New York, 1971, 288.

46 Lasansky, 'Urban Editing', 340–2; Lasansky, *Renaissance Perfected*, 142–3.

47 G Giovannoni, *Vecchie città ed edilizia nuova*, Turin, 1931; Choay, *Invention*, 131–5; Lasansky,'Urban Editing', 342; Castignoli *et al*, *Topografia*, 668–74; S Casiello (ed), *La cultura del restauro*, Venice, 205, 293, 314.

48 Giambruno, *Per una storia*, 133; Lasansky, 'Urban Editing', 340–1; Lasansky, *Renaissance Perfected*, 111–37, 160.

49 Choay, *Invention*, 221; Giambruno, *Per una storia*, 31–8; Irace, 'Le due città', in *Bergamo e il suo territorio*; S Angelini (ed), *Disegni di viaggio di Luigi Angelini*, Bergamo, 1982; W Barbero, G Gambirasio and V Zanella (eds), *Luigi Angelini. Ingegnere architetto, catalogo della mostra*, Milan, 1984.

50 Giambruno, 21–30; Caderna, *Mussolini urbanista*.

51 The 'old' title was *Königliche Generalkonservatorium der Kunstdenkmäle und Altertümer Bayerns*. R Gerwarth, 'The Past in Weimar History', *Contemporary European History*, 15, 1, 2006, 1–22; Hubel, *Denkmalpflege*, 101; Düwel and Gutschow, *Städtebau*, 70; *AR*, March 1980, 68; W Speitkamp, 'Das Erbe der Monarchie', *DKD*, 50, 1992, 10–19.

52 R Koshar, 'Building Pasts', in J Gillis (ed), *Commemorations*, 223; Clemen, *Die deutsche Kunst*, 1933, viii; J Hagen, 'Historic Preservation in Nazi Germany', *Journal of Historical Geography*, October 2009, 690–715.

53 Düwel and Gutschow, *Städtebau*, 108; T Scheck, *Denkmalpflege und Diktatur im Deutschen Reich zur Zeit des Nationalsozialismus*, Berlin, 1995; S Fleischner, *'Schöpferische Denkmalpflege', Kulturideologie der Nationalsozialismus und Positionen der Denkmalpflege*, Berlin, 1999.

54 E Lindvai-Dirksen, *Reichsautobahn, Mensch und Werk*, Berlin, 1937. Vinken, *Zone Heimat*, 154–5. B Pusback, *Stadt als Heimat*, Cologne, 2006, 15; Düwel and Gutschow, *Städtebau*, 108–9, 112–4.

55 J Tietz, *Das Tannenberg-Nationaldenkmal*, Berlin, 1999. The monument was designed by architects W and J Krüger.

56 See for example http://www.bundesarchiv.de/oeffentlichkeitsarbeit/bilder_dokumente/01831/index-1.html.de.

57 B Pusback, *Stadt als Heimat*, Cologne, 2006, 52–66.

58 G Vinken, *Zone Heimat*, Berlin, 2010, 123–201, 249; *DKD*, 1935, 105–10; Pusback, *Stadt als Heimat*, 121–4, 151–6; Düwel and Gutschow, *Städtebau*, 106.

59 H G Hiller von Gärtringen, *Sturm auf das Stadtbild*, in Baldrick and Clay (eds), *Iconoclasm*, 207, 221; http://de.wikipedia.org/wiki/Entstuckung. A Wiese, 'Entschandelung und Gestaltung als Prinzipien nationalsozialistischer Baupropaganda', *DD*, 1/2011, 34–41; M Dorn, *Die Semlowerstrasse in Stralsund*, Berlin, 1940; H Schulz, *Altstadtsanierung in Kassel*, Kassel, 1983; U von Pelz, *Stadtsanierung im Dritten Reich*, Dortmund, 1994.

60 Pusback, *Stadt als Heimat*, 225–68; Vinken, *Zone Heimat*, 150.

61 Düwel and Gutschow, *Städtebau*, 92–3; Pusback, *Stadt als Heimat*, 166–72; M Diefenbacher (ed), *Bauen in Nürnberg 1933–1945*, Nuremberg, 1995; Brockmann, *Nuremberg*; W Luebbeke, in Scheurmann, *Zeitschichten*, 132–7; *Mythos Burg*, 364–5; R Kuhn, *Past Finder: Nürnberg*, Düsseldorf, 2010, 20–1; B Friedel and G Grossmann, *Nuremberg Imperial Palace*, Regensburg, 2006; J Glatz, 'Rudolf Esterer', in Scheurmann, *Zeitschichten*, 132–7; B Friedel, 'Der Wiederaufbau der Nürnberger Burg', in M Bauernfeind and R Käs

(eds), *Wiederaufbau in Nürnberg*, Nuremberg, 2009, 24–211; H Berger, *DKD*, 1989, 1, 92–7; Hubel, *Denkmapflege*, 106, 110.

62 W Schlemmer, 'Ein Modell der Nürnberger Altstadt', *Nürnberger Schau*, 1/2, Feb 1940.

63 W Lindner and E Böckler, *Die Stadt, ihre Pflege und Gestaltung*, Munich, November 1939.

64 Hagen, *Preservation*, 185–215.

65 A M Cosme, *AR*, December 1991, 24–55.

66 C D Eby, *The Siege of the Alcazar*, New York, 1965; J Esteban-Chapapria, 'The Conservation of Spain's Architectural Heritage', *Future Anterior*, Winter 2008, 35–53; Stubbs and Makaš, *Architectural Conservation*, 94–5: the parador programme was imitated in Portugal – 'pousadas'.

67 S Muthesius, *Art, Architecture and Design*, 9; *DD*, 1–2005, 97; Piotr Paszkiewicz, 'The Russian Orthodox Cathedral of Saint Alexander Nevsky in Warsaw', *Polish Art Studies*, 1992; T Zarebska, in R Kain, *Planning for Conservation*, 75–96. E Zarebska, 'Kalisz', in W Kalinowski (ed.), Zabytki i Urbanistyky i Architektury, vol 1, Warsaw, 1986. 'Functional Warsaw' proposals of 1934 for rebuilding of the squalid and chaotic capital by Jan Chmielewski and Szymon Syrkus: B Malisz, *Planning Perspectives*, 2–1987, 254–69.

68 W F J der Uyl, *Amersfoorts Luisterrijk Stadsherstel*, Amersfoort 1953.

69 www.samfundetsterik.se/foreningen/historik/index.php; F Aalund, 'Physical Planning and Preservation of Historic Buildings and Settlements in Denmark', *Interplan*, 14, October 1996 (www.tegnestuen–raadvad.dk/Cultural%20heritage%20planning%20in%20DK.pdf); S de Boer, 'Diffusion or Diversity in Cultural Heritage Preservation?', *Institutional Dynamics in Environmental Governance*, 2006, 161–81. Brazil: Stubbs and Makaš, *Architectural Conservation*, 598.

70 H Albright, *The Birth of the National Park Service*, Salt Lake City, UT, 1985; S Fox, *John Muir and his Legacy*, Boston, 1981; A Runte, *National Parks, The American Experience*, Lincoln, NE, 1987; G R Willey and J A Sabloff, *A History of American Archeology*, London, 1974; *Monumentum*, vol 2, 1968; Murtagh, *Keeping Time*, 33–5; C H Wyrick, 'Stratford and the Lees', *JSAH*, 1971, 71; E Axelrod (ed), *The Colonial Revival in America*, New York, 1985; W B Rhoads, *The Colonial Revival*, New York, 1977; S Chamberlain, *A Small House in the Sun*, New York, 1936.

71 Murtagh, *Keeping Time*, 35–7, 96–102; *GRKG*, 135. Ford and history: interview in *Chicago Tribune*, May 25th, 1916.

72 *GRKG*, 135; A L Koch and H Dearstyne, *Colonial Williamsburg*, Williamsburg, 1949; M Whiffen, *The Public Buildings of Williamsburg*, Williamsburg, 1958; R Handler and A Gable, *The New History in an Old Museum: Creating the Past at Colonial Williamsburg*, Durham, NC, 1997; Colonial Williamsburg Inc, *The Official Guidebook of Colonial Williamsburg*, Williamsburg, 1951, 7; A Downing, *JSAH*, 1975, 290; W A R Goodwin, 'The Far-visioned Generosity of Mr Rockefeller' (1930), reproduced in *Colonial Williamsburg Journal*, Winter 2000–1; Murtagh, *Keeping Time*, 33–7, 62–3, 95–6, 99–102; C B Hosmer, *Preservation Comes of Age: From Williamsburg to the National Trust*, Charlottesville, 1981; Lindgren, 'A Spirit that Fires the Imagination', 121–3; Ivor N Hume superintended archaeological work on the site.

73 *GRKG*, 135, 274–6; F D Nicholls, *JSAH*, 1959, 163.

74 C Lounsbury, *JSAH*, 1990, 373–89.

75 *Monumentum*, vol 13, 1976; Murtagh, *Keeping Time*, 36–7.

76 R R Weyeneth, 'Ancestral Architecture', in Page and Mason, *Giving Preservation a History*, 257–81; R N Cote, *City of Heroes, the Great Charleston Earthquake of 1886*, Mount Pleasant, SC, 2006; S E Yuhl, *A Golden Haze of Memory, the Making of Historic Charleston*, Chapel Hill, 2005; Margaret Mitchell, *Gone With the Wind*, Macmillan, 1936; Quigley, *Shifting Grounds*; R W Richardson, *Historic Districts of America: The South*, Bowie, MD, 1987.

77 Yuhl, *Golden Haze*, 43; M D McInnis, *The Politics of Taste in Antebellum Charleston*, Chapel Hill, 2005; A J Stannis, *Creating the Big Easy*, Athens, GA, 2006; N R Walker, 'Savannah's Lost Squares', *JSAH*, December 2011, 512–31.

78 Wayland Kennet, *Preservation*, London, 1972, 40; *First Report of the Royal Fine Art Commission for Scotland*, Edinburgh, 1960, 5. The RFAC in England and Wales was established at the instigation of the Ministry of Works' Permanent Secretary, Lionel Earle: A J Youngson, *Urban Development and the Royal Fine Art Commissions*, Edinburgh, 1990; S A Kohler, *The Commission of the Arts, A Brief History 1910–1971*.

79 R Hurd, 'Design for Today', in *Scotland 1938*, Glasgow, 1938, 120–7; R W Finn, *Scottish Heritage*, London, 1938; Sir John Stirling Maxwell, *Shrines and Homes of Scotland*, London, 1937; Glendinning *et al*, *Scottish Architecture*, ch 8. In 1938, the young architect Robert Hurd, author of several Edinburgh Old Town

conservation schemes, defined the essence of 'traditional' architecture: a 'free yet basically Scottish character . . . bold simplicity threaded by an odd streak of vanity'.

80 Lord Bute, *A Plea for Scotland's Heritage* (National Trust for Scotland), Edinburgh, May 1936.

81 D Walker, *Transactions of the Ancient Monuments Society*, vol 38, 1994, 33–7; R Hurd, *Scotland Under Trust*, London, 1939; R J Prentice, *Conserve and Provide*, Edinburgh, c.1971; A Pittaway, 'A National Awakening', *St Andrews Studies*, III, St Andrews, 1993; R G Cant, 'The Founding of the St Andrews Preservation Trust, 1937–8', in St Andrews Preservation Trust, *Three Decades of Historic Notes*, St Andrews, 1991; *History Scotland Magazine*, April 2005; C McWilliam, 'Ian Lindsay', *EAA Yearbook*, 1967; C Sinclair, *An Clachan, Official Guide*, Glasgow, 1938.

82 Charlotte Square protection: the Edinburgh (Charlotte Square) Scheme Order 1930, *Scotsman*, 11 January 1935, 124; *Scotsman*, 6 February 1935, 10; *Scotsman*, 30 March 1935, 14; R Hurd, *The Face of Modern Edinburgh*, Edinburgh, 1939; G A S Powers, *Sir Frank Mears*, Edinburgh, 1983; Royal Commission on the Ancient Monuments of Scotland, file NMRS MSS.60, Scottish Town Survey, September 1936; Johnson and Rosenburg, *Renewing Old Edinburgh*; A Gale, *Scotland in Trust*, Spring 2006, 228.

83 C W Ellis, *Portmeirion, The Place and its Meaning*, Portmeirion, 1963; C W Ellis, *Portmeirion Still Further Illustrated and Explained*, Portmeirion, 1966; *Cornerstone*, vol 26, 2, 2005, 32ff.

84 J B Priestley, *English Journey*, London, 1934; A Mee, *Enchanted Land* (King's England Series), London, 1936; B Hillier, *John Betjeman, The Biography*, London, 2006.

85 *Architects' Journal*, 31 August 1982, 35.

86 Big city slum redevelopments: see e.g. L H Keay, *RIBA Journal*, 23 November 1935, 57–75; *RIBA Journal*, 22 February 1936, 393; *Transactions of the Ancient Monuments Society*, new series, vol 20, 1975, 49–84. Civic sector's expansion: 1917, for example, saw the founding of the National Federation of Women's Institutes.

87 Cant, 'Founding of the St Andrews Preservation Trust'.

88 R Byron, *AR*, May 1937, 217–24; E Beresford Chancellor, *Disappearing London*, London, 1927; *Architects' Journal*, 31 August 1982; *AR*, November 1970, 308.

89 Robert Byron, *How We Celebrate The Coronation: A Word To London's Visitors*, London, 1937.

90 *AJ*, 31 August 1982.

91 J Taylor, *A Dream of England*, Manchester, 1994; C Madge, 'Anthropology at Home', *The New Statesman and Nation*, 30 January 1937. For the link to the growth of aerial photography and archaeology of English landscapes, see also K Hauser, *Bloody Old Britain*, New York, 2008.

92 M. Lock, *Civic Diagnosis, a Blitzed City Analysed*, Hull, 1943; Ministry of Health, *New Houses for Old*, c.1938; M O Ashton, 'Tomorrow Town', in Welter and Lawson, *The City after Patrick Geddes*.

93 W Harvey, *The Preservation of St Paul's Cathedral and Other Famous Buildings: A Textbook on the New Science of Conservation*, London, 1925.

Chapter 7

1 J Düwel and N Gutschow, *Fortgewischt sind alle überflüssigen Zutaten*, Berlin, 2008, 145 and 213; H Beer, *Nürnberger Erinnerungen 14*, Nuremberg, 2005, 152.

2 M Schmidt, *Der Untergang des alten Dresden*, Dresden, 2005, 13; F Taylor, *Dresden: Tuesday 13 February 1945*, London, 2005.

3 Clemen, *Die deutsche Kunst*, 104–5; R Koshar, 'Building Pasts', in Gillis, *Commemorations*, 1994, 273; G Packer, 'Embers', *The New Yorker*, 1 Feb 1990, 32–4; D Irving, *The Destruction of Dresden*, New York, 1963.

4 W Sebald, *On The Natural History of Destruction*, London, 1999; S Lindquist, *A History of Bombing*, London, 2002. Post-1945 accusations: see for example Richard Peter's photograph cycle, e.g. *Dresden – eine Kamera klagt an*, Dresden, 1949.

5 Muthesius, *Poland*, 11, 17.

6 Also bombed in May 1940 was Middelburg in the Netherlands, in which 458 historic buildings were destroyed; *GRKG*, 344–5.

7 Rotterdam and Coventry: Lambourne, *War Damage*, 43–8. German bombing of London arguably provoked by RAF escalation: H W Koch, 'The Strategic Air Offensive against Germany: the Early Phase', *Historical Journal*, 34, 1, 1991, 117–41.

8 *Cornerstone*, vol 29, 1, 2008, 34–9; R Winstone, *Bristol Blitzed*, Bristol, 1973 and 1976, 26.

9   S Sillars, *British Romantic Art and the Second World War*, London, 1991; D Mellor (ed), *A Paradise Lost*, London 1987, 34–9.

10 J M Richards, *The Bombed Buildings of Britain*, London, 1942, 5, 8; Woodward, *In Ruins*, 212–21.

11 P J Larkham, 'Planning for Reconstruction after the Disaster of War: Lessons from England in the 1940s', *Perspectivas Urbanas*, vol 6, 2008, 3–14; *Bombed Churches as War Memorials*, London, 1945; Lambourne, *War Damage*, 182–3; SPAB, *The Treatment of Ancient Buildings in Wartime*, nd.

12 *Cornerstone*, vol 29, 1, 2008, 34–9; Improvements and Town Planning Committee of the Corporation of London, *The City of London, A Record of Destruction and Survival*, London, 1951; *GRKG*, 344–5.

13 M Tarnassia, *Firenze 1944–45, Danni di Guerra*, Livorno, 2007.

14 *St Lorenz* [journal], no 47, July 2002, 34–47; Briggs, *Goths and Vandals*, 232.

15 Kann, *Leningrad*; S Heym, S de Ponte and H Neumann, *Das Cuvilliés-Theater*, Munich, 2008; C Scott-Clark and A Levy, *The Amber Room, The Untold Story of the Greatest Hoax of the Twentieth Century*, London, 2004.

16 Lambourne, *War Damage*, 128–9, 139–49; H Probert, *'Bomber' Harris: His Life and Times*, London, 2006; A C Grayling, *Among the Dead Cities: The History and Moral Legacy of the World War II Bombing of Civilians in Germany and Japan*, New York, 2006; C Webster and N Frankland, *The Strategic Air Offensive Against Germany, 1939–1945*, London, 1961; Sir A Harris, *Bomber Offensive*, London, 1947 (2005 edition, p.52); J Friedrich, *The Fire: the Bombing of Germany, 1940–1945*, Munich, 2002.

17 Harris, *Bomber Offensive*, 52; H Beseler and N Gutschow, *Kriegsschicksale deutscher Architektur*, 2 vols, Wiesbaden, 2000.

18 Lutz Wilde, *Bomber gegen Lübeck*, Lübeck, 1999; Düwel and Gutschow, *Städtebau*, 139–40; Beseler and Gutschow, *Kriegsschicksale*, 10–13; Lambourne, *War Damage*, 52–3.

19 Richards, *Bombed Buildings*, 164; Lambourne, *War Damage*, 53–6, 141–3.

20 L P Lochner (ed, trans), *The Goebbels Diaries 1942–1943*, New York, 1948, entry for 27 April 1942, 189–90.

21 Lambourne, *War Damage*, 56–8, 91–3; *Picture Post*, 4 July 1942, 196; *Bauwelt*, 21–2, 1943, 1–7.

22 Richards, *Bombed Buildings*, 139.

23 *Kriegsschicksale*, vol 1, x; Wilde, *Bomber gegen Lübeck*.

24 M Bauernfeind and R Käs (eds), *Wiederaufbau in Nürnberg*, Nuremberg, 2009; P Heyl, *Toyland: Bomber über Nürnberg*, Nuremberg, 2004; G Althaus, 'Mit Gott unter einem Dach', in *Verein zur Erhaltung der St Lorenzkirche in Nürnberg*, 37, July 1992, 51ff.

25 Lambourne, *War Damage*, 145; Hagen, *Preservation*, 219–225; *Kriegsschicksale*, vol 2, 1469–72.

26 Düwel and Gutschow, *Fortgewischt*, 145 and 213.

27 Beseler and Gutschow, *Kriegsschicksale*, vol 1, x; House of Lords proceedings, 9 Feb 1944, col 737–46.

28 Lambourne, *War Damage*, 104–7; *Völkischer Beobachter*, 24 December 1944, 3; *Graz Tagespost*, 4 April 1944; S Harries, *Nikolaus Pevsner: The Life*, London, 2011, 338; Beseler and Gutschow, *Kriegschicksale*, vol 1, xi. 'Domplombe': P Börger, *Das Moderne Köln*, Cologne, 2006, 115–16; Dehio, *Denkmalschutz und Denkmalpflege*, translation courtesy of Jonathan Blower, 2012.

29 Stubbs and Makaš, *Architectural Conservation*, 260; Muthesius, *Poland*.

30 Kann, *Leningrad*.

31 DD, 1–2005, 98; Lambourne, *War Damage*, 125; *Siena 1939–1945, proteggere l'arte*, Siena, 1945.

32 See e.g. N Ferguson, *The House of Rothschild, The World's Banker, 1849–1998*, New York, 1998; www.kasteeldehaar.nl/uk/0303_uk.htm; J Luijt, *Van Vicus tot Vinex, een geschiedenis van Vleuten, De Meern en Haarzuilens*, Vleuten, 2011; information from castle tour guides, 2009.

33 A B Vihoude, *The Bergenhus Castle after the Explosion on 20 April 1944*, Bergen, 1994, 10–11.

34 Central Commission for Germany/British Committee on the Preservation and Restoration of Works of Art . . . in Enemy Hands, *Works of Art in Germany (British Zone of Occupation)*, London, 1946; Briggs, *Goths and Vandals*, 235; R M Edsel, *The Monuments Men*, Nashville, 2009; Lambourne, *War Damage*, 68–77.

35 Tarnassia, *Firenze*; Garnero, *Conservazione*; S Crane, 'Digging Up the Present in Marseille's Old Port: Toward an Archaeology of Reconstruction', *JSAH*, September 2004, 297–319; S Crane, *Mediterranean Crossroads*, 2011; S Crane, 'Mutable Fragments: Destructive Preservation and the Postwar Rebuilding of Marseille', *Future Anterior*, Summer 2005.

36 Ellis, *AR*, January 1993; *MH*, 189, 1993, 85ff; Wakeman, *Modernising the Provincial City*, 61; Grodecki, 'Denkmalschutzgesetzgebung', 13; Sutcliffe, *Autumn*, 300–7, 310–11; Lambourne, *War Damage*, 146.

471

37 T Hausman, 'Something Old, Something New', DOCOMOMO 2004 Conference, *Proceedings*, 229; Garnero, *Conservazione*, 141–9; B Vital-Durand, *Domaine Privé*, Paris, 1996.

38 G Thum, *Uprooted: How Breslau became Wrocław during the Century of Explusions*, Princeton, NJ, 2011 (German edition, 2003), xxv; Beseler and Gutschow, *Kriegsschicksale*, vol 1, xi. Lambourne, *War Damage*, 146.

39 *Cornerstone*, vol 29, 1, 2008, 34–9; *Builder*, 19 September 1941, 253; Koshar, 'Building Pasts', 223.

40 W H Godfrey, *Our Building Inheritance*, London, 1944, 20; A Reiach and R Hurd, *Building Scotland*, Edinburgh, 1941.

41 Düwel and Gutschow, *Städtebau*, 124–5, 130, 147–8; Ecclesiological Society, *The Continuity of the English Town*, London, 1943, 150–5, 161; T Sharp, *English Panorama*, 1936; J Paton Watson and P Abercrombie, *A Plan for Plymouth*, Plymouth, 1943; *Resurgam*, London, 1944.

42 J M Richards, *The Castles on the Ground*, London, 1946; *Town and Country Planning*, Spring 1947, 37.

43 Hubel, *Denkmalpflege*, 114; Düwel and Gutschow, *Städtebau*, 139–44.

Chapter 8

1 *Builder*, 19 September 1941, 253.

2 W Crofts, *Coercion or Persuasion*, London 1989; 'Art and the State', *Collected Writings of J M Keynes*, vol 28, London, 1982, 345.

3 R Stern, 'The Ruin Film', *AA Files*, 54, Summer 2006; Hagen, *Preservation, Tourism and Nationalism*, 253–4; *DD*, 2/2001, 109–15; Thum, *Uprooted*.

4 Huse, *Denkmalpflege*, 182–7, 193–7; *DD*, 2001, 109ff, 125ff; Falser, *Zwischen Identität und Authentizität*, 71–98. On Cologne devastation, see also Hermann Classen's photo–book, *Gesang im Feuerofen* (Düsseldorf, 1947).

5 Vinken, 'Gegenbild, Traditionsinsel, Sonderzone', 196–7; M R Kelsall and S Harris, *A Future for the Past*, Edinburgh, 1961, 11.

6 L Riall, *Garibaldi*, Yale, 2007, 7.

7 Jokilehto, *History*, 223–37; C D Valsassina, *Restauro – Made in Italy*, Milan, 2006, 19–56. Examples of 19th-century texts on preservation of art works: G S Suardo, *L'arte del restauro*, Milan, 1866; V Schaedler-Saub, 'Teoria e metodologia del restauro', in Falser, Lipp, Tomaszewski, *Conservation and Preservation*, 81–96.

8 S Casiello, *La cultura del restauro*, 339–56; Jokilehto, *History*, 231; F Meraz, 'Brandi's Theory of Restoration', DOCOMOMO 2008 Conference, 16 September 2008; Stubbs and Makaš, *Architectural Conservation*, 17.

9 *ICCROM Newsletter*, 33, June 2007, 8; Jokilehto, *History*, 227; R Bonelli, *Architettura e restauro*, Venice, 1959.

10 *Pasticcio*: Wikipedia: http://en.wikipedia.org/wiki/pastiche; J Fouriasté, *Les trentes glorieuses ou la révolution invisible de 1946 à 1975*, Paris, 1979.

11 L Verpoest, 'Raymond M Lemaire', in T Patricio, K van Balen and K de Jonge, *Conservation in Changing Societies*, Leuven, 2006, 19–26; *Monumentum*, vol 26, 2, 1983, 85–90; C Perier d'Ieteren, in *Geschichte der Restaurierung*, vol 1, 17; *Mythos Burg*, 285.

12 R Tubbs, *Living in Cities*, Harmondsworth, 1942, 32–3.

13 *Arquitectura popular em Portugal*, 1961; P C Goodwin, *Brazil Builds, Architecture New and Old, 1652–1942*, New York, 1943.

14 Interview with D Insall, 2006.

15 *Guardian*, 21 November 2008, 46; F Tindall, *Memoirs and Confessions of a County Planning Officer*, Ford, 1998; interview with F Tindall, 20 May 1992; *Country Life*, 10 August 1972, 318.

16 *Builder*, 23 February 1962, 401; Matthew 1964 lecture: Glendinning, *Modern Architect*, 282; C McWilliam, 'A Sentimental Subject', *EAA Year Book*, 10, 1966.

17 H Atkinson, 'A New Picturesque?', *Edinburgh Architecture Research*, 131, 2008, 26–37; B Appleyard, *The Pleasures of Peace*, London, 1989; N Pevsner, *AR*, December 1944, 139; 'Norwich', *Builder*, 11 August 1961, 245–6.

18 C McWilliam, *Scotsman*, 11 July 1968; K D Murphy, *JSAH*, March 2002, 68–89.

19 Casiello, *Cultura del restauro*, 329–56; Giambruno, *Per una storia*, chs 8–13; *GRKG*, 236–7; M Casciato, 'Neo-Realism in Italian Architecture', in S W Goldhagen and R Legault, *Anxious Modernisms*; D Y Ghirardo, *Italy: Modern Architectures in History*, London, 2011.

20 S Los, *Carlo Scarpa, an Architectural Guide*, Verona, 2004; A C Schultz, *Carlo Scarpa – Layers*, Fellbach, 2007; DOCOMOMO Italia Journal, special issue, *Franco Albini in Genova*, 2004.

21 Pane: Casiello, *Cultura del restauro*, 357–70; S Ascari, in Giambruno, *Per una storia*, 121–30.

22 Falser, *Zwischen Identität und Authentizität*, 74–7; *Revolution und Tradition: 150 Jahre Freies Deutsches Hochstift*, catalogue to exhibition at Goethehaus Museum, November 2009 to February 2010; Huse, *Denkmalpflege*, 186 and 198–201; W Dirks, *Mut zum Abschied*, Frankfurt, 1947; Von Buttlar *et al*, *Denkmalpflege Statt Attrappenkult*, 163.

23 R Dölling (ed), *The Conservation of Historical Monuments in the Federal Republic of Germany*, Munich, 1974, 31ff; *GRKG*, 351–3.

24 J Paul, *Deutsche Kunst und Denkmalpflege*, 1980, 1–2, 64–76; *GRKG*, 319.

25 Hubel, *Denkmalpflege*, 122; Huse, *Denkmalpflege*, 187; *Report of the Select Committee on the House of Commons (Rebuilding)*, London, 1944; Lambourne, *War Damage*, 96, 175; G D Rosenfeld, *Munich and Memory*, Berkeley, CA, 2000, 119–22.

26 L. Campbell, *Coventry Cathedral*, Oxford, 1998.

27 Egon Eiermann: Annemarie Jaeggi (ed), *Egon Eiermann (1904–1970). Die Kontinuität der Moderne* (catalogue to exhibition, Karlsruhe and Berlin), 2004; Lambourne, *War Damage*, 180–1. Cagliari: A Sanna, 'Reinforced Concrete and Limestone', *Proceedings of the 3ʳᵈ International Congress on Construction History*, 2009.

28 S Crane, 'Mutable Fragments', *Future Anterior*, Summer 2005, 1–10; P Dieudonné (ed), *Villes reconstruites du dessin au destin* (2 vols), Paris, 1994; J M Diefendorf (ed), *Rebuilding Europe's Bombed Cities*, New York, 1990; S Kopp, F Boucher and D Pauly (eds), *L'Architecture de la reconstruction en France*, Paris, 1982; J–P Azema *et al* (eds), *Reconstructions et modernization, la France après les ruines*, Paris, 1991; D Voldman, *La Reconstruction des villes françaises de 1940 à 1944*, Paris, 1997.

29 P Longuet, 'The Reconstruction of Dunkirk', *DOCOMOMO 2004 Conference Proceedings*, 211–17; *GRKG*, 344–5.

30 M Burkunk, *40 Jaar: 1956–1966*, Amsterdam, 1996; J Brinkgreve, *Alarm in Amsterdam*, Amsterdam, 1956; *Bouw*, 49, 5 December 1964; Dobby, *Conservation and Planning*, 77–8; Tillema, *Schetsen uit de Geschiedenis van de Monumentenzorg in Nederland*, The Hague, 1975; Amsterdam City Planning Department, *Amsterdam: Planning and Development*, 1983, 58; R Roegholt (ed), *Levend Amsterdam*, Amsterdam, 1987.

31 D Chapman, *Planning History*, vol. 21, 3, 1999; A Harrison and R Hubbard, *Outline Plan for Valletta and the Three Cities*, 1945; Dölling, *Conservation of Historic Monuments*; Vinken, *Stadt als Heimat*, 123–201; J M Diefendorf, 'Reconstruction Law and Building Law in Postwar Germany', *Planning Perspectives*, 1, 1986, 107–29; Huse, *Denkmalpflege*, 188; Düwel and Gutschow, *Städtebau*, 149–52; N Gutschow, *DKD*, 1980, 1–2, 41–9; Hubel, *Denkmalpflege*, 120; Bauernfeind and Käs, *Wiederaufbau in Nürnberg*, 214–21; Rosenfeld, *Munich and Memory*; Heym, de Ponte and Neumann, *Cuvilliés-Theater*; S Glatz, 'Rudolf Esterer', in Scheurmann, *Zeitschichten*, 142–7; *GRKG*, 168–77, 315.

32 Bauernfeind and Käs, *Wiederaufbau in Nürnberg*; Museen der Stadt Nürnberg, *Nürnberg Baut Auf!*, Nuremberg, 2009, transcript of speech by O. Schneider at opening of Nürnberg Baut Auf! Exhibition, Fembo-Haus, Nuremberg, 21 January 2009; Huse, *Denkmalpflege*, 190–3; E Kusch, *The Immortal Nuremberg*, Nuremberg, 1953; *DD*, 2007, 2, 164–5; *DD*, 1/2010, 79–81; Hagen, *Preservation, Tourism and Nationalism*, 225–35 and 260; *GRKG*, 348–9; Hubel, *Denkmalpflege*, 120; Düwel and Gutschow, *Städtebau*, 152.

33 Wilde, *Bomber gegen Lübeck*; Dumouchel, 'Federal Republic of Germany, West German Improvement Programmes'; J Petzsch, *Deutsche Volkszeitung*, 3 October 1974, 8.

34 Kennet, *Preservation*, 200; J Gould, *AR*, March 2007; J P Watson and P Abercrombie, *A Plan for Plymouth*, Plymouth, 1943; English Heritage, *Conservation Bulletin*, 56, 24–5; S Essex and M Brayshay, 'Boldness Diminished', *Urban History*, vol 35, 3, 437–61; M Brayshay (ed), *Post-war Plymouth*, Plymouth, 1983, 22–33; J M Richards, *The Bombed Buildings of Britain*, London, 1947 (enlarged edition); Architectural Press, *Bombed Churches as War Memorials*, Cheam, 1945; Lambourne, *War Damage*, 179–84; N Tiratsoo, 'The Reconstruction of Blitzed British Cities', *Contemporary British History*, Spring 2000, 27–44 .

35 T Sharp, *Cathedral City*, London, 1944; T Sharp, *Exeter Phoenix*, London 1946, 28, 35, 90, 98; P Abercrombie and R Nickson, *Warwick*, London, 1949; Corporation of Coventry, *The Future of Coventry*, 1944; T Sharp, *Oxford Replanned*, London, 1948; Canterbury City Council, *Canterbury – Conservation Area Appraisal*, 2010.

36 English Heritage, *Conservation Bulletin*, 56, 24–5; *GRKG*, 442; *Transactions of the Ancient Monuments Society*, new series, vol 30, 1986, 224; *The City Churches, Final Report of the Bishop of London's Commission*, London, 1946.

37 L Esher, *A Broken Wave*, London, 1981, 102–3, 114–19; Corporation of London, *Report . . . on Postwar Reconstruction in the City of London*, London, 1944.

38 As early as 1946, one architect commentator argued that, while mass production was essential, amenities must also be conserved: G Barry, *Builder*, 5 July 1946, 6; I B Whyte (ed), *Man-Made Future*, London, 2007.

39 See e.g. V G Childe, *Skara Brae, Orkney, Official Guide*, Edinburgh, 1933/1950.

40 *Independent*, 20 November 2008; *Guardian*, 21 November 2008, 46; D Mitchell, *Arup Journal*, May 1968 and July 1968; *Builder*, 22 January 1965, 257; D Insall, *Thaxted, An Historical and Architectural Survey for the County Council of Essex*, London, 1966; interview with D Insall, 2006.

41 M L Catoni (ed), *Il patrimonio culturale in Francia*, Milan, 2007; A Dale, 'Listing and Preserving', *AR*, August 1965, 100; A Dale *et al*, *Historic Preservation in Foreign Countries*, Washington, DC, 1983, 9–36; M F Sorlin, 'The French System for Conservation', in P Ward (ed), *Conservation and Development*, Newcastle, 1968; A Dobby, *Conservation and Planning*, London, 1978, 72–4; Wakeman, *Modernising the Provincial City*, 75–6, 88–9, 114.

42 Dale, 'Listing and Preserving', 101–2.

43 Dale, 'Listing and Preserving', 102–3.

44 Kennet, *Preservation*, 86; J Harvey, 'The Origin of Listed Buildings', *Transactions of the Ancient Monuments Society*, 37, 1993, 3–12.

45 *RIBA Journal*, December 1945, 59; Harvey, 'Origins'; A Acworth and A Wagner, '25 Years of Listing', *AR*, November 1970, 308–10.

46 *Transactions of the Ancient Monuments Society*, vol 38, 1994, 197–8; *Architects' Journal*, 31 March 1982, 39; *Builder*, 28 July 1961.

47 Kennet, *Preservation*, 43–4; F Kelsall, 'Not as Ugly as Stonehenge', *Architectural History*, 52, 2009, 1–30; A Saint, 'How Listing Happened', in M Hunter (ed), *Preserving the Past*, Stroud, 1996, 115–33.

48 National Trust for Scotland, *Newsletter*, 14 April 1956, 10; Civic Trust Circular, 'Joint Action for Conservation', January 1969; Dobby, *Conservation and Planning*, 77–9; Dale, 'Listing and Preserving', 103. For 1961 *Monumentenwet*, etc.: J A C Tillema, *Schetsen uit de Geschiedenis van de Monumentenzorg in Nederland*, The Hague, 1975.

49 Ministry of Town and Country Planning, *Report of the Second Annual Conference of Investigators*, Oxford, 27 March 1949.

50 National Trust: Murtagh, *Keeping Time*, 39–50; D E Finley, *History of the National Trust for Historic Preservation*, Washington, DC, 1965; E D Mulloy, *The History of the National Trust for Historic Preservation, 1963–1973*, Washington, DC, 1976.

51 L Huxtable, *The Architecture of New York*, vol 1, 1964; N Silber, *Lost New York*, New York, 1967; M Sawin (ed), *James Marston Fitch, Selected Writings*, New York, 2007; B Mackintosh, *The Historic Sites Survey*, Washington, DC, 1985; R Mason, *The Once and Future New York*, Minneapolis, MN, 2009.

52 *With Heritage so Rich: A Report of a Special Committee on Historic Preservation*, New York, 1965; Advisory Council on Historic Preservation, *The National Historic Preservation Act of 1966*, Washington, DC, 1986; S K Blumenthal (ed), *Federal Historic Preservation Laws*, Washington, DC, 1990; B Chittenden, *A Profile of the National Register of Historic Places*, Washington, DC, 1984; Dobby, *Conservation and Planning*, 81–5; Dale, 'Listing and Preserving', 277–9; Murtagh, *Keeping Time*, 62–7; Silber, *Lost New York*; *Monumentum*, 12, 1976; Irwin Unger, *The Best of Intentions: the Triumphs and Failures of the Great Society under Kennedy, Johnson, and Nixon*, Doubleday, 1996; R R Garvery and T B Morton, *The United States Government in Historic Preservation*, Washington, DC, 1973; R M Greenberg and S A Marusin (eds), National Register of Historic Places, vols 1, 2, Washington, DC, 1976 and 1979.

53 A Gowans, *JSAH*, 1965, 252–3; National Trust for Historic Preservation, *Reusing Old Buildings*, Washington, DC, 1984; R Tubesing, *Architectural Preservation in the United States, 1941–1975, a Bibliography of Federal, State and Local Government Publications*, New York, 1978; National Trust for Historic Preservation, *Historic Preservation Tomorrow*, Charlottesville, VA, 1967.

54 Stubbs and Makaš, *Architectural Conservation*, 533; A Falkner, '*Without Our Past?*', Toronto 1977; Heritage Canada, *Annual Report 1973–74*, Ottawa 1974; Dale, 'Listing and Preserving', 278–9; A Raines, 'Pursuits of Deeper Purpose: the Reconstruction of Fortress Louisbourg', in *Mirror of Modernity, Docomomo*

E–Proceedings 2, December 2009, 29–34; E d'Orgeix, 'Twentieth Century Architecture in Quebec City', in *Mirror of Modernity*, 23–8.

55 Murtagh, *Keeping Time*, 77.

56 Ethan Carr, *Mission 66, Modernism and the National Park Dilemma*, Amherst, 2007; F Tilden, *Interpreting our Heritage*, Chapel Hill, NC, 1957, 34; H M Albright *et al*, *National Park Service*, Las Vegas, 1987.

57 *Monumentum*, vol 12, 1976.

58 Raines, 'Fortress Louisbourg'; T MacLean, *Louisbourg Heritage*, Sydney, Nova Scotia, 1995.

59 S Mannheim, *Walt Disney and the Quest for Community*, Farnham, 2002; B Dunlop, *Building a Dream: the Art of Disney Architecture*, New York, 1996, 25–33 and 117–23; M Kelleher, 'Images of the Past: Historical Authenticity and Inauthenticity from Disney to Times Square', *CRM*, Summer 2004, 6–19; K A Marling, *The Architecture of Reassurance: Designing the Disney Theme Parks*, New York, 1997; J A Adams, *The American Amusement Park Industry*, Boston, 1991; A Bryman, *Disney and his Worlds*, London, 1995.

60 Bannockburn: Glendinning, *Modern Architect*, 290–1; *Scotsman*, 19 April 2000, 18; R J Prentice, *Conserve and Provide*, Edinburgh, 1971; in 2010, the complex was, in turn, 'listed' by the English government: www.english–heritage.org.uk/about/news/the–shakespeare–centre–is–listed.

61 B Alnquist, *The Vasa Saga*, Stockholm, 1965; M Geijer, 'Objectivity and Aesthetics', *Future Anterior*, Winter 2008, 21–31.

62 Glendinning, *Modern Architect*, 130–1.

63 *Imperial Calendars*, 1966–1973; J Gibson, *The Thistle and the Crown*, Edinburgh, 1985; R Samuel, *Guardian*, 4 October 1984, 16; Sir W Holford, *The Built Environment*, Tavistock Pamphlet 11, London, 1965.

64 Kennet, *Preservation*, 75: Kennet and his wife Elizabeth had in 1956 written a book on *Old London Churches*.

65 Kennet, *Preservation*, 71; *Builder*, 15 September 1961, 493; *Builder*, 19 April 1968, 102; *Builder*, 1 November 1968, 69; Boulting, 'The Law's Delays', 29–33; Ministry of Housing and Local Government, Circular 61/1968.

66 In England, among architects and planners, the 1940s witnessed an upsurge in anti-rural rhetoric: the small-town plans by Thomas Sharp insisted that the English were a nation of town-dwellers, and another author, Thomas Burke, claimed that, 'when England is at war, her men are not fighting for England as a whole. Many of them are fighting for an English village, but far greater numbers are fighting for a street in an English town. For the townsman, there is a distinct dearness in the High Street, a secret sweetness in the old bus stop, as strong as the countryman finds in his hill and his stream': Thomas Burke, *The English Townsman*, London, 1946.

67 C W Ellis, *Britain and the Beast*, London, 1938; I Nairn, *Your England Revisited*, London, 1964; I Nairn, *Outrage*, London, 1955; *AR*, December 1956; I Nairn, *Britain's Changing Towns*, London, 1967, 53; *The Times*, 18 August 1983; *AR*, December 1975, 328; Architectural Press, *Dartmoor, Building in the National Parks*, London, 1955.

68 Tindall, *Memoirs and Confessions*; interview with F Tindall, 1992; interview with J Reid, 1987; D Walker, *Country Life*, 10 August 1972, 318; Insall & Associates, *Thaxted*, 23–4: the Thaxted plan did include some echoes of the old Anti–Scrape rhetoric, notably in its criticism of 'bogus' neo-Georgian new buildings and 'drab' council housing.

69 The Gowers Committee argued, in the words of Christopher Hussey, that the houses were a 'living element in the social fabric of the nation, uniting visibly the present with national history': *Houses of Outstanding Historic or Architectural Interest, Report*, London, 1950; Kennet, *Preservation*, 46.

70 Duke of Bedford and G Mikes, *How to Run a Stately Home*, London, 1971; H Montgomery-Massingberd, C S Sykes, *Great Houses of England and Wales*, London, 1994, 155–6.

71 *International Federation for Housing and Town Planning, Newsletter*, November 1954.

72 C Buchanan, *Traffic in Towns* (shortened edition), London, 1963, 104–10, 243; Esher, *Broken Wave*, 52–4; G A Jellicoe, *Motopia*, London, 1961.

73 J M Richards, 'The Failure of the New Towns', *AR*, July 1953; RIBA, Symposium on 'The Living Town', 22 May 1959; P J Marshall, *Rebuilding Cities*, London, 1966; *Daily Telegraph*, 17 July 1993, 17; J Gold, 'A SPUR to action? The Society for the Promotion of Urban Renewal, "anti–scatter" and the crisis of city reconstruction, 1957–1963', *Planning Perspectives*, April 2012, 199–224.

74 Alison Smithson (ed), *Team 10 Primer*, London, 1965; Silver, *Lost New York*, 17–19; I de Wolfe, *The Italian Townscape*, London, 1963; G Cullen, *Townscape*, London, 1961; I de Wofle, *Civilia*, London, 1971; Taylor Woodrow, *Urban Renewal: Fulham Study*, London, 1964; T Crosby, *Architecture, City Sense*, London, 1965;

Esher, *RIBA Journal*, 1 November 1966; *RIBA Journal*, August 1960, 388–403. For example, *Planning Bulletin* [PB] *1, Town Centres: Approach to Renewal* (1962), insisted on a balance of comprehensiveness and preservation of each town's '*character*'; and PB4, *Town Centres – Current Practice*, of 1963, combined advocacy of multi-level traffic infrastructure with insistence that historic towns must not be 'torn apart'.

75   National Archives, file PRO HLG 117/24 (on CHAC 1964); J Yelling, *Planning Perspectives*, 14, 1, 1999, 1–18; *Architects' Journal*, April 1971, cover; Barbican Art Gallery, *Barbican: This was Tomorrow*, Gallery Guide, 2002; J Tetlow and A Goss, *Homes, Towns and Traffic*, London, 1965, 172–3; Department of the Environment, *New Housing in a Cleared Area*, London 1971; J Pendlebury, 'The Postwar Rise of Conservation in England', in *Mirror of Modernity*, 42–3.

76   National Archives, file PRO HLG 118/203, correspondence of May 1961, April 1963, April 1965: for example, in 1961/5 correspondence between MHLG's head civil servant, Dame Evelyn Sharp, and colleagues regarding ongoing urban redevelopments in Coventry (at Spon End and Hillfields), Sharp cautioned that 'we are not yet ready nationally to move on to large-scale urban renewal', and thus 'Coventry must consider improvement as well as clearance'.

77   W Burns, *New Towns from Old*, London, 1963; J Holliday (ed), *City Centre Redevelopment*, London, 1973; J Pendlebury, 'Alias Smith and Burns', *Planning Perspectives*, 16, 2001, 115–41; Newcastle Corporation Minutes, 16 March 1960, 1006–23; Esher, *Broken Wave*, 176–80; T Dan Smith, *An Autobiography*, Newcastle, 1970, 54; T Aldous, *Battle for the Environment*, London, 1972, 173–5.

78   J Pendlebury, Masterclass lecture, Edinburgh College of Art, 4 February 2010.

79   H Lottman, *How Cities are Saved*, New York, 1976, 164–83; Dunlop, *Building a Dream*, 120–1.

80   H G Hiller von Gaertringen, 'Sturm auf das Stadtbild', in Baldrick and Clay, *Iconoclasm*, 233–4; *Frankfurter Allgemeine Zeitung*, 5 July 2011, 33.

81   J Göderitz, R Rainer and H Hoffmann, *Die Gegliederte und Aufgelockerte Stadt*, Tübingen, 1957; also Gabi Dolff-Bonekämper and F Schmidt, *Das Hansaviertel*, Berlin, 1999; F Urban, *JSAH*, September 2004, 354–69; Düwel and Gutschow, *Städtebau*, 216–21; *Die Alte Stadt*, 3/1999, 219ff; Niedersächsische Sozialministerium and Stadt Göttingen, *Städtebauliche Erneuerung in Niedersachsen: Göttingen*, Hannover and Göttingen, 1990 (on *Sanierungsgebiete* overseen by the Lower Saxony Regional Development Company).

82   Glendinning and Muthesius, *Tower Block*, 20, 62, 80, 120, 266, 312: Hollamby was a member of the LCC's controversial faction of 'Socialist-Realist' communist-sympathising architects; Esher, *Broken Wave*, 157; *AR*, November 1967, 358; S Pepper, *Housing Improvement*, London, 1971, 72ff (in Leeds, a prototype reconditioned house was opened by junior housing minister W Deedes in November 1955); PRO HLG 118/203; D L Smith, *Amenity and Urban Planning*, London, 1974, 104–11; MHLG, *New Houses for Old*, London, 1954; MHLG, *Moving from the Slums*, London, 1956; Central Housing Advisory Committee, *Conversion of Existing Houses*, London, 1945; *Builder*, 21 October 1955 on Scotland.

83   Glendinning and Muthesius, *Tower Block*, 312–13; Esher, *Broken Wave*, 178; Newcastle Corporation Minutes, 7 December 1960, 615; 3 February 1965, 799; 6 December 1967; 7 May 1969, 1243; Pepper, *Housing Improvement*, 93.

84   S Laing, *Representations of British Working-class Life*, London, 1986, 184; *Builder*, 8 March 1963, 513; 18 August 1966, 115.

85   MHLG, *The Deeplish Study*, London, 1966: Roy Worskett did the drawings in the Deeplish Study. Rochdale had been already targeted in a 1965 Civic Trust investigation of improvement and redevelopment (Civic Trust, *Area Regeneration: a Pilot Study for Rochdale*, 1965).

86   Central Housing Advisory Committee, *Our Older Homes, A Call for Action*, London, 1966; J Yelling, *Planning Perspectives*, 14, 1, 1999, 1–18; *AR*, November 1967, 358.

87   Kennet, *Preservation*, 91; Scottish Development Department/Scottish Housing Advisory Committee, *Scotland's Older Houses*, Edinburgh, 1967; Scottish Development Department, *Towards Better Homes*, Edinburgh, 1973; Scottish Housing Adivsory Committee, *Modernising our Homes*, Edinburgh, 1947; D S Robertson, *Urban Studies*, vol 29, 7, 1992, 1115–36.

88   Choay, *Invention*, 142, 222.

89   Dale, *Historic Preservation in Foreign Countries*, 26–32; Wakeman, *Modernising the Provicial City*, 114.

90   Lottman, *How Cities are Saved*, 59–60; D I Scargill, *Urban France*, London, 1983, 125–8; D Repellin, 'Restoration in Old Lyon', in P Harrison (ed), *Civilising the City*, Edinburgh, 1990; O Godet, 'The Loi Malraux and the Urban Conservation Revolution in 1960s France', in *Mirror of Modernity*, 36–40.

91 M Minost, *AR*, December 1970, 359–64; *MH* (special issue), 'Centres et quartiers anciens', Paris, 1977; R Kain (ed), *Planning for Conservation*, 199–234; Sutcliffe, *Autumn*, 230, 308–14.

92 D Wildeman, *DKD*, 1/1971, 17–35; *Monumentum*, 1972, 124–5; Scargill, *Urban France*, 122–33; Dale, 'Listing and Preserving', 100–1; Dobby, *Conservation and Planning*, 74–7; see also Sutcliffe, *Autumn*, 315–19 for Les Halles row 1967–70.

93 Dale, 'Listing and Preserving', 107; *Monumentum*, vol 26, 3, 1983, 85–100; Lottman, *How Cities are Saved*, 49–53.

94 Dobby, *Conservation and Planning*, 78–9; Fried, *Planning the Eternal City*, New Haven, CT, 1973; Giambruno, *Per una storia*, 145–54 and 185–96; M MacKeith, *Period Home*, vol 2, no. 3, 1982. The Siena piano regolatore (1953–8) was authored by Piero Bottoni, Aldo Lucchini, and Luigi Piccinato: Giambruno, *Per una storia*, 131–44.

95 Lottman, *How Cities are Saved*, 91–108; R T Cerrino, 'Saverio Muratori', in Welter and Lawson, *The City after Patrick Geddes*.

96 *JSAH*, March 1989, 96–7; Dale, 'Listing and Preserving', 277–8; *AR*, October 1966; Silver, *Lost New York*; R Plunz, *A History of Housing in New York City*, New York, 1990, 291–2, 314–15.

97 Citizens Association of Georgetown, *The Future of Georgetown*, Washington, DC, c.1924; T E Beauchamp, *Georgetown Historic District*, Washington, DC, 1998; D Davis and S Hart, *Georgetown: 1751 to the Present*, Washington, DC, 1965; Foundation for the Preservation of Historic Georgetown. *A Walking Guide to Historic Georgetown*, Washington, DC, 1971; Historic Georgetown, Inc. *22nd Annual Report*, 1974; D M Williams, 'Georgetown: The Twentieth Century, A Continuing Battle', *Records of the Columbia Historical Society*, 48, 1972, 783–96.

98 B Greenfield, 'Marketing the Past', in Page and Mason, *Giving Preservation a History*, 176–84; *College Hill – a Demonstration Study, of Historic Area Renewal*, Providence, RI, 1959; 'New Life for Yesterday's City', *Architectural Forum*, January 1960; HUD, *Historic Preservation through Urban Renewal*, 1963; American Institute of Architects, *Open Space for People*, Washington, DC, 1970; D H Schmidt, 'Urban Triage', *Planning Perspectives*, October 2011, 569–89.

99 Choay, *Invention*, 145; *MH*, May 1995; *GRKG*, 286–7; D'Orgeix, *Mirror of Modernity*.

100 *AR*, November 1961, September 1967 and November 1968; *Builder*, 24 July 1964, 162; P Powell, *Monumentum*, vol 11/12, 1975, 55–7.

101 Kennet, *Preservation*, 106; Betjeman, *The Future of the Past*, 67; Boulting, 'The Law's Delays', 26; Civic Trust, *Civic Trust Awards*, 1970; L Borley (ed), *Dear Maurice*, East Linton, 1998, 147.

102 Civic Trust, *Urban Redevelopment*, London, 1962; Civic Trust, *Pride of Place*, London, 1972; Aldous, *Battle for the Environment*, 147–8; P Shapely, 'Planning, Housing and Participation in Britain', *Planning Perspectives*, January 2011, 75–90.

103 Kennet also argued that the most effective agitators were long-established organisations such as the York Civic Trust, who 'time and time again . . . have raised half the money necessary to preserve an old building or group and have then turned to the owners, as often as not the city council, and assumed with a friendly smile that the owners would naturally wish to pay the other half.' Kennet, *Preservation*, 153ff; City and Royal Burgh of Edinburgh, *Development Plan Review*, Edinburgh, 1965; Advisory Committee on City Development, report, Edinburgh, 1943; 'The Little Houses of Scotland', *RIAS Quarterly*, 95, 1954; R J Naismith, The Preservation of the Street in Old Scottish Towns, *RIAS Quarterly*, 96, 1954; I G Lindsay, *Georgian Edinburgh*, Edinburgh, 1948; E MacRae, *The Heritage of Greater Edinburgh*, Edinburgh, 1947; E MacRae, *City of Edinburgh, The Royal Mile, Second Report*, Edinburgh, 1947; 'Stirling's Historic Houses', *Builder*, 6 March 1953; D Walker, *Transactions of the Ancient Monuments Society*, vol 38, 1994.

104 G Scott-Moncrieff, *Scotland's Dowry*, Edinburgh, 1956; M Kelsall and S Harris, *A Future for the Past*, Edinburgh, 1961, 11; C McWilliam, *Scottish Townscape*, London, 1975, 203; I G Lindsay and M Cosh, *Inveraray and the Dukes of Argyll*, Edinburgh, 1973; D Walker, 'The HBC for Scotland', in Borley, *Dear Maurice*; Scottish Development Department, Urban Planning Directorate, Perth Case Study Technical Team, Agenda for Meeting, 14 August 1966 (and papers for 'teach-in' at MHLG, London, 21 March 1967).

105 Lecture by Lord Holford, in R Matthew, J Reid, M Lindsay (eds), *The Conservation of Georgian Edinburgh*, Edinburgh, 1972; C Bloch and M D West, *Conservation in the Edinburgh New Town and Georgian Bath*, Edinburgh, 1980, 19; McWilliam, *Scottish Townscape*.

106 B Little, 'Looking at the Conservation Reports', *Architect and Building News*, 19 June 1969, 61.

107   Colin Buchanan and Partners, *Bath, a Study in Conservation*, London, 1968; Viscount Esher, *York, a Study in Conservation*, London, 1968; D Insall & Partners, *Chester*, London, 1968; G S Burrows, *Chichester, a Study in Conservation*, London, 1968; Bloch and West, *Conservation*, 20; Kennet, *Preservation*, 69–70; D Insall and DoE, *Conservation in Action*, London 1982; D Linstrum, *Architects' Journal*, 4 July 1984, 25–9; Little, 'Looking at the Conservation Reports'.

108   R Crossman, *The Diaries of a Cabinet Minister*, vol 1, London, 1975, 516–17, 525; *RIBA Journal*, November 1966, 500–4.

109   D Smith, *Amenity and Urban Planning*, London, 1974, 111; Roy Worskett, *The Character of Towns*, London, 1969; Kennet, *Preservation*, 66; *Architects' Journal*, 18 January 1967, 130–200; P Ward (ed), *Conservation and Development in Historic Towns and Cities*, Newcastle, 1968; Boulting, 'The Law's Delays', 28. Follow–on activities included a guide to urban townscape design by ministry planner Roy Worskett (*The Character of Towns*, 1969), an MHLG 'coffee table' book, *Preservation and Change* (1967, with cover photograph of Powell & Moya's Cripps Building in Cambridge) and a Rowntree-funded conference on Historic Towns at York University in March/April 1968, covering a wide range of aspects of urban conservation.

110   Sir Kenneth Clark, in J H Waterhouse (ed), *Ruskin's Influence Today*, London, 1945, 15; *Architects' Journal*, 31 March 1982, 39, 46; *Builder*, 9 August 1957, 246; see also F Kelsall, *Architectural History*, 52, 2009, 1–30; George Scott-Moncrieff, *Edinburgh*, Edinburgh, 1947.

111   M Girouard, *The Victorian*, July 2004, 13; *The Victorian*, March 2006; I Grant, *Victorian Society Annual*, 1976, 3–5; 1977, 3–6; 1991, iv; 1991; *Builder*, 24 October 1958, 679; *The Times*, 11 July 1992, 19; *Daily Telegraph*, 6 July 1992, 19.

112   Harries, *Pevsner*, 484–500, 567–82, 639; N Pevsner, *The Englishness of English Art*, London, 1955; S Games (ed), *Pevsner, Art and Architecture*, London, 2002; S Games, *Pevsner, the Early Life*, London, 2010; S Muthesius, 'Nikolaus Pevsner', in H Dilly (ed), *Altmeister moderner Kunstgeschichte*, Berlin, 1990; *AR*, October 1983, 4; *Building Design*, 2 September 1983, 14–15; *Victorian Society Annual*, 1976, 5 and 1982–3, 5; *Builder*, 2 September 1955, 390; *Architectural Review*, April 1954, 227; 'Pevsner's Plea', *Hampstead and Highgate Express*, 30 June 1967.

113   R Hill, C Cunningham, A Reid, *Victorians Revalued*, London, 2010, 7; *Builder*, 6 October 1961, 615; Boulting, 'The Law's Delays', 26–7.

114   Hill, Cunningham and Reid, *Victorians Revalued*, 84.

115   A Briggs, *Victorian Cities*, London, 1963; J Robinson, *Outlook*, Autumn 1972, 5; B Trinder, *History Workshop*, 2, 1976; K Hudson, *Handbook for Industrial Archaeologists*, London, 1967; K Hudson, *Industrial Archaeology, an Introduction*, London, 1963; *Beamish, the Making of a Museum*, Newcastle, 1991.

116   L T C Rolt, *Railway Adventure*, London, 1953; M Bairstow, *The Keighley and Worth Valley Railway*, Halifax, 1991; W G Smith, *1247, Preservation Pioneer*, Kettering, 1991; D Pearce and M Binney (eds), *Off the Rails*, London, 1977.

117   *JSAH*, December 1993, 511; J Plank, *Crombie Taylor: Modern Architecture, Building Restoration and the Rediscovery of Louis Sullivan*, Richmond, CA, 2010.

118   *Builder*, 7 July 1961; Corporation of Glasgow Planning Committee, *Park Circus Area*, April 1967. What was unusual about Glasgow, in a UK context, was the single-minded way in which people had carried on building this rich classical architecture throughout the 19th century – including the furious eclecticism of Alexander 'Greek' Thomson – and had largely ignored the neo-Gothic passions of England. Ian Nairn argued in 1967 that Glasgow was 'one of the finest examples of an architecture of completely fulfilled promise – in fact, a dignified old age in some cases took classicism right up to the First World War': I Nairn, *Britain's Changing Towns*, London, 1967, 53; J Macaulay, *AR*, June 1965, 407; N Taylor, *Architect and Building News*, 6 May 1964; *Prospect*, vols 17 and 18, 1960.

119   Lord Esher, *Conservation in Glasgow: a Preliminary Report*, Glasgow, June 1971.

120   J Summerson, *Victorian Architecture, Four Studies in Evaluation*, New York and London, 1970, 12.

Chapter 9

1   J Betjeman, in A Fergusson, *The Sack of Bath*, Salisbury, 1973, 7.

2   J Pendlebury, Masterclass lecture, Edinburgh College of Art, February 2010.

3   B Feilden, *Conservation of Historic Buildings*, London, 1982.

4   *Architects' Journal*, 22 November 2007, 43; http://en.wikipedia.org/wiki/List_of_riots#1960s.

5 M Nicholson, *The Environmental Revolution*, London, 1970; E F Schumacher, *Small is Beautiful*, London, 1973; T Aldous, *Battle for the Environment*, London, 1972; E Partridge (ed), *Responsibilities to Future Generations*, Buffalo, NY, 1981; J Porritt, *Seeing Green*, Oxford, 1984.

6 I Nairn, *AR*, December 1975, 328–37; N Fairweather, *New Lives, New Landscapes*, London, 1970; S Zipp, 'The Battle of Lincoln Square', *Planning Perspectives*, October 2009, 409–33; Plunz, *History of Housing in New York City*, 308–12; G Lang and M Wunsch, *Genius of Common Sense*, Boston, MA, 2009; H Lefebvre, *Critique de la vie quotidienne*, Paris, 1947. In Northern Ireland, the upsurge in bomb attacks and arson created a different combination of 'threats' to the historic built environment, with much 'traditional' 19th-century working-class housing being targeted for attack and dereliction, rather than cherished as elsewhere in Western Europe: see e.g. R Wiener, *The Rape and Plunder of the Shankill*, Belfast, 1975.

7 W J Siedler, E Niggemeyer and G Angress, *Die gemordete Stadt*, West Berlin, 1964; Düwel and Gutschow, *Städtebau*, 209; A Mitscherlich, *Die Unwirtlichkeit unserer Städte*, Frankfurt/Main, 1965.

8 *Architects' Journal*, 22 November 2007, 46; S Ankers, D Kaiserman and C Shepley, *The Grotton Papers, Planning in Crisis*, London, 1979; L Wright, *Architects' Journal*, December 1977, 324; A Ravetz, *RIBA Journal*, April 1972, 144–51, and May 1973, 210–11.

9 H Klotz *et al*, *Keine Zukunft für unsere Vergangenheit?*, Giessen, 1975; F Bollerey and K Hartmann, *Denkmalpflege und Umweltgestaltung*, Munich, 1975; K Ohlenmacher, *Sind die Dörfer zum Sterben verurteilt?*, Limburg, 1975; D Sturdy, *How to Pull a Town Down*, London, 1972; C Amery and D Cruickshank, *The Rape of Britain*, London, 1975, 10; M T Will, *Architectural Conservation in Europe: a Selected Bibliography*, Monticello, Ill, 1980.

10 Initiative Marburg Stadtbild, *Marbach im Abbruch*, Marbach, 1972; Erwin Schleich, *Die Zweite Zerstörung Münchens*, Stuttgart, 1978; lecture by Rolf Sachsse at Denkmal-Werte-Dialog symposium, Dortmund, March 2012; A Fergusson, *The Sack of Bath, a Record and an Indictment*, Salisbury, 1973; Bloch and West, *Conservation*; R Worskett, R Redston and H Gunton, *Saving Bath*, Bath, 1978; *AR*, May 1973, 280; P Smithson, *Bath, Walks within the Walls*, Bath, 1971; interview with R Worskett, 1992; *AR*, July 1972, 2.

11 *AR*, December 1977, 324; *RIBA Journal*, May 1973; County Council of Essex, *A Design Guide for Residential Areas*, December 1973.

12 The mildly nationalistic tone of Taylor's argument, contrasting French *dirigisme* with English individualism and freedom, was amplified over a decade later in E R Scoffham's study of social housing, which hailed the English 'tradition of xenophobic housing' and argued that 'towns and technology are not English': N Taylor, *The Village in the City*, London, 1973; *RIBA Journal*, June 1973, 296–301; E R Scoffham, *The Shape of British Housing*, London, 1984, 231.

13 C Buchanan, *The State of Britain*, London, 1972.

14 G Millerson, 'Dilemmas of Professionalism', *New Society*, 4 June 1964; P Levin, 'Decisions and Decisionmaking', *Public Administration*, Spring 1972, 19–43; A Etzioni and R Remp, *Technological Shortcuts to Social Change*, New York, 1973; R Ellis and D Cuff (eds), *Architects' People*, New York, 1989, 260–81; J G Davies, 'The Councillor's Dilemma', *Official Architecture*, February 1972, 112–14; P Saunders, 'They Make the Rules', in G Parry and P Morriss (eds), *Political Sociology Year Book*, London, 1974, 332; M Gladstone, *The Politics of Planning*, London, 1976; J M Simmie, *Citizens in Conflict*, London, 1974.

15 R Goodman, *After the Planners*, London, 1972, 236–7; *ARse*, No. 3, May 1970; *Anarchy*, 117, November 1970 (see e.g. A Thomas, 'Motorway Madness', 344–8).

16 Laing, *Representations of Working-class Life*; H Perkin, *The Structured Ground*, Brighton, 1981, ch 11; Town Planning Institute, Summer School, 195, report on meetings by F J Osborn; M Young and P Wilmott, *Family and Kinship in East London*, London, 1957; *Architects' Journal*, 22 November 1956, 723; P J Marshall, *RIBA Journal*, April 1959, 193–201; *Architect and Building News*, 30 August 1951, 236; *RIBA Journal*, January 1963, 9–13; A and P Smithson, *Urban Structuring*, London, 1967; *Architectural Review*, November 1957, 333–6; B Rudofsky, *Architecture without Architects*, New York, 1964.

17 B Evans, *AR*, April 1985, 73–7; *Architects' Journal*, 15 June 1979, 1011 ff; G de Carlo, *An Architecture of Participation*, Melbourne, 1972.

18 M Broady (ed), *Planning for People*, London, 1968, 113; *Town and Country Planning*, September 1969, 396ff; Committee on Public Participation in Planning, *People and Planning* ('Skeffington Report'), London, 1969, 3, 17; Aldous, *Battle for the Environment*, 267; C Ward, *Tenants Take Over*, 1974; H Casson, *Observer*, 16 December 1984, 52; K Cupers, 'The Expertise of Participation', *Planning Perspectives*, January 2011,

29–54; S Harmann, 'Modernism was Hollow', *Planning Perspectives*, January 2011, 55–73; Shapely, 'Planning, Housing and Participation'.

19 Dale, *Historic Preservation in Foreign Countries*, 36.

20 Falser, *Zwischen Identität und Authentizität*, 112–25; Landeskonservator Rheinland, Arbeitsheft 14, *Wir Verändern ein Stückchen Bonn*, Cologne, 1975.

21 C McKean, *Fight Blight*, London, 1977; N Wates, *The Battle for Tolmers Square*, London, 1976. Berkeley: P Allen, 'The End of Modernism?' *JSAH*, September 2011, 354–74.

22 *Architectural Heritage Society of Scotland Magazine*, Winter 1999, 9–11.

23 Newcastle: *Architectural Review*, April 1977, 225–6; Esher, *Broken Wave*, 179–83; G Priest and P Cobb (eds), *The Fight for Bristol*, Bristol, 1980; *Transactions of the Ancient Monuments Society*, new series, vol 30, 1975, 22; Feilden, *Conservation of Historic Buildings*, v.

24 Worskett, Reston and Gunton, *Saving Bath*.

25 B Anson, *I'll Fight You For It*, London, 1981; C Amery, M Girouard and D Cruickshank, 'Save the Garden', *AR*, July 1972, 26–32; *Twentieth Century Architecture*, vol 7, 2004, 13–14; *AR*, July 1972, 16–32; *Cornerstone*, vol 29, 1, 2008, 42–3; *The Times*, 5 May 2009, 50; G Stamp, *Spectator*, 21 June 1980, 15.

26 R Hillier, *A Place Called Paddington*, Sydney, 1970; Robert Burns, 'Pushing Paddington', *Nation*, 2 December 1967, 13–15.

27 J Balk, *Jordaan – een toekomst voor het overleden*, Amsterdam, 1975; Lottman, *How Cities are Saved*, 218–19; *Architects' Journal*, 30 August 1978.

28 The Eisenheim initiative included a complex research project undertaken by Fachshochschule Bielefeld under Günter's direction, based in a local *Haus der Initiativen* in Werrastrasse 1 (1974–80). W Wolff, *Neuer Wert aus alten Häusern*, Berlin, 1966; J Günter, *Leben in Eisenheim*, Weinheim, 1980; Projektgruppe Eisenheim, *Rettet Eisenheim*, Bielefeld, 1973; Falser, *Zwischen Identität und Authentizität*, 104–22; *Bauwelt*, 63, 1972, 162–5; Deutscher Werkbund, *Werkbund Zeit-Forum*, 1972, issue 12, 1; J Dumouchel, *European Housing Rehabilitation*, Washington, DC, 1978, 55–64; Neue Heimat, *Monatshefte*, October 1981 (special Hamburg issue).

29 Glendinning and Muthesius, *Tower Block*, chs 30–1.

30 Kennet, *Preservation*, 90–1; *Architects' Journal*, 22 August 1973, 433; P Balchin, *Housing Improvement and Social Inequality*, Farnborough, 1979, 227; A E Holmans, *Housing Policy in Britain*, London, 1987; S Merrett, *State Housing in Britain*, London, 1979, 262; Ministry of Housing, *Design Bulletins* 21, 22; *Building*, 21 July 1967, 134; 10 May 1968, 122; 26 April 1968, 86; 17 March 1972, 67–74; *AR*, November 1967, 371–2.

31 See also *Building*, 30 August 1968, 79. Exeter: *Architects' Journal*, 10 June 1970; Department of the Environment, *Area Improvement Notes 5*, London, 1972. Lambeth Borough Council rehab show house, 80 Vassall Road: *Building*, 17 March 1972, 68–9.

32 *Municipal and Public Journal*, 27 December 1969, 3191–4; *Architects' Journal*, 1 July 1970, 15–35.

33 Barnsbury: see for example North Terrace, rehabilitated 1969–70, *Architects' Journal*, 22 August 1973, 428–44; *Building*, 17 May 1968, 93–4. Liverpool: *Architects' Journal*, 18 July 1984, 36ff; Esher, *Broken Wave*, 243–4. C Hague, *Housing Studies*, 5, 1990, 242–56; A Richardson, *Participation*, London, 1983; Scottish Development Department, *Tenant Participation and Housing Cooperatives*, Edinburgh, 1977. The movement towards community-based housing rehab could draw on the nearly century-old tradition of cooperative housekeeping, with its strong Arts and Crafts links in Britain. But the international tradition of radical communitarianism was an equally strong influence: by 1970, there were at least 50 communes of squatters in Britain: L F Pearson, *The Architectural and Social History of Cooperative Living*, Basingstoke, 1988; J Hands, *Housing Cooperatives*, London, 1975.

34 N Wates, *AR*, April 1985, 58; McKean, *Fight Blight*, 106–8; *Guardian*, 12 August 1985, 9, and 26 August 1985, 7; *Architects' Journal*, 20 February 1985, 27, and 24 April 1985, 34; *New Society*, 3 April 1987, 12ff; *Sunday Times*, 14 December 1986, 29.

35 Smith, *Amenity and Urban Planning*, 75–128; interview with R Worskett, 1992; Department of the Environment, *Area Improvement Notes*, 1, 1971, and 4, 1972.

36 Smith, *Amenity and Urban Planning*, 90; M Hook, *Architects' Journal*, 10 June 1970.

37 *Architects' Journal*, 1 July 1970, 15–17; Scottish Housing Advisory Committee, *Scotland's Older Houses*, Edinburgh, 1967; Scottish Development Department, *The New Scottish Housing Handbook*, Bulletin 2, 1969; interview with D Whitham, 2006; Building Research Station, *Damage to Buildings in Scotland in the Gale of January 14 and 15 1968*, Garston, 1968; *Evening News* (Edinburgh), 2 January 1969, 7; J English,

Urban Studies, 381–6; *Architects' Journal*, 1 July 1970, 17–19; *Scottish Affairs*, Winter 1995, 47–63; Robertson, 'Home Improvement', 1115–36; Smith, *Amenity and Urban Planning*, 90.

38 *Architectural Heritage XXI: Mirror of Modernity*, Edinburgh, 2011 (articles by Whitham, Robinson, Young, 59–108; interview with T Crombie and R Young, March 2010; *Scottish Daily Express*, 26 June 1969, 13; *Architects' Journal*, 30 May 1990, 30–1; *Architects' Journal*, 10 January 1973, 61–2; Merrett, *State Housing*, 115.

39 Ministry of Housing, Circular 64/1969; *Official Architecture*, March 1969, 287; Department of the Environment, *Better Homes, the Next Priorities*, London, 1973; A D Thomas, 'Area Based Renewal', University of Birmingham, 1979, vi; D McLennan, *Urban Housing Rehabilitation*, December 1983; Department of the Environment, Circular 13/1975, February 1975; *Architects' Journal*, 5 February 1975, 275–83, 19 February 1975, 383–92 and 17 November 1976, 925.

40 C Couch, *Urban Renewal Theory*, Basingstoke, 1990, 45; SLASH, *Rehabilitation and Modernisation Advisory Notes* 1, 1974, 2, 1975, 3, 1978; Scottish Local Authorities Special Housing Group, *The SSHA Housing Management Handbook*, Edinburgh, 1981; Scottish Development Department, *Housing in Clydeside*, Edinburgh, 1971; *Architects' Journal*, 5 October 1988, 69; K Powell, in Lowenthal and M Binney, *Our Past Before Us*, London, 1981, 143–57.

41 Scargill, *Urban France*, 130–2.

42 Senator für Bau-und Wohnungswesen, Info, *Strategien für Kreuzberg*, Berlin, 1978; Büro für Stadtsanierung, *Sanierung – für wen?*, Berlin, 1971; Düwel and Gutschow, *Städtebau*, 248–53; J P Kleihues, *AD Profile, Post-War Berlin*, London, 1982; J P Kleihues, C Baldus and U Frohne (eds), *750 Jahre Architektur und Städtebau in Berlin*, Stuttgart, 1987; www.udk–berlin.de/haemer; G Schlusche, *Die internationale Bauausstellung Berlin – eine Bilanz*, Berlin, 1997; *AR*, September 1984, 28–41.

43 M T R Will, 'Austria', *Period Home*, vol 2 5, 40–1.

44 The block was bounded by Ottakringer Strasse, Thalia Strasse and Eisner Strasse.

45 Dumouchel, *European Housing Rehabilitation*; M T Will, *Period Home*, vol 2, 4, 1982; G Kokkelink and H D Theen, *Bewertungsfragen der Denkmalpflege*, Hannover, 1976; C Berg *et al*, *Was ist ein Baudenkmal?*, Cologne, 1983.

46 S Dennis (ed), *Preservation Law Update*, Washington, DC, 1986; N Williams, E H Kellog and F Gilbert, *Readings in Historic Preservation*, Brunswick, NJ, 1983; C Creiff (ed), *Lost America*, 2 vols, 1971/2; J H Kay, *Lost Boston*, Boston, 1980; *The Old House Journal*, vol 1, 1, October 1973; R Keune (ed), *The Historic Preservation Yearbook*, Bethesda, MD, 1965; *The Brown Book, a Directory of Preservation Information*, Washington, DC, 1983; *Preservation: towards an Ethic in the 1980s*, Washington, DC, 1980; W J Murtagh, 'The National Register of Historic Places', *American Preservation*, February/March 1979; *Preservation News*, September 1980; Murtagh, *Keeping Time*, 74–7; R Longstreth, *The Buildings of Main Street*, Washington, DC, 1987; National Main Street Center, *National Main Street Training Manual*, Washington, DC, 1981; Old House Journal Editors, *The Old House Journal 1988 Catalog*, Brooklyn, NY, 1987; Preservation Action, *Blueprint for Lobbying*, Washington, DC, 1984; R Tubesing, *Architectural Preservation and Urban Renovation*, New York, 1982; National Park Service, *Local Preservation: A Selected Bibliography*, Washington, DC, 1988; National Park Service, *What are the Historic Preservation Tax Incentives?* Washington, DC, 1988; National Park Service, *What is the National Historic Preservation Act?*, Washington, DC, 1987.

47 Goodman, *After the Planners*; *American Preservation*, October/November 1977 and October/November 1978; R H McNulty and S A Kliment (eds), *Neighborhood Conservation*, New York, 1976/9; P Myers and G Binder, *Neighborhood Conservation*, Washington, DC, 1977; A Ziegler *et al*, *Revolving Funds for Historic Preservation*, Pittsburgh, 1975.

48 D Gamston, *The Designation of Conservation Areas*, York, 1975; S Cantacuzino, *Architectural Conservation in Europe*, London, 1975, 12; Aldous, *Battle for the Environment*, 149–75; *A Future for Old Buildings (Journal of Planning and Environmental Law Occasional Papers)*, 1977; Civic Trust, *Conservation in Action*, London, 1992, 63.

49 M Talbot, *Reviving Buildings and Communities*, Newton Abbot, 1986, 116.

50 Scottish Civic Trust (SCT), *Newsletter*, 1, 1991 and 13, 1997, 8–13; SCT, *Yearbook*, 6, 1980; SCT, *Environment Scotland*, Issues 1, 1969–70, 3, 1971, and 4, 1972.

51 I G Lindsay, *Georgian Edinburgh*, Edinburgh, 1948; G Scott-Moncrieff, *Edinburgh*, Edinburgh, 1947; W Holford and Partners, *City of Edinburgh High Buildings Policy*, 1965; A Rowan and J Cornforth, *Country Life*, 18 December 1969; A J Youngson, *The Making of Classical Edinburgh*, Edinburgh, 1966; RFACS, *Report*

for 1964–1966, Edinburgh, 1967, 13; *Building*, 25 August 1967; P Daniel, *Two Hundred Summers in a City, Edinburgh 1767–1967, Souvenir Programme*, Edinburgh, 1967.

52 *Architectural Heritage XXI: Mirror of Modernity*, Edinburgh, 2011 (articles by Gerrard, Hodges and Knight); D Rodwell, 'The French Connection', in P Harrison (ed), *Civilizing the City*, Edinburgh, 1990; C McWilliam, lecture of 10 December 1985 to Scottish Georgian Society; *New Edinburgh Review*, August 1970; A Davey, B Heath and D Hodges, *The Care and Conservation of Georgian Houses*, Edinburgh, 1978; *Architectural Review*, December 1970, 357–9.

53 Scottish Civic Trust, *Newsletter*, 13, 1997; New Lanark Conservation Trust, *The Story of New Lanark*, nd; Lanark County Council, *A Future for New Lanark*, 1973.

54 Esher, *Broken Wave*, 243–4; Cupers, 'The Expertise of Participation'.

55 Scottish Development Department Working Group on Urban Conservation, minutes 1971–9, courtesy of D Walker; M Magnusson (ed), *Echoes in Stone*, Edinburgh, 1983; R T Rowley and M Breakell (eds), *Planning and the Historic Environment*, Oxford, 1977; Gamston, *Designation*; Teignbridge UDC Planning Committee minutes, 28 April 1975 (Teignbridge Local History Library, Devon); *Official Architecture*, February 1970. Percentages of list categories as at 1969: in Scotland, Category A was 10% of A and B together; in England, Grade 1 was 4% of Grades 1 and 2 combined. Bath: Worskett *et al*, *Saving Bath*.

56 *JSAH*, May 1978, 108–11.

57 L Bravo, 'Area Conservation as Socialist Standard-Bearer: A Plan for the Historical Centre of Bologna in 1969', in *Mirror of Modernity*, 44–53; L Benevolo, *Le avventure della città*, Bari, 1974; S Cantacuzino and S Brandt (eds), *Saving Old Buildings*, London, 1980, 2–19; A Rossi, *The Architecture of the City*, Cambridge, MA, 1984 (first published in Italian, 1966); A Colquhoun, 'Typology', *Perspecta*, 12, 1967, 71–4; Lottman, *How Cities are Saved*, 201–12; *Edilizia Popolare*, November–December 1983 and January–February 1984 (special issue on housing and rehab in Venice).

58 Bravo, 'Area Conservation'; Giambruno, *Per una storia*, 169–78 and 185–96; *JSAH*, May 1978, 108–11; P L Cervellati and F Fontana, *Bologna, il volta della città*, Modena, 1975; P L Cervellati, R Scannavini and C de Angelis, *La nuova cultura delle città*, Milan, 1977; *Bologna Centro Storico* (exhibition catalogue), Bologna, 1970; Bologna, *Politica e metologia dello restauro nei centri storici*, Bologna, 1973; A Debold-Kritter and P Debold, *Bauwelt*, 33, 1974, 1112–32; H Bodenschatz, *Die Alte Stadt*, 1999, 3, 205ff.

59 Esher, *Broken Wave*, 296; S Games, *RIBA Journal*, February 1980, 36ff; B Auger, *Architectural Review*, August 1976, 77–9.

60 The 1980 Biennale display comprised a 'street' made up of 20 facades by 20 'great architects'. Falser, *Zwischen Identität und Authentizität*, 153–61; M Glendinning, *Architecture's Evil Empire?*, London, 2010, 47–51; Düwel and Gutschow, *Städtebau*, 248–54; *AR*, October 1982.

61 The new *Heimat* concept first started to emerge in the wake of Helmut Kohl's election victory in 1982 and was cautiously elaborated in the so-called *Historikerstreit* of 1986, for and against a more relativistic interpretation of Third Reich crimes: Falser, *Zwischen Identität und Authentizität*.

62 W Pehnt, *Karljosef Schattner*, Stuttgart, 1988; *Karljosef Schattner, ein Führer zu seinen Bauten*, Munich, 1998; *AR*, March 1982, 62–72; E Mulzer, *Nürnberger Bürgerhäuser*, Nürnberg 1954; E Mulzer, *Der Wiederaufbau der Altstadt von Nürnberg 1945 bis 1970* (Erlanger Geographische Arbeiten 31), Erlangen, 1972; www.altstadtfreunde–nuernberg.de/index.php?id=55.

63 *Bauwelt*, 72, 1981, 2128–64.

64 The sponsoring organisation was the *Koelnische Lebensversicherung*: W Queck (ed), *Die Steipe, Trier*, Trier, 1972; Doelling, *Conservation*, 76.

65 The architects for the 'new' Leibnizhaus were Wilfried Ziegemeier and Hubertus Pfitzner. In a further complication, the 'new' Leibnizhaus was to have been built on a site created by demolition of a surviving *Gründerzeit* mansion, the Noltehaus, but when this was listed at the last minute, following citizen protests in 1975, the Noltehaus was spared and the new Leibnizhaus site was simply shifted one plot along the street: Falser, *Zwischen Identität und Authentizität*, 156–7.

66 Düwel and Gutschow, *Städtebau*, 246–7; Hubel, *Denkmalpflege*, 280; *Frankfurter Allgemeine Zeitung*, 15 January 2011, Z1–2.

67 Conservation officials argued that the town's bomb-gutted Romanesque churches could be legitimately restored, as significant fabric survived. 1985 also saw the designation of two key reconstructed churches as UNESCO World Heritage Sites; Hubel, *Denkmalpflege*, 280–1; *GRKG*, 320–1; Falser, *Zwischen Identität und Authentizität*, 133–53; W Achilles, H-G Borck *et al*, *Der Marktplatz zu Hildesheim, Dokumentation des Wiederaufbaus*, Hildesheim, 1989.

68 M Sawin (ed), *James Marston Fitch, Selected Writings*, New York, 2007; *JSAH*, May 1983, 196; J H King, 'The National Trust', *American Preservation*, October/November 1978; Stubbs and Makaš, *Architectural Conservation*, 452–5 (e.g. on Quincy Market/Faneuil Hall Marketplace regeneration project in Boston, 1973–6, by architect Ben Thompson and developer James Rouse).

69 L Ward, *Things*, 5, Winter 1996, 7–37; J Cornforth, *Country Houses in Britain: Can they Survive?*, London, 1974, 1, 18, 129.

70 Many redundant country houses were safeguarded in ingenious conversion and subdivision schemes pioneered by Kit Martin (son of Sir Leslie of Festival Hall LCC fame); in a sign of the effectiveness of this campaign, a 1982 report by Binney and Martin focused mainly on insensitive alterations rather than on demolitions: M Binney, K Martin, *The Country House: To Be or Not To Be?*, SAVE, London, c.1982; Cornforth, *Country Houses*; *Grapevine* (Scottish Civic Trust), 33, December 1980, 1; R Strong, M Binney and J Harris, *The Destruction of the Country House*, London, 1974; M Binney and M Watson-Smith, *The SAVE Britain's Heritage Action Guide*, London, 1991.

71 D Walker, 'The Adaptation and Restoration of Tower Houses' in R Clow (ed), *Restoring Scotland's Castles*, Glasgow, 2000, 1–29; D McCrone and A Morris, in T M Devine (ed), *Scottish Elites*, Edinburgh 1994, 171; N Fairbairn, *Scottish Field*, September 1985, 39.

72 The progress of this unique and internationally little-known movement of individual, private restorations was charted in two unpublished government reports of 1982–5 by Walker: D Walker, 'The Restoration of Ruined Castles', January 1982 Scottish Development Department paper, revised 1985 (courtesy of D Walker); see also R Fawcett, 'The Portfolio of Monuments within the Estate: the Inspectorate View', Scottish Development Department paper, 1991 (courtesy of D Walker); Bruce Walker, *Scots Magazine*, February 1982; E W Proudfoot (ed), *Ancient Monuments, Historic Buildings and Planning*, CBA Scotland Occasional Papers No. 1, 1984; Clow, *Castles*, 161–192; R Fawcett and A Rutherford, *Renewed Life for Scottish Castles*, York, 2011; H Fenwick, *Scots Magazine*, July 1979.

73 *Architects' Journal*, 8 August 1984, 1–6; M Binney and M Watson-Smyth, *The Origins of SAVE*, London, 1991, 10–17; *Architects' Journal*, December 1975, 7–9, and 18 July 1984, 20; *Building Design*, 5 July 1985, 13; M Binney, *Our Vanishing Heritage*, London, 1984; M Binney, *Change and Decay*, London, 1977; SAVE, *Satanic Mills*, London, 1979; M Girouard, D Cruickshank and R Samuel, *The Saving of Spitalfields*, London, 1989; Talbot, *Reviving Buildings and Communities*, 71–7.

74 A Artley and J M Robinson, *The New Georgian Handbook*, London, 1985, 44, 87; G Gladstone and V Mather, 'The National Trust Navy', *Harpers and Queen*, July 1979, 84–7; *Sunday Times*, 16 November 1986, 53, and 12 March 1995; G Young, *Conservation Scene*, London, 1977; C McWilliam, lecture of 10 December 1985, Edinburgh.

75 *Times*, 1 May 1984; R Scruton, *The Aesthetics of Architecture*, London, 1979; R Scruton, *The Meaning of Conservatism*, London, 1980; A Potts, *History Workshop Journal*, vol 12, 1981, 159–62; C Aslet and A Powers, *The British Telephone Box –Take it as Red*, 1987; *Independent*, 19 January 1988; *Country Life*, 30 April 1987, 118; *Daily Telegraph*, 16 May 1992, 15.

76 L Krier, *Art and Design*, April 1985.

77 AD Profile, *Prince Charles and Architectural Debate*, London, 1989; C Jencks, *The Prince and the Architects and New Wave Monarchy*, London, 1988; Prince Charles, *A Vision of Britain*, London, 1989; *Perspectives on Architecture*, vol 1, 1, April 1994; *Sunday Times*, 10 September 1989, B2; *Daily Telegraph*, 3 December 1987, 17 and 14 August 1998; *Building Design*, 21 August 1998; *Architects' Journal*, 12 February 1992; *Architects' Journal*, 23 May 1984, 32, 30 May 1984, 22, and 8 May 1985, 19; *Building Design*, 26 August 1983, 1; *Evening Standard*, 10 June 1988, 6.

78 *Architects' Journal*, 20 February 1985, 27, 47; E M Farrelly, *Architectural Review*, March 1987, 27–31; N Wates and C Knevitt, *Building Design*, 21 November 1986, 21.

79 See e.g. E Harris, C Cecil and A Bronovitskaya (SAVE), *Moscow Heritage at Crisis Point*, London/Moscow, 2009.

80 Civic Trust, *Conservation in Action*; Doelling, *Conservation*, 71; M Hanna, 'Cathedrals at Saturation Point', in D Lowenthal and M Binney (eds), *Our Past Before Us*, London, 1981; *Telegraph Magazine*, 11 April 1992; MacKeith, 'Italy'. Rothenburg: Hagen, *Preservation, Tourism and Nationalism*, 245, 285–6.

81 Barcelona Pavilion: *GRKG* 359; Düwel and Gutschow, *Städtebau*, 282–6; Cantacuzino, *Architectural Conservation*, 67–70; Scottish Civic Trust, *Newsletter* 13, 1997; Ursula von Petz, 'City Planning Exhibitions in Germany, 1910–2010', *Planning Perspectives*, vol 25, 3, 2010, 375–82. G Gordon, 'Management and

Conservation in the Historic City', in G Gordon (ed), *Perspectives of the Scottish City*, Aberdeen, 1985. In Italy, 1970–82 saw an innovative attempt to renovate the dilapidated town of Gorgonza as a pre-structured 'tourist destination': S Cantacuzino, *New Uses for Old Buildings*, London, 1975; S Cantacuzino *et al*, *Saving Old Buildings*, London, 1980.

82 Scottish Development Department Historic Buildings and Urban Renewal file, December 1977 correspondence, courtesy of D Walker; *Glasgow 1990, the 1990 Story*, Glasgow, 1990; *Architects' Journal*, 6 May 1987, 39ff, and 30 May 1990; *Glasgow Herald*, 10 and 17 August 1982, 19 April 1983, and 24 August 1983, 14; *Architectural Review*, November 1974; *Glasgow Herald Weekend*, 11 February 1989, 17; *Scotland on Sunday*, 18 August 1996, 18. The Promenade, Halifax (Canada): W Bernstein and R Cawker, *Contemporary Canadian Architecture*, Don Mills, 1982, 57–63.

83 National Heritage Act 1983, ch 47, sections 32–8; *English Heritage Conservation Bulletin*, 49, Summer 2005, 'English Heritage: The First 21 Years'; D Pearce, *Conservation Today*, London, 1989, 8, 228–38; M Ross, *Planning and the Heritage*, London, 1991, 46–52; *Glasgow Herald*, 8 November 1984, 11.

84 R Hewison, *The Heritage Industry*, London, 1987; D Lowenthal, *PSAS*, 1990, 2; J Simpson, 'Whither Conservation?', in P Harrison (ed), *Civilizing the City*, Edinburgh, 1990.

85 Montagu Committee, *Report*, 1980, 4; Pearce, *Conservation Today*, 238; Lowenthal and Binney, *Our Past Before Us*, 14, 122–3, 210; *Building Design*, 25 April 1986.

86 *Scottish Review*, 27, August 1982; *Guardian*, 14 March 1979.

Chapter 10

1 E Hruška, *JSAH*, 38, 1979, 157. See also L Deiters, 'Historic Monuments', *JSAH* , 38, 1979, 145.

2 *JSAH*, 38, 1979, 158–75.

3 H Magirius, 'Denkmalpflege in der GDR', *DD*, 2/2001, 127–33; H Berger, 'Tendenzen der Dekmalpflege in der GDR', *DKD*, 1/1991, 3–4.

4 J Kirchner, 'Denkmalpflege und Stadtplanung in der GDR', in Scheurmann, *Zeitschichten*, 148–51; Magirius, 'Denkmalpflege in der GDR'; Magirius, 'Eine fast vergessene Zeitschicht', in I Scheurmann and H R Meier, *Echt, Alt, Schön, Wahr*, Munich and Berlin, 2006, 132–45.

5 E Hruška, *JSAH*, 38, 1979, 157. See also L Deiters, 'Historic Monuments', *JSAH*, 38, 1979, 145: 'Monuments in our country are the witnesses of political, cultural, economic and historical development. Because of their historical, social and artistic value, their preservation is of great interest for the socialist society . . . The chain of achievements that remains evident in the monuments, teaches us to comprehend this historical development, and helps us to achieve our own political goals.' Deiters grouped together, under the heading of socialist monuments, early industrial archaeological sites, *Altstadt* preservation in a town (Mühlhausen) associated with the 16th-century Peasants' War and the new Buchenwald memorial of 1954–8 to the 'victims of fascism'. See also G Piltz, *Streifzug durch die deutsche Baukunst*, Berlin, 1973; G Piltz, *Streifzüge durch die deutsche Architektur*, Leipzig/Berlin, 1973 (new edition of 1966 book).

6 P Kann, *Leningrad, a Guide*, Moscow, 1988, 167; *GRKG*, 282.

7 K Kalinowski, 'Der Wiederaufbau der historischen Stadtzentren', in *DKD*, 1989, 111; J Zachwatowicz, *Conservation in Poland*, Warsaw, 1965, 49–51.

8 M Derzhovina, N T Morozova and H D Monakhova (eds), *Volgograd*, Moscow, 1979, 85, 99, 106.

9 Choay, *Invention*, 75; A J Schmidt, *Impact of Perestroika*, 286.

10 S Massie, *Pavlovsk, The Life of a Russian Palace*, Boston, 1990; Kann, *Leningrad*; Dobby, *Conservation and Planning*, 88–9; http://news.bbc.co.uk/2/hi/europe/3025833.stm; Schmidt, *Impact of Perestroika*, 352; Stubbs and Makaš, *Architectural Conservation*, 274–5.

11 Dale, 'Listing and Preserving', 104; D Snyder, 'The Future has a Dubious Past', DOCOMOMO 2004 *Conference Proceedings*, New York, 267–73; A Ciborowski, *Warsaw, a City Destroyed and Rebuilt*, Warsaw, 1969.

12 Zamek Krolewski w Warszawie, *Jan Zachwatowicz w Stulecie Urodzin*, Warszawa, 2000; Kalinowski, 'Wiederaufbau', 102–13; S Lorentz, 'Reconstruction of the Old Town Centers of Poland', in Colonial Williamsburg, *Historic Preservation Today*, Charlottesville, 1966, 43–61; Kalinowski, 'The Development of Polish Architecture', *JSAH*, 38, 1979, 129–32; Polski Komitet Narodowy ICOMOS, *XXX ICOMOS 1965–1995*, Warsaw, 1995, 53–78.

13 M Murawski, 'A-political buildings: Ideology, Memory and Warsaw's Old Town', in *Mirror of Modernity*, 13–20; Dale, 'Listing and Preserving', 104; Dobby, *Conservation and Planning*, 85–8; *DD*, 2/2007, 189;

National Gallery, *Venice, Canaletto and his Rivals*, London, 2010.

14   Dobby, *Conservation and Planning*, 86; Lorentz, *Reconstruction*, 52–62; Ciborowski, *Warsaw*; Zamek Krolewski w Warszawie, *Jan Zachwatowicz*.

15   Zachwatowicz, *Conservation in Poland*, 49–51; Zachwatowicz, 'Protection of Historic Monuments'; Kalinowski, 'Development of Polish Architecture', *JSAH*, 38, 1979, 133; Polish National Committee of ICOMOS, *Problems of Heritage and Cultural Identity in Poland*, Warsaw, 1984 (ch by J Stankiewicz), 25–31.

16   J Friedrich, *Neue Stadt in Altem Gewand: Der Wiederaufbau Danzigs 1945–1960*, Cologne, 2010; Pusback, *Stadt als Heimat*; Lorentz, 'Reconstruction of the Old Town Centers', 63–70 and 75; Zachwatowicz, 'Protection of Historic Monuments'; Kalinowski, 'Development of Polish Architecture', 129ff.

17   Friedrich, *Neue Stadt*, 235–47 and figs 95–6; Pusback, *Stadt als Heimat*, 268; Thum, *Uprooted*, 352–4, 361–7; Choay, *Invention*, 76; J Zachwatowicz, *Conservation in Poland*, Warsaw, 1965, 49–51. Postwar rebuilding in Elbing/Elbląg: *GRKG*, 468–9.

18   Thum, *Uprooted*, 128; G Hryniewicz-Lamber, 2004 DOCOMOMO Conference proceedings, 235; Lorentz, 'Reconstruction of the Old Town Centers', 68–9; C Scott-Clark and A Levy, *The Amber Room, The Untold Story of the Greatest Hoax of the Twentieth Century*, London, 2004; P Bruhn, *Das Bernsteinzimmer in Zarskoje Selo bei Sankt Petersburg*, Berlin, 2004.

19   Magirius, 'Denkmalpflege in der DDR', 125–40; *DKD*, 1/1991.

20   H Nadler, 'Von den Anfängen der Denkmalpflege in Sachsen nach dem Kriege', *DKD*, 1/ 1991, 60–1; Magirius, 'Denkmalpflege in der DDR', 127, 130, 134; *DKD*, 1/1991, 63–6; Düwel and Gutschow, *Städtebau*, 219–21; F Urban, *Neo-Historical East Berlin*, Farnham, 2009.

21   Scheurmann, *Zeitschichten*, 224–37; H Magirius and U Aust, *DKD*, 1/1991, 24–5 and 67–78; Nadler, 'Von den Anfängen', 52–7; Deiters, 'Historic Monuments', 147; *DD*, 2/2007, 189.

22   B Rymaszewski, *Generations Should Remember*, Oświęcim, 2003.

23   J E Young (ed), *The Art of Memory: Holocaust Memorials in History*, Munich, 1994.

24   Scheurmann, *Zeitschichten*, 178–83; W Sabrow (ed), *Erinnerungsorte der DDR*, Munich, 2009; W Wippermann, *Denken statt Denkmälen*, Berlin, 2010, 102–5; V Knigge, *The History of the Buchenwald Memorial*, Weimar, 2000; Magirius, 'Denkmalpflege in der DDR', 131; R Koshar, *Germany's Transient Pasts*, Chapel Hill, 1998.

25   *Architectural Review*, October 1970, 304.

26   www.fsz.bnmc.hu/hungary/landscape/bptour; Dobby, *Conservation and Planning*, 91; *Architectural Review*, August 1965, 104; *Restauro*, March 1972; Stubbs and Makaš, *Architectural Conservation*, 237.

27   J Vohrna, F Soukup *et al*, *SURPMO*, Prague, nd (*c.*1985); Stulc, '19th Century Prague', 149–50; Hruška, 'Architectural Heritage', 155–7.

28   *Frankfurter Allgemeine Zeitung*, 28 August 1992, 9.

29   *Project Russia*, 43, 1/2007; V A Shkvarikov, *Essays on the Planning and Building of Russian Towns*, Moscow, 1954; O A Shvidkovsky, 'The Historical Characteristics of the Russian Architectural Heritage', *JSAH*, 38, 1979, 148; B L Altshuller *et al*, *Pamietniki Arkhitektury Moskovskoi oblasti*, Moscow, 1975.

30   Deiters, 'Historic Monuments', 146; 'Bericht der ehemaligen Arbeitsstellen', 48; Hubel, *Denkmalpflege*, 128; Snyder, 'Future has a Dubious Past'; D Kobielski, *Warszawa k Lotu Ptaka*, Warszawa, 1971; Stolica, *Nowa Warszawa w ilustracjach*, Warsaw, 1955.

31   Glendinning, *Modern Architect*, 263–7; G Hryniewicz-Lamber, 'Late Modern Buildings', 278; Düwel and Gutschow, 207.

32   Khrushchev: A Schmidt, *Impact of Perestroika*, 341, 348, 352; M Bourdeux, *Patriarch and Prophets*, New York, 1970, 124–40; J Ellis, *The Russian Orthodox Church: a Contemporary History*, Bloomington, 1986, 29–31; M Brudny, *Reinventing Russia*, Cambridge, MA, 2000, 44; Cracraft, *Petrine Revolution*, 9; W Prigge, *Ikone der Moderne*, Dessau, 2006; Piltz, *Streifzug durch die deutsche Baukunst*, 85, 127–9.

33   Shvidkovsky, 'Historical Characteristics', 148; Dobby, *Conservation and Planning*, 88; Stubbs and Makaš, *Architectural Conservation*, 309.

34   D Rodwell, 'Comparative Approaches to Urban Conservation in Central and Eastern Europe', *The Historic Environment*, November 2010, 116–42.

35   *ÖZKD*, 41, 1987, issue 314, 151; 'Berichte der ehemaligen Arbeitstellen', 37.

36   S Patkauskas, *Archeologiniai tyrimai Trakuose*, Vilnius, 1982; A Medonis, *Vilnius*, Vilnius, 1977, 17. Even in the Vilnius Old Town, propaganda could still override conservation, as in the bizarre 1960s' part-demolition of the 17th-century Radvila Palace during the shooting of a Mosfilm drama on the 'German

wartime destruction of Vilnius'. F Tomps (ed), *Pool ajandit restaureerimist Eestis*, Tallinn, 2009, 11–33 and 47. A new Estonian Scientific Restoration Workshop was established in 1950 by the State Architecture and Building Committee; a key leader was architect Ernst Ederberg. R Zobel, *Tallinna keskaegsaed kindlustused*, Tallinn, 1980; M Kalm, *Eesti 20. sajandi arhitektuur*, Tallinn, 2001, 425–33.

37 J Krastins, I Strautmanis, *Riga – Complete Guide to Architecture*, Riga, 2004, 24–7, 34, 59–62. The 1958 and 1963 Technical University and apartment blocks were by O Tilmanis; the 1969–70 'Red Riflemen' museum was designed by D Z Driba and G Lusis-Grinbergs (following a 1964–5 competition inaugurated by Soviet Prime Minister Nikolai Kosygin); the statue of a group of Riflemen was by sculptor V Albergs.

38 Berger, 'Tendenzen der Denkmalpflege', 4–5; 'Berichte der ehemaligen Arbeitstellen', 25; Magirius, 'Denkmalpflege', 134; Hubel, *Denkmalpflege*, 127–36.

39 H G Hiller von Gaertringen, 'Sturm auf das Stadtbild', in S Baldrick and R Clay (eds), *Iconoclasm*, Farnham, 2007, 235; *AR*, November 1970, 306; Magirius, *Denkmalpflege*.

40 Dobby, *Conservation and Planning*, 88; *Project Russia*, 43, 2007, 64; Brudny, *Reinventing Russia*, 45 and 67–8; G H Hosking, *Rulers and Victims*, Cambridge, MA, 2006, 357; Colton, *Moscow*, 555.

41 Schmidt, *Impact of Perestroika*, 350; Shvidkovsky, 'Historical Characteristics', 151; Dobby, *Conservation and Planning*, 88. The redevelopment proposals were drawn up by Workshop No. 1 of the Institute for the General Plan. Grishin's intervention was endorsed by veteran Soviet architect Mikhail Posokhin, who argued ingeniously that swathes of the historic centre should be preserved both 'as memorials to past ages and to events during the Revolution'. See also A Latur, *Moskva 1890–2000*, Moscow, 2009.

42 Schmidt, *Impact of Perestroika*, 335–6, 355; F Feldbrugge (ed), *Soviet Administrative Law*, Dordrecht, 1989; F Feldbrugge *et al*, *Encyclopaedia of Soviet Administrative Law*, Dordrecht, 1989; Brudny, *Reinventing Russia*, 68; B G Fyodorov *et al*, *Hroniki unichtozhenia staroi Moskvy*, Moscow, 2006, 10; *MH*, 179, 1992, 49–53, 67–70, 104–8; Shvidkovsky, 'Historical Characteristics', 148–51; *Project Russia*, 43, 2007, 64–6; *GRKG*, 472–3; Düwel and Gutschow, 211.

43 Fitch, *JSAH*, 1966; Hruška, 'Architectural Heritage', 157–8.

44 *AR*, November 1970, 305; Glendinning, *Modern Architect*, 262–8 and 356–77; J Haspel, 'Das Brandenburger Tor', in Scheurmann, *Zeitschichten*, 94–101.

45 *JSAH*, 38, 1979, 123ff.

46 PKZ, *Les ateliers de restauration des monuments historiques*, Warsaw, 1981; Dobby, *Conservation and Planning*, 86–7.

47 Urban, *Neo-historical East Berlin*, 143–62; Berger, *Tendenzen der Denkmalpflege*, 7; Brudny, *Reinventing Russia*, 139–142.

48 Deiters, 'Denkmalpflege', 134; 'Berichte der ehemaligen Arbeitstellen', 25 and 42; Deiters, 'Historic Monuments', 142; G Piltz, *Streifzug durch die deutsche Baukunst*, 49; Berger, 'Tendenzen', 7; *AR*, November 1970, 304 and 306.

49 Berger, 'Tendenzen der Denkmalpflege'.

50 Urban, *Neo-Historical East Berlin*, 35–66.

51 B Heckart, 'The Battle of Jena', *Journal of Urban History*, 32, 4, 2006, 546; 'Berichte der ehemaligen Arbeitstellen', 19 and 26; Berger, 'Tendenzen der Denkmalpflege', 7; Hubel, *Denkmalpflege*, 133; Magirius, 'Denkmalpflege in der DDR', 137.

52 The Nikolaiviertel was questioned by some contemporary critics as 'Old-Berlin stage-scenery': Urban, *Neo-Historical East Berlin*, 99–141, 163–171, 186–205; Berger, 'Tendenzen der Denkmalpflege', 6; F Urban, *Planning Perspectives*, 23, 1, 2008, 1–28; Düwel and Gutschow, *Städtebau*, 256–60.

53 Schmidt, *Impact of Perestroika*, 347; TASS: 'G Somov, 'Novoe v istoricheskom okruzhenii', *Stroitelstvo i Arkhitektura Moskvy*, February 1981, 20–3; *Project Russia*, 43, 1/2007: Gutnov acclaimed the old centre of Moscow as 'a custodian of historical time, as an architectural chronometer whose face always has room for one more feature from our own time'; A Kurg, M Laanemets (eds), *Keskonnad, Projektid, Kontseptsioonid: Tallinna Kooli Arhitektid 1972–1985* (Museum of Estonian Architecture), Tallinn, 2008, 208.

54 Porumbescu etc.: Glendinning, *Modern Architect*, 407–9. Postwar publications on vernacular architecture in Romania: see e.g. G Ionescu, *Arhitectura Populara in Romania*, Bucharest, 1971; F Stanculescu *et al*, *Arhitectura Populara Romaneasca* (multi–vol series), Bucharest, 1956 onwards. Albania, role of L M Ugolini: Stubbs and Makaš, *Architectural Conservation*, 389.

55 S Cantacuzino, 'Reconstruction in Bucharest', *ICOMOS Information*, 2, 1987, 9–18; M de B Cavalcanti, 'Ceaușescu's Bucharest', *Planning History*, vol 16, 3, 1994, 18–24; N Ceaușescu, *On the Way of Building up*

the Multilaterally Developed Socialist Society, vol 14, Bucharest, 1977, 145; D C Giurescu, The *Razing of Romania's Past*, London, 1990, 23, 26, 38, 61.

56 *New York Times*, 3 December 1989; A Budișteanu, *Arhitectura*, 28, 1–2, 1980, 12–17; A Budișteanu, 'Sistimatizarea si modernizarea', *Revista economica*, 24, 1988, 3–6.

Chapter 11

1 R Lemaire, lecture transcript, *Proceedings of First International Congress on Architectural Conservation*, Basel, 28 March 1983.

2 Forrest, *International Law and the Protection of Cultural Heritage*, 63–132; 1954 Hague Convention: www.icomos.org/hague; C E M Pearson, *Designing Unesco*, Abingdon, 2010.

3 Martin and Krautzberger, *Handbuch Denkmalschutz*, 19.

4 National Trust for Historic Preservation, *Historic Preservation Today*, Charlottesville, VA, 1966, 40–1.

5 *ICCROM Newsletter*, 35, October 2009, 30.

6 P Gazzola, *The Past in the Future*, Rome (ICCROM/University of Rome), 1st edition, 1969 (revised 1975), 8, 28–34, 55, 105–8, 125–32.

7 *Builder*, 7 January 1966, 19; T Spencer, 'The Race To Save Abu Simbel is Won', *Time Magazine*, 2 December 1966; cf. Skopje earthquake rebuilding: R Golik, T Arsovski (eds), *Skopje 1963–1973*, Skopje, 1973.

8 *Builder*, 29 August 1958, 357.

9 *ICCROM Newsletter* 35, 1–2.

10 Gazzola, *Past in the Future*, 91–2; J Jokilehto, *Built Environment*, vol 33, 3, 280; *ICCROM Newsletter* 35, 3; P Philippot, 'Histoire et actualité de la restauration', in *Geschichte der Restaurierung*, vol 1, Worms, 1991, 7–13; J Jokilehto, 'Les fondaments des principes modernes en conservation', in *Geschichte der Restaurierung*, vol 1, 29–33.

11 *ICCROM Newsletter* 33, June 2007, 8; H Spande (ed), *Conservation Legacies of the Florence Flood of 1966*, London, 2009; *DOC Toscana*, Special Issue, 'Anno 66', November 2006.

12 H Lottman, *How Cities are Saved*, New York, 1976, 91–8; Dobby, *Conservation and Planning*, 79–80; Sir Ashley Clarke, 'The Preservation of Venice', *Transactions of the Ancient Monuments Society* new series 17, 1970, 51–62; S Cantacuzino, *Architectural Conservation in Europe*, London, 1975, 28–35.

13 Kain, *Planning for Conservation*, 235–58; Craig A Mauzy, *Agora Excavations 1931–2006*, Athens, 2006; American School of Classical Studies, *The Stoa of Attalus II*, Princeton, NJ, 1992. The 1950s' Agora campaign also included the restoration of a small Byzantine church (1954–5, with a new domed roof by Travlos), funded by the Samuel H Kress Foundation, and a verdant landscaping plan, executed from 1954, by Pittsburgh landscape architect Ralph E Griswold.

14 Ministry of Culture, *The Acropolis at Athens, Conservation, Restoration and Research*, Athens, 1983; C Bouras and K Zambas, *Die Arbeiten . . . auf der Athener Akropolis*, Athens, 2002; C Bouras, 'Strict and Less Strict Adherence to the Principles of Anastelosis of the Ancient Monuments in Greece', *Acropolis Restoration News* (YSMA), 9, Athens, 2009, 2–8; A Loukaki, *Living Ruins, Value Conflicts*, Farnham, 2008.

15 *Acropolis Restoration News*, 7, July 2007, 2–5; *The Study for the Restoration of the Erechtheion, 1977*.

16 Lottman, *How Cities are Saved*, 109–10; 'Back to Sanity in Jerusalem', *AR*, July 1972, 41–3; A Kutcher, *The New Jerusalem, Planning and Politics*, London, 1973; Pullan and Sternberg, 'Holy Basin', *Planning Perspectives*, April 2012, 238–42; W Pullan, M Gwiazda, M Dumper and C Larkin, *The Struggle for Jerusalem's Holy Places*, London, 2012. For a detailed recent account of post-1967 Israeli policies and debates, see B Slae, R Kark and N Shoval, 'Post-war Reconstruction and Conservation of the Historic Jewish Quarter in Jerusalem, 1967–75', *Planning Perspectives*, vol 27, no 3, July 2012, 369–92.

17 Gazzola, *Past in the Future*, 112–13.

18 *ICCROM Newsletter*, 35, October 2009, 1; F Valderrama, *A History of UNESCO*, Paris, 1995; J Jokilehto, Masterclass at ECA, 16 March 2006.

19 Gazzola, *Past in the Future*; *ICCROM Newsletter* 35, 3.

20 Text extracts from Venice Charter: 'our duty to hand [monuments] on in the full richness of their authenticity' . . . 'permanent conservation' . . . 'traditional setting . . . must be kept. No new construction, demolition of modification which would alter the relations of mass or colour must be allowed' . . . 'restoration . . . must stop at the point where conjecture begins' . . . 'reconstruction . . . ruled out a priori' . . . 'additions cannot be allowed except in so far as they do not detract from the interesting parts of the building, its traditional setting . . .'; *Monumentum*, 1, 1967; *RIBA Journal*, April 1961, 230–40.

21 L Verpoest, 'Raymond M Lemaire', lecture to 'Conservation in Changing Societies' conference, Leuven, 2006 (and proceedings, eds Patricio, van Balen and de Jonge, 2006).

22 Polski Komitet Narodowy ICOMOS, *XXX ICOMOS*, 11–54; Dobby, *Conservation and Planning*, 100.

23 Verpoest, 'Lemaire', 19–26; Gazzola, *Past in the Future*, 28–34; Conference on the Role and Works of Piero Gazzola, Verona, 2008; *ICCROM Newsletter*, October 2009; 'An Interview with Raymond Lemaire', *Monumentum*, vol 26, 2, 1983, 84–91.

24 ICOMOS, 2nd General Assembly, Oxford, July 1969: *Report on the Activities of the National Committees*; Gazzola, *Past in the Future*, 118–21.

25 The only non–European/American participants were Ghana, Israel, Japan and Iran. In addition, there were observers from UNESCO, the Council of Europe, the Rome Centre (Plenderleith), Europa Nostra, IFLA (International Federation of Landscape Architects), UIA (Union Internationale des Architectes), and AIEST (the International Association of Scientific Experts on Tourism). ICOMOS, 2nd General Assembly, 1969: *Report on the Activities of the National Committees*; Dobby, *Conservation and Planning*, 100.

26 B Feilden, *Conservation of Historic Buildings*, Oxford, 1982.

27 These included 1965 conferences in Palma (on inventorisation) and Vienna (on the difficulties of 'reanimation' of monuments), a 1966 conference at Bath on the principles and methods of 'reanimation' and a synthesis conference at Avignon in 1968. Gazzola, *Past and the Future*, 15, 54–9, 112–13; Council of Europe, *The Preservation and Development of Ancient Buildings and Historical or Other Sites*, 1963; Dobby, *Conservation and Planning*, 99–100.

28 See e.g. Friends of the Earth, *The Stockholm Conference: Only One Earth*, London, 1972; *How Do You Want To Live?* HMSO, 1972.

29 E F Schumacher, *Small is Beautiful*, London, 1973; D Bell, *The Historic Scotland Guide to International Conservation Charters (Technical Advice Note 8)*, Edinburgh, 1997, 16; Brundtland Report, *Our Common Future*, 1987.

30 D Bell, *The Historic Scotland Guide to International Conservation Charters (Technical Advice Note 8)*, Edinburgh, 1997, 27.

31 Gazzola, *Past in the Future*, 112–17; Dobby, *Conservation and Planning*, 100–5; Cembureau Photonews, *Modern Interventions in Historic Towns*, 1975; Lottmann, *How Cities are Saved*.

32 J Kirschmann, 'Eine Zukunft für unseren Vergangenheit', in Scheurmann, *Zeitschichten*, 166–71; Council of Europe, Committee on Monuments and Sites, *European Programme of Pilot Projects*, Strasbourg, 1973; Düwel and Gutschow, *Städtebau*, 239–43; Y Friedman, *Votre ville est à vous*, Strasbourg, 1975; G Korff, 'Denkmalisierung', *Die Denkmalpflege*, 2005, 2, 133–44; Falser, *Zwischen Identität und Authentizität*, 104; J Tietz, 'Welche Vergangenheit für unsere Zukunft?', *DD*, 2/2005, 145–50. Anti-Modernist interpretation: M Hardy (ed), *The Venice Charter Revisited: Modernism, Conservation and Tradition in the 21st Century*, Newcastle, 2009.

33 Countess Dartmouth, *What is our Heritage?*, London, 1975; D Pearce, *Conservation Today*, London, 1989, 2–8; Falser, *Zwischen Identität und Authentizität*, 99–107; Cantacuzino, *Architectural Conservation*; Dobby, *Conservation and Planning*, 100–113; H Perkin, *The Structured Ground*, Brighton, 1981.

34 M Parent, 'The Heritage Tomorrow', in Council of Europe, *A Future for our Past*, 1985.

35 Council of Europe/Civic Trust, *European Charter of the Architectural Heritage*, 1975.

36 ICOMOS German National Committee, *Colloque sur la conservation des petites villes historiques*, Rothenburg ob der Tauber, May 1975, 1975 (Conference Rapporteur General: Prof. Gerd Albers, Professor of City and Regional Planning, TU Munich); German Commission for UNESCO, *Protection and Cultural Animation of Monuments, Sites and Historic Towns in Europe*, Bern, 1980.

37 Bell, *Conservation Charters*, 14 and *passim*.

38 D Rodwell, 'The Unesco World Heritage Convention 1972–2012, Reflections and Directions', *The Historic Environment*, April 2012, 64–85; F Leblanc, 'An Inside View of the Convention', *Monumentum*, 1984, 17–32; interview with H Stovel, October 2006; C Cameron and M Rössler, 'Voices of the Pioneers: UNESCO's World Heritage Convention 1972–2000', *Journal of Cultural Heritage Management*, vol 1, 1, 2011; UNESCO Oral Archives Initiative: www.unesco.org/archives/new2010/en/oral_archives.html.

39 L Pressouyre, *The World Heritage Convention, Twenty Years Later*, UNESCO, Paris 1993; C Cameron, 'The Evolution of the Concept of OUV', in Stanley-Price and King, *Conserving the Authentic*, 127; B von Droste, 'The Concept of Outstanding Universal Value and its Application: "From the Seven Wonders of the Ancient World to the 1000 World Heritage Places Today"', *Journal of Cultural Heritage Management*, vol

1, 1, 2011; J Jokilehto, 'World Heritage: Observations on Decisions related to Cultural Heritage', *Journal of Cultural Heritage Management*, vol 1, 1, 2011; M Batisse and G Bolla, '"L'invention du "patrimoine mondial"', *Les cahiers de l'histoire*, Paris, 2003; *Patrimoine mondial de l'Unesco*, Ittingen, 2010; T Patricio, K van Balen and K de Jonge (eds), *Conservation in Changing Societies*, 2006, 39–48; S M Titchen, *Outstanding Universal Value*, 1995; S M Titchen, 'On the Construction of Outstanding Universal Value', *Conservation and Management of Archaeological Sites*, 1, 4, 1996, 235–42.

40 Choay, *Invention*, 221; O Martin and G Piatti (eds), *World Heritage and Buffer Zones (UNESCO World Heritage Papers*, 25), Paris, 2009; H Dagslund, *Bryggen Guide*, Bergen 2004.

41 Programme of First International Congress on Architectural Conservation, University of Basel, March 1983.

42 Choay, *Invention*, 149; *AR*, May 1972, 263–4.

43 Glendinning, *Architecture's Evil Empire?*, 48–60; Dobby, *Conservation and Planning*, 100.

44 Interview with H Stovel, October 2006.

45 ICOMOS Australia Inc, *The Burra Charter*, 1999; M Walker and P Marquis-Kyle, *The Illustrated Burra Charter* (updated edition), Burwood, Vic., 2004; N Gutschow, 'Restaurierung und Rekonstruktion . . . im Kontext Südasiens', *DKD*, 49, 2, 1991, 156–9.

Chapter 12

1 J Jokilehto, Masterclass lecture, Edinburgh College of Art, 16 March 2006.

2 See for example www.denkmalwerte.org; H R Meier and I Scheurmann (eds), *DENKmalWERTE*, Berlin/Munich, 2010; *DD*, 1/2011, 3 (foreword).

3 Stubbs and Makaš, *Architectural Conservation*. See also J H Stubbs, *Time Honored: A Global View of Architectural Conservation*, Hoboken, NJ, 2009.

4 Mostar, etc.: M Coward, *Urbicide*, London, 2009, 1–7; C Grodich, 'Reconstituting Identity and History in Postwar Mostar', *City*, 6, 2002, 61–82; Z Zderic, R Radic and A Kindij, 'Mostar Old Bridge Rehabilitation', *Arch 07: 5th International Conference on Arch Bridges, 2007, Proceedings*, 695–703; Lambourne, *War Damage*, 205–6. E Palazzo (ed), *Esempi di Architettura*, no 7, 2009.

5 Choay, *Invention*, 144–5, 159, 222–4.

6 M Rössler, lecture to Royal Society of Arts, Edinburgh, 3 December 2010 'Culture and Heritage, European Perspectives'; T Fejerdy, 'Approaching 40 Years Old', in Stanley-Price and King, *Conserving the Authentic*, 138; World Heritage: J Jokilehto, Masterclass lecture, Edinburgh College of Art, 16 March 2006. Malbork: see e.g. R Rząd, 'Pomniki malborskie', *Rocznik Gdański*, 53, 1993, 95 ff; R Rząd, 'Malbork 1945, Relacja z obrony zamku', *Militaria i Fakty*, 2002; R Rząd, *Zamek w Malborku 1882–1945*, Malbork, 1996. For a typically blatant example of the interrelationship of World Heritage and tourist commercialism, see e.g. *South China Morning Post*, 19 July 2011, special report ('Go Hangzhou': on West Lake UNESCO designation).

7 Choay, *Invention*, 145–8, 158–9, 163; J Harris and R Williams (eds), *Architecture, Art and Urban Style within the Global Politics of City Branding*, Liverpool, 2011; P Harrison (ed), *Civilizing the City*, Edinburgh, 1990, 64, 80; R Maitland, 'Tourism and Changing Representation in Europe's Historic Capitals', *Revista di Scienze del Turismo*, 2/2010, 73–119; L Kolbe, 'Central and Eastern European Capital Cities', *Planning Perspectives*, vol 22, 1, 2007, 79–111.

8 Temple Bar Properties, *Temple Bar Lives*, Dublin, 1991; Temple Bar Properties, *Development Programme for Temple Bar*, Dublin, 1992; P Quinn (ed) *Temple Bar, the Power of an Idea*, Dublin, 1996; Irish Architectural Archive, *Temple Bar 15*, Dublin, 2006; R Koolhaas and B Mau, *S, M, L, XL*, 1995, 1248.

9 Rebecca Madgin, 'Reconceptualising the Historic Urban Environment', *Planning Perspectives*, January 2010, 29–48; M Binney, *Bright Future: the Re–Use of Industrial Buildings*, London, 1990; M Bianchini and M Parkinson, *Cultural Policy and Urban Regeneration: the West European Experience*, Manchester, 1994; Urban Task Force, *Towards an Urban Renaissance*, London, 1999, 251–2; M Stratton, *Industrial Buildings: Conservation and Regeneration*, London, 2000. Expansion of Beamish as Greenfield-style complex in north-east England, targeted both at regional economic 'regeneration' and at safeguarding the 'distinct identity' of industrial north-east England from the losses of 'our families, our neighbourhoods, our region' (in the words of founder Frank Atkinson): G Muirhead, *Beamish, the North of England Open Air Museum*, Norwich, 2003.

10 'Glocalisation': see Barry Wellman and Keith Hampton, 'Living Networked On and Offline', *Contemporary Sociology*, November 1999, 648–54; K Hampton and B Wellman, 'The Not So Global Village of Netville', in B Wellman and C Haythornthwaite (eds), *The Internet in Everyday Life*, Oxford, 2002, 345–71; N Al Sayyad, *Hybrid Urbanism*, New York, 2001. Koolhaas and consumerism: CRONOCAOS preservation tour, part 4, www.designbom.com/weblog/cat/9/vie=w/11428/rem–koolhaas–oma–cronocaos. 2010: M Glendinning, *Architecture's Evil Empire?*, London, 2010; Sir Simon Jenkins, 'Building the Big Society', Royal Society of Edinburgh, 22 November 2010. Launch of BBC conservation game-show format series: 'Restoration' *Architects' Journal*, 14 August 2003, 20–1; P Wilkinson, *Restoration: Discovering Britain's Hidden Architectural Treasures*, London, 2003 (won by a swimming pool in Manchester, with 52 per cent of the 282,000 votes cast); M Slocombe, 'Restoration on TV: Friend or Foe?', *Cornerstone*, Winter 2011, 4. 2010 advocacy of popularised interpretation by incoming National Trust head, Sir Simon Jenkins: 'It's the Dead Hand We're Trying To Remove, that [strips heritage] of Character and Personality and Reduces It to Art History.' M Kelleher, 'Images of the Past: Historical Authenticity and Inauthenticity from Disney to Times Square', *CRM*, Summer 2004, 6–19. Medina: A A Kaki, Urban Fabric for Al–Madinah, Jeddah, 2005; M S Makki, Medina, Saudi Arabia, London, 1982; S Bianca, 'Madinah Al-Muniwarah: Planning and Urban Design Concepts for the Central Area', in *MIMAR 12: Architecture in Development*, Singapore, 1984; A Al-Alaidarous, Conservation of Historic Sites of Al-Madinah (MSc project, Edinburgh College of Art), 2011.

11 J Jokilehto, Masterclass lecture, Edinburgh College of Art, 16 March 2006.

12 V Caldelli, G Meucci, *La Torre Pendente, il restauro del secolo*, Pisa, 2005; *Newsweek*, 18 June 2001, 58–9; J Jokilehto, in Falser, Lipp and Tomaszewski, *Conservation and Preservation*, 35; Polish National Committee of ICOMOS, *Problems of Heritage*, 32–6.

13 G Ashworth. 'Pluralising the Past – Heritage Policies in Plural Societies, *Edinburgh Architecture Research*, 30, 2006; Puccio, Raymond Lemaire International Centre for Conservation, *Conservation in Changing Societies*, Leuven, 2006. T Mann and Lübeck: R Pommer, in Scheurmann, *Zeitschichten*, 7. C Sauer, *The Morphology of Landscape. University of California Publications in Geography*, 22, 1925, 19–53; D C Harvey, 'Heritage Pasts and Heritage Presents', *International Journal of Heritage Studies*, December 2001, 319–38; Polish National Committee of ICOMOS, *Problems of Heritage*, 8–19.

14 J R Gillis, *Commemorations*, Princeton, NJ, 1994, 17.

15 M Halbwachs, *La memoire collective*, Paris, 1950, 235; A Forty and S Küchler (eds), *The Art of Forgetting*, Oxford, 1999; G Lucas and V Buchli (eds), *Archaeologies of the Contemporary Past*, London, 2001; M Proust, *The Remembrance of Things Past*, vol 3, *Time Regained*, New York, 1981 (originally published 1927); M Crinson, *Urban Memory*, London, 2004; A Steenkamp, 'Apartheid to Democracy', *ARQ*, vol 10, 3–4, 2006, 249–54; Knigge, in Scheurmann, *Zeitschichten*.

16 Benedict Anderson, *Imagined Communities, Reflections on the Origin and Spread of Nationalism*, London, 1983/91. See also A Gruszecki, 'Cultural and National Identity and the Protection of Foreign Fortifications in Poland', in Polish National Committee of ICOMOS, *Problems of Heritage*, 48: '. . . a link with Polish history and Polish landscape, in many cases also with Polish blood, is a good basis for the social approval and protection of foreign fortifications, even those originally directed against Poland, which, however, constitute an important contribution to the development of human thought on a supranational scale'.

17 Choay, *Invention*, 166–71; Glendinning, *Architecture's Evil Empire?*; Koshar, *Germany's Transient Pasts*, 292–3.

18 In some cases, the commemoration was still quite elitist in character: for example, the artistic evocation of deindustrialisation in Bernd and Hilla Becher's depopulated photographs of defunct industrial sites. S Lange, *Bernd and Hilla Becher, Life and Work*, Cambridge, MA, 2006.

19 *SAH News*, September–November 2009, 3–4; www.msbluestrail.org; B L Savage, *African-American Historic Places*, Washington, DC, 1994; US Department of the Interior, National Park Service, *Guidelines for Identifying . . . America's Historic Battlefields*, Washington, DC, 1992; N Kaufman, *Place, Race and Story: Essays on the Past and Future of Historic Preservation*, New York, 2009.

20 C Chu, 'Heritage of Disappearance', *Traditional Dwellings and Settlements Review*, vol 18, 2, 2007, 43–55.

21 'Dark Tourism': G Ashworth and R Hartmann, *Horror and Human Tragedy Revisited: The Management of Sites of Atrocities for Tourism*, New York, 2005; R E Kelly, 'America's World War II Home Front Heritage' and M G Moersi, 'Presenting Race and Slavery at Historic Sites', *CRM*, Summer 2004, 34–50 and 92–5; Liz Sevcenko, 'Sites of Conscience: Imagining Reparations', in *Change over Time*, vol 1, 1, Spring 2011, 6–33; V Knigge, *The History of the Buchenwald Memorial*, Weimar, 2000; *Die Neukonzeption der Gedenkstätte*

Buchenwald, Weimar, 2001; N G Finkelstein, *The Holocaust Industry, Reflections on the Exploitation of Jewish Suffering*, London, 2000; *DD*, 2/2011 (themed issue on Jewish heritage). Weathermen: see e.g. www.complex.com/city–guide/2011/02/look–up–18–west–11th–street.

22 I Gutman, M Bernbaum and Y Gutman, *Anatomy of the Auschwitz Death Camp*, Bloomington, IN, 1998; R L Rubenstein and J K Roth, *Approaches to Auschwitz*, Louisville, 1987; Rymaszewski, *Generations Should Remember*. A composite committee-design of tomb and sarcophagus designs emerged from a 1957–9 international competition chaired by Henry Moore (and completed in 1967).

23 *Independent Magazine*, 15 August 1992, 41; H Hübner, *Das Gedächtnis der Stadt*, Berlin, 1997; E Francais and H Schulze (eds), *Deutsche Erinnerungsorte*, Munich, 2001; Knigge, in Scheurmann, *Zeitschichten*.

24 N Foster, *Rebuilding the Reichstag*, New York, 2000; Jacob Baal-Teshuva, *Christo and Jeanne-Claude*, Cologne, 2005.

25 Wippermann, *Denken statt Denkmälen*; Akademie der Künste Berlin, *Streit um die Neue Wache*, Berlin, 1993; S Thoma, *Vergangenheitsbewältigung am Beispiel der Auseinandersetzungen um die Neue Wache*, Berlin, 1995; S F Kellerhoff, 'Todesstreifen soll wieder aufgebaut werden', *Welt Online*, 4 February 2009.

26 Andreas Huyssen, 'The Voids of Berlin', *Critical Inquiry*, vol 24, 1, Autumn, 1997, 57–81; G Dolff-Bonekämper, 'Sites of Hurtful Memory', *Getty Conservation Institute Newsletter*, 17.2, Summer 2002.

27 Interview with H Stovel, October 2006. Nara: Falser, Lipp and Tomaszewski, *Conservation and Preservation*, 115–32; K E Larsen and N Marstein (eds), *Conference on Authenticity in relation to the World Heritage Convention, Preparatory Workshop, Bergen, Norway*, Trondheim, 1994; K E Larsen (ed), *Nara Conference on Authenticity, Japan 1994*, Trondheim, 1995.

28 Interview with H Stovel, October 2006.

29 In his papers to the conference and a preparatory workshop in Bergen, Jokilehto argued that conservation was not only about 'keeping the material, but also recognising this spirit, this "non-physical" essence and authenticity of the heritage, and its relation with society', Larsen, *Nara Conference on Authenticity*; Stanley-Price and King, *Conserving the Authentic*, 152.

30 J Jokilehto, 'Considerations on Authenticity and Integrity in the World Heritage Context', *Edinburgh Architecture Research*, 30, 2006, 7–12; UNESCO, *Convention for the Safeguarding of the Intangible Cultural Heritage*, Paris, 2003; Stanley-Price and King, *Conserving the Authentic*, 151–2.

31 B Szmygin, in Falser, Lipp and Tomaszewski, *Conservation and Preservation*, 98; interview with H Stovel, October 2006.

32 ICOMOS UK, 'Conservation Philosophies, Global or Local?', conference call for papers, York, 2010.

33 J Jokilehto, Masterclass lecture, Edinburgh College of Art, March 2006; B Feilden, 'A Profile of Jukka Ilmari Jokilehto', *Journal of Architectural Conservation*, March 2004, 49–52; Stanley-Price and King, *Conserving the Authentic*, 7.

34 Jokilehto, 'Considerations on Authenticity and Integrity', 7.

35 A McCleery *et al*, *Intangible Cultural Heritage in Scotland*, Edinburgh, 2008; Jokilehto, in Falser, Lipp and Tomaszewski, *Conservation and Preservation*, 31.

36 National reflections of these international attempts to widen the definitions of authenticity: e.g. Historic Scotland, *The Stirling Charter*, Edinburgh, 2000; English Heritage, *Conservation Principles, Policies and Guidance for the Sustainable Management of the Historic Environment*, London, 2008.

37 C Cameron, 'The Evolution of the Concept of Outstanding Universal Value', in Stanley-Price and King, *Conserving the Authentic*, 134.

38 *Economist*, 28 August 2010, 44–5.

39 Caisteal Tioram: *Scotsman*, 22 August 2004, 15; *Country Life*, 20 November 1997, 42; N Ashurst, 'The Siege that Saved Tioram', *Cornerstone*, vol 27, 4, 2006, 50–2; G Bennett, *Guardian*, 27 April 1995, 5; *Times Magazine*, 16 July 1994, 27.

40 Establishment in England of Department of Heritage: *Architects' Journal*, 22 April 1992, 13; D Sudjic, *Observer*, 3 April 2005, 6; *Architects' Journal*, 17 January 2002, 4; English Heritage, *Building in Context: New Development in Historic Areas*, London, 2001; *Building Design*, November 2001; *Cornerstone*, vol 29, 1, 2008, 45. Similar clashes at a national level: within England, for example, the later 1990s and 2000s saw a shift to an aggressively 'modernising' government administration, with the state heritage organisation, English Heritage, coming under pressure from the advocates of 'iconic' neo-modernism, such as London mayor Ken Livingstone, who denounced it as 'the English Taliban'.

41 G Debord, *La société du spectacle*, Paris, 1967; F Jamieson, *Postmodernism, or, The Cultural Logic of Late Capitalism*. Durham, NC, 1991.

42 *DOCOMOMO-International Journal*, 27 ('The History of DOCOMOMO'), 2002, 8–9; interview with H Stovel, October 2006; T Prudon, *Preservation of Modern Architecture*, Hoboken, NJ, 2008. One 2009 DOCOMOMO conference, 'Mirror of Modernity', even examined the postwar history of the Conservation Movement itself as a kind of 'hidden' or 'complementary' modernity.

43 DOCOMOMO 2008 International Conference, Rotterdam: lecture by B Bergdoll and debate session, 'Reconstruction of Modern Movement buildings'; A Canzani, 'La pellicola dell' analogia', *Scienza e beni culturali*, 20, 2004.

44 In 2010, DOCOMOMO's biennial conference was held in the UNAM campus in Mexico City, showpiece of Mexico's exotically 'regional' modernism and inscribed on the World Heritage List in 2007, joining the prehistoric pyramids of Teotihuacan and the exotic water-gardens of Xochimilco (inscribed 1987): http://whc.unesco.org/en/list/1250; www.international.icomos.org/risk/2007/pdf/Soviet_Heritage_09_1–3_Cohen.pdf. Modernism and World Heritage: see also H-J Henket (ed), *The Modern Movement and the World Heritage List*, Rotterdam, 2009.

45 R Koolhaas (ed), *Content*, Cologne, 2004, 454–65; 'Estudio de preservación para Pekín', *El Croquis*, 134–5, 2007, 38–43.

46 Echoing Korn's 1940s' praise of the 'purity' of ruined Hamburg facades, Harrap argued that 'the ruin had to be de-ruinated to rediscover its state of innocence': ECA Masterclass lecture, March 18, 2010. Staatliche Museen zu Berlin, *Das Neue Museum Berlin*, Berlin, 2009; F P Hesse, 'Neues vom Neuen Museum', *DD*, 1/2000, 19–32; English Heritage, *Conservation Bulletin*, 59, Autumn 2008, 9–13; S Los, *Carlo Scarpa*; P Davey, 'Diocesan Dialogue', *AR*, November 2007, 37–42; *Detail*, January 2008, 16; Glendinning, *Architecture's Evil Empire?* 124.

47 *Acropolis Restoration News*, 1, July 2001 and 9, July 2009.

48 *SPAB in Scotland News*, no 1, 1996, 2; letter from Andrea Schuler, Europa Nostra, to P Buchanan, Historic Scotland, 10 January 2001 (per Richard Emerson).

49 English Heritage, *Conservation Bulletin*, 59, Autumn 2008, 6–9.

50 *Project Russia*, 43, January 2007; B Mitchneck, 'The Heritage Industry, Russian-style', *Urban Affairs Review*, vol 34, 1, 1998, 28–51.

51 http://en.wikipedia.org/wiki/Yury_Luzhkov; *Izvestia*, 19 May 2004. In a pattern that became familiar elsewhere, the Moscow facsimile movement then rapidly spread to more far-fetched projects, such as that for the reconstruction of the Kolomenskoye palace of Tsar Alexis (demolished in the 18th century), or for the construction of Catherine II's unbuilt palace at Tsaritsyno.

52 Krastins and Strautmanis, *Riga*; G T Tielemann, *Geschichte der Schwarzen Häupter in Riga*, Amsterdam, 1970 (reprint); www.omf.lv/eng/news/eka.html; LDKVR, *The Palace of the Grand Dukes of Lithuania*, Vilnius, 2010.

53 Frankfurt and Palast der Republik: Falser, *Zwischen Identität und Authentizität*; *Frankfurter Allgemeine Zeitung*, 26 January 2008, 46 and 1 February 2008, 33; *DD*, 1/2008; *DD*, 2/2007, 164–5; N Maak, 'Frankfurts neue Altstadt', *Frankfurter Allgemeine Zeitung*, 25 March 2011, 35 and 42, 15 January 2011, Z1–2. Mainz Marktplatz: *GRKG*, 458–60. Berlin: http://berliner–schloss.de/start.php?navID=31&typ+main.

54 'Meinungsttreit', *DKD*, 1/1991, 79ff; DD, 2/2002, 183–8; F Loeffler and H Magirius, *Frauenkirche Dresden* (Schnell Art Guide), English edition, 1994; *Frankfurter Allgemeine Zeitung*, 31 October 2005, 9; *Dresdner Neueste Nachrichten*, special edition, 'Frauenkirche Dresden, Geschichte vom Wiederaufbau', October 2005; *GKRG*, 326–8. The Hotel Dresdner Hof was by architect Wolfgang Levin of Berlin.

55 *Frankfurter Allgemeine Zeitung*, 31 October 2005, 9; *Dresdner Hefte*, 92, 4/2007, 78–95.

56 S Hertzig, 'Die historische Rekonstruktion des Dresdner Neumarktes', *DD*, 1/2000, 5–9; Hubel, *Denkmalpflege*, 286. Area 1 (immediately north-west of the Frauenkirche, adjoining late 1980s' GDR blocks) was largely complete at the time of the inauguration in 2005, and the other site parcels soon followed.

57 *Neumarkt Kurier*, 2/2007, 5 ('Zum Beispiel: Dresdner Neumarkt'). GHND member Prof. Hans Joachim Neidhardt pleaded that the new-old facsimile Neumarkt was a 'clear gesture of opposition to the levelling processes of globalisation'; he argued that, although the new Gewandhaus project was a 'high–quality piece of architecture . . . the design had no relationship to Dresden, and so could be built anywhere in the world.

In this particular spot, it's far more important to show some restraint, than to rush off in pursuit of global fashion.'

58 Deutsches Architekturmuseum, *Architektur in Deutschland: Rekonstruktion – alles bleibt anders* (edited by I Flagge and A Götz), Munich/Frankfurt, 2004; J Habich *et al*, *Denkmalpflege statt Attrappenkult*, Basel, 2011; U Hassler and W Nerdinger (eds), *Das Prinzip Rekonstruktion*, Munich, 2010; M Braum and U Baus (eds), *Rekonstruktion in Deutschland*, Basel, 2009; C Welzbacher, *Durchs wilde Rekonstrutktistan*, Berlin, 2010; Von Buttlar *et al*, *Denkmalpflege statt Attrappenkult*, 19–41, 57; G Moersch, 'Ist Rekonstruktion erlaubt?', in *Schloss, Palast, Haus Vaterland*, Berlin, 1998, 62–73. See also J Janowski, 'Restoring the Spirit and the Spirit of Restoration: Dresden's Frauenkirche as model for Bamiyan's Buddhas', in *Ethos Logos Pathos, Abstract Book 2011*, American Institute for Conservation, Washington, DC, 2011, 6.

59 D B Abramson, 'The Aesthetics of City-scale Preservation Policy in Beijing', *Planning Perspectives*, April 2007, 129–66 (especially 140); *El Croquis*, 134–5, 2007, 39–43; F Feng, *Developing Heritage*, MSc thesis, Edinburgh College of Art, 2010, 17; Liang Sicheng, 'Why We Study Chinese Ancient Architecture', *Serials of Construction Skills*, vol 7, 1, 1945; D Lu, *Remaking Chinese Urban Form*, London, 2006; Q Wang, 'Theory and Method in Urban History Conservation', Master thesis, Zheijiang University, 2003; D F d'Ayala and H Wang, 'Conservation Practice of Chinese Timber Structures', *Journal of Architectural Conservation*, vol 12, 2006, 7; *GKRG*, 210; Zhao Xiaofeng, 'Contemporary Chinese Conservation in Action', 24 March 2011 Masterclass lecture at Edinburgh College of Art.

60 *South Asia Studies*, 10, 1994, 37–44; Bhaktapur, *GRKG*, 212–13.

61 Seoul: Jong-Deok Choi, 'The Palace, the City and the Past', *Planning Perspectives*, April 2010, 193–213. Gyeongbokgung: in the latter case of the 1911 invasion, largely involving extensive structures built as recently as the regency of Daewongun in the late 1860s and early 1870s. Baekje: *Korea Herald*, 27 August 2010; Song Woong-ki, *Korea Herald*, 31 August 2010; Rössler, Edinburgh lecture, 3 December 2010.

Epilogue

1 F Nietzsche, *Unzeitgemässe Betrachtungen*, vol 2 (*Vom Nutzen und Nachteil der Historie für das Leben*), Leipzig, 1874, para. 219.

2 Dehio Kaiserrede: Huse, *Denkmalpflege*, 124–49.

3 Jokilehto, *History*, 1999, 19 (cited in D Rodwell, *Conservation and Sustainability*, Chichester, 2007, 205–6). For the often convoluted debates on the relationship of 'sustainability' and conservation, see e.g. J Carroon, *Sustainable Preservation*, New Jersey, 2010; aata.getty.edu/nps; *Conservation Perspectives*, Spring 2011, 'Sustainability and Heritage'; City of Regensburg, *Earth, Wind, Water, Fire: Environmental Challenges to Urban World Heritage*, Regensburg, 2009; R Oxley, *Survey and Repair of Traditional Buildings: A Conservation and Sustainable Approach*, Shaftesbury, 2003; J M Teutonico and F Matero (eds), *Managing Change*, Los Angeles, CA, 2003; English Heritage, *Sustainability and the Historic Environment*, London, 1996.

4 Riegl perceptible substratum: Huse, *Denkmalpflege*, 124–49; Niall Ferguson, *Civilisation: The West and the Rest*, London, 2011.

5 Edgar Allan Poe, *The Coliseum*, 1833.

– Bibliography –

Owing to space constraints, this bibliography is focused almost exclusively on book references; for article references, see the notes, or the bibliographies of certain other overview accounts (such as Jokilehto's *History of Architectural Conservation*). Also mostly excluded are references to individual conservation charters and specialist reports of international conservation organisations such as ICOMOS, ICCROM or UNESCO World Heritage Committee: for a full coverage of those, readers should consult the websites of the relevant organisations.

J A Adams, *The American Amusement Park Industry*, Boston, MA, 1991

Advisory Council on Historic Preservation, *The National Historic Preservation Act of 1966*, Washington, DC, 1986

Akademie der Künste Berlin, *Streit um die Neue Wache*, Berlin, 1993

H Albright, *The Birth of the National Park Service*, Salt Lake City, UT, 1985

T Aldous, *Battle for the Environment*, London, 1972

J Alfrey and T Putnam, *The Industrial Heritage*, London, 1992

J Allibone, *George Devey, Architect 1820–1886*, Cambridge, 1991

B Alnquist, *The Vasa Saga*, Stockholm, 1965

G Alpin, *Heritage: Identification, Conservation and Management*, Oxford, 2002

B L Altshuller *et al*, *Pamietniki Arkhitektury Moskovskoi oblasti*, Moscow, 1975

American School of Classical Studies, *The Stoa of Attalus II*, Princeton, NJ, 1992

C Amery and D Cruickshank, *The Rape of Britain*, London, 1975

C Amiel and J Pinies, *La cité des images*, Carcassonne, 1999

Benedict Anderson, *Imagined Communities, Reflections on the Origin and Spread of Nationalism*, London, 1983/91

C Applegate, *A Nation of Provincials, The German Idea of Heimat*, Berkeley, CA, 1990

B Appleyard, *The Pleasures of Peace*, London, 1989

D Appleyard (ed), *The Conservation of European Cities*, Cambridge, MA, 1979

Architectural Association of Ireland, *Architectural Conservation: An Irish Viewpoint*, Dublin, 1975

Architectural Press, *Bombed Churches as War Memorials*, London, 1945

G C Argan, *La creazione dell'Istituto Centrale del Restauro*, Rome, 1989

A Artley and J M Robinson, *The New Georgian Handbook*, London, 1985

J Ashurst, *Conservation of Ruins*, Amsterdam, 2007

G J Ashworth and J E Turnbridge (eds), *Building a New Heritage: Tourism, Culture and Identity in the New Europe*, London, 1994

G Ashworth and R Hartmann, *Horror and Human Tragedy Revisited: The Management of Sites of Atrocities for Tourism*, New York, 2005

P Astrinidou (ed), *Restoration of Byzantine and Post-Byzantine Monuments*, Thessaloniki, 1986

E Avrami, R Mason and M de la Torre, *Values and Heritage Conservation*, Los Angeles, CA, 2000

E Axelrod (ed), *The Colonial Revival in America*, New York, 1985

C Aymonino *et al*, *Roma Capitale 1870–1911*, Venice, 1984

J-P Azema *et al* (eds), *Reconstructions et modernization, la France après les ruines*, Paris, 1991

E Bacher, *Kunstwerk oder Denkmal? Alois Riegls Schriften zur Denkmalpflege*, Vienna, 1995

D Baker, *Living with the Past*, Bradford, 1983

N Balanos, *Les monuments de l'Acropole*, Paris, 1938

P Balchin, *Housing Improvement and Social Inequality*, Farnborough, 1979

P N Balchin, *Urban Development in Renaissance Italy*, Hoboken, NJ, 2008

S Baldrick and R Clay, *Iconoclasm*, Farnham, 2007

G Baldwin Brown, *The Care of Ancient Monuments*, Cambridge, 1905

J Balk, *Jordaan – een toekomst voor het overleden*, Amsterdam, 1975

A Barbacci, *Il restauro dei monumenti in Italia*, Rome, 1956

W Barbero, G Gambirasio and V Zanella (eds), *Luigi Angelini. Ingegnere architetto, catalogo della mostra*, Milan, 1984

M D Bardeschi (ed), *Il Monumento e suo doppio*, Florence, 1981

T F Barnard, *Herder's Social and Political Thought*, Oxford, 1965

D Barthel, *Historic Preservation*, New Brunswick, 1996

M Batisse and G Bolla, 'L'invention du "patrimoine mondial"' (*Les cahiers de l'histoire*), Paris, 2003

M Bauernfeind and R Käs (eds), *Wiederaufbau in Nürnberg*, Nuremberg, 2009

G Beard, *The Work of Christopher Wren*, Edinburgh, 1982

G Beard, *The Work of John Vanbrugh*, London, 1986

A Bell (ed), *Lord Cockburn, 1779–1979*, Edinburgh, 1969

A S Bell (ed), *The Scottish Antiquarian Tradition*, Edinburgh, 1981

D Bell, *The Historic Scotland Guide to International Conservation Charters (Technical Advice Note 8)*, Edinburgh, 1997

L Beltrami, *Indagine e documenti riguardanti la Torre Principale del Castello di Milano*, Milan, 1905

L Benevolo, *Le avventure della città*, Bari, 1974

F Berce, *Les premiers travaux de la Commission des Monuments Historiques*, Paris, 1979

F Berce, *Des monuments historiques au patrimoine du dix-huitieme siècle*, Paris, 2005

S Bergeon, G Brunel and E Mognetti (eds), *La conservation restauration en France*, Rome, 1999

I Bernali, *Historia de la arquelogia en Mexico*, Mexico City, 1979

H Beseler and N Gutschow, *Kriegsschicksale deutscher Architektur*, 2 vols, Wiesbaden, 2000

M Bianchini and M Parkinson, *Cultural Policy and Urban Regeneration: The West European Experience*, Manchester, 1994

I Bignamini and C Hornsby, *Digging and Dealing in 18th Century Rome*, New Haven, CT and London, 2010

A Biörnstad, *Skansen under hundra år*, Wiken, 1991

Bishop of London's Commission, *The City Churches, Final Report of the Bishop of London's Commission*, London, 1946

J R Blanco, *De varia restauratione: teoria e historia de la restauración arquitectónica*, Madrid, 2001

S K Blumenthal (ed), *Federal Historic Preservation Laws*, Washington, DC, 1990

F N Bohrer, *Photography and Archaeology*, London, 2010

C Boito, *I restauratori*, Florence, 1884

C Boito, *Questioni pratiche di belle arti: restauri, concorsi, legislazione, professione, insegnamento*, Milan, 1893

C Boito (A A Crippa ed), *Il nuovo e l'antico in architettura*, Milan, 1970 edition

F Bollerey, K Hartmann, *Denkmalpflege und Umweltgestaltung*, Munich, 1975

Bologna, *Politica e metologia della restauro nei centri storici*, Bologna, 1973

R Bonelli, *Architettura e restauro*, Venice, 1959

R Bonelli, *Scritti sul restauro e sulla architettonica*, Rome, 1995

P Boniface and P J Fowler, *Heritage and Tourism in the Global Village*, London, 1993

H Boockmann, *Die Marienburg im 19. Jahrhundert*, Berlin, 1982

Nicola Borger-Keweloh, *Die mittelalterlichen Dome im 19. Jahrhundert*, Munich, 1986

A Borowiec, *Destroy Warsaw!* Westport, CT, 2001

L Boturini, *Historia de la America septentrional*, 1746

C Bouras and K Zambas, Die *Arbeiten . . . auf der Athener Akropolis*, Athens, 2002

M C Boyer, *The City of Collective Memory*, Cambridge, MA, 1994

V Brand (ed), *The Study of the Past in the Victorian Age*, Oxford, 1998

C Brandi, *Teoria del restauro*, Rome, 1963 (English edition: *Theory of Restoration*. Rome, 2005)

C Brandi, *Teoria generale della critica*, Turin, 1974

C Brandi, *Il restauro, teoria e pratica 1939–1986*, Rome, 1995

M Braum and U Baus (eds), *Rekonstruktion in Deutschland*, Basel, 2009

M Brayshay (ed), *Post-war Plymouth*, Plymouth, 1983

D Brett, *The Construction of Heritage*, Cork, 1996

M S Briggs, *Goths and Vandals*, London, 1952

British-American Peace Centenary Committee, *The British Home of the Washingtons: Sulgrave Manor and its Associations*, London, 1912

M Broady (ed), *Planning for People*, London, 1968

S Brockmann, *Nuremberg, the Imaginary Capital*, New York, 2006

B Brolin, *Architecture in Context*, New York, 1980

G Bruce, *Some Material Good*, Edinburgh, 1975

M Brudny, *Reinventing Russia*, Cambridge, MA, 2000

P Bruhn, *Das Bernsteinzimmer in Zarskoje Selo bei Sankt Petersburg*, Berlin, 2004

C Buchanan, *Traffic in Towns* (shortened edition), London, 1963,

Colin Buchanan and Partners, *Bath, a Study in Conservation*, London, 1968

C Buls, *L'esthétique des villes*, Brussels, 1893

C Buls, *La restauration des monuments anciens*, Brussels, 1903

M Burian, *La protection des monuments historiques en Tchécoslovaquie*, Prague, 1957

E Burke, *A Philosophical Enquiry into the Origin of our Ideas of the Sublime and the Beautiful*, 1757

W Burns, *New Towns from Old*, London, 1963

Büro für Stadtsanierung, *Sanierung – für wen?* Berlin, 1971

Lord Bute, *A Plea for Scotland's Heritage* (National Trust for Scotland), Edinburgh, May 1936

Robert Byron, *How We Celebrate The Coronation: A Word To London's Visitors*, London, 1937

R Cabrera Castro, *Teotihuacan*, INAH, Mexico City, 2003

H Cain, *Gipsabgüsse*, Nuremberg, 1995

V Caldelli and G Meucci, *La Torre Pendente, il restauro del secolo*, Pisa, 2005

D Caldwell and E Caldwell (eds), *Rome: Continuing Encounters between Past and Present*, Farnham, 2011

Canadian Association for Conservation of Cultural Property, *Code of Ethics and Guidance for Practice*, Ottawa, 1986

S Cantacuzino, *Architectural Conservation in Europe*, London, 1975

S Cantacuzino, *New Uses for Old Buildings*, London, 1975

S Cantacuzino and S Brandt (eds), *Saving Old Buildings*, London, 1980

G Carbonara, *Avvicinamento al restauro*, Naples, 1997

G Carbonara (ed), *Trattato di restauro architettonico*, Turin, 1996–2008 (11 vols)

M Carboni, *Cesare Brandi*, Rome, 1992

A Carmel and A J Eisler, *Der Kaiser reist ins Heilige Land*, Stuttgart, 1999

Ethan Carr, *Mission 66, Modernism and the National Park Dilemma*, Amherst, 2007

S Casiello (ed), *La cultura del restauro*, Venice, 2005

F Castignoli, C Cecchelli, G Giovannoni and M Zucca, *Topografia e urbanistica di Roma*, Rome, 1958

M L Catoni (ed), *Il patrimonio culturale in Francia*, Milano, 2007

P Cavvadias and G Kawerau, *Die Ausgrabung der Akropolis vom Jahre 1885 bis zum Jahre 1890*, Athens, 1906

A Cederna, *Mussolini urbanista*, Rome, 1979

Cembureau Photonews, *Modern Interventions in Historic Towns*, 1975

Central Commission for Germany/British Committee on the Preservation and Restoration of Works of Art . . . in Enemy Hands, *Works of Art in Germany (British Zone of Occupation)*, London, 1946

M G Cerri, D Fea and L Pittarello, *Alfredo d'Andrade*, Turin, 1981

P L Cervellati and F Fontana, *Bologna, il volta della città*, Modena, 1975

P L Cervellati, R Scannavini and C de Angelis, *La nuova cultura della città*, Milan, 1977

C Ceschi, *Teoria e storia del restauro*, Rome, 1970

K Chamberlain, *War and Cultural Heritage*, Leicester, 2004

M Charlesworth (ed), *The Gothic Revival 1720–1870, Literary Sources and Documents*, 2002 (3 vols), Hastings, 2002

R Chapman, *The Sense of the Past in Victorian Literature*, London, 1986

B Chittenden, *A Profile of the National Register of Historic Places*, Washington, DC, 1984

F Choay, *L'allégorie du patrimoine*, Paris, 1992

F Choay, *The Invention of the Historic Monument*, Cambridge, 1998

F Choay, *La conference d'Athènes*, Paris, 2002

A Ciborowski, *Warsaw, a City Destroyed and Rebuilt*, Warsaw, 1969

City Plan Commission, Providence, RI, *College Hill – a Demonstration Study, of Historic Area Renewal*, Providence, RI, 1959

Civic Trust, *Pride of Place*, London, 1972

G Clarke, *Roman House, Renaissance Palaces: Inventing Antiquity in 15ᵗʰ Century Italy*, Cambridge, 2003

P Clemen, *Die deutsche Kunst und die Denkmalpflege*, Berlin, 1933

R Clow (ed), *Restoring Scotland's Castles*, Glasgow, 2000

H Cocclosis and P Nijkamp (eds), *Planning for our Cultural Heritage*, Aldershot, 1995

C C Collins and G R Collins, *Camillo Sitte and the Birth of Modern City Planning* (2 vols), New York, 1965

Colonial Williamsburg Inc, *The Official Guidebook of Colonial Williamsburg*, Williamsburg, 1951

Committee on Public Participation in Planning, *People and Planning* ('Skeffington Report'), London, 1969

Comune di Roma, *L'invenzione dei Fori imperiali*, Rome, 2008

P Connerton, *How Societies Remember*, Cambridge, 1989

A Conti, *Storia del restauro e della conservazione delle opere d'arte*, Milan, 1988

J Cornforth, *Country Houses in Britain: Can they Survive?* London, 1974

Corporation of London, *Report . . . on Postwar Reconstruction in the City of London*, London, 1944

C Corvisier, *Le Chateau de Coucy*, Paris, 2009

L A Coser (ed), *Maurice Halbwachs, On Collective Memory*, Chicago, IL, 1992

R N Cote, *City of Heroes, the Great Charleston Earthquake of 1886*, Mount Pleasant, SC, 2006

Council of Europe, *The Preservation and Development of Ancient Buildings and Historical or Other Sites*, 1963

Council of Europe, Committee on Monuments and Sites, *European Programme of Pilot Projects*, Strasbourg, 1973

Council of Europe/Civic Trust, *European Charter of the Architectural Heritage*, 1975

Council of Europe, *A Future for our Past*, 1985

Council of Europe, *Compendium of Basic Texts of the Council of Europe in the Field of Cultural Heritage*, Strasbourg, 1998

Council of Europe, *European Cultural Heritage* (2 vols), Strasbourg, 2002

M Coward, *Urbicide*, London, 2009

M Crinson, *Urban Memory*, London, 2004

G Croci, *The Conservation and Structural Restoration of Architectural Heritage*, Southampton, 1978

W Crofts, *Coercion or Persuasion*, London, 1989

D Crowley, *National Style and Nation State*, Manchester, 1992

G Cullen, *Townscape*, London, 1961

M Culot *et al*, *Le bombardement de Bruxelles par Louis XIV*, Brussels, 1992

C Cunningham, *The Acropolis*, Milton Keynes, 1979

P J H Cuypers and F Luyten, *Le Chateau de Haar a Haarzuylen*, Utrecht, 1910

A Dakin, M Glendinning and A MacKechnie (eds), *Scotland's Castle Culture*, Edinburgh, 2011

A Dale *et al*, *Historic Preservation in Foreign Countries*, Washington, DC, 1983

C Damas, *Le bombardement de Bruxelles a 1695*, Bruxelles 1951

A Davey, B Heath and D Hodges, *The Care and Conservation of Georgian Houses*, Edinburgh, 1978

D d'Ayala and E Fodde (eds), *Structural Analysis of Historic Construction*, London, 2008

G Debord, *La société du spectacle*, Paris, 1967

G Dehio, *Was wird aus dem Heidelberger Schloss werden?* Strasbourg, 1901

G Dehio, *Handbuch der deutschen Kunstdenkmäler* (first volume of series: *Mitteldeutschland*, 1ˢᵗ edition, Berlin, 1905)

G Dehio, *Geschichte der deutschen Kunst*, Berlin, 1930

A del Bufalo, *Gustavo Giovannoni*, Rome, 1982

C Dellheim, *The Face of the Past*, Cambridge, 1982

C Delon, *Notre capitale Paris*, Paris, 1888

F R de Montalembert, *Du vandalisme et du catholicisme dans l'art*, Paris, 1839

D Dendooven and J Dewilde, *The Reconstruction of Ieper*, Ieper, 1999

497

W Denslagen, *Architectural Restoration in Western Europe*, Amsterdam, 1994

W Denslagen, *Romantic Modernism: Nostalgia in the World of Conservation*, Amsterdam, 2011

W Denslagen and N Gutschow (eds), *Architectural Imitations: Reproductions and Pastiches in East and West* , Maastricht, 2005

A Desgodetz, *Les édifices antiques*, Paris, 1682

Deutsches Architekturmuseum, *Architektur in Deutschland: Rekonstruktion – alles bleibt anders* (edited by I Flagge and A Götz), Munich/Frankfurt, 2004

E de Vattel, *The Law of Nations*, Philadelphia, PA, 1844

I de Wolfe, *The Italian Townscape*, London, 1963

S Diaz-Berrio Fernandez, *La conservación del patrimonio cultural en México*, Mexico City, 1990

M Diefenbacher (ed), *Bauen in Nürnberg 1933–1945*, Nuremberg, 1995

J M Diefendorf (ed), *Rebuilding Europe's Bombed Cities*, New York, 1990

P Dieudonné (ed), *Villes reconstruites du dessin au destin* (2 vols), Paris, 1994

A Dobby, *Conservation and Planning*, London, 1978

M Dobson, *The Making of the National Poet*, Oxford, 1992

R Dölling (ed), *The Conservation of Historical Monuments in the Federal Republic of Germany*, Munich, 1974

C d'Onofrio, *Gli obelischi di Roma*, Rome, 1967

M Dorn, *Die Semlowerstrasse in Stralsund*, Berlin, 1940

J A Douglas, *The Redemption of St Sophia*, London, 1919

C M Dow, *The State Reservation at Niagara, a History*, Albany, 1914

Dresden Frauenkirche Foundation, *The Frauenkirche in Dresden*, Dresden, 2005

Duke of Bedford and G Mikes, *How to Run a Stately Home*, London, 1971

C R Dulau, *Pierrefonds Castle*, Paris, 1999

A Duménil and P Nivet (eds), *Les reconstructions en Picardie*, Amiens, 2003

B Dunlop, *Building a Dream: The Art of Disney Architecture*, New York, 1996

J Düwel and N Gutschow, *Städtebau in Deutschland im 20. Jahrhundert*, Stuttgart, 2001

J Düwel and N Gutschow, *Fortgewischt sind alle überflüssigen Zutaten*, Berlin, 2008

M Dvořák, *Katechismus der Denkmalpflege*, Vienna, 1915/1918

M Dvořák, *Gesammelte Aufsätze zur Kunstgeschichte*, Munich, 1929

Ecclesiological Society, *The Continuity of the English Town*, London, 1943

R M Edsel, *The Monuments Men*, Nashville, 2009

C W Ellis, *Britain and the Beast*, London, 1938

C W Ellis, *Portmeirion, the Place and its Meaning*, Portmeirion, 1963

English Heritage, *Monuments of War*, London, 1998

English Heritage, *Building in Context: New Development in Historic Areas*, London, 2001

English Heritage, *Conservation Principles, Policies and Guidance for the Sustainable Management of the Historic Environment*, London, 2008.

L Ennen, *Der Dom zu Köln*, Cologne, 1880

Ente bolognese manifestazioni artistiche, *Bologna Centro Storico* (exhibition catalogue), Bologna, 1970

C Erder, *Our Architectural Heritage*, UNESCO, Paris, 1986

L Esher, *York, a Study in Conservation*, London, 1968

L Esher, *Conservation in Glasgow: A Preliminary Report*, Glasgow, 1971

L Esher, *A Broken Wave*, London, 1981

J Evans, *A History of the Society of Antiquaries*, Oxford, 1956

Exchequer Committee, *Houses of Outstanding Historic or Architectural Interest*, report, London, 1950

A Falkner, *Without Our Past' A Handbook for the Preservation of Canada's Architectural Heritage*, Toronto, 1977

M Falser, *Zwischen Identität und Authentizität*, Dresden, 2008

M S Falser, W Lipp and A Tomaszewski (eds), *Conservation and Preservation, Interactions between Theory and Practice*, Florence, 2010

J Favier (ed), *Reconstructions et Modernisations*, Paris, 1991

J Fawcett (ed), *The Future of the Past*, London, 1976

B Feilden, *Conservation of Historic Buildings*, Oxford, 1982

B Feilden, *Between Two Earthquakes, Cultural Property in Seismic Zones*, Rome, 1987

B Feilden and J Jokilehto (ICCROM), *Management Guidelines for World Cultural Heritage*, Rome, 1993

F Feldbrugge *et al, Encyclopaedia of Soviet Administrative Law*, Dordrecht, 1989

A Fergusson, *The Sack of Bath, a Record and an Indictment*, Salisbury, 1973

P Ferriday, *Lord Grimthorpe 1816–1905*, London, 1957

M G Filetici, F Giovanetti, F M Tufano and E Palletino (eds), *I restauri dell'Acropoli d'Atene*, Rome, 2003

N G Finkelstein, *The Holocaust Industry, Reflections on the Exploitation of Jewish Suffering*, London, 2000

D E Finley, *History of the National Trust for Historic Preservation*, Washington, DC, 1965

M F Fischer (ed), *Wie lange dauern die Werke?* Munich, 1990

L F Fisher, *Saving San Antonio*, Lubbock, 1996

J M Fitch, *Historic Preservation: The Curatorial Management of the Built World*, Charlottesville, VA, 1990

C Forrest, *International Law and the Protection of Cultural Heritage*, London, 2010

M Forsyth (ed), *Structures and Construction in Historic Building Conservation*, Oxford, 2007

M Forsyth (ed), *Understanding Historic Building Conservation*, Oxford, 2007

M Forsyth (ed), *Materials and Skills in Historic Building Conservation*, Oxford, 2008

A Forty and S Küchler (eds), *The Art of Forgetting*, Oxford, 1999

N Foster, *Rebuilding the Reichstag*, New York, 2000

Foundation for the Preservation of Historic Georgetown, *A Walking Guide to Historic Georgetown*, Washington, DC, 1971

A Fradeletto, *Il Campanile di San Marco riedificato* (*Studi, ricerche, relazioni a cura del Comune di Venezia*), Venice, 1912

E Francais and H Schulze (eds), *Deutsche Erinnerungsorte*, Munich, 2001

F Francioni, *The 1972 World Heritage Convention, a Commentary*, Oxford, 2008

E A Freeman, *Principles of Church Restoration*, London, 1846

Y Friedman, *Votre ville est à vous*, Strasbourg, 1975

J Friedrich, *The Fire: The Bombing of Germany, 1940–1945*, Munich, 2002

J Friedrich, *Neue Stadt in Altem Gewand*, Cologne and Vienna, 2010

W Frodl, *Idee und Verwirklichung: das Werden der staatlichen Denkmalpflege in Österreich*, Vienna, 1988

E Frodl-Kraft, *Gefährdetes Erbe: Österreichs Denkmalschutz und Denkmalpflege, 1918–1945*, Vienna, 1997

J Frycz, *Restauracja i Konserwacja Zabytków Architektury w Polsce w Latach 1795–1918*, Warsaw, 1975

B G Fyodorov *et al, Hroniki unichtozhenia staroi Moskvy*, Moscow, 2006

A Galeazzi (ed), *Ripristino architettonico*, Fiesole, 1999

D Gamboni, *The Destruction of Art*, London, 1997

S Games (ed), *Pevsner, Art and Architecture*, London, 2002

D Gamston, *The Designation of Conservation Areas*, York, 1975

S Garnero, *Conservazione e restauro in Francia 1919–39*, Florence, 2006

R R Garvery and T B Morton, *The United States Government in Historic Preservation*, Washington, DC, 1973

P Gazzola, *The Past in the Future*, Rome (ICCROM/University of Rome), 1st edition, 1969 (revised 1975)

A Gebessler (ed), *Schutz und Pflege von Baudenkmälern in der BRD*, Cologne, 1980

P Geddes, *City Development*, Birmingham, 1904

E Gentile, *Fascismo di pietra*, Rome, 2007

German Commission for UNESCO, *Protection and Cultural Animation of Monuments, Sites and Historic Towns in Europe*, Bern, 1980

Germanisches Nationalmuseum, *Mythos Burg*, Nuremberg, 2006

Getty Institute, *Historical and Philosophical Issues in the Conservation of Cultural Heritage*, Los Angeles, CA, 1996

D Y Ghirardo, *Italy: Modern Architectures in History*, London, 2011

A M Ghisalberti (ed), *La vita di Cola di Rienzo*, Florence, 1928

M Giambruno (ed), *Per una storia del restauro urbano*, Novara, 2007

C Giannattasio, *Il restauro urbanistico in Francia 1962–2002*, Naples, 2003

M Gilbert, *Jerusalem in the Twentieth Century*, London, 1996

J R Gillis (ed), *Commemorations*, Princeton, NJ, 1994

D Gilmour, *Curzon*, London, 1994

G Giovannoni, 'Vecchie città ed edilizia nuova' (*Nuova Antologia* 48), Rome, 1913

G Giovannoni, *Sistemazione edilizia del Quartiere del Rinascimento in Roma*, Rome, 1919

G Giovannoni, *Vecchie città ed edilizia nuova*, Turin, 1931

G B Giovenale, *La Basilica di S. Maria in Cosmedin*, Rome, 1926

M Girouard, *The Return to Camelot*, New Haven, CT and London, 1981

D C Giurescu, *The Razing of Romania's Past*, London, 1990

M Gladstone, *The Politics of Planning*, London, 1976

W J Glaser, *Making Lahore Modern, Constructing and Imagining a Colonial City*, Minneapolis, 2008

M Glendinning, *Architecture's Evil Empire?: The Triumph and Tragedy of Modernism*, London, 2010

M Glendinning, R MacInnes and A MacKechnie, *A History of Scottish Architecture*, Edinburgh, 1996

W H Godfrey, *Our Building Inheritance*, London, 1944

L Grassi, *Camillo Boito*, Milan, 1959

R B Gratz, *The Battle for Gotham*, New York, 2010

A C Grayling, *Among the Dead Cities: The History and Moral Legacy of the World War II Bombing of Civilians in Germany and Japan*, New York, 2006

R M Greenberg and S A Marusin (eds), *National Register of Historic Places*, vols 1, 2, Washington, DC, 1976 and 1979

M Greenhalgh, *The Survival of Roman Antiquities in the Middle Ages*, London, 1989

X Greffe, *La valeur economique du patrimoine*, Paris, 1990

T Guha-Thakurta, *Monuments, Objects, Histories: Institutions of Art in Colonial and Postcolonial India*, New York, 2004

I Gutman, M Bernbaum and Y Gutman, *Anatomy of the Auschwitz Death Camp*, Bloomington, IN, 1998

N Gutschow and B Klain, *Vernichtung und Utopie: Stadtplanung Warschau 1939–1945*, Hamburg, 1994

W Haas, *Der Dom zu Speyer*, Königstein, 1988

J Habich *et al*, *Denkmalpflege statt Attrappenkult*, Basel, 2011

J Hagen, *Preservation, Tourism and Nationalism*, Aldershot, 2006

M Halbwachs, *La mémoire collective*, Paris, 1950

M Hall, *Towards World Heritage*, Farnham, 2011

L Hammer, *Die Geschichtliche Entwicklung des Denkmalrechts in Deutschland*, Tübingen, 1995

R Handler and A Gable, *The New History in an Old Museum: Creating the Past at Colonial Williamsburg*, Durham, NC, 1997

G Hans, *Denkmalschutz in Baden im 19. und 20 Jahrhundert*, Freiburg, 1985

J F Hanselmann, *Die Denkmalpflege in Deutschland um 1900*, Frankfurt/M, 1996

M F Hansen, *The Eloquence of Appropriation*, Rome, 2003

M Hardy (ed), *The Venice Charter Revisited: Modernism, Conservation and Tradition in the 21th Century*, Newcastle, 2009

P M Harman, *The Culture of Nature in Britain 1680–1860*, New Haven, CT and London, 2009

S Harries, *Nikolaus Pevsner: The Life*, London, 2011

Sir A Harris, *Bomber Offensive.* London, 1947

J Harris and R Williams (eds), *Architecture, Art and Urban Style within the Global Politics of City Branding*, Liverpool, 2011

P Harrison (ed), *Civilising the City*, Edinburgh, 1990

R Harrison (ed), *Manual of Heritage Management*, Oxford, 1994

V Hart and P Hicks (eds), *Palladio's Rome*, New Haven, CT and London, 2009

U Härtl, *Die Neukonzeption der Gedenkstätte Buchenwald*, Weimar, 2001

J Harvey, *Conservation of Buildings*, London, 1972

W Harvey, *The Preservation of St Paul's Cathedral and Other Famous Buildings: A Textbook on the New Science of Conservation*, London, 1925

A Haskell, *Caring for our Built Heritage*, London, 1993

F Haskell and N Penny, *Taste and Antique: The Lure of Classical Sculpture*, New Haven, CT, 1981

U Hassler and W Nerdinger (eds), *Das Prinzip Rekonstruktion*, Munich, 2010

A Hastings, *The Construction of Nationhood*, Cambridge, 1997

D Hayden, *The Power of Place: Urban Landscapes as Public Heritage*, Cambridge, MA, 1995

M F Hearn, *The Architectural Theory of Viollet le Duc*, Cambridge, MA, 1990

O Hederer, *Friedrich von Gärtner*, Munich, 1976

Henry Moore Institute, *Angkor Wat: From Temple to Text*, Leeds, 2010

D T Herbert (ed), *Heritage, Tourism and Society*, London, 1995

Heritage Canada, *Preservation Pays*, Ottawa, 2002

G Hermann, *Das Biedermeier im Spiegel seiner Zeit*, Berlin, 1913

J B Hernandez and J J I Tresseras, *Gestión del patrimonio cultural*, Barcelona, 2001

R Hewison, *The Heritage Industry*, London, 1987

R Hewison, *Ruskin on Venice*, New Haven, CT and London, 2010

R Hill, *God's Architect*, London, 2007

R Hill, C Cunningham and A Reid, *Victorians Revalued*, London, 2010

B Hillier, *John Betjeman, The Biography*, London, 2006

HMSO, *How Do You Want To Live?*, HMSO (Department of the Environment), London, 1972

E Hobsbawm and T Ranger (eds), *The Invention of Tradition*, Cambridge, 1983

M Holleran, *Boston's 'Changeful Times': Origins of Preservation and Planning in America*, Baltimore, MD, 1998

R Home, *Of Planting and Planning*, London, 1997

V Horta, *L'entourage des monuments: principles generaux*, Paris, 1933

C B Hosmer, *Presence of the Past*, New York, 1965

C B Hosmer, *Preservation Comes of Age: From Williamsburg to the National Trust*, Charlottesville, VA, 1981

A Hubel, *Denkmalpflege*, Stuttgart, 2006

H Hübner, *Das Gedächtnis der Stadt*, Berlin, 1997

HUD, *Historic Preservation through Urban Renewal*, Washington, 1963

K Hudson, *Industrial Archaeology, an Introduction*, London, 1963

M Hufford (ed), *Conserving Culture: A New Discourse on Heritage*, Urbana, OH, 1994

Victor Hugo, *Notre-Dame de Paris*, 1831

M Hunter, *Preserving the Past: The Rise of Heritage in Modern Britain*, Stroud, 1996

R Hurd, *Scotland Under Trust*, London, 1939

R Hurd, *The Face of Modern Edinburgh*, Edinburgh, 1939

N Huse, *Denkmalpflege: deutsche Texte aus drei Jahrhunderten*, Munich, 1984

M Hutter and I Rozzo (eds), *Economic Perspectives on Cultural Heritage*, New York, 1997

L Huxtable, *The Architecture of New York*, vol 1, Garden City, NY, 1964

ICCROM, *The First Decade, 1959–1969*, Rome, 1969

ICCROM/Getty Conservation Institute, *International Directory in Training in Conservation of Cultural Heritage*, Rome, 1994

ICOMOS, 2nd General Assembly, Oxford, July 1969: *Report on the Activities of the National Committees*

ICOMOS, *Monuments and Sites: Finland*, Helsinki, 1999

ICOMOS, *Pillage en Europe*, Paris, 2000

ICOMOS Australia Inc, *The Illustrated Burra Charter*, Sydney, 1992

ICOMOS Canada, *Preserving our Heritage: Catalog of Charters and Other Guides*, Quebec, 1990

ICOMOS German National Committee, *Colloque sur la conservation des petites villes historiques*, Rothenburg ob der Tauber, May 1975, 1975

Improvements and Town Planning Committee of the Corporation of London, *The City of London, A Record of Destruction and Survival*, London, 1951

J Ingamells, *A Dictionary of British and Irish Travellers in Italy*, New Haven, CT, 1997

D Insall, *Thaxted, An Historical and Architectural Survey for the County Council of Essex*, London, 1966

D Insall, *Living Buildings*, Mulgrave, Vic., 2008

D Insall & Partners, *Chester*, London, 1968

Institut de cooperation intellectuelle (ICI), *Le conservation des monuments d'art et d'histoire*, Paris, 1933

Institut für Denkmalpflege der DDR, *Denkmale der Geschichte und Kultur*, Berlin, 1976

Institut für Denkmalpflege der DDR, *Martin Luther, Stätten seines Lebens und Wirkens*, East Berlin, 1983

G Ionescu, *Arhitectura Populara in Romania*, Bucharest, 1971

W Irwin, *The New Niagara*, University Park, PA, 1996

Y R Isar, *The Challenge to our Cultural Heritage*, Washington, DC, 1986

M Iversen, *Alois Riegl, Art History and Theory*, Cambridge, MA, 1993

J B Jackson, *The Necessity for Ruins*, Amherst, 1980

J Jacobs, *The Death and Life of Great American Cities*, New York, 1961

S Jacobs, *Historic Preservation in Europe*, Ithaca, NY, 1966

F Jamieson, *Postmodernism, or, The Cultural Logic of Late Capitalism*, Durham, NC, 1991

J Johnson, L Rosenburg, *Renewing Old Edinburgh*, Glendaruel, 2010

J Jokilehto, *A History of Architectural Conservation*, Oxford, 1999/2004

M Jones, *Fake: The Art of Deception*, London, 1990

R Kain (ed), *Planning for Conservation*, London, 1981

R Kain (trans.), *Alois Riegl, Problems of Style*, Princeton, NJ, 1992

A Kaldellis, *The Christian Parthenon*, Cambridge, 2009

F E Kaplan (ed), *Museums and the Making of 'Ourselves'*, London, 1994

D Karg, *Zum 200. Geburtstag von Ferdinand von Quast*, Berlin, 2008

M Karjalainen, *Alvar Aalto Library in Vyborg*, Helsinki, 2010

N Kaufman, *Place, Race and Story: Essays on the Past and Future of Historic Preservation*, New York, 2009

E H Kellog and F Gilbert, *Readings in Historic Preservation*, Brunswick, NJ, 1983

J M Kelly, *The Society of Dilettanti*, New Haven, CT and London, 2010

M R Kelsall and S Harris, *A Future for the Past*, Edinburgh, 1961

N Kelvin (ed), *The Collected Letters of William Morris* (2 vols), Princeton, NJ, 1984–7

H Kendall, *Jerusalem, the City Plan*, London, 1948

Wayland Kennet, *Preservation*, London, 1972

C Kiadó, *Historical Monuments and their Protection in Hungary*, Budapest, 1984

M Kneubühler, *Handbook on the European Heritage Days*, Strasbourg and Brussels, 2009

V Knigge, *The History of the Buchenwald Memorial*, Weimar, 2000

A Knöpfli, *Schweizerische Denkmalpflege: Geschichte und Doktrinen*, Zurich, 1972

A L Koch and H Dearstyne, *Colonial Williamsburg*, Williamsburg, 1949

G Kokkelink and H D Theen, *Bewertungsfragen der Denkmalpflege*, Hannover, 1976

Kongl: Mayst:tz Placat och Påbudh, *Om gamble monumenter och antiquiteter*, Stockholm, 1666

R Koolhaas (ed), *Content*, Cologne, 2004

R Koolhaas, B Mau, *S, M, L, XL*, New York, 1994/1995

S Kopp, F Boucher and D Pauly (eds), *L'Architecture de la reconstruction en France*, Paris, 1982

W Kos and C Rapp (eds), *Alt-Wien, die Stadt die niemals war*, Vienna, 2004

R Koshar, *Germany's Transient Pasts*, Chapel Hill, NC, 1998.

S Kostof, *The Third Rome*, Berkeley, 1973

J Krastins, I Strautmanis, *Riga – Complete Guide to Architecture*, Riga, 2004

R Krautheimer, *Rome, Profile of a City*, Princeton, NJ, 1980

H E Kubach and W Haas, *Der Dom zu Speyer* (3 vols), Munich, 1972

A Kutcher, *The New Jerusalem, Planning and Politics*, London, 1973

B Ladd, *Urban Planning and Civic Order in Germany, 1860–1914*, Cambridge, MA, 1990

B Ladd, *The Ghosts of Berlin*, Chicago, 1997

N Lambourne, *War Damage in Western Europe*, Edinburgh, 2001

G la Monica, *Ideologia e prassi del restauro*, Palermo, 1974

R Lanciani, *La distruzione di Roma antica*, Milan, 1971 edition

Landesamt für Denkmalpflege im Regierungspräsidium Stuttgart, *Traum und Wirklichkeit, Vergangenheit und Zukunft der Heidelberger Schlossruine*, 2005

S Lange, *Bernd and Hilla Becher, Life and Work*, Cambridge, MA, 2006

F La Regina, *Come un ferro rovente: cultura e prassi del restauro architettonico*, Naples, 1992

K E Larsen, *Architectural Preservation in Japan*, Trondheim, 1994

K E Larsen (ed), *Proceedings, Conference on Authenticity in Relation to the World Heritage Convention, Nara, Japan 1994*, Trondheim, 1995

K E Larsen and N Marstein (eds), *Conference on Authenticity in Relation to the World Heritage Convention, Preparatory Workshop, Bergen, Norway*, Trondheim, 1994

D M Lasansky, *The Renaissance Perfected*, University Park, PA, 2004

S E Lawrence (ed), *Piranesi as Designer*, New York, 2007

A J Lee (ed), *Past Meets Future: Saving America's Historic Environments*, Washington, DC, 1992

T Lemain, *Forgery, The History of a Modern Obsession*, London, 2011

E Lemire (ed), *The Unpublished Lectures of William Morris*, Detroit, 1969

J-M Leniaud, *L'utopie française: essai sur le patrimoine*, Paris, 1992

P Leon, *Les Monuments Historiques*, Paris, 1913

P Leon, *La guerre at l'architecture*, Paris, 1918

502

P Leon, *La vie des monuments français*, Paris, 1951

J D Le Roy, *Les ruines des plus beaux monuments de la Grèce*, Paris, 1758

P M Letarouilly, *Edifices de Rome Moderne*, Liège, 1849

M J Lewis, *The Politics of the German Gothic Revival: August Reichensperger*, Cambridge, MA, 1994

J M Lindgren, *Preserving the Old Dominion*, Charlottesville, VA, 1993

J M Lindgren, *Preserving Historic New England*, New York, 1995

W Lindner and E Böckler, *Die Stadt, ihre Pflege und Gestaltung*, Munich, 1939

S Lindquist, *A History of Bombing*, London, 2002

I G Lindsay, *Georgian Edinburgh*, Edinburgh, 1948

M Lock, *Civic Diagnosis, a Blitzed City Analysed*, Hull, 1943

F Loeffler and H Magirius, *Frauenkirche Dresden* (Schnell Art Guide), English edition, 1994

S Loew, *Modern Architecture in Historic Cities, Planning, Policy and Building in Contemporary France*, London, 1998

R Longstreth, *The Buildings of Main Street*, Washington, DC, 1987

P V Lottin de Laval, *Manuel complet de Lottinoplastique*, Paris, 1857

H Lottman, *How Cities are Saved*, New York, 1976

A Loukaki, *Living Ruins, Value Conflicts*, Farnham, 2008

D Lowenthal, *The Past is a Foreign Country*, Cambridge, 1985

D Lowenthal, *The Heritage Crusade and the Spoils of History*, Cambridge, 1998

D Lowenthal and M Binney (eds), *Our Past Before Us*, Oxford, 1981

D Lu, *Remaking Chinese Urban Form*, London, 2006

G Lucas and V Buchli (eds), *Archaeologies of the Contemporary Past*, London, 2001

M Lupano, *Marcello Piacentini*, Rome, 1991

K Lynch, *What Time is this Place?* Cambridge, MA, 1972

R Macaulay, *Pleasure of Ruins*, London, 1984

C McKean, *Fight Blight*, London, 1977

B Mackintosh, *The Historic Sites Survey*, Washington, DC, 1985

R H McNulty and S A Kliment (eds), *Neighborhood Conservation*, New York, 1976/9

E MacRae, *The Heritage of Greater Edinburgh*, Edinburgh, 1947

C McWilliam, *Scottish Townscape*, London, 1975

M Maderna, *Camillo Boito*, Milan, 1995

M Magnusson (ed), *Echoes in Stone*, Edinburgh, 1983

C Magny, *La beauté de Paris*, Paris, 1911

U Mainzer (ed), *Paul Clemen – zur 125. Wiederkehr seines Geburtstages*, Cologne, 1991

A Mallwitz, *Olympia und seine Bauten*, Darmstadt, 1972

A Malraux, *Le musée imaginaire*, Paris, 1965

A Malraux, *La grande pitié des monuments de France*, Paris, 1998

P Mandler, *The Rise and Fall of the Stately Home*, New Haven, CT and London, 1997

S Mannheim, *Walt Disney and the Quest for Community*, Farnham, 2002

P Marconi, *Guiseppe Valadier*, Rome, 1964

P Marconi, *Arte e cultura della manutenzione dei monumenti*, Rome, 1990

S Marks (ed), *Concerning Buildings: Studies in Honour of Sir Bernard Feilden*, Oxford, 1996

K A Marling, *The Architecture of Reassurance: Designing the Disney Theme Parks*, New York, 1997

T Marosovic, *The Preservation of Urban and Architectural Heritage*, Split, 1983

D Martin and M Krautzberger (eds), *Handbuch Denkmalschutz und Denkmalpflege*, Munich, 2004

O Martin and G Piatti (eds), *World Heritage and Buffer Zones (UNESCO World Heritage Papers*, 25), Paris, 2009

R Mason, *The Once and Future New York*, Minneapolis, MN, 2009

S Massie, *Pavlovsk: The Life of a Russian Palace*, Boston, 1990

G Massimi, *S Maria in Cosmedin*, Rome, 1953

R Matthew, J Reid and M Lindsay (eds), *The Conservation of Georgian Edinburgh*, Edinburgh, 1972

Craig A Mauzy, *Agora Excavations 1931–2006* (The American School of Classical Studies at Athens), Athens, 2006

O Mazzei, *Alfonso Rubbiani*, Bologna, 1979

H R Meier and I Scheurmann (eds), *DENKmalWERTE*, Berlin/Munich, 2010

D Mellor (ed), *A Paradise Lost*, London 1987

P Merimée, *Correspondence générale, 1864–5*, Toulouse, 1958 edition

P M Messenger, *The Ethics of Collecting Cultural Property*, Albuquerque, NM, 1989

K E Meyer, *The Plundered Past*, New York, 1977

MHLG, *New Houses for Old*, London, 1954

MHLG, *The Deeplish Study*, London, 1966

G Miarelli Mariani (ed), *Il restauro dei monumenti nei paesi europei*, Rome, 2001

S Michalski, *Public Monuments, Art in Political Bondage*, London, 1998

C Miele (ed), *From William Morris*, New Haven, CT and London, 2005

M Miles, *Art as Plunder*, Cambridge, 2008

Ministero della Pubblica Istruzione, *La ricostruzione del patrimonio artistico italiano*, Rome, 1950

Ministry of Culture and Sciences, *International Meeting on the Restoration of the Erechtheion*, Athens, 1976

Ministry of Culture, *The Acropolis at Athens, Conservation, Restoration and Research*, Athens, 1983

Ministry of Health, *New Houses for Old*, London, *c.*1938

C Mires, *Independence Hall in American Memory*, Philadelphia, 2002

A Mitscherlich, *Die Unwirtlichkeit unserer Städte*, Frankfurt/Main, 1965

R Mohr de Perez, *Die Anfänge der Staatlichen Denkmalpflege in Preussen*, Worms, 2001

G Moller, *Denkmäler der deutschen Baukunst* (3 vols), Darmstadt, 1815–21

H Montgomery-Massingberd and C S Sykes, *Great Houses of England and Wales*, London, 1994

C Morgan and I Orlova, *Saving the Tsar's Palaces*, Clifton, 2005

E D Mulloy, *The History of the National Trust for Historic Preservation, 1963–1973*, Washington, DC, 1976

E Mulzer, *Der Wiederaufbau der Altstadt von Nürnberg 1945 bis 1970*, Erlangen, 1972

A Muñoz, *Il restauro del tempio della Fortuna Virile*, Rome, 1925

A Muñoz Cosme, *La conservación del patrimonio arquitectónico español*, Madrid, 1989

R Munro, *Archaeology and False Antiquities*, London, 1905

J Murtagh, *Keeping Time: The History and Theory of Preservation in America*, New York, 1988/1997

Museen der Stadt Nürnberg, *Das Albrecht Dürer Museum*, Nuremberg, 2006

Museum of the City of New York, *The American Style: Colonial Revival and the Modern Metropolis*, exhibition, New York, 2011

S Muthesius, *Art, Architecture and Design in Poland*, Königstein, 1994

S Muthesius, *The Poetic Home: Designing the 19th-Century Domestic Interior*, London, 2009

I Nairn, *Outrage*, London, 1955

I Nairn, *Your England Revisited*, London, 1964

National Park Service, *Local Preservation: A Selected Bibliography*, Washington, DC, 1988

National Trust for Historic Preservation, *Historic Preservation Today: Essays Presented to the Seminar on Preservation and Restoration, Williamsburg, Virginia, September 8–11, 1963*, Charlottesville, VA, 1966

National Trust for Historic Preservation, *Historic Preservation Tomorrow: Revised Principles and Guidelines for Historic Preservation in the United States*, Charlottesville, VA, 1967

National Trust for Historic Preservation, *Preservation: Toward an Ethic in the 1980s*, Washington, DC, 1980

National Trust for Historic Preservation, *Reusing Old Buildings*, Washington, DC, 1984

W Nerdinger (ed), *Geschichte der Rekonstruktion: Konstruktion der Geschichte*, Munich, 2010

L H Nicholas, *The Rape of Europa*, New York, 1995

J Nivet, *Sainte Croix d'Orleans*, Orleans, 1984

P Nora, *Les lieux de mémoire*, Paris, 1984–92

W Noth, *Die Wartburg und ihre Sammlungen*, Leipzig, 1972

R O'Byrne, *The Irish Georgian Society: A Celebration*, Dublin, 2008

A Orbaşli, *Tourists in Historic Towns*, London, 2000

A Orbaşli, *Architectural Conservation*, Oxford, 2008

I Ordieres Diez, *Historia de la restauración monumental en España*, Madrid, 1995

M Ottosen, *Dansk byggningsrestaurerings historie, en indføring*, Århus, 1984

M Page and R Mason, *Giving Preservation a History*, New York, 2004

Palmer/RAE Associates, *European Cities and Capitals of Culture*, Brussels, 2004

R Pane, *Attualità e dialettica del restauro*, Chieti, 1987

J-P Panouille, *The City of Carcassonne* (Editions du Patrimoine, CNMHS), Paris, 2001

Parks Canada, *Guide to Canada's National Historic Parks*, Ottawa, 1985

C C Parslow, *Rediscovering Antiquity*, Cambridge, 1995

J Paton Watson and P Abercrombie, *A Plan for Plymouth*, Plymouth, 1943

T Patricio, K van Balen and K de Jonge (eds), *Conservation in Changing Societies*, Leuven, 2006

A Peacock, *The Heritage Game: Economics, Policy and Practice*, New York, 2008

D Pearce, *Conservation Today*, London, 1989

W Pehnt, *Karljosef Schattner*, Stuttgart, 1988

N Pevsner, *Some Architectural Writers of the Nineteenth Century*, Oxford, 1972

C Pharr, *The Theodosian Code*, Princeton, NJ, 1952

R D Pickard, *Conservation in the Built Environment*, Essex, 1996

R D Pickard, *Policy and Law in Heritage Conservation*, London, 2001

G Piltz, *Streifzug durch die deutsche Baukunst*, Berlin, 1973

G Piranesi, *Antichità romane*, Rome, 1756

G Piranesi, *Trattato della magnificenza e architettura de'Romani*, Rome, 1961

M Pisani, *Architetture di Marcello Piacentini. Le opere maestre.* Rome, 2004

H Plenderleith, *The Conservation of Antiquities and Works of Art*, Oxford, 1971

PKZ, *Les ateliers de restauration des monuments historiques*, Warsaw, 1981

Polish National Committee of ICOMOS, *Problems of Heritage and Cultural Identity in Poland*, Warsaw, 1984

Polish National Committee of ICOMOS, *XXX ICOMOS 1965–1995*, Warsaw, 1995

M Pollak, *Vom Erinnerungsort zur Denkmalpflege*, Cologne, 2009

D Poulot, *Musée, nation, patrimoine, 1789–1815*, Paris, 1997

D Poulot, *Patrimoine et musée*, Paris, 2001

G A S Powers, *Sir Frank Mears*, Edinburgh, 1983

L Pressouyre, *La convention du patrimoine mondial vingt ans après*, Paris, 1993

N Price, M Talley and A Vaccaro (eds), *Historical and Philosophical Issues in the Conservation of Cultural Heritage*, Los Angeles, 1996

Prince Charles, *A Vision of Britain*, London, 1989

H Probert, *'Bomber' Harris: His Life and Times*, London, 2006

G Proietti (ed), *L'eccellenza del restauro italiano nel mondo*, Rome, 2005

Projektgruppe Eisenheim, *Rettet Eisenheim*, Bielefeld, 1973

L V Prott and P J O'Keefe, *Law and Cultural Heritage*, Oxford, 1984

M Proust, *The Remembrance of Things Past*, vol 3, *Time Regained*, New York, 1981 (originally published 1927)

T Prudon, *Preservation of Modern Architecture*, Hoboken, NJ, 2008

A W N Pugin, *Contrasts*, 1836

B Pusback, *Stadt als Heimat*, Cologne, 2006

Uwe Puschner, Walter Schmitz and Justus H Ulbricht (eds), *Handbuch zur Völkischen Bewegung' 1871–1918*, Munich, 1999

S Quaedvlieg-Mihailović and R G Strachwitz (eds), *Heritage and the Building of Europe*, Berlin, 2004

P Quigley, *Shifting Grounds: Nationalism and the American South, 1848–1865*, Oxford, 2011

P Quinn (ed) *Temple Bar, the Power of an Idea*, Dublin, 1996

A M Racheli, *Restauro a Roma, 1870–1990*, Venice, 1995

Sir T Raleigh (ed) *Lord Curzon in India, 1898–1905*, London, 1906

Raymond Lemaire International Centre for Conservation, *Conservation in Changing Societies*, Leuven, 2006

L Réau, *Histoire du vandalisme*, Paris, 1959

Regia Stamperia, *Le antichità di Ercolano esposte*, 8 vols, 1755–92

A Reiach and R Hurd, *Building Scotland*, Edinburgh, 1941

S Rentzhog, *Friluftsmuseerna, en skandinavisk idé erövrar världen*, Stockholm, 2007

Resurgam, London, 1944

D M Reynolds (ed), *Remove not the Ancient Landmark: Public Monuments and Moral Values*, Amsterdam, 1996

W B Rhoads, *The Colonial Revival*, New York, 1977

J M Richards, *The Bombed Buildings of Britain*, London, 1942

R W Richardson, *Historic Districts of America: The South*, Bowie, MD, 1987.

A Richmond, A Bracker (eds), *Conservation: Principles, Dilemmas and Uncomfortable Truths*, Oxford, 2009

A Riegl, *Der moderne Denkmalskultus, sein Wesen, seine Entstehung*, Vienna, 1903 (published 1929 in collected works; English translation: *Oppositions*, Autumn 1982)

A Riegl, 'Neue Strömungen in der Denkmalpflege' (*Mitteilungen der K. K. Zentralkommission*, III (iv)), Vienna, 1905

A Riegl, *Gesammelte Aufsätze*, Augsburg, 1929

C Rifkind, *Main Street: The Face of Urban America*, New York, 1977

O Rigaud, *Reims à l'époque de l'Art Déco*, Paris, 2006

O Rigaud and M Bedarida, *Reims, reconstructions 1920–1930*, Reims, 1988

Riksantikvarienämbetet, *Protection of the Architectural Heritage of Sweden*, Stockholm, 1975

J Rivera, *Historia de las restauraciones de la Catedral de Leon*, Valladolid, 1903

A G Robinson, *Old New England Doorways*, New York, 1919

J M C Rodríguez, *La restauración monumental en España*, Sevilla, 2000

D Rodwell, *Conservation and Sustainability in Historic Cities*, Chichester, 2007

B Roeck, *Florence 1900: The Quest for Arcadia*, New Haven, CT and London, 2009

G D Rosenfeld, *Munich and Memory*, Berkeley, CA, 2000

L Ross, E Schnaubert and C Hansen, *Die Akropolis von Athen nach den neuesten Ausgrabungen*, Berlin, 1839

M Ross, *Planning and the Heritage*, London, 1991

A Rossi, *The Architecture of the City*, Cambridge, MA, 1984 (first published in Italian 1966)

M S Roth, C Lyons and C Merewether, *Irresistible Decay: Ruins Reclaimed*, Los Angeles, CA, 1997

Y Rowan and U Baram, *Marketing Heritage: Archaeology and the Consumption of the Past*, Walnut Creek, CA, 2004

C Rowell and J M Robinson, *Uppark Restored*, London, 1996

R T Rowley and M Breakell (eds), *Planning and the Historic Environment*, Oxford, 1977

A Rubbiani, *Di Bologna riabbellita*, Bologna, 1913

R L Rubenstein and J K Roth, *Approaches to Auschwitz*, Louisville, KY, 1987

A Runte, *National Parks, The American Experience*, Lincoln, NE, 1987

J Ruskin, *The Seven Lamps of Architecture*, 1849

J Ruskin, *The Stones of Venice*, 1851–3

R Rząd, *Zamek w Malborku 1882–1945*, Malbork, 1996

W Sabrow (ed), *Erinnerungsorte der DDR*, Munich, 2009

St Andrews Preservation Trust, *Three Decades of Historic Notes*, St Andrews, 1991

C Sauer, *The Morphology of Landscape. University of California Publications in Geography*, Berkeley, 1925

B L Savage, *African-American Historic Places*, Washington, DC, 1994

M Sawin (ed), *James Marston Fitch, Selected Writings*, New York, 2008

D I Scargill, *Urban France*, London, 1983

S Scarrocchia, *Alois Riegl, teoria e prassi della conservazione dei monumenti*, Bologna, 1995

S Schama, *Landscape and Memory*, London, 1995

I Scheurmann (ed), *Zeitschichten: Erkennen und Erhalten – Denkmalpflege in Deutschland*, Munich and Berlin, 2005

I Scheurmann and H R Meier (eds), *Echt, Alt, Schön, Wahr – Zeitschichten der Denkmalpflege*, Munich and Berlin, 2006

D Schindler and J Toman, *The Laws of Armed Conflicts*, Leiden, 2004

K F Schinkel, *Grundsätze zur Erhaltung alter Denkmaler*, Berlin, 1815

G Schlusche, *Die internationale Bauaustellung Berlin – eine Bilanz*, Berlin, 1997

B Schmid, *Die Wiederherstellung der Marienburg*, Königsberg, 1934

A Schmidt (ed), *The Impact of Perestroika on Soviet Law*, Dordrecht, 1990

L Schreiner, *K F Schinkel und die erste westfälische Denkmäler-Inventarisation*, Münster, 1968

H Schück, *Kgl. Vitterhets historie och antikvitets akademien*, Stockholm, 1932–1944

C Schultz, *Carlo Scarpa – Layers*, Fellbach, 2007

H Schulz, *Altstadtsanierung in Kassel*, Kassel, 1983

P Schultze-Naumburg, *Kulturarbeiten* (5 vols), Munich, 1902–9

E F Schumacher, *Small is Beautiful*, London, 1973

Schweizerischer Verband für Konservierung und Restaurierung, *Geschichte der Restaurierung in Europa*, vol 1, Worms, 1991

Schweizerischer Verband für Konservierung und Restaurierung, *Geschichte der Restaurierung in Europa*, vol 2, Worms, 1993

G G Scott, *A Plea for the Faithful Restoration of our Ancient Churches*, Oxford, 1850

G G Scott, *Personal and Professional Recollections*, London, 1879

C Scott-Clark and A Levy, *The Amber Room, The Untold Story of the Greatest Hoax of the Twentieth Century*, London, 2004

Scottish Historical Landscapes, London, 2002

W Sebald, *On The Natural History of Destruction*, London, 1999

K Sekino, *The Conservation of Ancient Buildings in Japan*, Tokyo, 1929

Select Committee on the House of Commons (Rebuilding), *Report of the Select Committee on the House of Commons (Rebuilding)*, London, 1944

S W Semes, *The Future of the Past: A Conservation Ethic for Architecture*, New York, 2009

I Serageldin, E Shluger and J Martin-Brown (eds), *Historic Cities and Sacred Sites*, Washington, DC, 2001

V Shacklock, *Architectural Conservation: Issues and Developments*, Shaftesbury, 2006

T Sharp, *English Panorama*, London, 1936

T Sharp, *Exeter Phoenix*, London, 1946

A S Shchenkov (ed) *et al*, *Pamyatniki arkhitektury v dorevolutsionnoi Rossii*, (*Architectural Monuments of Pre-Revolutionary Russia*) Moscow, 2002

W J Siedler, E Niggemeyer and G Angress, *Die gemordete Stadt*, West Berlin, 1964

N Silber, *Lost New York*, New York, 1967

S Sillars, *British Romantic Art and the Second World War*, London, 1991

U Singh, *The Discovery of Ancient India*, Delhi, 2004

M-A Sire, *La France du patrimoine*, Paris, 2005

C Sitte, *Der Städtebau nach seinen künstlerischen Grundsätzen*, Vienna, 1889

H C Smith, *Sulgrave Manor and the Washingtons*, New York, 1933

A Smolar-Meynart (ed), *Autour du bombardement de Bruxelles de 1695*, Brussels, 1995

M Snodin, *Horace Walpole's Strawberry Hill*, New Haven, CT and London, 2009

SPAB, *The Treatment of Ancient Buildings in Wartime*, n.d.

H Spande (ed), *Conservation Legacies of the Florence Flood of 1966*, London, 2009

R E Spear and P Sohm (eds), *Painting for Profit, The Economic Lives of 17th Century Italian Painters*, New Haven, CT and London, 2010

Staatliche Museen zu Berlin, *Das Neue Museum Berlin*, Berlin, 2009

F Stanculescu *et al*, *Arhitectura populara romaneasca* (multi-vol series), Bucharest, 1956 onwards

N Stanley-Price and J King, *Conserving the Authentic*, Rome, 2009

N Stanley-Price, M K Talley and A M Vaccaro (eds), *Historical and Philosophical Issues in the Conservation of Cultural Heritage: Readings in Conservation*, Los Angeles, CA, 1996

A J Stannis, *Creating the Big Easy*, Athens, GA, 2006

R E Stipe, *Historic Preservation in Other Countries*, vols 1–5, Washington, DC, 1982–90

R E Stipe, *A Richer Heritage*, Chapel Hill, NC, 2003

Sir J Stirling-Maxwell, *Shrines and Homes of Scotland*, London, 1937

H Stovel, *Risk Preparedness: A Management Manual for World Cultural Heritage*, Rome, 1998

M Stratton, *Industrial Buildings: Conservation and Regeneration*, London, 2000

R Strong, M Binney and J Harris, *The Destruction of the Country House*, London, 1974

J Stuart and N Revett, *The Antiquities of Athens*, published 1762–87

J H Stubbs, *Time Honored: A Global View of Architectural Conservation*, Hoboken, NJ, 2009

J H Stubbs, *Architectural Conservation in Europe and the Americas*, Hoboken, NJ, 2009

J Stubbs, M Makos and M Bouchenaki, *Architectural Conservation in Europe and the Americas*, Chichester, 2011

D Sturdy, *How to Pull a Town Down*, London, 1972

H Stynen, G Charlier and A Buellens, *Het verwoeste gewest 15/18: Mission Dhuicque*, Bruges, 1985

G S Suardo, *L'arte del restauro*, Milano, 1866

J Summerson, *Victorian Architecture, Four Studies in Evaluation*, New York/London, 1970

A Sutcliffe, *The Autumn of Central Paris*, London, 1970

M Swenarton, *Artisans and Architects*, London, 1989

M Talbot, *Reviving Buildings and Communities*, Newton Abbot, 1986

J Tamási (ed), *Heritage Protection within the Compass of Legal Regulation*, Budapest, 2001

M Tarnassia, *Firenze 1944–45, Danni di Guerra*, Livorno, 2007

C Taylor, *The Ethics of Authenticity*, Cambridge, MA, 1991

J Taylor, *A Dream of England*, Manchester, 1994

N Taylor, *The Village in the City*, London, 1973

P W Taylor, *Respect for Nature: A Theory of Environmental Ethics*, Princeton, NJ, 1986

A M Tazzer, *La restauración arquitectónica*, Mexico City, 1991

Temple Bar Properties, *Temple Bar Lives*, Dublin, 1991

Temple Bar Properties, *Development Programme for Temple Bar*, Dublin, 1992

S Thoma, *Vergangenheitsbewältigung am Beispiel der Auseinandersetzungen um die Neue Wache*, Berlin, 1995

E Thomas, *Monumentality and the Roman Empire: Architecture in the Antonine Age*, Oxford, 2007

M W Thompson, *Ruins: Their Preservation and Display*, London, 1981

P Thompson, *The Work of William Morris*, Oxford, 1991 (first published 1967)

G Thum, *Uprooted: How Breslau became Wroclaw during the Century of Expulsions*, Princeton, NJ, 2011

F Tilden, *Interpreting our Heritage*, Chapel Hill, NC, 1957

J A C Tillema, *Schetsen uit de Geschiedenis van de Monumentenzorg in Nederland*, The Hague, 1975

F Tindall, *Memoirs and Confessions of a County Planning Officer*, Ford, 1998

A Tomaszewski (ed), *Values and Criteria in Heritage Conservation*, Florence, 2008

F Tomps (ed), *Pool ajandit restaureerimist Eestis*, Tallinn, 2009

F Tönnies, *Gemeinschaft und Gesellschaft*, Leipzig, 1887

F Torchio, *Siena 1939–1945, proteggere l'arte*, Siena, 1945

P Torsello, *Restauro architettonico*, Milan, 1984

P Tournikiotis (ed), *The Parthenon and its Impact in Modern Times*, Athens, 1994

R Towse and A Khakee (eds), *Cultural Economics*, Berlin, 1992

M Trachtenberg, *Building in Time from Giotto to Alberti and Modern Oblivion*, New Haven, CT and London, 2010

E N Trubetskoi, *Saint Sophia: Russia's Hope and Calling*, London, 1916

S Tschudi-Madsen, *Restoration and Anti-Restoration*, Oslo, 1976

R Tubbs, *Living in Cities*, Harmondsworth, 1942

R Tubesing, *Architectural Preservation in the United States, 1941–1975, a Bibliography of Federal, State and Local Government Publications*, New York, 1978

R Tubesing, *Architectural Preservation and Urban Renovation*, New York, 1982

J E Tunbridge and G J Ashworth, *Dissonant Heritage: The Management of the Past as a Resource in Conflict*, Chichester, 1996

A M Tung, *Preserving the World's Great Cities*, New York, 2001

M M Turina, *La memoria de los piedras. Anticuarios, arqueologos y coleccionistas de anitiguedades en la España de los Austrias*, Madrid, 2010

N Tyler, T J Ligibel and I R Tayler, *Historic Preservation: An Introduction to its History, Principles and Practice*, New York, 2009

J Tyrwhitt, *Patrick Geddes in India*, London, 1947

UNESCO, *Final Act of the Intergovernmental Conference on the Protection of Cultural Property in the Event of Armed Conflict* (The Hague), Paris, 1954

UNESCO, *World Culture Report*, Paris, 1998

UNESCO, *Convention for the Safeguarding of the Intangible Cultural Heritage*, Paris, 2003

United States Conference of Mayors, *With Heritage so Rich: A Report of a Special Committee on Historic Preservation*, New York, 1965

J Unrau, *Ruskin and St Mark's*, London, 1984

R Unwin, *Town Planning in Practice*, London, 1909

F Urban, *Neo-Historical East Berlin*, Farnham, 2009

Urban Task Force, *Towards an Urban Renaissance*, London, 1999

N Urbanova and M Sukajlova, *Slovakia's Cultural Heritage*, Bratislava, 1996

J Urry, *Consuming Places*, London, 1995

US Department of the Interior, National Park Service, *Guidelines for Identifying . . . America's Historic Battlefields*, Washington, DC, 1992

G Valadier, *Narrazione artistica dell'operato finora nel ristauro dell' Arco di Tito*, Rome, 1822

F Valderrama, *A History of UNESCO*, Paris, 1995

L Vale and T Campanella, *The Resilient City*, Oxford, 2005

C D Valsassina, *Restauro – Made in Italy*, Milano, 2006

A Vance, *William Morris: His Art, His Writings, and His Public Life*, London, 1897

S van Riel, *Monumenti e centri storici*, Firenze, 1996

VEB Verlag für Bauwesen, *Geschichte der Denkmalpflege*, Berlin (GDR), 1989

VEB Verlag Zeit im Bild Dresden, *Monument Preservation in the German Democratic Republic*, Dresden, 1982

Verhandlungen der Heidelberger Schloss-Konferenz, Karlsruhe, 1902

G Vinken, *Zone Heimat*, Munich, 2010

E E Viollet-le-Duc, *Dictionnaire raisonné de l'architecture française* (10 vols), Paris, 1854–69

E E Viollet-le-Duc, *On Restoration* (translation), London, 1875

M R Vitale, *Restauri in Francia, 1970–2000*, Palermo, 2001

L Vitet, *Histoire des anciennes villes de France*, Dieppe and Paris, 1833

D Voldman, *La Reconstruction des villes françaises de 1940 à 1944*, Paris, 1997

M von Bunsen, *John Ruskin*, Leipzig, 1903

A von Buttlar, G Dolff-Bonekämper, M Falser, A Hubel and G Mörsch, *Denkmalpflege Statt Attrappenkult*, Basel, 2011

B von Droste, H Plachter and M Rössler (eds), *Cultural Landscapes of Universal Value*, Jena, 1995

U von Pelz, *Stadtsanierung im Dritten Reich*, Dortmund, 1994

F von Quast, *Mittheilungen über Alt und Neu Athen*, Berlin, 1834

W H Wackenroder and L Tieck, *Herzensergiessungen eines kunstliebenden Klosterbrüders*, 1796

R Wagner-Rieger and W Krause (eds), *Historismus und Schlossbau*, Munich, 1975

R Wakeman, *Modernising the Provincial City*, Cambridge, MA, 1997

M Walker and P Marquis-Kyle, *The Illustrated Burra Charter* (updated edition), Burwood, Vic., 2004

P Ward (ed), *Conservation and Development*, Newcastle, 1968

B Ward-Perkins, *From Classical Antiquity to the Middle Ages*, Oxford, 1984

D Watkin, *The Rise of Architectural History*, London, 1980

J P Watson and P Abercrombie, *A Plan for Plymouth*, Plymouth, 1943

C Webster and N Frankland, *The Strategic Air Offensive Against Germany, 1939–1945*, London, 1961

V Welter and J Lawson (eds), *The City after Patrick Geddes*, Bern, 2000

M Wheeler and N Whiteley (eds), *The Lamp of Memory: Ruskin, Truth and Architecture*, Manchester, 1992

M Whiffen, *The Public Buildings of Williamsburg*, Williamsburg, 1958

R Wiener, *The Rape and Plunder of the Shankill*, Belfast, 1975

P Wilkinson, *Restoration: Discovering Britain's Hidden Architectural Treasures*, London, 2003

M T Will, *Architectural Conservation in Europe: A Selected Bibliography*, Monticello, Ill, 1980

G R Willey and J A Sabloff, *A History of American Archeology*, London, 1974

S A Williams, *The International and National Protection of Movable Cultural Property*, New York, 1978

R G Wilson, *The Colonial Revival House*, New York, 2004

J J Winckelmann, *Anmerkungen über die Baukunst der Alten*, Leipzig, 1762

J J Winckelmann, *Geschichte der Kunst des Alterthums*, Dresden, 1764

W Wippermann, *Denken statt Denkmälen*, Berlin, 2010

M Wohlleben (ed), *Georg Dehio und Alois Riegl, Konservieren nicht Restaurieren*, Zürich, 1988

G Wolff, *Zwischen Tradition und Neubeginn: zur Geschichte der Denkmalpflege in der 1. Hälfte des 19. Jahrhunderts*, Frankfurt/Main, 1992

W Wolff, *Neuer Wert aus alten Häusern*, Berlin, 1966

A C Wood, *Preserving New York*, New York, 2008

C Woodward, *In Ruins*, London, 2001

Roy Worskett, *The Character of Towns*, London, 1969

Wu Hung, *Ruins in Chinese Art and Visual Culture*, London, 2011

J E Young (ed), *The Art of Memory: Holocaust Memorials in History*, Munich, 1994

A J Youngson, *The Making of Classical Edinburgh*, Edinburgh, 1966

A J Youngson, *Urban Development and the Royal Fine Art Commissions*, Edinburgh, 1990

S E Yuhl, *A Golden Haze of Memory, the Making of Historic Charleston*, Chapel Hill, NC, 2005

J Zachwatowicz, *Conservation in Poland*, Warsaw, 1965

Zamek Krolewski na Wawelu, *Wawel Narodowi Przywrocony*, Krakow 2005

Zamek Krolewski w Warszawie, *Jan Zachwatowicz w Stulecie Urodzin*, Warszawa 2000

P Zanker, *The Power of Images in the Age of Augustus*, Ann Arbor, MI, 1988

J Zielinski, *Warsaw, Ruined and Rebuilt*, Warsaw, 1997

– Image credits –

Although every effort has been made to trace and contact copyright holders, this has not been possible in every case; we apologise for any that have been omitted. Should the copyright holders wish to contact us after publication, we would be happy to include an acknowledgement in subsequent reprints.

Named copyright holders

Abbé L Verguet 3.12a
Acropolis Restoration Service 1.8
Allom Lovell Architects, Brisbane 11.11
American School of Classical Studies at Athens 6.6a, b
L Angelini 6.10b, c, d
Anta Architecture 9.17c
Antony Foto 6.2b
Architectural Press 6.24; 7.3a, b, c; 9.1a; 9.4b, c; 9.10a; 9.15a; 12.3a
ASSIST Architects 9.10b, c
Bayerisches Hauptstaatarchiv 2.1a
Bayerisches Landesamt für Denkmalpflege 7.6b
Bayerisches Nationalmuseum 1.12b
N Beseler 7.4a, b, c
Blair Castle Collection 2.14a, b
BNVE Roma 6.8a
British School in Rome (public domain image) 1.13a
F Brouty, J Fayard & Cie 7.2
Brownstone Revival Coalition 8.24a, b
Colonial Williamsburg Foundation 6.19a, b
Compton Russell Ltd 9.2
Coventry Evening Telegraph 0.2a
Cultural and Town Planning Board of Dragør Municipality 11.19f
Denkmalpflege des Kantons Basel-Stadt Fotoarchiv 8.2a, b
Department of the Environment 9.8a, b, d; 11.8a, b; 11.9a, b, c
Deutsche Herold 9.4a
Deutscher Kunstverlag 5.2a; 5.4a; 9.11a
Deutsches Architektur-Museum 8.3b
Dienst der publieke werken, Amsterdam 11.9d
DOCOMOMO International 12.11a
Ebury Press/Random House 9.18b
Editura Meridiane 10.16a
Eesti Arhitekturi Muuseum 10.9b, d
F Florin 8.13
Flugblatt 'Pro Palast' 12.18a
Fonds Ministère de la Culture et des Communications 8.26a, b

Foreign Languages Publishing House, Pyongyang 10.17a, b
Frank Lloyd Wright Home and Studio Foundation 8.30a
Frankfurter Goethe-Haus Museum 8.5a, b
Friends of Cast-Iron Architecture 8.24
Fromm Verlag 0.1a
Germanisches Nationalmuseum 3.8a; 3.22a; 4.7a; 5.1a, b
Glasmuseum Rheinbach 3.8b
Grossherzoglich-Badischen Finanzministerium 5.6
Heinze-Greenberg 5.13d
Heritage Canada 8.16c
Karl Holtz 12.12a
Huntington Library, San Marino, CA 1.1
ICCROM/J A Flores 11.6b, c
Institut für Denkmalpflege der DDR 3.22c, d
Izdanie Mosgorispolkoma 10.15
Katholieke Universiteit Leuven 11.7
Kinderbuchverlag Berlin 10.13c, d, e
Kirchenkreis Lübeck 7.11a
Osbert Lancaster 8.18
Langen Müller Herbig 9.1b
Leeds City Museum 2.3
Lukas-Verlag 7.7a, b
LVR – Amt für Denkmalpflege im Rheinland 5.4d
MAP Paris 6.5
Meijer Pers b.v. 11.9e
Ministry of Housing and Local Government 8.21c; 8.22a, b; 8.28a, b, c, d, e; 9.3a, b, c, d
Mintis, Vilnius 10.10c
Monumental 1.7b, 3.14c
Musée Carnavalet 3.1c
Musée de la Ville de Bruxelles 2.2a
Musée Paul Dupuy, Toulouse (cliché Daniel Molinier) 3.2b
Museen der Stadt Nürnberg 3.10a, b; 6.13a, b; 6.14a, b, c; 7.6a; 8.12a, b
Museum of Estonian Architecture 10.9b, 10.9d
National Museum in Krakow 5.7b
National Trust for Historic Preservation 8.16a, b
P Neff Verlag 5.4c
Newcastle County Borough Council 8.21d
Niedersächsisches Landesamt für Denkmalpflege 5.5d
North Western Gas Board 9.8c
Norwich City Council 8.3b
La Parisienne de Photographie 3.1c, 3.11a, 3.13b
Perfect Harmony Ltd 9.18d
G Petrie 1.11b
PKZ Warsaw 10.12a
Progress Publishers, Moscow 10.2
Projektgruppe Eisenheim 9.7c, d
Anne Raines 3.21e; 8.17a, b; 9.19b, c
Peter Robinson 9.10d
M Roerbye 3.16
Rosebery Estates 3.5a
F Rouff, Paris (by courtesy of O Rigaud) 6.2c
St Albans Cathedral 4.4b, c
P Sandby 3.20a
SAVE 9.17a; 9.18a

Scherz Publishers 8.1
Scottish Development Department 8.15; 9.14a, b
Stadt Göttingen 8.20a, b
Städtische Galerie Dresden 7.1
State Art Publishers 7.9b
State Political Publishing House of the Lithuanian SSR 10.1
Stichting Congres 1975 Amsterdam 8.27a
Stichting Kasteel de Haar 4.7c, d
Stiftung Gedenkstätte Buchenwald-Mittelbau-Dora/M Schuck 10.7b
Super-Illu 12.19c
SURPMO/Polytechna Foreign Trade Corporation 10.12b
Temple Bar Properties 12.3b
Trustees of the American School of Classical Studies at Athens 6.6a, b
Ufficio Fori Imperiali 1.5
Universe Books, New York 11.3b
Ure Smith 9.5a, b
US Department of Housing and Urban Development 9.13
VEB F A Brockhaus Verlag 10.13a
Victorian Society 8.29a, b
J J Waltz 5.1c, d
Wawel Royal Castle 5.7 a, c
Zamek Krolewski 10.4c
Zöllner Foto, Erfurt 10.7a

Images reproduced from publications

W Abelshauser, *Wirtschaft in Westdeutschland*, Stuttgart, 1975 7.5a
N Arkhipov and A Raskin, *Petrodvorets*, Leningrad, 1959 7.9b
Bauwelt, H6, 1978, 187 12.16
M Briggs, *Goths and Vandals*, London, 1952 3.19c
J Britton, *History and Antiquities of the Cathedral Church of Salisbury*, London, 1814 2.11c
The Builder, 29 September 1860 3.20b
Cartoon Portraits and Biographical Sketches of Men of the Day, 1873 4.3
Coventry Evening Telegraph, 18 May 1962 0.2a
Die Denkmalpflege, 1901 5.12a
Die Denkmalpflege, 1903 5.12b
W Denslagen, *Architectural Restoration in Western Europe*, Amsterdam, 1994 5.4b
W Denslagen and N Gutschow, *Architectural Imitations*, Maastricht, 2005 6.2b
Domenica del Corriere, 1912 6.3b, and 3 March 1935 6.8a
A Dorn, *Die Semlowersrasse in Stralsund*, Berlin, 1940 6.12b
W Durth and N Gutschow, *Träume in Trümmern*, Braunschweig, 1988 7.11b
C Flatow, *6+ Berlin*, Berlin, 1972 9.7a
P Geddes, *City Development*, 1904 5.16b, c, d
W Hearne and T Byrne, *Antiquities of Great Britain*, 1786–1807 2.11b
L'Illustrazione Italiana, 20 July 1902 6.3a
W Lindner and E Böckler, *Die Stadt: Ihre Pflege und Gestaltung*, 1939 6.16a, b
Lustige Blätter, no 34, 1904 5.6b
Magazine of Art, January 1891 4.2a
C Martin, *L'Art de Bâtir les Villes*, Paris, 1902 5.11b, c
Old Edinburgh International Exhibition, 1886 5.13a, b, c
Pennant's *Tours in Scotland*, 1769, 218 2.14d
W Pinder, *Deutsche Dome des Mittelalters*, Duesseldorf, 1910 1.9b, 1.12a, 3.7b

Piranesi, *Vedute di Roma*, 1750/71 3.6a, b

E Ross, E Schaubert, C Hansen, *Der Tempel der Nike Apteros*, Berlin, 1839 3.16b

J Ruskin, *The Seven Lamps of Architecture* (1883 edition) 4.2b, c

P Schulze-Naumburg, *Kulturarbeiten*, 1906 5.3a, b

Vetusta Monumenta, 1826 2.8

Völkischer Beobachter, July 1943 7.5b

Het Vrije Volk, October 1955 8.10

V Welter, *Biopolis*, Cambridge, MA, 2002, 10 5.16a

Internet public domain sources

HABS public domain 3.21a, b, c, d; 5.9a, b, c; 6.19 c, d; 6.21a, b, c; 8.25a, b

Wikimedia Commons, Abu_Simbel_relocation_by_Zureks.jpg:Zureks 11.2

Wikimedia Commons, Creative Commons: Itziaram, 2007 4.9

Wikimedia Commons, http://commons.wikimedia.org/wiki/File:General_view_of_the_ruins_from_Tribune_Building,_Booksellers_Row_in_the_centre,_by_Lovejoy_%26_Foster.jpg 6.1

Wikimedia Commons, http://en.wikipedia.org/wiki/File:Disneyland_Main_Street.jpg (photographer, A A Si) 8.17c

Wikimedia Commons, Belgie_ieper_1919_ruine.jpg 6.2a

Wikimedia Creative Commons: Attribution Share-Alike Licence Conditions 2.5: Voytek S 12.14

– Index –

Note: Page numbers followed by 'f' refer to figure captions and followed by 'n' refer to notes.

WITHDRAWN
UST
UNIVERSITY OF ST. THOMAS LIBRARIES
Libraries